Spatial Analysis for Radar Remote Sensing of Tropical Forests

SAR REMOTE SENSING

A SERIES

Series Editor
Jong-Sen Lee

Polarimetric SAR Imaging
Theory and Applications
Yoshio Yamaguchi

Imaging from Spaceborne and Airborne SARs, Calibration and Applications
Masanobu Shimada

Radar Scattering and Imaging of Rough Surfaces
Modeling and Applications with MATLAB®
Kun-Shan Chen

Spatial Analysis for Radar Remote Sensing of Tropical Forests
Gianfranco (Frank) De Grandi and Elsa Carla De Grandi

For more information about this series, please visit: https://www.routledge.com/SAR-Remote-Sensing/book-series/CRCSRS

Spatial Analysis for Radar Remote Sensing of Tropical Forests

Gianfranco (Frank) De Grandi and
Elsa Carla De Grandi

CRC Press
Taylor & Francis Group
Boca Raton London New York

CRC Press is an imprint of the
Taylor & Francis Group, an **informa** business

First edition published 2021
by CRC Press
6000 Broken Sound Parkway NW, Suite 300, Boca Raton, FL 33487-2742

and by CRC Press
2 Park Square, Milton Park, Abingdon, Oxon, OX14 4RN

CRC Press is an imprint of Taylor & Francis Group, LLC

Library of Congress Cataloging-in-Publication Data
Names: De Grandi, G. F. (Gianfranco), author. | De Grandi, Elsa Carla, author.
Title: Radar remote sensing of tropical forests: spatial analysis
techniques / by Gianfranco D. De Grandi and Elsa Carla De Grandi.
Description: First edition. | Boca Raton, FL: CRC Press/Taylor & Francis
Group, LLC, 2021. | Series: SAR remote sensing | Includes
bibliographical references and index. | Summary: "This book is based on
authors' extensive involvement in large Synthetic Aperture Radar (SAR)
mapping projects, targeting the health of an important earth ecosystem,
the tropical forests. It highlights past achievements, explains the
underlying physics that allow the radar practitioners to understand what
radars image, and can't yet image, and paves the way for future
developments including wavelet-based techniques to estimate tropical
forest structural measures combined with InSAR and Lidar techniques.
As first book on this topic, this composite approach makes it appealing for
students, learning through important case studies; and for researchers
finding new ideas for future studies"– Provided by publisher.
Identifiers: LCCN 2020049160 | ISBN 9780367259402 (hardback) |
ISBN 9780429290657 (ebook)
Subjects: LCSH: Rain forests–Remote sensing. | Synthetic aperture radar. |
Forest surveys–Data processing. | Spatial analysis (Statistics)
Classification: LCC SD247 .D43 2021 | DDC 577.34–dc23
LC record available at https://lccn.loc.gov/2020049160

ISBN: 978-0-367-25940-2 (hbk)
ISBN: 978-0-367-74266-9 (pbk)
ISBN: 978-0-429-29065-7 (ebk)

Typeset in Times
by SPi Global, India

Cover design by Alessandro De Grandi

Contents

Preface

"Scusatemi, se da sol, mi presento. Io sono il prologo...", Ruggero Leoncavallo, Pagliacci[i] ("Please? Will you allow me? Ladies! Gentlemen! Excuse me if I appear thus alone. I am the Prologue.")

This book is a four-handed piano sonata. Wolfgang Amadeus Mozart wrote masterpieces in this musical form (but the book is certainly not a masterpiece). The sonata is performed by two pianists, and the book is written by two authors (strangely, the second author bears the same surname as the first, though they never met).

In the sonata, the score is split into two parts, "secondo" and "primo", each one encoding the notes and instrumental designations for each pianist (see the introductory two pages of Mozart's sonata in D major in the figure). Accordingly, in the book, there are two parts.

Usually, the "secondo" part is written in the bass clef and is intended as an accompaniment to the "primo", this being where action is taking place, and where the main themes of the composition are developed. The first part of the book is, in a certain sense, like an accompaniment to the main theme and it is written in a lower key. It is a look into the past, a look back over one's shoulders.

Somewhere in a box hidden in a dusty drawer one can find black-and-white pictures portraying grandfathers and grandmothers, or perhaps how we were as babies playing on the beach. Those pictures, maybe worn by the passing of time, were probably long forgotten and of no use to anybody. Yet they are important just because they are the only surviving shadows of the past, which would otherwise be forlorn.

We like to believe that the first part of the book is like one of those worn-out black-and-white pictures, perhaps considered as of no use to anybody, but important because they are memories. Memories are important in the social sciences, such as history. But we believe they are also important in science and technology, where the pace of progress is so much faster, where the stuff our achievements are made of (algorithms, mathematics) is sometime very soft, and consequently also the rate of oblivion is so much faster.

Memories are not only recording media of what is no more, but also vehicles of emotions. It is our hope that the first part of the book may convey to the reader (especially to young researchers) memories and emotions. Memories will be carried by the description of the underpinning technical aspects of the SAR projects that are recounted in the Part I chapters. Emotions, we hope, will be roused by the awareness of the role of the human component, the efforts and the hard work of the many people around the world (Europe, Japan, USA), who, in a truly international and synergistic way, made those projects a reality.

The "primo" part is where all the trills, pirouettes and good humor are. In the same way, Part II of the book is where the reader will possibly find the more interesting and novel matters (indeed, with some technical pirouettes too). In this part, the arrow of time is reversed. Instead of being the historical recount of past deeds, we present here some techniques that have been the subject of our recent research. In doing so, we hope also to have sown the seed for the curiosity of some readers and for the blossoming of future applications based on these techniques.

The piano sonata has a structure, with three main sections: the exposition, the development and the recapitulation. The book has a structure too, albeit here our analogy becomes a bit thinner.

The book is like a set of pearls (well, perhaps beads) threaded on a string, a fil-rouge. The fil-rouge of Part I is time, which gives the beat to a succession of developments, all based on the same multiresolution image-processing paradigm. From the initial conception (ESA–JRC CAMP 1990) of the idea that by gluing together several little tiles one could retain the information content (if the tiles were high resolution radar images), the reader is taken through successive implementations of the concept (GRFM 2000–ALOS K&C 2012) with increased image quality and thematic content.

The beads of Part II are spatial random fields, and wavelets are the fil-rouge. The first bead in Chapter 5 had no "hair" (to borrow from the language of our friends the astrophysicists). We merely computed by use of a numerical model the energy scattered back by a homogeneous slab of forest. The field is a constant, and it establishes the link between the observable and the target properties. Thus it sets the groundwork for the more interesting fields in the following chapters.

The fields in Chapter 6 have grown "hair", as they contain variations (textural patterns) that reflect the forest spatial structure. Here, wavelets are the yardsticks by means of which we measure those ripples of space. Some fields contain variations of backscattered energy. Other fields, derived from InSAR data, have grown "vertical hair". These contain information on the top of canopy spatial arrangement, on the bumps and troughs generated by the layout of neighboring trees of different heights. Here, wavelets will provide statistical measures of the ripples of the canopy layout.

In Chapter 7, attention is moved to the corners of the field, to discontinuities (edges) that appear at the boundary between two different targets, such a forest and grassland. Wavelets here provide a localization of these discontinuities and measures of their shape and strength.

In Chapter 8, fields go to the hairdresser. Variations (speckle and texture) may be a nuisance and not providing valuable information for pattern recognition algorithms that require conditions of regularity in the image. Here, wavelets take advantage of the measures of discontinuities, developed in the previous chapter, for building smooth approximations of backscatter fields and, at the same time, preserving their contours.

In Chapter 9, the beads, the fields, are again different, now holding InSAR coherence. Here, the "hair" is phase noise, and wavelets, taking advantage of the availability of many fields corresponding to the same scene, are able to reconstruct a series of better estimates of the true coherence values in stationary areas, with preservation of non-stationary discontinuities.

Chapter 10 is an outlier, because the fil-rouge is broken. There are no wavelets here. Information is not provided by the content of one field, but by the difference between two fields portraying the height of the top of the canopy at two dates. The outcome is a measure of forest volume loss or gain in a given time frame. The remarkable difference with respect to the previous approaches is that here the atomic quantity of interest is no more a single resolution element within the field, but the entire field, which is treated as an object characterized by average properties related to the gain or loss of forest volume.

Now we have revealed what is in the book, let us explain what is not there. The book is not about tropical forests. Forests are the stage sets, and not the main characters in the drama. Therefore the reader will not find a chapter dedicated to a general-purpose review of tropical forests' ecology and importance in global climate change studies. Albeit, there is some up-to-date descriptive material of the forest characteristics concerning the sites where the remote sensing experiments took place. References to related literature on the subject of tropical forest can be found in those sections.

The book is also not on SAR signal processing, instrument design or SAR image understanding. As the more famous (and with a better voice) Frank used to sing in "My Way", the book is just about "the way we did it".

So, what is the book good for? We will lean on the shoulder of a giant to provide a justification.

In 1976, the famous mathematician Freeman Dyson wrote a seminal and far-reaching paper in the journal Energy [1], proposing the theoretical possibility of controlling the amount of carbon dioxide in the atmosphere by growing trees. The paper starts by framing the problem in terms of known quantities in carbon dynamics, such as the current amount of carbon in the atmosphere, the carbon mass added per year by the burning of fossil fuels, the amount of carbon absorbed by oceans and the biosphere. The following disclaimer is made at the outset: "The present carbon content of the biosphere is only roughly known."

At the time of this book's publication, global estimates of the tropical and boreal forest biophysical parameters, and their evolution over time, are still a crucial variable in modeling carbon dynamics. It is our hope that the experiments and techniques described in our book might make a small contribution in this direction, if not in a quantitative manner, but at least in a qualitative one, just by viewing the problem from a different angle.

Raise the curtain then, in Leoncavallo's words: "Mark well, therefore, our souls, rather than the poor players' garb we wear, rather than our mathematical doodlings, for we are men of flesh and bone, like you breathing the same air of this orphan world. This, then, is our design. Now give heed to its unfolding. On with the show! Begin!"

REFERENCE

Dyson, F. (1977). Can we control the carbon dioxide in the atmosphere? *Energy*, vol. 2, pp. 287–291.

Acknowledgements

The first author would like to thank...

At the outset, I would like to express my gratitude to the Institution of the European Commission (EC), where I spent some 40 years of my professional life.

When I first joined the lab at the EC Joint Research Centre for thesis work during my graduate studies, I soon realized that not only very few people spoke in my mother tongue (Italian), but I also used to meet in the corridors and conference rooms a flurry of people visiting from all over the world. Indeed, my first visiting scholar was from China, and my first PhD student (now a star at JPL) was from Canada.

The international spirit that oozed throughout the lab was instrumental in making this young student set foot on the steps that turned out to be his lifelong groundwork: science holds universal values and has no boundaries, and sharing knowledge and experience pays back with great intellectual richness. Correlation increases variance in human affairs as in speckle filters.

A catapult into the world was not the only asset endowed by the Institution. Other benefits included the freedom of pursuing one's own scientific interests, and plenty of hard and soft resources to achieve the intended goals. Last but not least, I also wish to express my gratitude for the moral and practical support from my Institution at the time when the BIG KILLER tried to avoid the pain of this book to the readers.

I hope in exchange to was a good ambassador of the EC whenever I was called on to represent it when lecturing, teaching or carrying out other work throughout the world.

A thought of deep gratitude goes to all my students, colleagues, collaborators, technicians and supporting personal who shared with me ideas, efforts and, most important, friendship. Much of the work presented in this book would have not been possible without their contributions. Naming all of them would easily fill the space allowed for the book. Let me cite just two cases, by recounting two anecdotes from my life that are in some way pertinent to the content of the book.

Swept by political winds from nuclear energy to environment, I had just shifted from counting neutrons to counting photons. Indeed, I was drawn into the "big attractor", the microwave team that Dr Alois Sieber, whom I acknowledge, had just founded. I dared to present as a poster at an Institute of Electrical and Electronic Engineers (IEEE) International Geoscience and Remote Sensing Symposium (IGARSS) conference a speckle filter, the first of my many dysfunctional ones (one is described in Chapter 9). My poster location began to look more and more like a cemetery, and I started to get a bit discouraged. Then a kind fellow with an American drawl stopped by, pondered a bit on my artwork and said: "Hey, this solution looks quite original, keep going young chap. By the way, I am Dr Jong Sen Lee of NRL (US Naval Research Laboratory, Washington, DC)." It was the beginning of a long relationship, of joint research efforts in the field of radar polarimetry (some are recounted in the book; see Chapter 6) and of friendship, (indeed it was more like a string quartet, with the addition of Dale Schuler and Tom Ainsworth).

With the second anecdote, I wish to remember and give thanks (in memoriam) to one of the pillars of synthetic aperture radar (SAR) polarimetry: Professor Wolfgang Boerner. I had conceived the idea of polarimetric texture, an idea that had been looked on as a bit odd by the SAR community. I was scheduled to deliver a talk on the subject, once more at an IEEE IGARSS conference. I started to discuss a few slides, and was becoming a bit nervous because I felt the audience was not following. Then a resounding voice from the floor interrupted my talk: "Pay attention now you all, because what Dr De Grandi is saying is very important." It was Prof. Boerner, as outspoken as always. It was again the beginning of a long relationship and friendship. Without his encouraging words, I would probably have stopped the line of research that underpins the content of this book.

One more exception to the list of concealed names in the list of thanks reveals the name of Professor Richard Lucas, because he played a central role in the developments of Chapter 6, and because I, as a father, owe him a debt of gratitude.

Part of the work described in this book was performed while I was member of the scientific staff at the EC Joint Research Centre. I also served as Principal Investigator of the Japan Aerospace Exploration Agency (JAXA), and of the German Aerospace Center (DLR). These Institutions are acknowledged.

I would like to dedicate the book with loving memory to my parents, who bestowed on me the right blend of Newton and Beethoven, a richness that made my life better. In addition, by symmetry, to my daughter and son, to whom I hope to have passed along the same richness.

Gratitude goes to my wife Marisa, who could not surf the waves while attending with love and care the secluded husband and daughter, busy playing with wavelets.

The second author would like to thank…

I would like to thank Professor Richard Lucas and Dr Pete Bunting for transmitting their enthusiasm for remote sensing at the start of my studies, and for supervising my BSc and MSc thesis.

A special thank you to my PhD supervisor, Edward Mitchard, for his support throughout my PhD studies, and for contributing his expertise to my professional development. Thank you to Professor Iain Woodhouse for providing knowledge and valuable experience in SAR remote sensing.

I would also like to thank those who contributed to improving my remote sensing work and knowledge. In particular, Professor Dirk Hoekman, for enabling me to carry out work on one of his study sites, for providing extensive light detection and ranging (LiDAR) and aerial photography datasets, and for his SAR expertise (see Chapter 6, Section 6.9).

I would like to acknowledge the data providers. In particular, the team at DLR (Dr Thomas Busche and Prof. Irena Hajnsek) for providing TanDEM-X data (TanDEM-X AO VEGE6702), used in Chapters 6 and 10.

My gratitude goes to Sarmap, who provided SARScape software to process SAR data during the course of my PhD. In particular, thank you to Dr Francesco Holecz, Dr Paolo Pasquali and the Sarmap team for teaching me about SAR and Sarscape during my early career internship.

Last but not least, I would like to dedicate this book to my parents and in particular to my father who has always been a great source of inspiration (and discussions on SAR) throughout my career. Thanks to my mother, who always supported me and always tried to bring out the positive side of things. Thanks for her efforts in moderating mine and my father's discussions on SAR. And finally, thanks to my brother who I hope will also become a true remote sensing geek in the future!

The authors are grateful to Irma Shangla Britton, senior editor at CRC Press, who convinced them to write a "true" book, and made their dream book come true. Many thanks also to the CRC Press and SPi Global staff for their support in producing the book.

"Grazie" to Keith Povey, our copy editor, who could make even equations speak good English.

The Authors

Gianfranco (Frank) De Grandi received a doctorate degree in nuclear engineering (with honors) from the Politecnico di Milano, Italy, in 1973. In 1977 he joined the European Commission Joint Research Centre (EC JRC), in Ispra, Italy as a scientific staff member. This lab was the base camp where he developed his main research activities across the next forty years, but also was the catapult to his many adventures on the international stage.

In 1985, he was visiting scientist at Bell Communications Research, Morristown, NJ, USA, where he participated in the design of Metrocore, one of the first research projects for GB-rate metropolitan area networks.

From 1986 to 1989, he headed the signal processing section of the Electronics Division, JRC, where he introduced VLSI (very large-scale integration) design technology and conducted research, in cooperation with Bellcore, on packet video, and with ITALTEL Italy on the European digital mobile phone network.

In 1989, he joined the Joint Research Centre's Institute for Remote Sensing Applications, where he began his research activity in radar remote sensing for earth observations, and in particular in SAR polarimetry.

From 1997 to 2001 he served as assistant professor with the Faculté de foresterie, de géographie et de geomatique, Laval University, Quebec, PQ, Canada.

He retired from active service with the European Commission in 2012, but continued to 2017 with his activities in science and education as a visiting scholar at the University of Wales at Aberystwyth, Ceredigion, UK.

His research interests spanned a wide gamut, including global-scale forest mapping using high resolution space-borne SAR, and wavelet techniques applied to spatial statistics of InSAR (interferometric synthetic aperture radar) digital elevation models (DEMs) for forest structure detection, image regularity characterization, the generation of edge-preserving smooth approximation of SAR images, multi-temporal speckle filtering, and polarimetric SAR backscatter two-point statistics.

Dr De Grandi collaborated with several laboratories in the USA, such as the Lawrence Livermore National Laboratory, and the Los Alamos National Laboratory for nuclear safeguards, Caltech Jet Propulsion Laboratory and the US Naval Research Laboratory for radar remote sensing.

He served as EC representative at the UN International Atomic Energy Agency (IAEA) for nuclear safeguards problems.

At the JRC, he was team leader of the Central Africa Mosaic Project (CAMP), a joint initiative of the EC and the European Space Agency (ESA), where he performed seminal work in wide-area high-resolution SAR (Synthetic Aperture Radar).

He was principal investigator with the Japan Aerospace Exploration Agency (JAXA), Global Rain Forest Mapping and Global Boreal Forest Mapping projects, of the JAXA Earth Observation Research Center (EORC) Kyoto & Carbon Initiative .He was an international collaborator in a NASA Carbon Cycle Science project, led by the University of Washington, WA, USA.

He was a science team member for the DLR TanDEM-X mission, and principal investigator of the DLR Tandem-X pre-operational project XTI-VEGE0330, aimed at the characterization of tropical forests' horizontal and vertical structure.

In 2002, Dr De Grandi was elected an IEEE Fellow, with the citation: "For seminal work in continental scale vegetation mapping using high-resolution SAR mosaics and innovative contributions in the area of information extraction from SAR data." He is a member of the IEEE Geoscience and Remote Sensing Society, and the IEEE Signal Processing Society.

He is first author in frequently cited peer-reviewed papers published in international journals, such as *IEEE Transactions on Geoscience and Remote Sensing* and Taylor & Francis *International Journal of Remote Sensing*.

When not playing with wavelets, he plays Chopin and Beethoven sonatas on a Yamaha grand piano, and sings Verdi's opera arias.

Elsa Carla De Grandi received a BSc degree in physical geography and an MSc (with distinction) in remote sensing and geography from Aberystwyth University, Aberystwyth, UK, in 2011 and 2012, respectively.

In 2017 she was awarded a PhD in Remote Sensing (Atmospheric & Environmental Science programme) from the University of Edinburgh. Her PhD focused on developing and testing novel SAR and InSAR methods for mapping deforestation and forest degradation in tropical forests.

Since 2019 Elsa has been working as an Earth Observation Engineer at GMV NSL (United Kingdom). She is also acting as project manager for several projects, and contributing to proposal writing, particularly on the use of SAR for environmental monitoring.

She has been a reviewer for *IEEE Transactions on Geoscience and Remote Sensing*. She has published peer-reviewed papers in international journals, and delivered talks on radar remote sensing of tropical forest at several conferences.

List of Abbreviations

ACF	autocorrelation function
ALOS PALSAR	ALOS Phased Array type L-band Synthetic Aperture Radar
ALOS	Advanced Land Observing Satellite
AMI	Active Microwave Instrument
AO	Announcement of Opportunity
ASAR	Advanced Synthetic Aperture Radar
ATSR	Along Track Scanning Radiometer
AVHRR	Advanced Very High Resolution Radiometer
Caltech	California Institute of Technology
CAMP	Central Africa Mosaic Project
CCF	cross-correlation function
CDF	cumulative distribution function
CEOS	Committee on Earth Observation Satellites
CFAR	constant false alarm rate
CHM	Canopy Height Model
CLGC	Contextual Local–Global Clustering
CoSSC	Co-registered Single-look Slant-range Complex
CWT	continuous wavelet transform
DEM	Digital Elevation Model
DFT	discrete Fourier transform
DLR	German Aerospace Center (Deutsches Zentrum für Luft-und Raumfahrt)
DN	digital number
DRC	Democratic Republic of the Congo
DSM	Digital Surface Model
DTM	Digital Terrain Model
DWT	discrete wavelet transform
EC	European Commission
ELBG	Enhanced Linde–Buzo–Gray non-contextual clustering algorithm
ENVISAT	Environmental Satellite
EORC	(JAXA) Earth Observation Research Center
EOS (AMSR-E)	Advanced Microwave Scanning Radiometer
ERS	European Remote Sensing satellite
ESA	European Space Agency
FBD	Fine Beam Dual mode
FBFP	fine beam fully polarimetric
FIR	finite impulse response
GB	gigabyte (one billion bytes)
GEM6	Goddard Earth Model 6
GRFM	Global Rain Forest Mapping
HFCL	Hansen Forest Cover Loss
HH	co-polarized – horizontal transmit, horizontal receive
HV	cross-polarized– horizontal transmit, vertical receive
IAEA	UN International Atomic Energy Agency
IGBP DISCover	International Geosphere–Biosphere Program Data and Information System global land cover data set
IID	independent identically distributed (*random variables*)
INPE	Instituto Nacional de Pesquisas Espaciais (Brazilian Space Agency)

InSAR	Interferometric Synthetic Aperture Radar
ISODATA	Iterative Self-Organizing Data Analysis Technique algorithm
JAXA	Japan Aerospace Exploration Agency
JERS-1	Japanese Earth Resources Satellite 1
JM	Jeffries–Matusita distance
JPL	(Caltech–NASA) Jet Propulsion Laboratory, California
JRC	[European Commission] Joint Research Centre
K & C	Kyoto & Carbon Initiative
KLT	Karhunen–Loève transform
LANDSAT	Land satellite: NASA/US Geological Survey *Landsat* series of Earth Observation satellites
LBG	Linde–Buzo–Gray (*method of vector quantization*)
LiDAR	Light Detection and Ranging
MDT	Multiple Discriminant Transform
METI	Ministry of Economy, Trade and Industry (Japan)
MITI	Ministry of International Trade and Industry (Japan) (now METI)
MLP	multilayer perceptron
MMU	minimum mapping unit
MODIS	MODerate Resolution Imaging Spectroradiometer
MPAC	multiscale modified Pappas adaptive clustering algorithm
MPDI	microwave polarization difference index
MRA	Multi-resolution analysis
MSE	mean square error
NASA	National Aeronautics and Space Administration
NASDA	National Space Development Agency of Japan (now JAXA)
NMP	nearest multiple prototype
NOAA	National Oceanic and Atmospheric Administration
NP	Nyman–Pearson theorem
PAC	Pappas adaptive clustering algorithm
PAF	German Processing and Archiving Facility
PALSAR	JAXA Phased Array type L-band Synthetic Aperture Radar; see ALOS
PCH	phase center height
PDF	probability density function
PMF	parametric matched filter
PRI	Precison Image Product
PSF	point spread function
RCS	radar cross-section
REDD+	Reducing Emissions from Deforestation and Forest Degradation (*and the role of conservation, sustainable management of forests, and enhancement of forest carbon stocks in developing countries*)
REIMP-CA	Regional Environmental Information Management Program on Central Africa
RGB	red/green/blue
RMS(E)	root mean square (error)
ROI	region of interest
RP	reflection plane
RRMSE	relative root mean square error
RVOG	random volume over-ground model
SAR	Synthetic Aperture Radar
SARscape	software for processing SAR data
SEOM	standard error of the object's mean
SLC	Single Look Complex

SMA	Spectral Mixture Analysis
SNR	signal-to-noise ratio
SPOT 4	Satellite Pour l' Observation de la Terre
SRF	spatial random field
SRTM	Shuttle Radar Topography Mission
SWBD	SRTM Water Bodies Database
TanDEM-X	TerraSAR-X Add-On for Digital Elevation Measurement
TerraSAR-X	German Earth-observation satellite
TM	Thematic Mapper
TREES	TRopical Ecosystem Environment observations by Satellite
TRFIC	Tropical Rain Forest Information Center (NASA)
TRMM	Tropical Rainfall Measuring Mission
UNFCCC	United Nations Framework Convention on Climate Change
UTM	Universal Transverse Mercator
VARMAP	variable sized cells holding class labels
VGT	SPOT VEGETATION instrument
VHR	Very High Resolution
VLSI	very large-scale integration
VQ	vector quantization
VV	vertical transmit, vertical receive polarization (co-polarized)
WASS	wavelet scaling signature
WFLATS	Wavelet Flatness Factor Signature
WRI	World Resources Institute
WSC	Wind Scatterometer
WSS	wide-sense stationary process
WT	wavelet transform

List of Figures and Tables

LIST OF FIGURES

LIST OF TABLES

Part I

Sarcheology: The Era of the Big Radar Mosaics

1 The Dawn of the SAR Mosaics Era

The ESA–JRC Central Africa Mosaic Project

1.1 RADAR MOSAICS: WHAT AND WHY

The first part of the book is about memories. Memories of the dawn of a new era, when the big leviathans, the big mosaics, started populating the world of SAR remote sensing.

Well, that sentence may look a bit too operatic ("Ritorna vincitor .... (Return victorious!)", Verdi, *Aida*[1]), a bit over the top. Perhaps the SAR mosaics we are talking about were not the first ones, or the biggest or the most notable form the scientific standpoint. Yet they were born of a wonderful and lasting professional and human experience. They were also witness of the power of synergism in science and technology, since assets from Europe, with ESA, the EC and many researchers from member states, from the USA, with Caltech NASA JPL, and Japan with JAXA, were put at stake to realize this endeavor.

It is just about this great human experience and professional endeavor that we would like to narrate in this chapter and the following ones. The reader may well at this point be wondering about the connection between the large mosaics and spatial analysis, the main theme of the book.

Indeed, the mosaics introduced a novel paradigm in SAR imagery spatial analysis. This new way of doing things was inherited in part from an approach in image processing called multi-resolution analysis (MRA) [1] (see also the Appendix for a wavelet implementation). MRA is a way of building a virtual microscope that allows for zooming in and out of an image, thus providing views that are useful on the scale of interpretation that is best correlated with the underlying phenomena. The large mosaic data set, being assembled from high spatial resolution elements, provides at the same time the capability of doing analysis of slow varying extended phenomena (e.g. flooded areas) and to zoom in on small-scale or rapidly occurring local phenomena (e.g. logging).

Historically, up to the early 1990s, requirements for large-scale – regional to continental – monitoring of terrestrial ecosystems has led to the extensive use of low spatial resolution optical data from orbital instruments [2–4]. Among the most popular instruments in this context, we cite the NOAA Advanced Very High Resolution Radiometer (AVHRR) and the ESA Along Track Scanning Radiometer (ATSR). For example, land cover maps of Central Africa at 1 km and 5 km resolution were compiled using these instruments [5,6].

The launch in 1991 of the ESA ERS-1 satellite marked a turning point in the capability of monitoring terrestrial ecosystems, because a new observational window had been opened in the electromagnetic spectrum, and improved spatial and temporal resolutions could be provided by microwave technology in space. Our team at the Joint Research Center by then had acquired experience in radar remote sensing, but we were mainly fiddling with the new capabilities of microwaves from space in geographically limited experiments in tropical areas. Typically, the trajectory followed in these trials was to set up an investigation focusing on a given ecological theme in a test site, then set up an acquisition plan and order a few scenes from ESA, or even a time series matching the purpose of the experiment.

However, one day a new idea came into the world (well, into our team), and the prophet was Jean Paul Malingreau, the team leader, who suggested: what about trying to get a high resolution blanket coverage of a whole ecosystem by stitching together many, many SAR scenes? This idealized concept was supported by the concrete technical fact that an orbital SAR could indeed acquire continuously in time and space over large earth areas and in all weather conditions (barring extreme precipitation events). being an active microwave sensor.

The idea set in motion a joint ESA–JRC effort for the development of an ERS–SAR mosaic covering the entire biogeographical domain of the tropical forests in Central Africa. The project, code named Central Africa Mosaic Project – CAMP) [7,8] called for the acquisition and compilation of 477 ERS-1 scenes acquired in the period from July 15 to August 28, 1994. The mosaic canvas (Figure 1.1) depicts the whole rain forest domain of Central Africa as well as the Northern and Southern transition zones. It contains parts of Cameroon, Nigeria, Gabon, Guinea Bissau, Republique Centr'Africaine, Congo Brazzaville, Zaire, Rwanda and Burundi. The land covered by the mosaic is over 3 million square kilometers (km^2).

As a matter of fact, humankind was in the business of building mosaics well before our efforts. If you have a chance to make a trip to Enna, in Sicily, you will have the privilege of seeing beautiful artworks, compiled around the fourth century AD, the mosaics of the villa Romana del Casale, Piazza Armerina, such as the one shown in Figure 1.2.

These mosaics, assembled from many little colored glass tiles (tesserae), were, ante litteram, the result of a process similar to what nowadays is called digital image processing. The principle is the same: the combination of many components with low information content (binary numbers in digital images, ceramic tiles in the Armerina mosaics) yields a product with high information content (an image that suggests a meaning, a story or feelings to the observer).

On the other hand, this process is not reversible. If you pick out one little tile from the Armerina mosaic you get only a color. Similarly, if you pick out a pixel from a digital image you get only a number.

The striking difference is that if we pick out a tile from our SAR mosaic, we do not destroy information, rather we augment what we have.

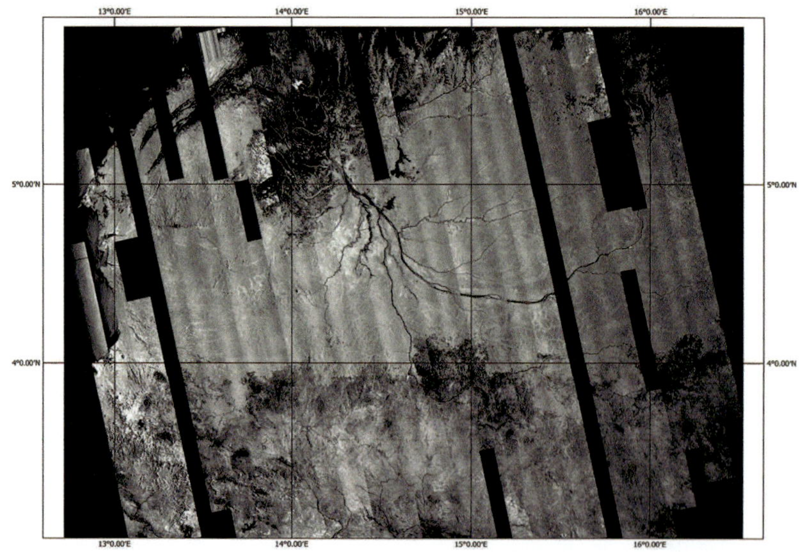

FIGURE 1.1 The Central Africa tropical forest domain depicted by the CAMP canvas (ESA–JRC 1992).

FIGURE 1.2 A mosaic at the Villa Romana del Casale, Piazza Armerina, Enna, Sicily (4th century A.D.).

To prove this statement, let us pick out a few tiles from the canvas in Figure 1.1.

The first tile (Figure 1.3) depicts the boundary between a mix of evergreen/semi-deciduous mesophyllous forest formations and the mixed seasonal savanna formations in the southern part of the Congo Basin. The evergreen cover and the more deciduous (trees) or senescent (grass) appear well delineated with high intensity contrast and rich textural details. The backscatter strong difference between the two vegetation classes is caused by the tile acquisition time, this corresponding to the dry season in the south (July–August).

The next example in our portraits room shows the boundary between the rain forest and the mixed savanna – gallery forest formations in the northern parts of the Congo Basin (Figure 1.4).

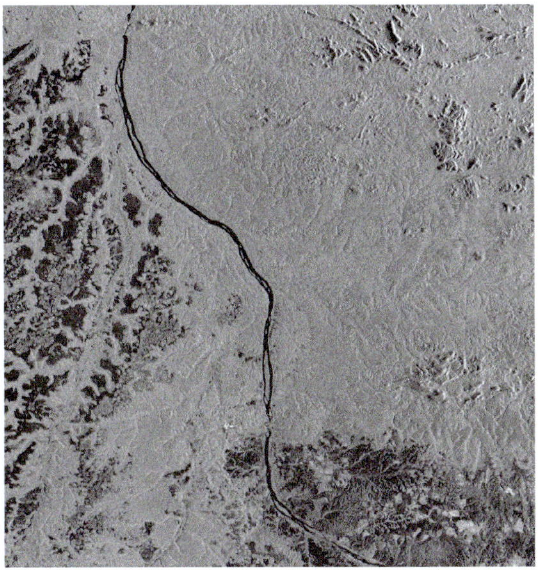

FIGURE 1.3 The boundary between the rain forest and the mixed savanna formations in the southern part of the Congo Basin.

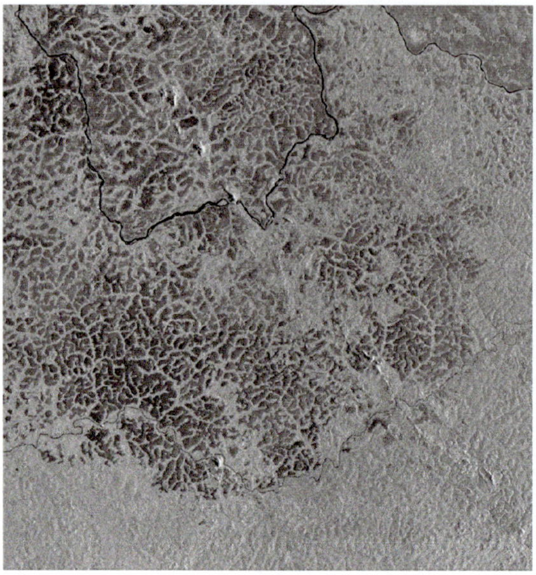

FIGURE 1.4 The boundary between the rain forest and the savanna in the northern part of the Congo Basin.

The boundaries are surprisingly sharp despite the acquisition time corresponding now to the rainy season.

In Figure 1.5 a ribbon of secondary formations following the older road network of Zaire is visible in contrast with the surrounding primary forest. These forest "galleries" correspond to the pattern of land management that since colonial times has essentially followed the road communication system.

The tile in Figure 1.6 takes us to the Congo River swamp forests. The swamp forests are an important ecosystem, since they host a great richness of flora and fauna. Moreover, they play an

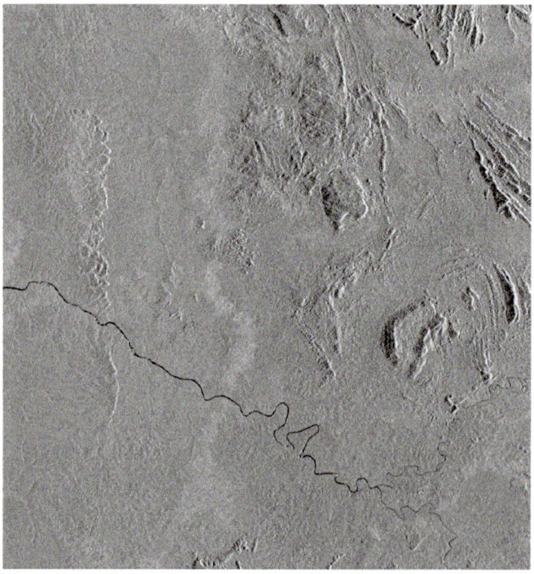

FIGURE 1.5 A ribbon of secondary formations (higher backscatter) emerges in contrast with the surrounding primary forest in this frame acquired in Zaire.

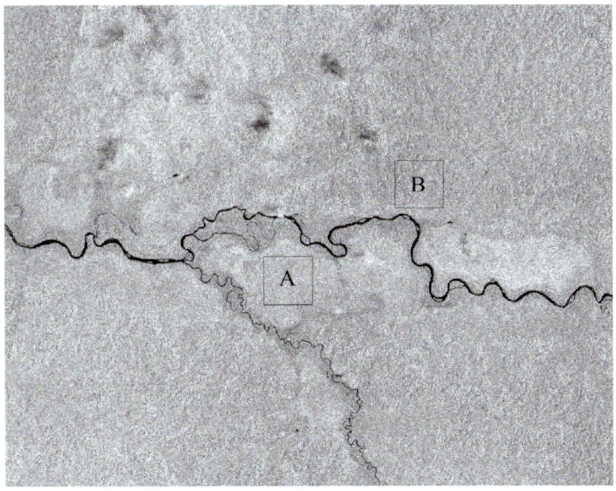

FIGURE 1.6 The Congo River swamp forests. The sample marked (A) is representative of the flooded swamp forest, while the sample (B) is located within the lowland rain forest.

important role in biosphere–atmosphere exchange processes. We shall see in later chapters how the smoother spatial distribution of the swamp forest upper canopy layer (area around marker A in Figure 1.6), in comparison to the more irregular structure of the primary forest (marker B), allows for the mapping of their extent. Moreover, the strong dihedral retro-reflection between the smooth water surface and the vegetation volume enables the detection of flooding conditions.

Shifting the view to the whole mosaic canvas, the visitor will indeed behold a unique and uniform cross-section of important tropical biomes, from the savanna and dry forest in the north, through the entire rain forest domain and again through the seasonal formations south of the Equator (savanna and edge of the Miombo woodland). The portrayed landscape crosses the Equator, and contains at the same time dry and rainy season acquisitions with a gradient of wetness between them. This allows the interpreter to carry out a range of observations with respect to the occurrence or to the lack of seasonal contrast between various vegetation formations.

The compilation of the large mosaic with its thematic richness was not trouble-free, though. Difficulties arose both in the area of digital signal transmission and processing, and in the area of radar science. We recollect among the major issues:

1. The spacecraft on-board memory capacity (6.5 Gbit) was very high for the solid state technology of the time (1990), but the AMI imaging SAR data, because of the high rate (105 Mb/sec), could not be stored in real time. Therefore, SAR raw data were acquired by a dedicated real-time downlink by the ESA Libreville ground station. Successively, they were brought by physical recording media to the German Processing and Archiving Facility (PAF) at DLR, where they were processed from raw signal data to ground range detected precision images (PRI, 12.8 m resolution). These products were finally shipped to our laboratory at the JRC by means of Exabyte tape cassettes.
2. The sheer volume of the high-resolution PRI data (62 Gbyte) required careful allocation and design of adequate storage and processing resources. This point must be made in the context of the computer technology of the time.
3. The geolocation and radiometric accuracy of the high-resolution data did set constraints on the capability of merging them into a seamless global image (spatially and radiometrically). In other words, the capability of fitting neighboring images into the mosaic jigsaw puzzle without discontinuities of the location and brightness of image features.
4. SAR imagery over the whole area is not acquired instantaneously. It is built up by sets of images acquired almost simultaneously with respect to the terrestrial ecosystem time constant

along one orbit, but at a time lag of approximately one month for two adjacent orbits. Changes in the vegetation or soil water content within this period could induce radiometric changes that needed to be accounted for if a smooth overall intensity was to be achieved at the margin between two orbits.

5. The SAR image intensity is highly affected by the local incidence angle. Therefore, since the rightmost margin of an image corresponds to far range (lower incidence angle) and the leftmost margin of the next neighboring image corresponds to near range (higher incidence angle), a discrepancy will always appear between images acquired in two neighboring orbits. These discrepancies will depend on the type of natural target, and on the scattering process (surface, volume). They cannot be corrected globally and a priori, but must be dealt with using an adaptive smoothing algorithm.

1.2 THE CAMP DATA PROCESSING MACHINE

The main tenet of the CAMP data processing was to achieve a hierarchy of products at different spatial and radiometric resolution, in such a way as to mimic the paradigm of multiresolution analysis. Four tiers of products can be located across the processing flow:

1. The original PRI ground range data sets at full resolution (12 m).
2. The mosaic tiles re-sampled at 100 m spacing.
3. Special purpose high-resolution products (such as speckle filtered amplitude imagery, texture products or classification maps).
4. Compact products with lower resolution and dynamic range for data distribution and documentation (e.g. CD ROMs).

The entire collated mosaic of the Central Africa domain is produced only at pixel spacing of 100 m, this being an optimal compromise between visual interpretability and data volume (a factor of 64 reduction with respect to the original data sets). This baseline product is assembled from the tiles in the second tier, and can be considered as a reference database, in which the user can navigate, zoom in at each point, and access higher resolution products. The full-resolution bespoke products in the third tier are then generated on request and according to the specific needs of the thematic analysis.

The most important passage in the data processing flow consists of the generation of the tiles in the second tier and their assemblage on to the mosaic canvas. Within this process, a hierarchy of products is defined: tiles, stripes, mosaic canvas. Products at one hierarchy level are composed of elements belonging to the previous level.

The mosaic construction starts with the selection of all frames that belong to a stripe; a stripe corresponds to a spatially contiguous set of images (with overlaps in latitude) acquired during one orbit of the satellite.

Each set belonging to a stripe is assembled using a two-step, successive approximation algorithm: (1) a coarse registration algorithm, based on the geographic coordinates of the PRI image corner points, as given in the SAR CEOS leader file; and (2) a fine registration algorithm that exploits the cross-correlation between two images in the overlap areas. Correlation between images in the long-track overlap area is high because the same speckle pattern is present in both images.

Radiometric balancing between tiles is performed in some exceptional conditions. Indeed, because of the stability of the ERS sensor, we assume there is a spatially and temporally invariant radiometric calibration error across the whole mosaic. Circumstances when this condition fails because of instrumental and processing errors are described in the next section.

The stripes are therefore generated in the SAR image reference frame (PRI ground-range projected and de-skewed images with no relief correction). This method has some advantages with respect to mapping each frame directly into a cartographic reference system. Basically, the stability of the spacecraft orbit, the low error sensitivity of the slant to ground range mapping process, and the cross-correlation between adjacent frames are exploited to achieve a fit of high accuracy among the elements.

On the other hand, precision geocoding and terrain correction based on the solution of the range Doppler equations was ruled out, since a digital elevation model of sufficient accuracy, such as SRTM (see next chapter), was not yet available over the whole Central African region.

For each stripe assembled in the ground-range projection, a number of control points corresponding to the corners of the individual tiles are available, since each PRI image contains its geodetic coordinates (latitude and longitude) in the leader file. The array of control points is used to perform a second-order polynomial least-squares fitting between the pixel coordinates in the image reference frame and the latitude/longitude coordinates in the geodetic reference frame. The associated error budget in rms is always less than 1 pixel (100 m).

Stripes are subsequently sorted by longitude and warped into the geographical reference frame using the coefficients computed by the least-squares fitting; the reference ellipsoid is taken to be the Goddard Earth Model 6 (GEM6), with an oblateness coefficient of 1/298.257.

Finally, the stripes are pasted with a simple left-to-right overlap algorithm into an incrementally growing canvas. Given the stripe geometry in the geographic reference frame, sentinel values are used to mask areas outside the backscatter values. The masked part of the stripe needs to be overlapped transparently.

1.3 RADIOMETRY

The foundations of the mosaic's radiometric properties lay on the ERS AMI radar instrument radiometric stability. Transponder measurements over two years [9] yielded a radiometric stability of 0.38 dB (standard deviation of the measured values), and a maximum variation of the measured RCS (radar cross-section) of $\pm 0.75\ dB$. These values were indeed halved after a correction of the ADC power loss estimate was performed.

Image processing in the mosaic assembly was tailored to reproducing at any point of the mosaic the capability of estimating the radar cross-section as closely as possible to the intrinsic accuracy of the instrument. This objective is achieved by proper calibration of each tile (range spreading loss, antenna pattern removal, effective ground scattering area correction), but otherwise with no further radiometric correction, e.g. to reduce intensity differences between neighboring tiles. In other words, accuracy of the RCS estimation was not traded off by a better visual appearance.

Indeed, the RCS will have temporal and spatial variations across the mosaic, these being caused by the scattering mechanism (e.g. RCS dependence on incidence angle) or to target seasonal or meteorological conditions. According to their cause, these changes can be localized or common to an entire orbit. In this case, even small radiometric differences (order of 0.5 dB) produce a visual banding effect in the mosaic image (see Figure 1.1).

The dichotomy between overall mosaic smoothness and RCS estimation accuracy never concedes a win–win strategy. The choice taken in the CAMP development was particularly suitable for applications based on the underlying physics of microwave scattering (for example, comparison with an RCS database) [10]. As we shall see in the next chapter, the dual strategy (global normalization of the image intensity) was adopted in other Africa mosaic compilations, this being the most reasonable choice for enabling unsupervised classification products.

The radiometric changes caused by variation in the underlying RCS over time and within the image will be discussed next.

1.3.1 Radiometric Changes in Time

The entire snapshot of the Central Africa Mosaic is not taken instantaneously, but in a time frame (two months) when changes can be expected both in the land cover, in the meteorological conditions, and in the sensor's nominal parameters. In more detail, two contiguous stripes (a set of consecutive ERS-1 frames along the same orbit) are acquired roughly 18 days apart, while the third stripe in a row (moving along longitude) is acquired three days apart with respect to the first. This temporal pattern gives rise to a number of effects that must be taken into account for the mosaic thematic interpretation.

Beyond localized and intermittent radiometric changes related to varying target properties, it was important to investigate the presence of systematic patterns in the stripes' time series. For this purpose, a target with a timewise stable radiometric response is needed. A good candidate is the primary rain forest.

Characterization of the primary forest radar cross-section using the C-band Wind Scatterometer (WSC) of the ERS-1 spacecraft is given in [29]. Findings in this study prove that the backscattering coefficient of the primary forest is very stable and, because of the scatterometer system's stability can be estimated typically with a mean value standard deviation of 0.22 dB. As to the time variations, diurnal changes of 0.3 dB and seasonal changes of 0.8 dB are observed. These results provide an excellent basis for interpreting the CAMP radiometric values in terms of time stability and incidence angle dependency.

To establish a comparison with the scatterometer reference data, the backscattering coefficient for the primary rainforest was estimated from the mosaic data, following the procedure outlined in [25]. For each stripe, homogeneous areas in the primary forest were selected visually at different range positions to take into account the local incidence angle dependency. Moreover, care was taken to select flat terrain areas to reduce the influence of topography.

Comparing values at the same range position (angular effects will be discussed in the next section), we obtained an inter-stripe pattern of variations. In most cases, the pattern follows an alternating sequence between adjacent stripes, this corresponding in time to a period of 18 days. The maximum difference is 0.8 dB.

The values were also sorted by time to investigate the presence of a trend in the time series, which would indicate, for example, a systematic drift in the sensor's electronics, say. No such trend was observed.

Still, the fact that no meaningful trend was observed in time, but radiometric differences are systematic within stripes points at a calibration problem, probably because of a power imbalance from imaging sequence to imaging sequence introduced by the ERS-1 internal calibration procedure. Indeed, the transmitted pulse replica, used by the PAF SAR processor to scale the raw data, may sometime be corrupted and not correlated to the true transmitted power at the time of the image acquisition [26]. Alternative ways to estimate the transmitted power were sought (e.g. estimates of the true transmitted power from historical log data available at ESA), but no accurate or robust solution could be found.

The lesson learnt was that the interpretation of the scattering coefficient wide-area time series in terms of system-related effects or scattering mechanism effects is problematic, because the values we are trying to estimate lie both within the range of natural variations in the primary forest and within the calibration uncertainty.

In contrast to the wide-area inter-stripe changes, several cases of localized radiometric changes caused by temporal variation of the natural target could be found in the mosaic. For example, changes in the herbaceous savanna are observed, these being ascribed to changes in the moisture content or to the removal of the absorbing grass layer following, for example, a fire event.

Indeed, these variations may be exploited in the overlap areas between contiguous frames to detect ecosystem short-term changes as a result of human activities. An interesting case is reported for an example in [24], where two images acquired at a time lag of 1 month over the Plateau Bateke, at the border between Gabon and the Congo Republic, show a sharp increase of the radar backscattering coefficient in the savanna. A number of contextual features lend support to the hypothesis that these radiometric changes could be caused by burn scars.

1.3.2 Within Tile Radiometric Changes in range

Within each image in the mosaic (tile) and in the range direction, the backscatter power can change because of three factors:

1. The effective scattering area on the ground corresponding to the fixed sampling rate in slant range changes with ground range.
2. The radar cross-section of a natural target depends on the incidence angle; assuming flat terrain and taking into account only the curvature of the earth ellipsoid, the ERS-1 nominal incidence angle varies approximately from 19.5 to 26.5.
3. If topography is present, the range of local incidence angles is much wider than the nominal one given above (spanning the gamut from shadow to layover state) and therefore the effect on the radar cross-section variation is enhanced.

The effect in (1) is corrected by a calibration procedure [11], that amounts to the calculation of the correction factor $\sin\alpha_i / \sin\alpha_{REF}$, where α_i is the local incidence angle for flat terrain and a function of the range distance.

Regarding the effect in 2), the angular dependence of the backscattering coefficient σ^0 over different vegetation types in Africa was characterized in an experiment [12] using data from the ERS-1 wind scatterometer (WSC). A linear model of the dependence of the backscattering coefficient on incidence angle in logarithmic scale $10\,log_{10}\sigma^0 = f(\vartheta)$ is proposed in this study. The measured rates of change are 0.06 dB/deg for the primary rain forest, 0.12 dB/deg for the woody savanna, and 0.15 dB/deg for Sahelian areas. This means that, in the typical range of incidence angles on flat terrain within an ERS-1 image, we can have a difference $\nabla\sigma^0$ of 0.42 dB from near range to far range, even for the stable primary forest target. For woody savanna, we would have a difference of 1 dB. This was verified by estimating values at near and far range for the primary forest in the ERS-1 PRI images, after correction for the effective ground scattering area.

Because of the σ^0 angular dependency, the same target in the overlap area of two consecutive stripes, being at far range in one stripe and at near range in the next, will yield slightly different σ^0 values. As we shall see in the next chapter, a mosaic-wide image intensity normalization, relying on inter-stripe overlap area differences, needs to perform a global optimization to avoid propagation of the errors due to the σ^0 angular dependency.

1.3.3 Quantization Noise

Nowadays remote sensing imagery is usually represented and processed with 32- or 64-bit words, using processors that also provide internally this kind of data bus width. However, at the time of the CAMP mosaic compilation, all data sets, from the original ESA PRIs down to the mosaic canvas, were represented in 16-bit integer digital numbers. Since the mosaic data were poised also to support applications where the radar cross-section estimation was of importance, the issue arose of how much quantizing the backscatter amplitude to 16-bit integer values might impact the estimator.

To answer the question, numerical analysis of quantization noise was carried out using the SAR backscatter one-point statistic over selected mosaic regions of interest, corresponding to stationary targets. Since this type of SAR data set characterization is not very common in the literature, we reckon it is worthwhile to summarize here selected results of the analysis.

For a linear quantizer using N levels and lower and upper limits t_{low} t_{up}, the mean squared signal (RMS) after quantization can be calculated as:

$$\langle s^2 \rangle = \sum_{i=1}^{N} a_i^2 \int_{a_i}^{a_i+a} p(x)\,dx \tag{1.1}$$

where a is the spacing between quantization levels, a_i is the *ith* quantization level, and $p(x)$ is the PDF of the random signal.

The quantization error is defined as the difference between the true signal value and the quantization level assigned to that value. The mean-squared quantization error is given by:

$$\langle \epsilon^2 \rangle = \int \epsilon^2 f(\epsilon) d\epsilon = \sum_i \int_{a_i}^{a_i+a} (v - a_i)^2 p(v) dv \tag{1.2}$$

Where $p(\nu)$ is the PDF of the signal, $f(\varepsilon)$is the PDF of the quantization error, a_i is the *ith* quantization level, and a is the level spacing.

From Equations (1.1) and (1.2) the RMS signal to quantization noise ratio $\langle S^2 \rangle / \langle \epsilon^2 \rangle$ was computed for samples of amplitude data extracted within ROIs of the PRI data sets (see Table 1.1)

The signal to noise (SNR) is inversely proportional to the number of looks. This happens because SNR depends on the shape of the data PDF (see Equations (1.1) and (1.2)). Therefore, for higher number of looks the PDF gets narrower (smaller variance) and the noise increases in the bins that are more populated.

On an absolute basis, it is instructive to compare the noise level introduced by quantization with the inherent noise level present in the SAR signal because of speckle statistics.

A measure of speckle noise is the ratio of the amplitude mean squared to variance. For an N-Look amplitude image the ratio is:

$$\frac{VAR(A)}{\langle A \rangle^2} = \frac{N\Gamma^2(N) - \Gamma^2\left(N + \frac{1}{2}\right)}{\Gamma^2\left(N + \frac{1}{2}\right)}$$

For a 3-Look image we have $S/N = 10.63$ dB, and for a 24-Look image $S/N = 19.8$ dB. Therefore, the quantization noise is by far less important than the intrinsic speckle noise at the 8-bit level.

Indeed, the major problem with 8-bit quantization of the SAR image is the impact on the first-order statistical properties of the compressed image rather than the noise level. Bit quantization of a discrete digital signal is equivalent to a nonlinear transformation in which, for each quantization bin, the linear transfer function between the input and output signals is replaced by a unit step function centered on the bin. Therefore, the statistical properties of the signal after the quantizer will be modified. In general, if a random variable $X(RV)$ is characterized by a PDF $f_X(x)$, the RV after a transformation $y = T(x)$ will have a modified PDF $f_Y(y)$. Therefore, statistical estimators, such as one-point mean value and variance, which rely on the PDF of the RV, will give different results if applied before or after the quantization.

To illustrate the problem, we report in Table 1.2 the mean intensity of 24-Look samples taken within the same ROIs as in Table 1.1, and the intensity computed from Equation (1.1), this corresponding to

TABLE 1.1
RMS Signal-to-noise Quantization Ratio for PRI Data Represented with 8 bit and 16-bit Words and Processed with 3 and 24 Looks

Classes	⟨A⟩	S/N 8b 3 looks	S/N 8b 24 looks	ESA PRI
Water	97.7	48.8	44.4	91.0
Savanna	207.6	57.1	52.6	94.0
Forest	382.5	64.9	60.5	99.0

TABLE 1.2
Mean Intensity Value of a 24-Look Amplitude Image, when Estimated before and after an 8-bit Quantizer. $\langle I_T \rangle$ is the mean Intensity Computed from the Theoretical PDF of the Original Image, $\langle I_Q \rangle$ is the Estimated mean Intensity after the Quantizer and ϵ is the Relative Error

Classes	$\langle I_T \rangle$	$\langle I_Q \rangle$	$\lvert \varepsilon \rvert$
Forest	159368	157816	0.97 %
Savanna	47969	47119	1.77 %
Water	13576	13126	3.31 %

an 8-bit quantization. The amplification of errors of the statistical estimator is inversely proportional to the backscatter intensity. Similar effects must be expected with higher order moments. Therefore care should be taken to evaluate texture measures using compressed 8 bit imagery.

NOTE

1. *Aida* is an opera in four acts by the Italian composer Giuseppe Verdi with libretto by Antonio Ghislanzoni.

REFERENCES

1. S.G. Mallat, "A theory for multi-resolution signal decomposition: The wavelet representation," *IEEE Transactions on Pattern Machine Inteligence*, vol. 11, p. 674, July 1989.
2. C. Justice, J.R.G. Townsend, B.N. Holben, and C.J. Tucker, "Analysis of the phenology of global vegetation using meteorological satellite data", *Inernational Journal of Remote Sensing*, vol. 6, 1271–1281, 1985.
3. E. Lambin, and D. Ehrlich, "Combining vegetation indices and surface temperature for land cover mapping at broad spatial scales", *International Journal of Remote Sensing*, vol. 16, 573–579, 1995.
4. C.J. Tucker, J.R. Townsend, and T.E. Goff, "African land cover classification using satellite data", *Science*, vol. 227, 369–375, 1985.
5. P. Mayaux, E. Janodet, C.M. Blair-Myers, and P. Legeay-Janvier, "Vegetation map of Central Africa scale 1/5M", *TREES Publications Series D, No. 1, EUR 17322 EN*, European Commission, Luxembourg, 1997.
6. N.T. Laporte, S.J. Goetz, C.O. Justice, and M. Heinicke, "A new land cover map of Central Africa derived from multi-resolution multi-temporal AVHRR data", *International Journal of Remote Sensing*, vol. 19, 3537–3550, 1998.
7. J.P. Malingreau, G.F. De Grandi, M. Leysen, P. Mayaux, and M. Simard, "*The ERS-1 Central Africa Mosaic: a New Role for Radar Remote Sensing in Global Studies of the Tropical Ecosystem*", *Proc. IEEE IGARSS'97*, pp. 1725, Aug. 3–8, 1997, Singapore.
8. G.F. De Grandi, J.P. Malingreau, M. Leysen, "The ERS-1 Central Africa mosaic: A new perspective in radar remote sensing for the global monitoring of vegetation", *IEEE Transactions on Geoscience Remote Sensing*, vol. 37, no. 3, pp. 1730–1746, May 1999.
9. H. Laur, J. Sanchez, E. Dwyer, and P. Meadows, "*ERS-1 SAR radiometric calibration*," in *Proc. CEOS Calibration Workshop*, Noodwijk, The Netherlands, Sept. 1993.
10. F.T. Ulaby, and M. C. Dobson, "*Handbook of Radar Scattering Statistics for Terrain*", Artech House, Inc. 1989.
11. H. Laur, P. Bally, P. Meadows, J. Sanchez, B. Schaettler, and E. Lopinto, "Derivation of backscattering coefficient σ^0 in ERS-1 SAR PRI products," *ESA Doc.* ES-TN-RS-PM-HL09, Issue 2, Rev. 4, May 1997.
12. P. L. Frison, and E. Mougin, "Use of ERS-1 wind scatterometer data over land surfaces," *IEEE Transactions on Geosci. Remote Sensing*, vol. 34, pp. 550–560, Mar. 1996.

2 The L-Band Breed

The GRFM Africa Radar Mosaic

2.1 THE GRFM PROJECT

The tabloids report that in 1996 a technical delegation from the National Agency for Space Development of Japan (NASDA, nowadays Japan Agency for Space Exploration – JAXA), visited the JRC to enquire whether, in the wake of the CAMP project, there might be grounds to set up a similar collaborative project using the Japanese JERS-1 radar. The delegation, to their disdain, was briefed by a weird fella (the first author), who talked about Einstein's relativity, about a relativistic mosaic, where time and space were bound together, and space was not absolute. My space is not the same as your space, because with my dilated space I can look at an entire ecosystem, and with your contracted space you can spot a subtle logging road.

Despite this commotion, the seed was planted for the launch of a new, and extended, global tropical vegetation radar mapping project: the Global Rain Forest Mapping project (GRFM) [1,2].

The scope of the project was to produce a pan-tropical map of the world's tropical forests, from South and Central America to Africa, Asia and Australia, using imagery acquired by the L-band radar aboard the JERS-1 spacecraft. Though GRFM was an initiative of NASDA, the overall effort was shared among several institutions and relied on an extensive international collaboration for data processing and for the science program. In particular, the European Commission Joint Research Center (JRC) and Caltech–NASA Jet Propulsion Laboratory (JPL) took the lead for the part related to tropical Africa. Next we shall expound this part of the project: GRFM Africa.

As in the CAMP approach [17], the major landmark of GRFM was the capability of generating regional-scale thematic information related to the vegetation's bio-physical and structural parameters from the aggregation of many high-resolution radar observations acquired with spatial and temporal continuity, and with measurable radiometric and geometric accuracy. In a sense, we could talk of dual variables in a Heisenberg principle: resolution and coverage. Contrary to classical remote sensing approaches, in GRFM large continental scale areas of some million km^2 are mapped using a resolution of tens of meters. Stretching the coverage resolution duality implied the solution of a number of challenging technical problems, which will be discussed below.

The GRFM project calls for the coverage of a continental-scale geographic area (a linear distance of some 6000 km in the case of the Africa mosaic) using a SAR sensor that acquires a scene within a swath of some 75 km and with a ground resolution of approximately 18 m. This is obtained by tiling together several acquisitions that are not taken instantaneously but still within a short time frame (two months). Moreover, because of the orbital characteristics of the JERS-1 spacecraft, contiguous swaths are imaged within two consecutive days, which assures a smooth time evolution through the whole coverage.

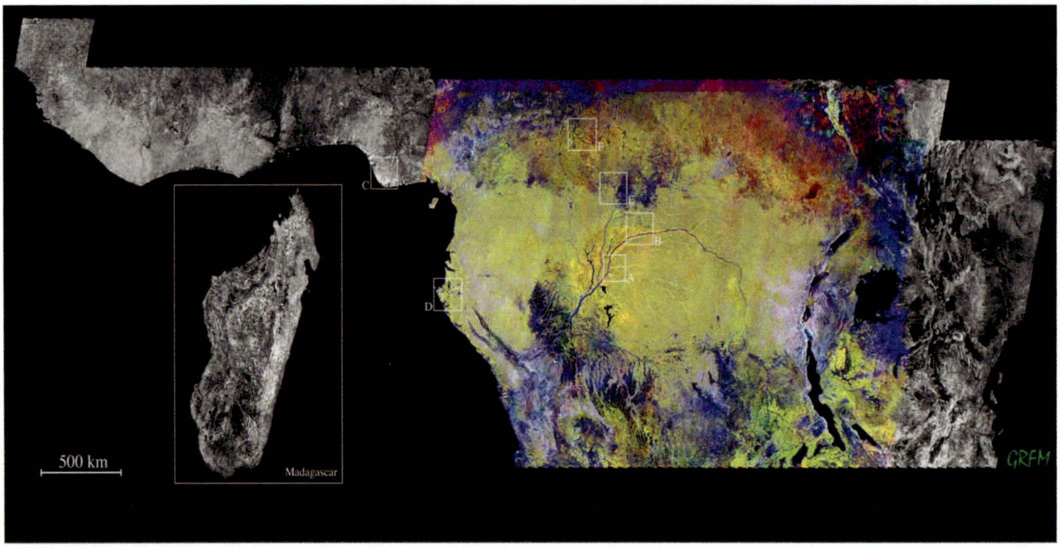

FIGURE 2.1 The GRFM Africa mosaic. The mosaic extends from the Western coast of Africa in Sierra Leone at 14°*W* to the Eastern coast in Kenya and Tanzania at 42°*E* in longitude and from 10°*S* to 10°*N* in latitude. In addition, the island of Madagascar is included (see inset). A blanket coverage of the whole area was acquired during January–March 1996 in the low water season of the Congo River. A second acquisition from 8°*E* to 36°*E* was also performed during the high-water season in October–November 1996. The area covered by the two acquisitions is shown in false color in the figure. Image courtesy of JAXA and EC JRC.

The resulting product [3] is a spatially continuous radar map (Figure 2.1), which can eventually be degraded to lower resolutions for studying scale-dependent natural phenomena, or to extract thematic information at the most suitable resolution. In the case of tropical ecosystems, the merit of the approach therefore lies in the possibility of estimating some geophysical parameters of interest (e.g. biomes distribution, deforestation) globally and at unprecedented resolution.

Previous approaches, relying on optical sensors, produced either wide area, continuous sampling but low resolution maps (e.g. the vegetation maps based on NOAA AVHRR 1 km data [4]), or high resolution but random coverage data sets [5–7] (e.g. LANDSAT TM data). In the latter case, some global geophysical parameters were then estimated by extrapolation using statistical sampling techniques.

The combination of continuous coverage, short acquisition time and high resolution is a landmark asset of the GRFM radar mosaic. Experience gained to date lets us confidently say that data sets such as the GRFM mosaic will add significantly to our knowledge of the tropical and other poorly documented earth ecosystems. However, as in the case of the precursor CAMP mosaic, the benefit comes at the cost of a number of technical hurdles. Most of these were discussed in the previous chapter, but we recap the main points here in common with the CAMP effort, thus highlighting the peculiarities of the GRFM project.

First, the end-to-end process from the satellite down link to the generation of several layers of lower resolution products entails a staggering data volume, and a high processing complexity. Experience in handling large data volumes was certainly a benefit of the CAMP project. Nevertheless, the GRFM project introduced one more shift in complexity, caused by the geographically distributed processing and the layers of low-resolution products (multi-temporal, multi-resolution).

To give a flavor, the GRFM Africa mosaic consists of some 3900 SAR scenes, which is tantamount to 312 Gigabytes of high-resolution ground range data, and 4.8 Gigabytes for one 100 m baseline product, without accounting for all the intermediate and special purpose lower resolution products.

This level of data volume was already posing some problems even for simple operations such as data ingestion at a time when the available off-the-shelf computer technology was mainly based on high-end Unix servers, such as Digital Equipment Corporation Vaxen.

Second, since the coverage is not obtained in one snapshot, perturbation of the imaging system's nominal values (e.g. the satellite attitude) affects the images that compose the mosaic differently according to their position (e.g. radiometric and geometric distortions). These effects must be taken into account and corrected for; moreover, the problem of parameters estimation is exacerbated by the data volume issue, because most manual techniques, such as tie-pointing, must be ruled out.

Third, to be validated, even a self-consistent measure of a physical parameter still requires a comparison with independent estimates. Validation of the image characteristics and, as a second step, of the thematic information that can eventually be extracted (e.g. a vegetation map), is indeed in our opinion one of the central issues in the generation and exploitation of wide area high resolution remote sensing data sets.

The validation of a measurement that depends on the geographic location and extends over a wide area requires a reference set of known and suitable accuracy, and dense enough to assure the correct sampling of the signal to validate. The problem is compounded by the fact that ad hoc ground experiments were at the time difficult to set up in areas of the world like Central Africa, which were plagued by a high degree of social and political unrest. Historical data sets were in this case the only source for benchmarks.

In the context of the GRFM Africa, post-processing the validation of the georeferencing accuracy was of particular relevance, because clearly it constitutes a prerequisite to the use of the GRFM radar maps for cartographic applications.

These peculiar connotations needed to be taken into account in the GRFM development process, and constituted the rationale for the development of suitable processing techniques and tools (see diagram in Figure 2.2). Selected topics related to these techniques and tools are the subjects of the following section.

2.2 THE GRFM AFRICA PROCESSING CHAIN

2.2.1 Input Datasets

The JERS-1 SAR data acquisitions of the GRFM Africa project include two blanket coverages of the Central Africa tropical region between 10° *S* and 10° *N*. The first coverage was acquired during January–March 1996 and extends between 8° *E* and 42° *E*; the second during October–November 1996 and extends between 8° *E* and 36° *E*. The two data sets will be dubbed in the following as the "low water" and "high water" data set, with reference to the perceived hydrological state of the main river network in the region, the Congo River and its tributaries. In addition, the low-water data set covers West Africa from 14° *W* to 8° *E* and the island of Madagascar (acquired January 1997).

The SAR raw data, corresponding to these acquisitions, were correlated at NASDA Earth Observations Center EOC. The output of this processing phase is a product defined by NASDA as a standard geocoded image in a Universal Transverse Mercator (UTM) projection and dubbed a Level 2.1 product. However, the rows and columns of the Level 2.1 raster file are aligned with the slant range and azimuth directions, and not with the Mercator reference system.

The low water acquisition comprises 2173 scenes, the high-water acquisition 1460 scenes in continental Africa and 263 scenes for the island of Madagascar. All the scenes were acquired on descending orbits.

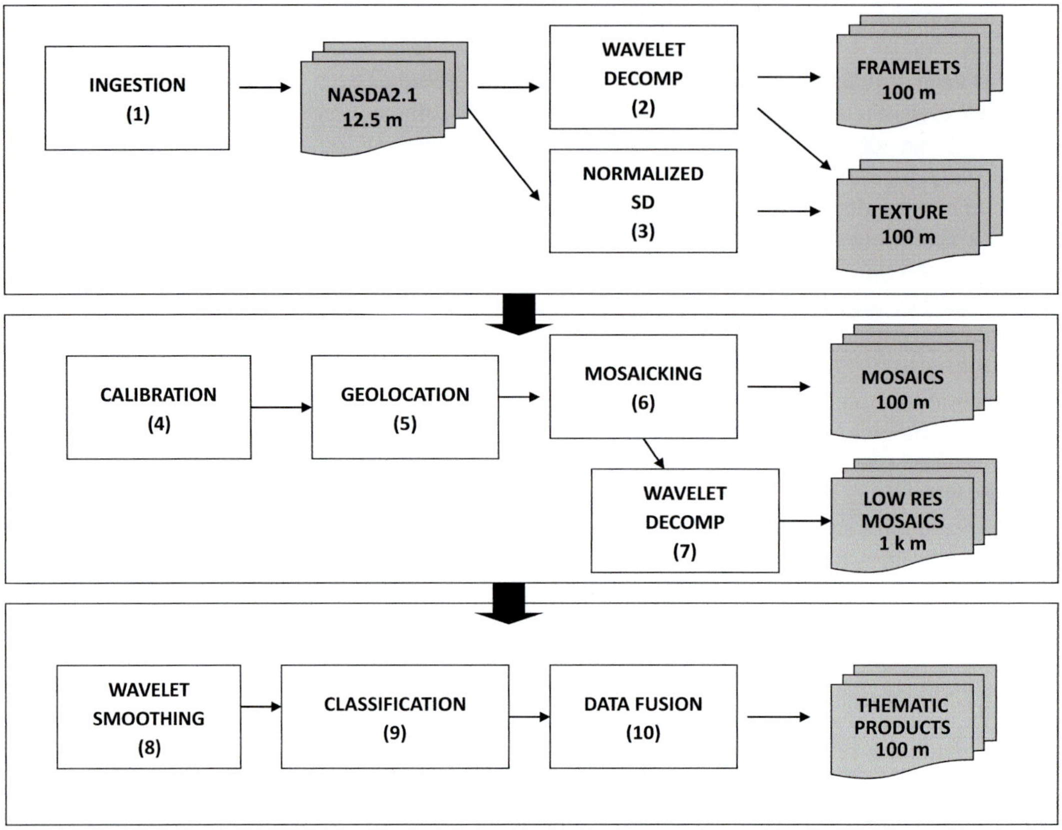

FIGURE 2.2 Flow diagram of the GRFM Africa processing chain.

2.2.2 Data Flow

Proper management of the data flow, in terms of hardware resource allocation, operation synchronization and housekeeping information is an issue of primary importance in a complex processing system such as the GRFM one.

Typically, the following data flow takes place. The NASDA level 2.1 high resolution images are ingested in batches into the system. The data ingestion step (Figure 2.2, block 1) is the most time critical in the overall chain of operations. The GRFM Africa dataset ingestion took approximately two calendar months.

The high-resolution data are then processed by the wavelet multiresolution decomposition block (block 2, Figure 2.2), which generates images and texture measures down-sampled at 100 m (see Section 2.3 and Appendix). Also, a texture measure based on the normalized standard deviation is derived from the high-resolution data (block 3, Figure 2.2). The 100 m products are the basic building blocks for all successive operations in building the mosaic; products at this resolution will be called "framelets".

The framelets are radiometrically calibrated (block 4, Figure 2.2), geolocated and tiled together (block 5 and block 6 in Figure 2.2; see Section 2.2). This step produces the baseline bi-temporal mosaics at 100 m resolution.

Larger pixel size (lower resolution) amplitude and texture measure datasets are generated again by wavelet decomposition from the baseline products (block 7, Figure 2.2). Finally, the multi-scale radiometric and texture products are used to support further analysis, data distribution, visual interpretation and automatic thematic information extraction (blocks 9 and 10, Figure 2.2).

Classification of GRFM mosaics is performed using algorithms that assume a piece-wise constant image model with additive white Gaussian noise (see Chapter 8). To adapt the SAR imagery to this type of algorithm, a pre-processing step is adopted that generates edge-preserving smooth approximations of the SAR imagery (block 8, Figure 2.2). This technique is based on detecting and following the maxima of a particular wavelet representation (with a non-orthogonal basis) and discussed in Chapter 8.

2.3 GEOLOCATION

2.3.1 The Block Adjustment Method

The geolocation of the individual scenes (frames) is a fundamental step in the compilation of the mosaic. Each image is acquired by the radar instrument in a viewing geometry that is dependent on the instantaneous position of the spacecraft with respect to the target on the ground, and the attitude of the sensor. Moreover, radar is a ranging device; the image is therefore acquired in a reference system where the coordinates are distances with respect to the line of sight from the instrument to the terrain (slant-range geometry).

To produce a cartographic product, the radar image must be projected into a geographic reference system; for example, a Mercator or an equiangular projection. Errors in the determination of the sensor's position and attitude, and errors in the cartographic projection become errors in the relative positions of the frames when they are tiled one next to the other in the mosaic compilation.

Moreover, these errors result in discrepancies between the positions of features in the image and the true position of these features on the geoid. The relative positioning of the frames is referred to as the mosaic internal geometric consistency, while absolute positioning with respect to a reference system anchored to the geoid is called absolute geometric consistency.

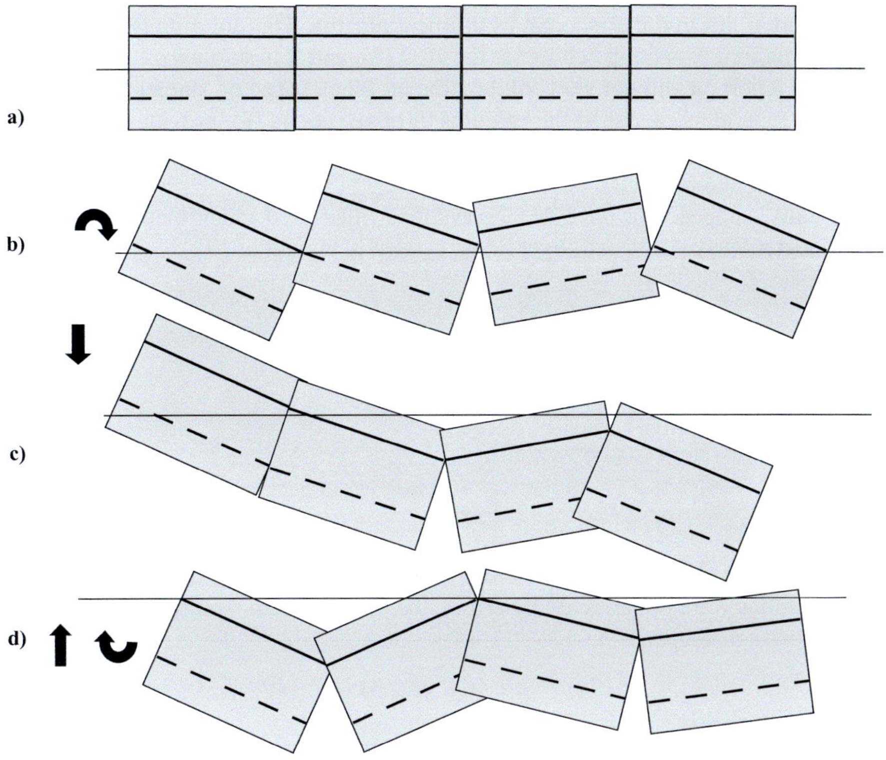

FIGURE 2.3 The problem of block adjustment for the geolocation of the SAR images.

To illustrate the problem let us consider a simple example. Suppose we have four images with two linear features (see Figure 2.3). In the cartographic system they should ideally all line up in a straight-line configuration (Figure 2.3a), where both linear features are perfectly matching.

However, suppose that, because of our imperfect knowledge of the sensor's position and attitude, the frames ended up being rotated with respect to their center point when projected into the geographic reference system. The situation is illustrated in Figure 2.3b. Clearly, the linear features would no longer match at the margin between one frame and the next.

The problem is to find a method that would allow us to minimize these relative positioning errors, based only on measures we can perform on the images themselves. Observations of the discrepancies between homologous features in neighboring scenes can provide us with useful information about the relative positions of the frames. The next problem is how to use these measures.

We assume further that every single image, once mapped into the geographic reference system, is internally consistent; that is, our imaging system does not introduce geometric deformations. In this case, we have only to be concerned about the translation and rotation of a rigid body.

Suppose that as a first trial we fix the position of the leftmost frame, and translate in a sequence from left to right the next ones in such a way as to make the upper linear features match at the boundaries (see Figure 2.3c). This strategy would have catastrophic results, because while errors relative to the first feature would indeed be minimized, errors relative to the second feature would not. Moreover, propagation of errors along the chain would result in a progressive displacement of the features from their true positions.

A better trial could be to rotate the frames and match the upper features instead of translating them. The result is shown in Figure 2.3d. This move looks better in term of error propagation, but still errors relative to the second feature will not be reduced.

What is needed is a global optimization technique that would find the best combination of translations and rotations with respect to the minimization of all the discrepancies among all features in the image (see Figure 2.4). The solution to the global optimization problem adopted for the GRFM mosaics compilation lies in the mathematics of linear algebra. This solution is mainly thanks to Dr Yrjo Rauste, at the time a visiting scholar at the JRC [8], and it is summarized next. In the following, we will also report on another viewpoint on the problem, based on statistical estimation theory.

A branch of linear algebra, called the calculus of observations [9], provides the groundwork for such an optimization technique Let us first introduce the principles of this matrix technique and then see how it is applied in our specific case.

Mathematically, suppose we have M observations (measured quantities) that are linked to N unknown parameters (to be adjusted) by a linear system of equations [9]:

$$\begin{aligned} a_{11}x_1 + \cdots + a_{1n}x_n &= b_1 \\ \cdots\cdots\cdots\cdots&\cdots\cdots\cdots \\ a_{m1}x_1 + \cdots + a_{mn}x_n &= b_m \end{aligned} \tag{2.1}$$

In matrix form the system can be written:

$$\underline{A}\underline{x} = \underline{b} \tag{2.2}$$

where $\underline{x}$ is an M vector of unknown parameters, $\underline{A}$ is an N by M matrix, and $\underline{b}$ is an N vector.

If the observations were not affected by errors, the linear system would be consistent; otherwise we would measure a discrepancy δ:

$$\underline{\delta} = \underline{A}\underline{x} - \underline{b} \tag{2.3}$$

We wish to determine the vector $\underline{x}$ that would minimize this discrepancy in a least squares sense:

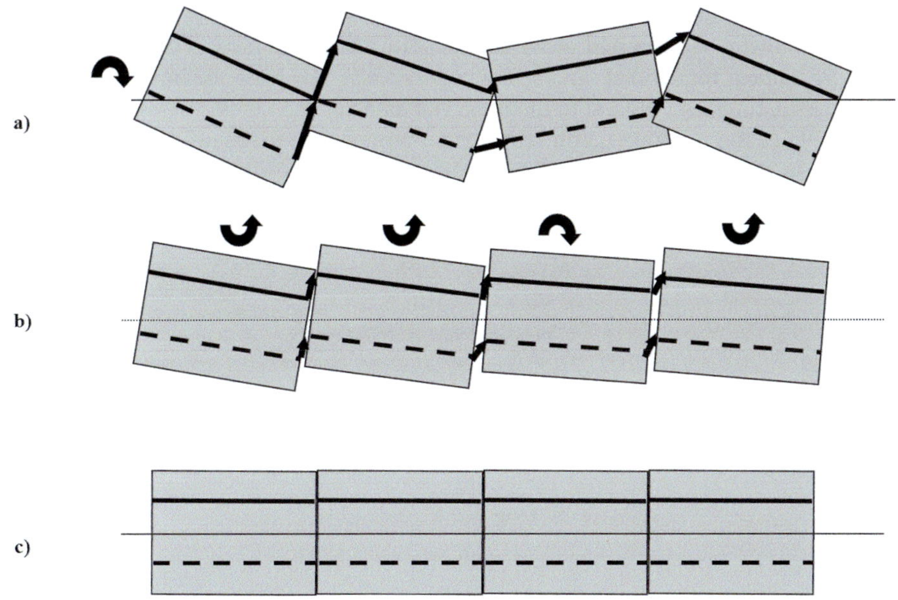

FIGURE 2.4 The solution of the block adjustment problem by linear least squares minimization of discrepancies.

$$\underline{\delta}^T\underline{\delta} = \underline{x}^T\underline{A}^T\underline{A}\underline{x} - \underline{x}^T\underline{A}^T\underline{b} - \underline{b}^T\underline{A}\underline{x} - \underline{b}^T\underline{b} \tag{2.4}$$

The condition for a minimum is:

$$\frac{\partial}{\partial \underline{x}}\left(\underline{\delta}^T\underline{\delta}\right) = \underline{0} \tag{2.5}$$

If $\underline{A}^T\underline{A}$ is non-singular, then the solution is:

$$\underline{x} = \left(\underline{A}^T\underline{A}\right)^{-1}\underline{A}^T\underline{b} \tag{2.6}$$

These are so-called normal equations and can be solved for the parameters vector $\underline{x}$ if the NxN matrix is non-singular. Weighting factors can also be introduced in the error criterion in order to emphasize the samples that are deemed more reliable. Then the error to be minimized becomes:

$$\underline{\delta}^T\underline{W}\underline{\delta} = \left(\underline{b} - \underline{A}\underline{x}\right)^T\underline{W}\left(\underline{b} - \underline{A}\underline{x}\right) \tag{2.7}$$

where $\underline{W}$ is a positive definite NxN weighting matrix. In this case, the normal equations are:

$$\underline{x} = \left(\underline{A}^T\underline{W}\underline{A}\right)^{-1}\underline{A}^T\underline{W}\underline{b} \tag{2.8}$$

Another way of looking at the problem is through least squares estimation theory [10]. Here we consider a deterministic signal s and observations perturbed by noise x. The signal is supposed to be described by a linear model that links s to a parameter θ. The least squares estimator (LSE) of θ finds the value $\hat{\theta}$ that makes s closest to the observed data x.

As an example of these optimization techniques, suppose we make noisy measurements of two physical parameters linked by a linear relationship (say biomass b versus backscatter σ in logarithmic scale). Let the noisy measurements be $(\beta_1,\sigma_1),(\beta_2,\sigma_2),(\beta_3,\sigma_3)$. The linear model linking the observations is $\sigma = m\beta + c$. The unknown (m,c) must satisfy the equations:

$$\begin{aligned} m\beta_1 + c &= \sigma_1 \\ m\beta_2 + c &= \sigma_2 \\ m\beta_3 + c &= \sigma_3 \end{aligned} \tag{2.9}$$

where in the matrix notation adopted above:

$$\begin{aligned} \underline{x} &= \left[m,c\right]^T \\ \underline{A} &= \begin{bmatrix} \beta_1 & 1 \\ \beta_2 & 1 \\ \beta_3 & 1 \end{bmatrix} \\ b &= \left[\sigma_1,\sigma_2\right]^T \end{aligned} \tag{2.10}$$

From Equation (2.6) the solution is:

$$\left[m,c\right]^T = \begin{bmatrix} \beta_1^2+\beta_2^2+\beta_3^2 & \beta_1+\beta_2+\beta_3 \\ \beta_1+\beta_2+\beta_3 & 3 \end{bmatrix}^{-1} \begin{bmatrix} \beta_1\sigma_1+\beta_2\sigma_2+\beta_3\sigma_3 \\ \sigma_1+\sigma_2+\sigma_3 \end{bmatrix} \tag{2.11}$$

Let us now go back to our geolocation problem in the mosaic compilation [3].

We assume that, if we consider the frames in a Mercator projection, only the scene center translations and one rotation need to be taken into consideration in the adjustment procedure. Therefore in the model linking the data to the unknown parameters the parameters to be estimated are the scene center translations and one rotation angle. This model relies on the confirmed assumption that there is no internal deformation in the NASDA level 2.1 products.

Working the problem in a conformal mapping projection minimizes the risk of introducing projection-dependent systematic deformations. Moreover, the rotation angles are kept small, and therefore the block adjustment can be simplified by linearizing the observation equations.

With the aid of an auxiliary reference system with origin at the scene center (Figure 2.5), the relationship in the Mercator reference system between a point P in the image and the unknown parameters can be derived:

$$\begin{aligned} N^P &= N^C + y\ \cos\alpha - x\ \sin\alpha \\ E^P &= E^C + x\ \cos\alpha + y\ \sin\alpha \end{aligned} \tag{2.12}$$

where $N^P E^P$ are the Mercator coordinates of point P; x, y are the image coordinates in the auxiliary reference system; $N^C E^C$ are the coordinates of the scene center; and α is the rotation angle of the auxiliary system with respect to the Mercator system.

What we measure are the discrepancies in the positions of homologous points in neighboring scenes. With reference to the linear system in Equation (2.3), the vector b will now contain the

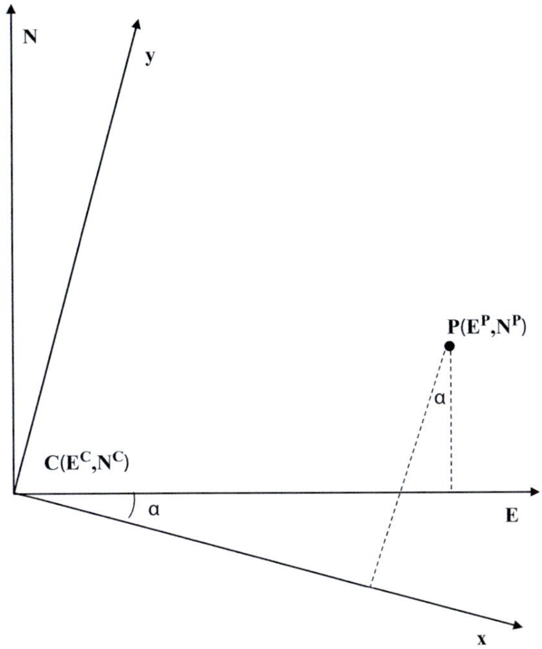

FIGURE 2.5 The auxiliary reference system for establishing the relationship between a point P in the image and the unknown parameters (scene center translation and rotation).

observed discrepancies before adjustment, the unknown scene parameters are in vector x, and we want to minimize in a least squares sense the residual discrepancies vector δ.

The observation equations based on tie-points between scenes are:

$$\delta N^P = N_1^P - N_2^P = N_1^C - N_2^C + y_1 \cos\alpha_1 - y_2 \cos\alpha_2 - x_1 \sin\alpha_1 + x_2 \sin\alpha_2 \tag{2.13}$$

$$\delta E^P = E_1^P - E_2^P = E_1^C - E_2^C + x_1 \cos\alpha_1 - x_2 \cos\alpha_2 - y_1 \sin\alpha_1 + y_2 \sin\alpha_2 \tag{2.14}$$

These equations are put into a suitable linear form by a Taylor's series and retaining only first order terms in the parameters: $N_1^C\ N_2^C\ E_1^C\ E_2^C\ \alpha_1\ \alpha_2$.

Thus, for example:

$$\delta N^P = f\left(N_1^C, N_2^C, \alpha_1, \alpha_2\right) = \delta N_0^P + \left.\frac{\partial f}{\partial N_1^C}\right|_0 dN_1^C - \left.\frac{\partial f}{\partial N_2^C}\right|_0 dN_2^C + \left.\frac{\partial f}{\partial \alpha_1}\right|_0 d\alpha_1 + \left.\frac{\partial f}{\partial \alpha_2}\right|_0 d\alpha_2 \tag{2.15}$$

where $dN_1^C dN_2^C$ are the translations in Northing of scenes 1 and 2, $\alpha_1\alpha_2$ are the rotations of the scenes with respect to their local reference systems, and δN_0^P is the initial Northing discrepancy before adjustment.

Thus the elements of the matrix $\underline{A}$ in the observation equations are the partial derivatives with respect to the parameters, and the vector $\underline{b}$ corresponds to the initial discrepancies before the adjustment.

Tie point measurement is based on image correlation performed at 100 m pixel spacing between adjacent scenes belonging to the same date mosaic, or between scenes on two different dates. For scenes acquired along the same orbit, the correlation peak is well defined even without high-contrast features; even homogeneous areas can be correlated successfully because the same speckle pattern is present in both scenes.

The points within one date and between strips always require the presence of a high-contrast feature that remains stable during the interval between the SAR acquisitions of the adjacent strips. The same applies to the points between dates, but here the overlap area to search for candidate features is larger because the scenes (same node or path–row position) cover the same area.

To increase the probability of a high correlation maximum, a simple interest operator was implemented. A small area around the center point of a correlation block is searched for candidate templates. The template that produces the highest value of the interest operator is then selected and used in the correlation maximum search.

Measurements related to tie points between two neighboring scenes assure only control over the internal consistency and the co-registration of the bi-temporal mosaic. However, in order to improve the absolute geolocation accuracy ground control points (GCPs) derived by external cartographic data were also introduced. In this case, the observation equations are:

$$\delta N^P = N^P_{meas} - N^P_{true} = N^C + y\ cos\alpha - x\ sin\alpha - N^P_{true} \tag{2.16}$$

$$\delta E^P = E^P_{meas} - E^P_{true} = E^C + x\ cos\alpha + y\ sin\alpha - E^P_{true} \tag{2.17}$$

Finally, information on the image location derived from the SAR ancillary data was also added to the LLS adjustment. This last piece of information is far less accurate than the one derived by tie points between images, but it is still useful to establish a bound on the amount that the images can be rotated and translated with respect to their original position.

The structure of the LLS normal equations coefficient matrix is block diagonal, with bandwidth:

$$w = \left(2s_{path} + 1\right) \cdot n_{dates} \cdot n_{par} \tag{2.18}$$

where s_{path} is the maximum number of scenes in a single satellite path, n_{dates} is the number of acquisitions in time (in the GRFM case $n_{dates} = 2$), and n_{par} is the number of parameters in the geometric model.

A conjugate gradient method [11] can be used efficiently for the solution of linear systems with a block diagonal coefficient matrix. In the case of the bi-temporal Africa mosaic, the matrix size is 10,875 × 10,875, a figure that clearly reflects the processing complexity.

2.3.2 Geolocation Validation

To validate the accuracy of the georeferencing, a sequence of block adjustments using different sources for the observations in the LSS were performed, and the relative residual mean square errors (RMSE) measured. The hierarchy of tests included: (1) observations based only on tie-points between neighboring scenes; (2) additional observations based on ground control points (GCP) located on the coastlines; and (3) additional observations based on GCP located in the interior of the continent. All three adjustments covered 3624 scenes and 62 006 tie-points.

Digital data from the World Vector Shoreline database were used for the GCPs along coastlines. The World Vector Shoreline data cover West Africa fairly well, plus the Western coastal zone of Africa down to Angola, and the part of the Tanzanian–Kenyan coast that is included in one of the mosaics.

Digital topographic maps scale 1:200000 in the Central African Republic and the Republic of Congo (Congo/Kinshasa) were used in test case 3. The maps were produced by IGN/France, mainly in the 1950s. They were scanned and digitized by I-Mage Consult, Namur, Belgium within the framework of the Regional Environmental Information Management Project on Central Africa (REIMP-CA), initiated under the PRGIE program of the World Bank.

TABLE 2.1
Residual mean Square Error (RMSE) Statistic in the GRFM Block Adjustment

Test case	Tiepoint *N*	Tiepoint *E*	Center *N*	Center *E*	GCP *N*	GCP *E*	∇N_{12}	∇E_{12}
1	33.2	34.8	477.5	751.7	503.0	1021.4	26.6	290.0
2	35.5	40.8	532.0	736.2	193.7	201.6	18.6	–69.3
3	40.1	39.7	543.3	758.0	173.2	166.3	24.3	–46.8

Results of the validation process are reported in Table 2.1. The table gives RMS errors in meters for various quantities of interest. The columns *Center N, Center E* are scene translations in Northing and Easting. The columns ∇N_{12} ∇ E_{12} give the average relative translations between the low-water and the high-water mosaics. The RMS error on GCPs in case 1 indicates the error computed when the 248 coastline GCPs are not included in the LLS adjustment but only used to estimate the error.

As to case 2, the inclusion of GCPs in the LLS degrades the tie-points RMSE by a few meters. On the other hand, it greatly reduces the discrepancy between the mosaic and external control data (from 500 m to less than 200 m in Northing and from 1000 m to 201 m in Easting).

In case 3. about 50 GCPs located in the Central African Republic and the Republic of Congo were added to the observations, but the RMSE statistics are not changed significantly. From this we can infer that these GCPs are compatible with the coastline GCPs at the scale of the mosaic pixel size (100 m).

In a one-dimensional least squares estimation problem it is possible to compute the variance (or the standard error s_e) of the estimated parameter, assuming that the input variables (observations) are random variables. Similarly, in our case, where the parameters such as the scene center are vectors in a two-dimensional space it is possible to compute the covariance matrix of the two coordinates. The distribution of the standard error in space is conveniently represented by the so-called error-ellipses; the semi-major axis of the ellipsis is the s_e of the parameter of interest in the direction where it reaches the maximum. The semi-minor axis is the s_e in the direction where it reaches the minimum.

In an experiment of scene geometry revision where the block adjustment of the whole bi-temporal mosaic included the GCPs, the semi-major axis of the error ellipse varied between 6 m and 55 m with a median at 11 m. The semi-minor axis varied between 6 m and 3 m with a median at 10 m. Larger error-ellipses corresponded to scenes where the number of tie points were reduced because of lack of inter-scene correlation.

2.4 WAVELET MULTIRESOLUTION DECOMPOSITION

A multi-resolution signal decomposition based on the wavelet transform is used to generate the GRFM Africa products [12]. Note that this image representation is not the same as the one presented in Chapter 6. Here, the multiresolution approximations are built using orthogonal wavelet bases (see Appendix), while the wavelets used to build texture measures in Chapter 6 are frames, a non-orthogonal and redundant basis. The wavelet coefficients at each level of the multiresolution approximation carry information on the details that were lost in passing to the next level of approximations. The frames act as differential operators, and their variance provides a statistic that approximates the structure function of order 2. The texture measures derived here with the multiresolution representation are of quite a different nature than the ones discussed in Chapter 6.

The rationale for using this approach instead of a classical down sampling method, such as block averaging, stems from several considerations.

Multiresolution is an intrinsic concept in the wide area radar mapping approach, where a high resolution sensor is used for wide area mapping; the wavelet pyramid gives the possibility of generating approximations of the original radar imagery at several spatial scales (at each stage in the pyramid the pixel size is doubled with respect to the previous level). Therefore it is possible to generate a set of products that best match the thematic analysis requirements for a certain application in terms of resolution, signal to noise ratio and data volume.

The wavelet coefficients at a scale 2^j carry information on the "details" of the signal that were present in the approximation at a finer scale 2^{j-1} and were lost in the approximation at scale 2^j. The wavelet coefficients can therefore be interpreted as texture measures at that scale and at the same time the wavelet decomposition provides the approximated image, and information on spatial structures that exist in the image at a certain scale. This fact opens up several possibilities for the automatic extraction of structural and textural information [13], a definite advantage over other multiresolution techniques.

The existence of fast, discrete wavelet transform algorithms, which require only a number of operations proportional to the size N of the original data, makes this approach ideal in the case of high data volume (in contrast, for example, fast Fourier transforms are typically $N \log N$ algorithms).

An overview in tutorial form of the principles underlying wavelet multiresolution analysis is given in the Appendix.

2.4.1 Multiresolution Products

The wavelet decomposition generates from each high resolution level image and for each scale, three products: (1) the smooth signal; (2) the wavelets coefficients (horizontal, vertical, oblique); and (3) the scalogram. The scalogram at scale 2^j as defined in [14] is the quadratic sum of the wavelet coefficients normalized by the squared smooth signal at that scale. Normalization is required to compensate for the wavelet amplitude modulation caused by multiplicative speckle noise in stationary areas. Scalograms are useful for constructing multi-scale texture maps [13].

It is important to notice that the low pass and high pass filters in the decomposition are applied after a square law detector. In this way, over stationary areas the intensity mean value is preserved after low pass filtering, and therefore an unbiased estimation of the radar backscattering coefficient is still possible at higher scales (lower resolution).

In parallel to the main pyramid another texture measure product is generated by computing the normalized standard deviation (sample standard deviation normalized by the sample mean) SD_n of the high-resolution amplitude image over contiguous blocks of 8 × 8 pixels.

This texture measure is intuitively proportional to the one-point signal variations (namely independent of the spatial ordering) within a resolution element of 100 m. For a homogeneous area with constant underlying radar reflectivity, they are entirely caused by speckle noise; SD_n will therefore assume the nominal value for a 3-look amplitude image with correlated, fully developed speckle. Deviations from this nominal value indicate that other variations are occurring in the signal within the 100 m resolution cell.

It is important to realize that the normalized standard deviation SD_n is complementary and carries different information with respect to the wavelet scalogram SW^j. The scalogram at scale 2^j contains the details that were lost from the finer scale (higher resolution) 2^{j-1} in approximating the signal, and therefore contains variations that were detectable at scale 2^{j-1} but were wiped out at scale 2^j. For example, SW^j at 100 m contains textural information at the 50 m scale. By contrast the SD_n measure contains texture at all scales from 12.5 m up to 100 m.

2.5 THE GRFM AFRICA MOSAIC SECOND EDITION

The availability in 2008 of the Shuttle Topographic Mission (SRTM) digital elevation model (DEM) [18] suggested a possible revision of the GRFM mosaic accuracy, both in the spatial domain (orthorectification) and in the radiometric domain (backscattering coefficient dependence on the local incidence angle).

Applying precision geocoding – for example, based on the range-Doppler algorithm – to each individual scene in the mosaic and then recompiling the entire dataset, would imply a high computational burden and possibly new errors in the recompilation process. Instead, a neat solution was adopted (mainly thanks to Jan Kropacek, then a Postdoc at the JRC) with a process that operated on, and corrected, the full assembled dataset. In a nutshell, the approach consisted of:

a. Generation of a virtual backscatter image obtained by illuminating the SRTM DEM in the radar range geometry. Each resolution element, in a grid corresponding to the DEM samples, carries a value corresponding to a constant radar reflectivity modulated by variations of the local incidence angle, because of elevation differences. The resulting patterns are also shifted with respect to the real position in the geographic grid, as a result of the radar viewing geometry.
b. Warping the original SAR backscatter image to the simulated image by co-registration based on cross-correlation with homologous features.
c. Ortho-rectification of the co-registered data set, by inversion of the SAR range geometry model (slant range to ground range repositioning of each resolution element).
d. Radiometric correction for local incidence angle dependency.

The procedure is documented in detail in [15]. We report here only some highlights.

The model underpinning the simulated image consists of a constant radar reflectivity modulated by incidence angle variation resulting from topography. Therefore, it lays out most of the variability in the real SAR image because of topographic features, e.g. surface slopes. As an example, a sample of the virtual image and the corresponding set of the GRFM mosaic is shown in Figure 2.6.

Evidently, the simulated image is rich in homologous features in areas where topographic variations are present. These features are used as control points in the co-registration procedure (based on cross-correlation on a grid of sampling windows). However, in flat areas, where these features are not present, control points are supplemented by structures from other sources, such as the SRTM Water Bodies Database (SWBD), and the GeoCover Landsat imagery [16] (Figure 2.7).

An example of the improvement in spatial accuracy as a result of the orthorectification is shown in Figure 2.8. Slopes around the sides of a mountaintop, when imaged by radar, appear contracted toward the radar looking position and dilated in the opposite direction. The geometric correction reinstates the correct length of the slopes and the correct position of the mountaintop.

Concerning the radiometric corrections, two steps were implemented: correction for the effective scattering area; and dependence of the backscattering coefficient on incidence angle.

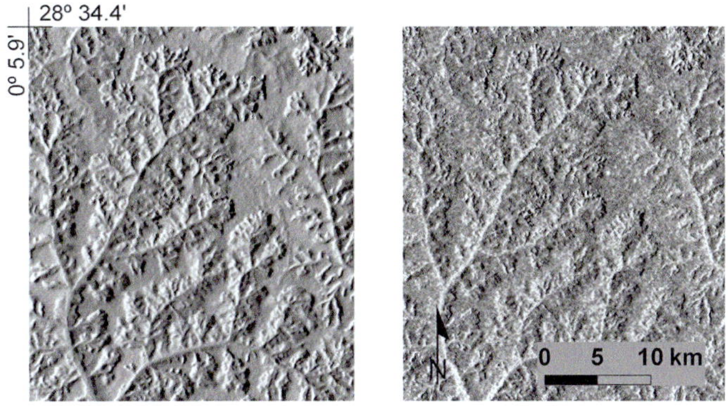

FIGURE 2.6 Simulated image, constructed by illuminating the SRTM DEM in the SAR slant range geometry (left frame), and the corresponding real SAR image of the GRFM mosaic (right frame).

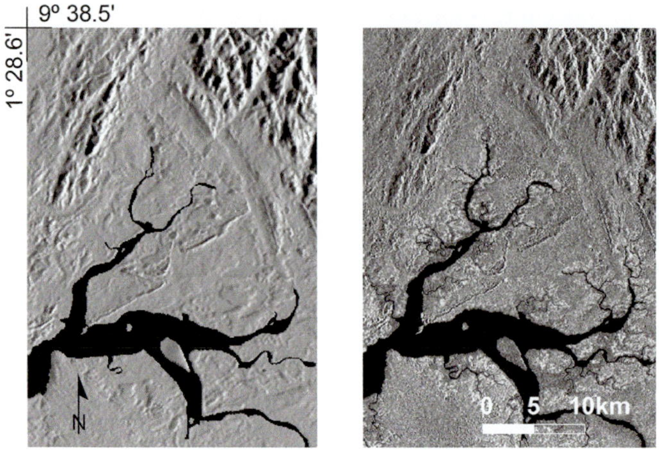

FIGURE 2.7 Simulated image compounded with a water body feature extracted from the SWBD vector data set (left frame), and the corresponding subset of the GRFM mosaic (right frame).

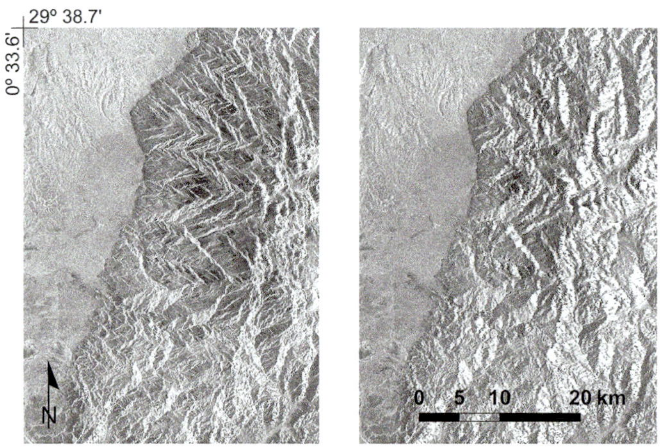

FIGURE 2.8 Example of the improved geometric accuracy. The terrain elevation distorts the lengths and position of the slopes on both sides of a mountaintop, when imaged in the radar slant geometry (left frame). The corrected scene after the orthorectification, where the length of the slopes is irrespective of the orientation (right frame).

REFERENCES

1. A. Rosenqvist, "The Global Rainforest Mapping Project by JERS-1 SAR", *International Archives of Photogrammetry and Remote Sensing.,* vol. 31, Part B7, pp. 594–598, ISPRS, Vienna, Austria, 1996.
2. A. Rosenqvist, M. Shimada, B. Chapman, A. Freeman, G.F. De Grandi, S. Saatchi, and Y. Rauste, "The global rain forest mapping project – A review", *International Journal of Remote Sensing*, special issue on "Global and Regional Land Cover Characterization from Satellite Data", vol. 21, no. 6–7, pp. 1375–1387, April 2000.
3. G.F. De Grandi, P. Mayaux, Y. Rauste, A. Rosenqvist, M. Simard and S. Saatchi, "The global rain forest mapping project JERS-1 radar mosaic of tropical Africa: development and product characterization aspects", *IEEE Transactions on Geoscience and Remote. Sensing,*vol. 38, no. 5, 2218–2233, 2000.
4. P. Mayaux, F. Achard, and J.P. Malingreau, "Global tropical forest area measurements derived from coarse resolution satellite imagery: A comparison with other approaches", *Environmental Conservation*, vol. 25, no. (1), pp. 37–52, 1998.

5. D. Skole, and C. Tucker, "Tropical deforestation and habitat fragmentation in the Amazon: Satellite data from 1978–1988", *Science*, vol. 260, pp. 1905–1910, 1993.
6. J.P. Malingreau, and C. Tucker, "Large scale deforestation in the Southern Amazon basin of Brazil", *Ambio*, vol. 17, pp. 49–55, 1988.
7. P. Mayaux, G.F. De Grandi, and J.P. Malingreau, "Central Africa forest cover revisited: A multi-satellite analysis", *Remote Sensing and Environment*, vol. 71, pp. 183–196, 2000.
8. Y. Rauste, T. Richards, G.F. De Grandi, G. Perna, E. Franchino, and A. Rosenqvist, "Compilation and validation of the GRFM Africa SAR mosaics using multi-temporal block adjustment", JERS-1 Science Program 99 PI Reports, NASDA EORC, pp. 99–108, March 1999.
9. L. Mirsky, "*An Introduction to Linear Algebra*", Dover, 1990, pp. 156–158, ISBN 0-486-66434-1.
10. S.M. Kay, *Fundamentals of Statistical Signal Processing*, Prentice Hall Intern. Editions, 1993, pp. 294–299, ISBN 0-13-042268-1.
11. W.H. Press, B.P. Flannery, S.A. Teukolsky, and W.T. Vetterling, *Numerical recipes in C: The art of scientific computing*, Cambridge University Press, pp. 52, 84–85, 1992.
12. S.G. Mallat, "A theory for multi-resolution signal decomposition: The wavelet representation", *IEEE Transactions on Pattern and Machine Intelligence*, vol. 11, pp. 674, July 1989.
13. M. Simard, G.F. De Grandi and K. Thomson, "Adaptation of the wavelet transform for the construction of multi-scale texture maps of SAR images", *Canadian Journal of Remote Sensing*, vol. 24, no. 3, pp. 264–285, September 1998.
14. M. Simard, G.F. De Grandi, K.P.B. Thomson, and G.B. Benie, "Analysis of speckle noise contribution on wavelet decomposition of SAR images", *IEEE Transactions on Geoscience and Remote Sensing*, vol. 36, no. 6, pp. 1953–1962, Nov. 1998.
15. J. Kropacek, G. De Grandi, and Y. Rauste, "Geo-referencing of continental-scale JERS-1 SAR mosaics based on matching homologous features with a digital elevation model: Theory and practice", *International Journal of Remote Sensing*, vol. 3, pp. 2413–2433, October 2011.
16. Tucker, C.J., Grant, D.M., and Dykstra, D., 2004, NASA's global orthorectified Landsat, data set, *Photogrammetric Engineering & Remote Sensing*, vol. 70, pp. 313–322.
17. G.F. De Grandi, J.P. Malingreau, and M. Leysen, "The ERS-1 Central Africa mosaic: A new perspective in radar remote sensing for the global monitoring of vegetation", *IEEE Transactions on Geoscience and Remote Sensing*, vol. 37, no. 3, pp. 1730–1746, May 1999.
18. Jarvis, A., Reuter, H., Nelson, A., and Guevara, E., 2008, Hole-filled SRTM for the globe version 4, CGIAR-CSI SRTM 90m Database, Available online at: /http://srtm.csi.cgiar.org.

3 The GRFM–CAMP Thematic Products

3.1 FROM BACKSCATTER TO A THEMATIC MAP

The many aspects and components of the CAMP and GRFM SAR mapping projects, described in the previous chapters, lay the foundation for the last – and possibly more important – phase: thematic information extraction. It is in this phase that quantities of interest for the application are derived from the remote sensing observations. While the radar mosaic data can lend support to many applications – such as deforestation and degradation measures, flooding extent, illegal logging detection, river mapping, flooding extent, burnt scar detection, vegetation biomass estimates and geomorphology characterization – the derived thematic products share some common important characteristics. These are:

1. The spatial resolution combined with large-area homogeneous coverage. Most classification maps are generated at 100 m pixel size and cover areas of several million square kilometers. For certain thematic classes (e.g. fragmented forest, gallery forests, plantations) this level of detail gives an unprecedented documentation encompassing entire ecosystems.
2. The qualification and quantification of new thematic classes. Scattering processes that occur in the microwave domain can be exploited to sense particular vegetation and soil properties, such as the structure of the upper canopy and the presence of standing water in the swamp forests.
3. Geolocation accuracy. The GRFM Africa mosaic absolute geolocation was obtained through a global optimization procedure based on external geographic reference data. Measured accuracy in the root mean square sense is 240 m.
4. Validation. This is a major and complicated issue in the production of large-scale high-resolution maps. Using suitable methods, data for the maps' accuracy are derived and constitute an important characterization of the products.

We shall focus our attention here on a tropical vegetation mapping exercise whose groundwork was provided by the CAMP and GRFM radar mosaics. The vegetation mapping problem is defined in terms of the spatial distribution of a number of vegetation classes in the Central Congo Basin. The resulting product will be referred to as the GRFM–CAMP vegetation map of Central Congo [1–7].

Central Africa contains the second largest area of contiguous moist tropical forest in the world. The GRFM–CAMP vegetation map of Central Congo therefore represents a valuable information resource for global climate change studies, and for supporting effective forest management policies. The map covers an area (15°*E*–22°*E*; 5°*N*–3°*S*) of approximately 56 million square kilometers, while keeping a resolution of 200 m for most classes.

A brief summary of the main vegetation formations in the Central Congo Basin follows. A detailed description of tropical vegetation can be found in [8,9]. The legend of the GRFM–CAMP map is tailored to these formations.

3.2 VEGETATION CLASSES

The climax ecosystem in the central part of the Congo Basin is the lowland rain forest (Figure 3.1). The upper stratum (35–45 m) of this formation is composed of a few evergreen or briefly deciduous species, whose age distribution is quite large. Age distribution makes the upper layer somewhat irregular, and the canopy, observed from high above, appears discontinuous and rolling. Leaves are dark and mesophyllous. The high canopy density precludes the development of grass and shrubs in the understory. The rain forest does not show seasonal behavior.

Within the area of the map, the largest formations of swamp forests are present in the floodplains. Large floodplains in the tropics are interesting ecosystems from several points of view [10]. From the local economy's perspective, they are significant as land reserves for agriculture, settlement areas for a rapidly increasing population, animal husbandry and forestry. In the ecological sense, they function as water storage, habitats for fauna and flora, and fishery. Moreover, they host a series of bio-chemical processes, such as nitrogen turnover, and methane emission. Since methane is a greenhouse gas, this process has a role in global climate change issues.

These formations are influenced by the presence of frequent floods, and soils with poor drainage. The upper canopy layer (up to 45 m in some cases), composed of a small number of species, is fairly homogenous compared to the lowland rain forest.

Where extended flooding and poor drainage conditions preclude the subsistence of trees, large swamp grasslands appear in the forest domain, usually at the periphery of swamp forests (Figure 3.2).

At the northern and southern boundaries of the rainforest domain, savannas (Figure 3.3) represent the para-climax vegetation and are mainly the result of human activity (fire being the main agent). Trees and shrubs are sparse, while grass forms a continuous layer.

In the midst of savannas, narrow corridors of tree formations (300–500 m in width) develop along river banks. These formations are called gallery forests.

FIGURE 3.1 The lowland rain forest – the climax ecosystem in the central part of the Congo Basin.

FIGURE 3.2 Swamp forests develop in the floodplains along the rivers. They are an important habitat for bio-diversity. Where extended flooding and poor drainage conditions preclude the subsistence of trees, large swamp grasslands appear in the forest domain.

FIGURE 3.3 Savanna and grassland.

Large plantations (rubber, oil palm and cotton) were established in colonial times in the forest domain. The plantations are characterized by a uniform cover and follow a geometrical layout.

A major feature of the Central African forest biome is the presence of corridors of secondary forest formations along the old road network (Figure 3.4). The vegetation found here is composed of a complex of secondary re-growth, fallow and food crops.

FIGURE 3.4 Secondary forest formations, and area cleared for food crops.

3.3 MAP COMPILATION METHODS

The vegetation map of Central Congo is compiled from observations by several orbital instruments. These are the synthetic aperture radars (SAR) on board the ESA ERS-1 (CAMP mosaic) and the NASDA JERS-1 satellites (GRFM mosaic) (C-band and L-band), and the imaging spectrometer VEGETATION on board SPOT 4.

This last instrument is specifically designed for global-scale vegetation monitoring. It acquires images in four spectral bands (VisBlue, VisRed, Near Infra-Red and Short-Wave Infra-Red) at 1 km spatial resolution. In the work presented here, band 1 is employed to detect image areas affected by cloud and haze, while bands 2–4 are exploited for vegetation classification. The VGT data (S1 products) were mosaicked using the nominal scene positions derived from orbital data, and then co-registered to the GRFM SAR mosaic using a polynomial warping procedure and manually selected tie-points.

The optical and microwave observations provide a different view of the same target, in terms of the probing wavelength, spatial resolution, acquisition time and viewing geometry. Clearly, then, a new augmented classification scenario can be defined by combining the observations.

Indeed, each instrument "sees" certain thematic classes while being "blind" to others. A synoptic view of the class visibility in connection with the radar and optical instruments is given in Figure 3.5.

A flowchart of the main building blocks and flow of operation of the classification method [6,11,12] is presented in Figure 3.6. Because of their different characteristics (spatial resolution, noise model, feature space) the radar and optical data sets are processed independently.

The radar data are first pre-processed to generate piecewise smooth approximations using the wavelet algorithm presented in Chapter 8. Also, a texture measure, based on local estimates of the normalized second moment (see Chapter 2) is smoothed.

The classification procedure for the radar data is based on a rule-based hierarchical classifier. A feature vector composed of the ERS amplitude data, the JERS-1 amplitude data and a texture

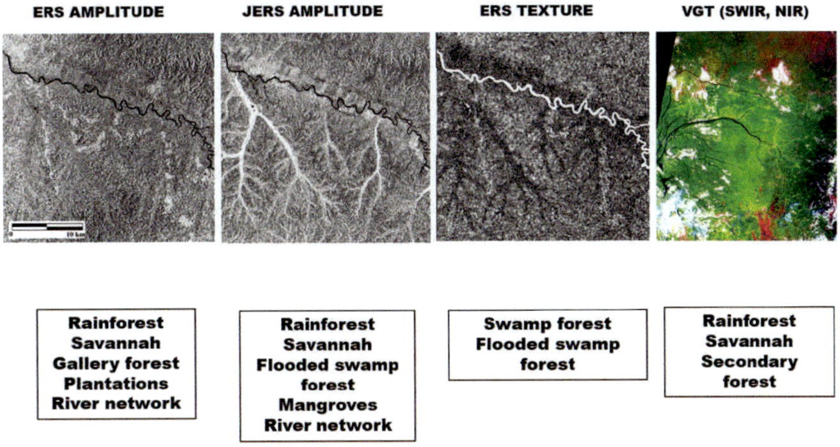

FIGURE 3.5 Class visibility provided by each remote sensing instrument.

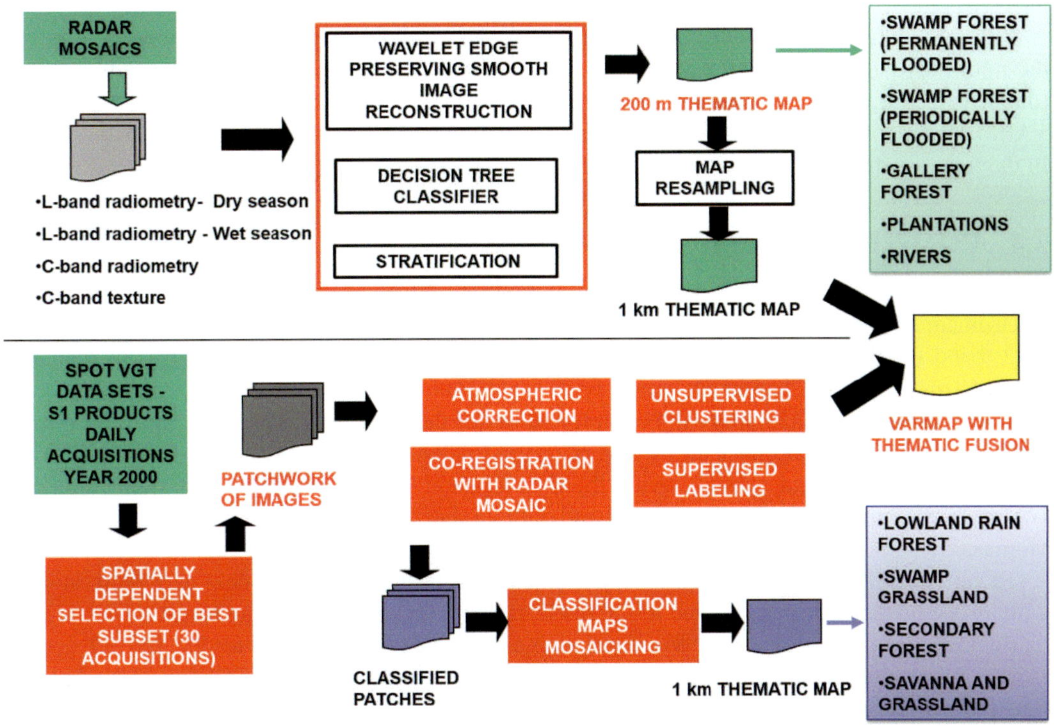

FIGURE 3.6 Flow diagram of the classification method.

measure derived from the high resolution (12.5 m pixel size) ERS amplitude data is used. The classification rules are defined by inference from a series of training sets in different areas of the Congo Basin. The training sets were selected by visual inspection using national or local maps [2].

For the VGT data two non-contextual clustering algorithms (Enhanced Linde-Buzo-Gray (ELBG) [13,14] and Fully Automatic Clustering System (FACS) [15]) and one contextual clustering procedure, termed Contextual Local–Global Clustering (CLGC) [16], are adopted. This approach presents significant advantages in terms of ease of use, sensitivity to initial parameters, and the preservation of small but genuine features.

The VGT data classification strategy, which focuses on detecting the class *secondary forest*, comprises the following steps:

1. Partition of the whole data set into three classes: "no cloud", "thin cloud" and "cloud". Band 1 (VisBlue) is used and the classification sequence is: ELBG (with 32 clusters), CLGC, and cluster aggregation by a photo interpreter. The partitions are used as data masks in the following steps. Raw data belonging to class "cloud" are discarded from further analysis. Subsequently, only VGT bands 2 to 4 are used.
2. Processing of the cloud-free image section. The classification scheme is: ELBG (with 16 clusters), CLGC, and cluster aggregation by a photo interpreter.
3. Processing of the region corresponding to thin cloud. The sequence is: FACS (input parameter is the mean squared error MSE value computed as output by ELBG in step 1), CLGC, and cluster aggregation by a photo interpreter. In this way, the VGT dataset is clustered according to a uniform MSE criterion, and a time-consuming trial-and-error strategy in choosing the initial number of templates is avoided when a large section of the dataset is involved with clustering.
4. The two VGT classification maps generated in steps 1 and 2 are overlapped.

Information fusion is achieved at the level of the classification maps derived independently from the microwave and the optical instruments. The derived classification product is a data structure (dubbed VARMAP [6]) composed of elementary variable sized cells holding class labels. The VARMAP supports the generation of thematic products at different scales according to the end use. In the final product, each class will inherit the original spatial resolution.

3.4 COMPLEMENTARITY OF RADAR SENSORS

The two radar sensors (JERS-1, ERS) "see" the target with a different viewing geometry and by electromagnetic waves of different wavelength (5.6 cm for the C-band ERS sensor and 23.5 cm for the L-band JERS-1 sensor). In turn, the wave target interaction is governed by different scattering mechanisms [1,2] (see Chapter 5).

These different mechanisms can be exploited in our thematic context: (1) to capture the structural properties of the upper canopy layers in the lowland rain forest and to map the low biomass vegetation (e.g. grasslands), using the C-band data; (2) to sense the water status under the swamp forest canopies and, using the L-band data, to discriminate vegetation classes with higher biomass.

To gain insight into the complementary characteristic of the sensors we deal here more in depth with a specific thematic case related to the lowland rain forest and the swamp forest.

Backscattering statistics of the two vegetation classes are reported in Tables 3.1 and 3.2.

The classes are labeled SF (swamp forest), LRF (lowland rain forest) and SFflo (flooded swamp forest). For each vegetation class the following quantities are reported: the backscattering coefficient σ, standard deviation of the amplitude values normalized by the mean value SD_n, the variance of the normalized standard deviation $VAR(SD_n)$, and the order parameter ν of the $K(\nu, I)$ distribution,

TABLE 3.1
ERS-1 Backscattering Statistics

Class	$\sigma^0(dB)$	SD_n	ν	No. pixels
SF	−5.55	0.3	71	405000
LRF	−5.49	0.338	10	195000

TABLE 3.2
JERS-1 Backscattering Statistics

Class	$\sigma^0(dB)$	SD_n	ν	No. pixels
SF	−7.21	0.310	28.0	554000
LRF	−7.10	0.327	13.0	554000
SFflo	−5.48	0.318	20.0	15000

which characterizes SAR backscattering statistics for this type of natural target (see also Chapter 6). The theoretical background and more details about this statistical characterization of the CAMP and GRFM data can be found in [3,17].

For the swamp forest areas, high values of the order parameter ν indicate that the Probability Density Function (PDF) tends to a generalized Rayleigh statistic with 6 degrees of freedom (nominal 3-Looks amplitude data). In this case, the fully developed speckle regime with constant underlying radar cross-section is therefore present.

For the lowland rain forest area, ν decreases for both sensors, and therefore the PDF deviates significantly from the Rayleigh distribution. However even if the trend is similar for both sensors, there are differences in the absolute values. The ν values for the JERS-1 data are always higher, meaning that the underlying radar cross-section fluctuates less.

The estimated backscattering coefficients for the lowland rain forest and the swamp forest are very close. Standard deviations of the estimators are typically 0.01 dB, taking into account that the normalized intensity auto-covariance function tends to zero after four lags; also, the estimated backscattering coefficients are computed for areas approximately at the same range positions.

Penetration depths for the ERS and JERS-1 sensor, computed by a wave scattering model and for both the swamp forest and the primary forest canopy, are approximately 54 cm (C-band) and 3.6 m (L-band). Therefore, major contributions to the backscattering are provided through the upper canopy layer volume scattering. This also explains why the two classes show very similar average radiometric characteristics. At C-band, the strong attenuation rules out any interaction with the lower canopy structure and the soil (see also Chapter 5).

Variation of the spatial distribution of the top canopy layers at the radar resolution (30 m) is what can account for the radar cross-section modulation and hence the different first order statistical properties of the observed signal. Indeed, visual interpretation of images taken from the airborne video camera shows that at the 30 m scale the swamp forest closed canopy looks much flatter and more homogeneous than the lowland rain forest canopy (see Figure 3.7).

In the lowland forest canopy we can therefore postulate a micro-relief effect at the top canopy level, which modulates the local incidence angle and hence the effective scattering area. This effect – which is inversely proportional to the penetration depth – will modulate the radar cross-section and hence the order parameter ν of the $K(\nu, I)$ distribution will be affected. Theoretical considerations of the effects of the canopy shape on the SAR backscatter statistics are given in [18] and in Chapter 6.

Therefore, an estimator of the order parameter ν (alternatively, of the normalized second moment) will be a suitable discriminator for the two thematic classes. The sensitivity to the forest canopy structural parameter will be higher for the ERS C-band data.

At L-band, HH polarization and 35° incidence angle the total loss through the dense swamp forest, modeled as a three-layer structure, will be approximately 15%. The radiation hitting the ground can be specularly retro-reflected – in the limiting case – by the presence of a water layer, and backscattered by a double bounce trunk–ground mechanism. This would account for a difference in backscattering coefficient values of approximately 2 dB between the non-flooded and the flooded conditions for the type of vegetation cover considered in our modeling exercise (see Chapter 5).

FIGURE 3.7 The upper canopy spatial patterns of the lowland rain forest (right) and the swamp forest as seen by an airborne survey system. The material is courtesy of the Nouabale-Ndoki project in Congo-Brazzaville.

This difference is, of course, a function of the vegetation layer's characteristics. Therefore the JERS-1 L-band can be used to map the water status even under the closed canopy of dense forest, and, if a time-series were available, to assess the flooded area extent in tropical floodplains [2].

In summary, we can say that, in the case of the dense lowland forest and the swamp forest at C-band, there is high sensitivity to the upper canopy spatial distribution at the sensor's resolution scale. This information is captured by a texture measure. On the other hand, because of the high absorption in the canopy layer, there is no direct mapping of the underlying extent of flooding.

This type of information is provided, in a complementary way, by the lower frequency JERS-1 backscatter data.

As an example of the considerations made so far, three images extracted from the layers of the composite mosaic are shown in Figure 3.8. The site is in the Central part of the Congo in DRC (Maringa River). In the ERS amplitude image the degradation corridors are clearly visible along the road network (light gray at the South of the river), while they are neither evident in the JERS amplitude image, nor in the ERS texture image.

On the other hand, the reticulated pattern of swamp forests is well delineated in the ERS texture (dark patches) and to a lesser extent in the JERS amplitude. The combination of those two channels allows us to refine the generic class "swamp forests" into "permanently and periodically flooded forests". Low values in the ERS texture image are characteristic of all flooded forests, while in the

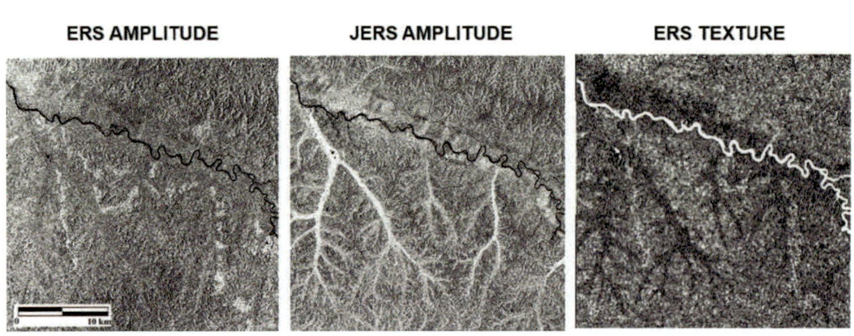

FIGURE 3.8 Three samples of the ERS amplitude, JERS amplitude and ERS texture datasets (Maringa River in DRC) illustrating how different thematic features may be detected exploiting the complementarity of the datasets.

JERS amplitude data, the permanently flooded forests appear brighter than the periodically flooded ones.

3.5 VALIDATION

Validation of regional scale vegetation maps poses challenging problems [2]. Indeed, the scale and the extent of such documents preclude any classical ground data collection at an acceptable cost. The strategy followed for the GRFM products, therefore, entailed a comparison with a compilation of national vegetation maps including the same thematic classes [19–22]. The main characteristics of the maps are described in Table 3.3.

When derived from Landsat images, the national maps were produced by visual interpretation, a procedure that tends to blur the spatial details. Regional vegetation maps of Central Africa, such as the maps produced by the TREES, the IGBP DISCover and the Landsat Pathfinder Projects [23–25], were not used in the comparison because they did not include a "Swamp Forest" class.

A contingency matrix, computed between the national vegetation maps and the radar-derived map, is reported in Table 3.4.

The overall correspondence between the two data sets is 76.3%. The main source of difference is related to the confusion between classes "lowland rain forest" and "swamp forest". Several factors can be called for to explain the discrepancies.

TABLE 3.3
Characteristics of the Maps used in the Validation Process

Country	Authors	Date	Scale	Data Source
Cameroon	CENADEFOR	1985	1:1000000	TM, ground data
Republic of Congo	Laudet	1967	1:2000000	Existing maps
Democratic Rep. Congo	SPIAF	1995	1:1000000	TM
Central African Republic	Boulvert	1985	1:1000000	TM, ground data

TABLE 3.4
Contingency Matrix between the Compilation of National Maps and the Radar-derived Classification

		National vegetation maps						
	Classes	Lowland Rain Forest	Swamp Forest	Swamp Grassland	Savanna	Water	Plantation	Total
	Lowland rain forest	**142**	30		15	1	1	189
	Swamp forest	17	**57**	2	2	1		79
	Swamp grassland		2	**8**				10
Radar-derived classification	Savanna	4	3		**40**	1		48
	Water		2			**12**		14
	Plantation		2				**8**	10
	Total	163	96	10	57	15	9	**350**

First, the narrow ribbons of swamp forest along the secondary river network are not delineated in the national maps, whereas they are well captured in the radar classification. Second, the boundary between the forest types is sometimes very fuzzy and, in the case of visual interpretations, different interpreters can assign pixels differently.

Another major discrepancy consists of pixels classified as "lowland rain forest" in the radar classification and as "savanna" in the national maps. Actually, pixels that generate this confusion between the two vegetation classes mainly correspond to the gallery-forests within the savanna domain, which do not appear in the traditional maps, because they are smaller than the minimum mapping unit of those documents.

3.6 TOUR OF RELEVANT FEATURES

The Central Congo map constitutes a big step forward in the quantitative assessment of the vegetation spatial extent and composition in the Congo floodplain. It provides information on the forest composition of a whole ecosystem and contains better spatial detail than previously compiled regional map. Of special relevance is the highly detailed mapping of the swamp forests, a very important faunistic and florensis habitat, but previously poorly documented. Moreover, the water condition of the floodplain – distinction between the permanently and periodically flooded forests – was never accessible before using optical data only.

The richness of thematic information in the map can only be fully appreciated by close examination of selected areas at full resolution. For the purpose, a few selected examples of important thematic features are presented next. Each area is identified in Figure 3.9 by a numbered box.

The first stop of the tour is the floodplain between the Congo and the Oubangi rivers (Box 1). The two most important vegetation classes here are the dense primary rain forest and the swamp forest. A color composite of ERS and JERS backscatter image is shown in Figure 3.10 and gives a view of the site as seen by the radar eyes. The reddish spots occur where the ERS texture is dominating, and it is an indicator of the canopy heterogeneity. Indeed this channel is the one that allows us to discriminate the lowland rain forest from the swamp forest based on the roughness of the canopy, as explained previously.

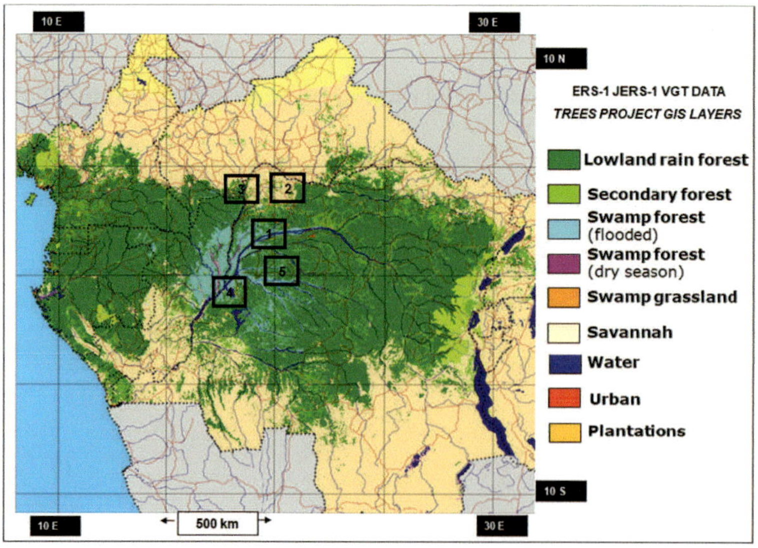

FIGURE 3.9 The CAMP–GRFM vegetation map of Central Congo. Courtesy EC JRC.

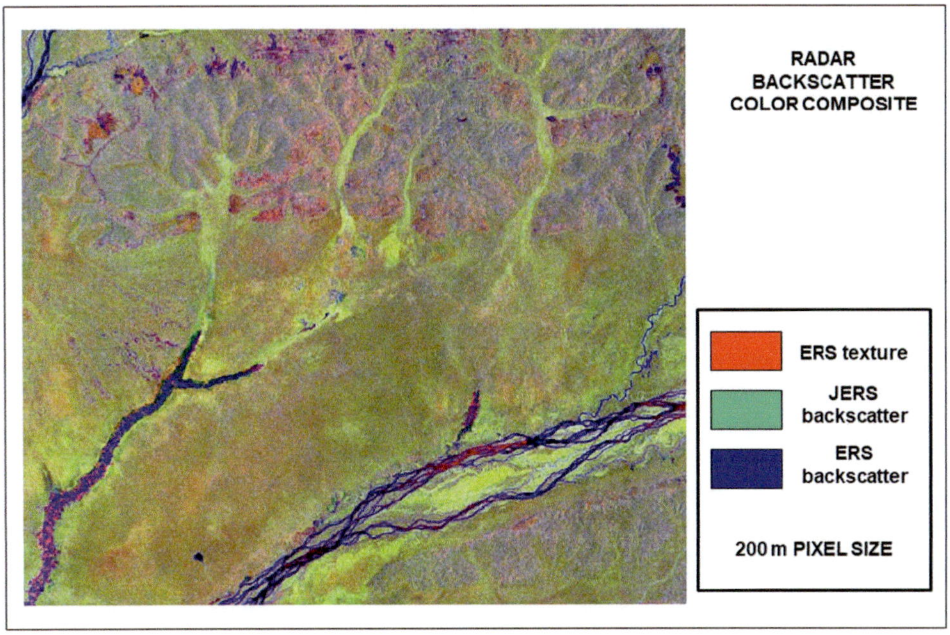

FIGURE 3.10 The floodplain between the River Congo and the River Oubangi as portrayed in a RGB composite image of ERS texture, JRS backscatter and ERS backscatter.

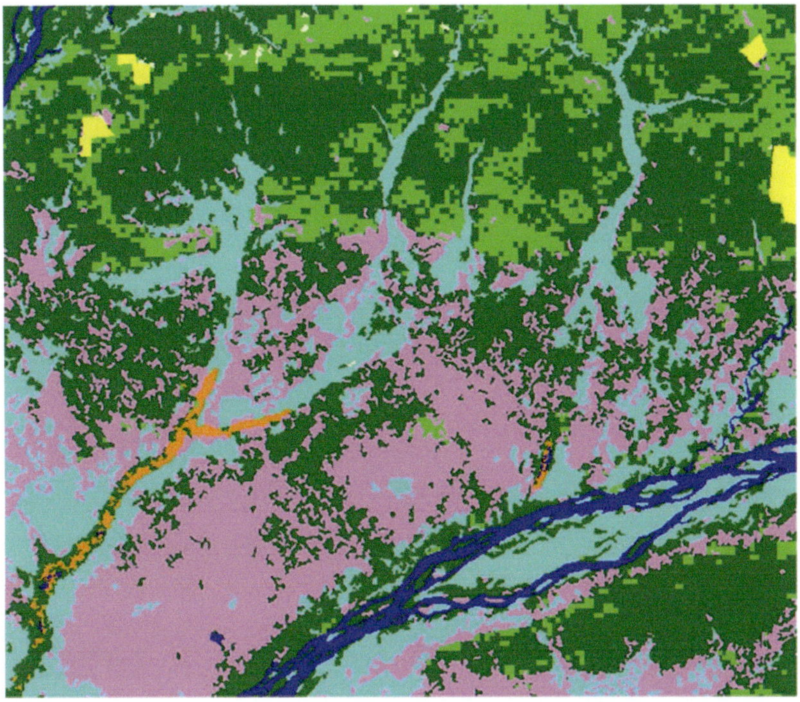

FIGURE 3.11 Classification map of the floodplain Congo/Oubangi Rivers. The thematic classes are listed in the legend of Figure 3.9. The map pixel size is 200 m for all classes except the "Secondary forest" class that is detected by the VGT instrument at 1 km pixel size.

Notice that for some vegetation patterns the backscatter is the same as in the river (barring the influence of texture). Observations from the optical instrument are used in the decision process to resolve these ambiguities in the radar feature space.

The corresponding classification map is shown in Figure 3.11. Notice the large patches of swamp forests (pink and azure for the flooded and non-flooded forest respectively); they are indeed the largest formations in the world. Towards North, we have the margin with the primary forest and also some degradation patterns, which are captured by the VGT data, and therefore appear at low resolution. Finally, some plantations are visible in yellow.

The next area (Box 2) straddles the savanna–forest margin to the north.

Savannas are formations that are mainly a result of human activity (fire being the main disturbance that causes the transition from forest to savanna). Also, some gallery forests are present. These formations are narrow corridors of trees developing along the river banks and within the savanna domain.

The radar color composite is shown in Figure 3.12 and the classification in Figure 3.13. The classification map clearly indicates how the forest is fragmented in this densely populated area. The fine strings of green are the gallery forest in the savanna domain.

The next area is again located at the forest savanna border near the city of Bangui (Box 3). The radar image is shown in Figure 3.14. In the corresponding map (Figure 3.15) one can observe how the savanna is intertwined with the forest. Also, some degradation patterns are visible in the forest.

Area 4 is located back at the core of the floodplain near Lake Tumba and the city of Mbandaka, a commercial and transportation center, with tanning and fishing industries.

Note how the river is extremely well delineated in the radar image (Figure 3.16). Indeed, thanks to the GRFM mosaic geolocation accuracy it was possible to derive by an expert photo-interpreter a GIS vector layer of the Congo Basin river network. This document strikingly showed how current geographic maps are inaccurate with respect to the current river spatial location.

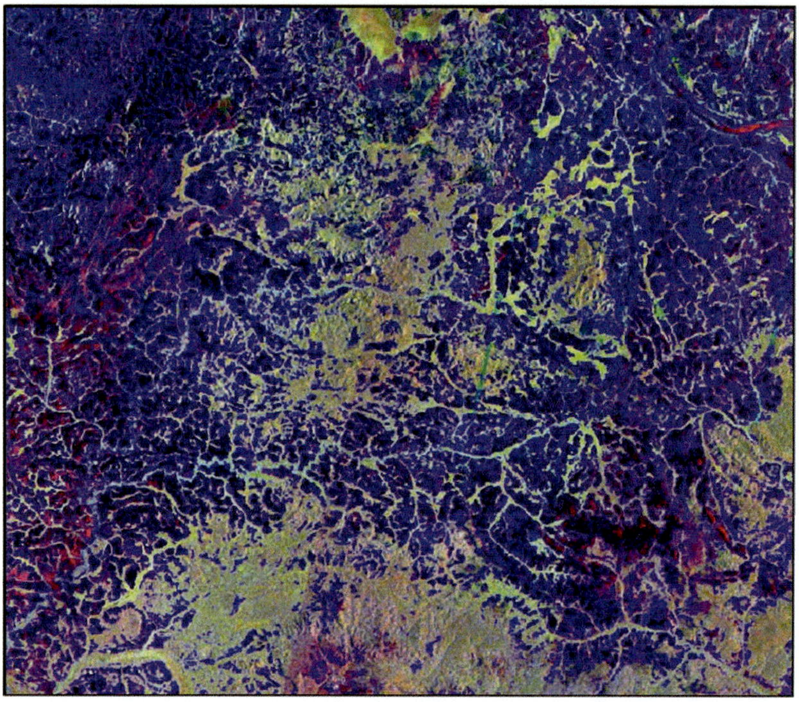

FIGURE 3.12 The forest–savanna margin in the north. RGB composite radar image (see legend in Figure 3.10).

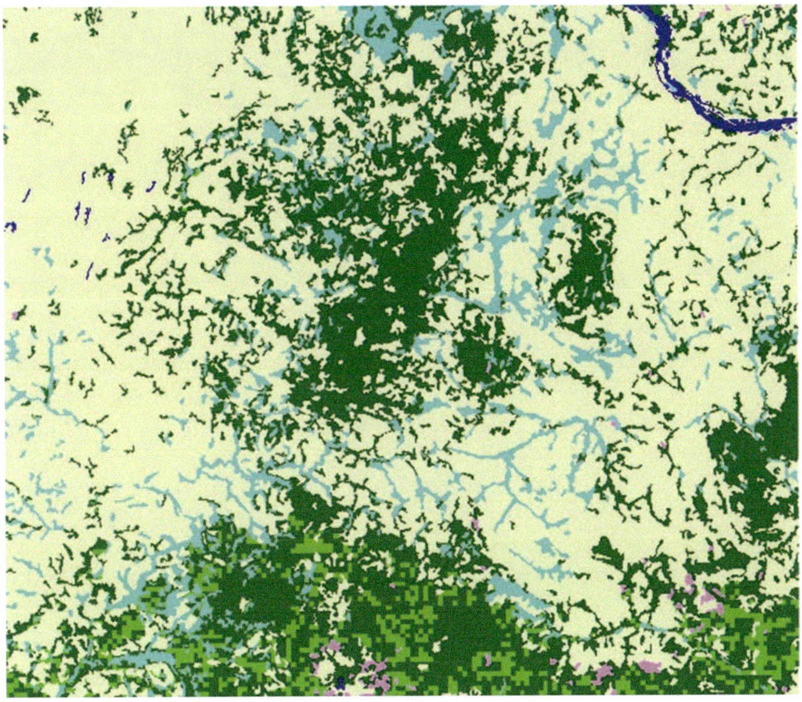

FIGURE 3.13 The vegetation map of the forest – savanna transition zone. The class legend is the same as in Figure 3.9.

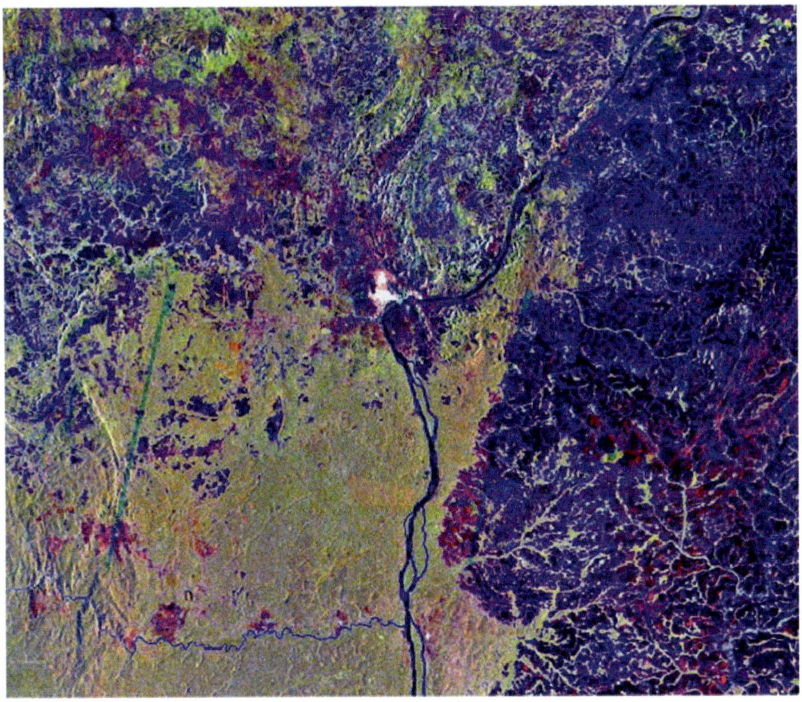

FIGURE 3.14 The forest-savanna border near the city of Bangui. RGB composite radar image (see legend in Figure 3.10).

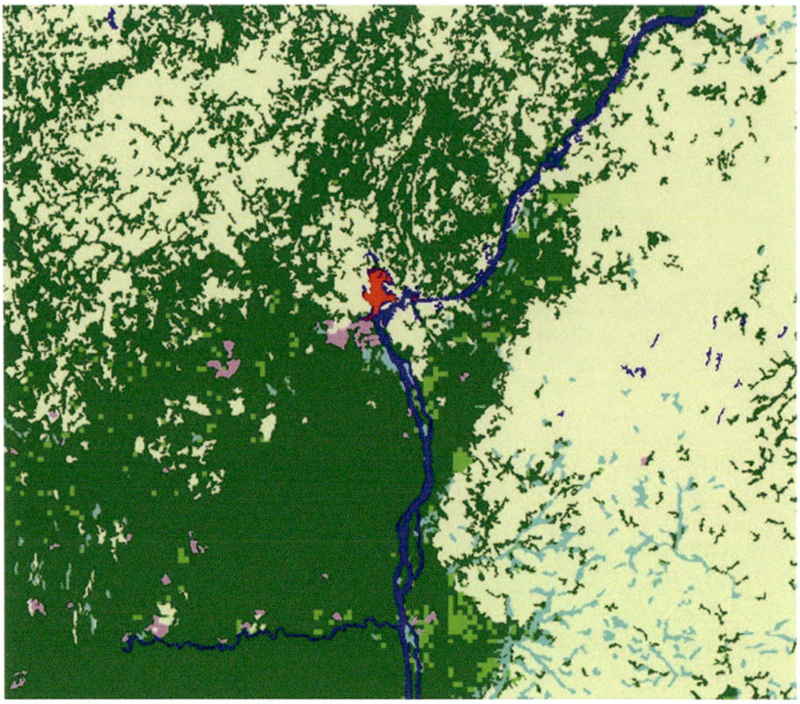

FIGURE 3.15 The classification map of the radar data set in Figure 3.14 (area in Box 3).

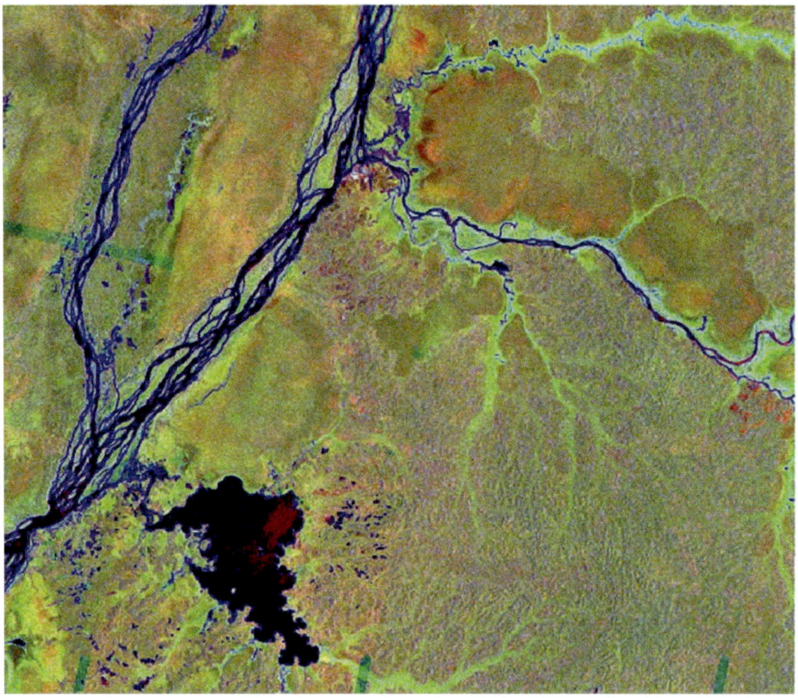

FIGURE 3.16 The Congo floodplain near Lake Tumba (Box 4). RGB composite radar image.

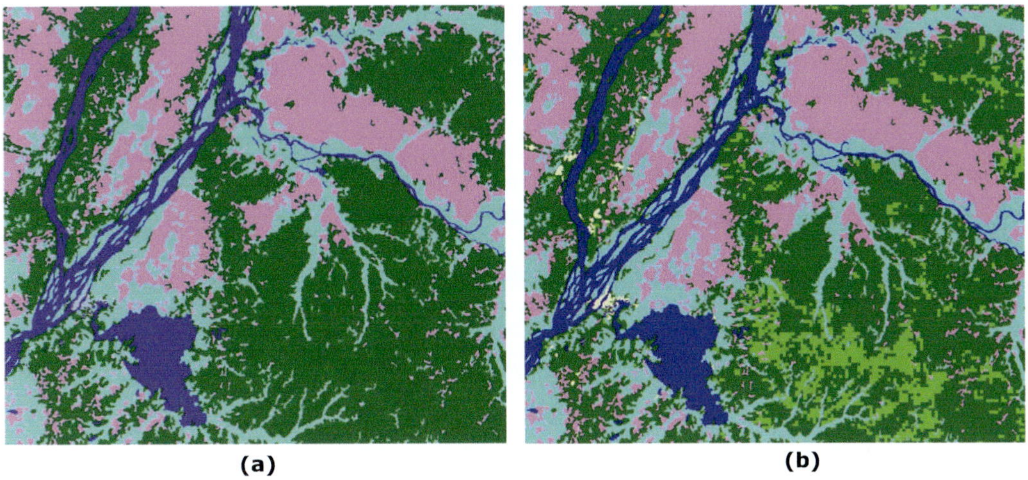

FIGURE 3.17 (a) The classification map of the Congo floodplain (area in Box 4) using only the radar imagery. The class "degraded forest" is missing in this product. This class (light green in b) can be detected by adding information from the SPOT VGT dataset, though only at a resolution of 1 km.

The thematic map derived only from the radar data sets is shown in Figure 3.17. This case illustrates well the variable resolution concept. In fact, in this map one class is lost –the class "degraded forest". This class cannot be uniquely defined by the radar observations, and is delineated in the amp by adding information from the optical instrument (SPOT VGT). Note that the degraded forest class is resolved only at 1 km.

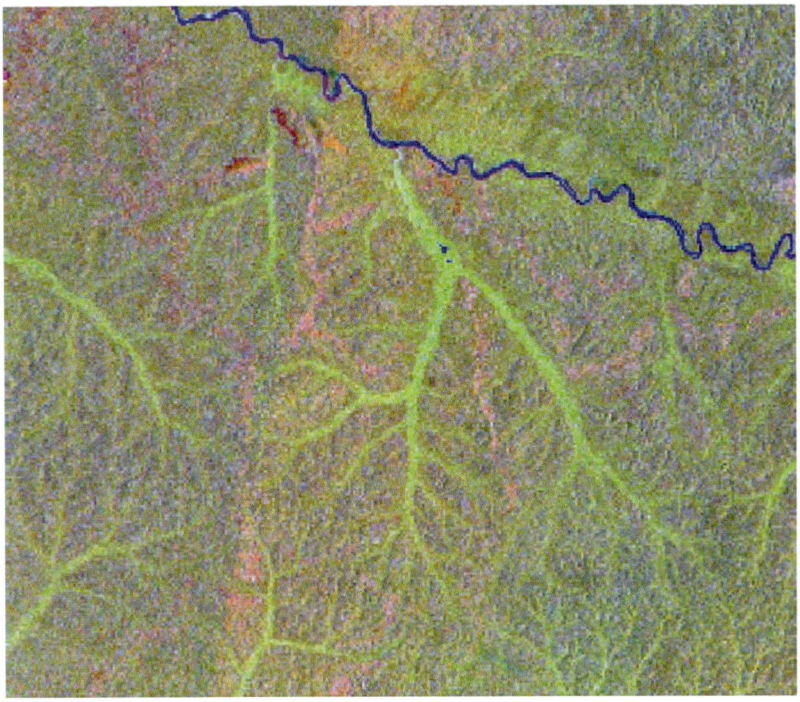

FIGURE 3.18 Network of swamp forests and degradation patterns (radar image).

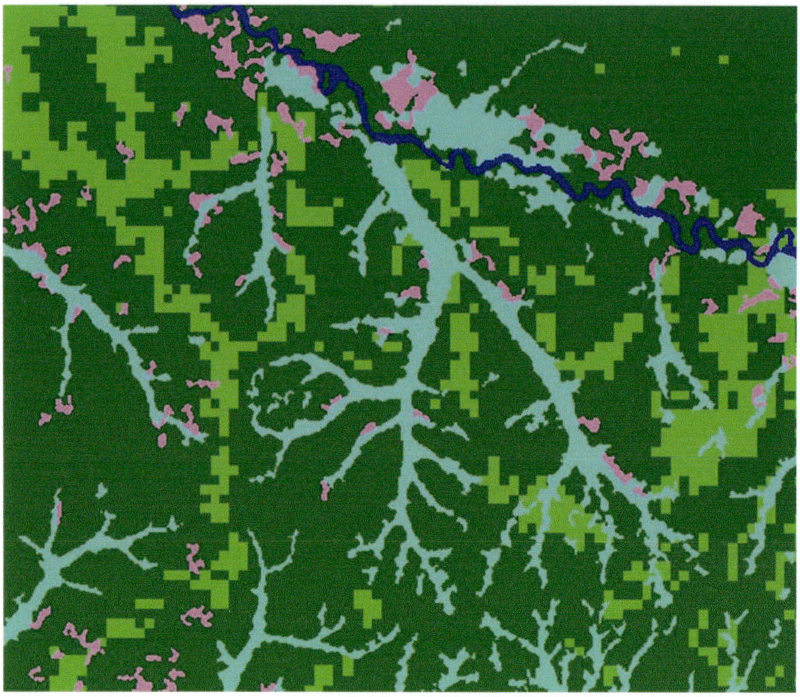

FIGURE 3.19 Network of swamp forests and degradation patterns (classification map).

The last example is related again to the floodplain, and it is particularly meaningful to illustrate the benefit of high resolution to capture the filamentary patterns of the swamp forests (see Figure 3.18 and Figure 3.19).

REFERENCES

1. Mayaux, P., De Grandi, G.F., and Malingreau J.P., 2000. "Central Africa forest cover revisited: A multi-satellite analysis", *Remote Sensing of Environment*, vol. 71, 183–196.
2. P. Mayaux, G.F. De Grandi, Y. Rauste, M. Simard, and S. Saatchi, "Regional scale vegetation maps derived from the combined L-band GRFM and C-band CAMP wide area radar mosaics of Central Africa", *International Journal of Remote Sensing* (GRFM special issue), vol. 23, no. 7, 1261–1282, April 2002.
3. G.F. De Grandi, P. Mayaux, J.P. Malingreau, A. Rosenqvist, S. Saatchi, and M. Simard, "New perspectives on global ecosystems from wide-area radar mosaics: Flooded forest mapping in the tropics", *International Journal of Remote Sensing*, special issue on "Global and Regional Land Cover Characterization from Satellite Data", vol. 21, no. 6–7, pp. 1235–1250, Apr. 2000.
4. G.F. De Grandi, P. Mayaux, A. Rosenqvist, Y. Rauste, S. Saatchi, M. Simard, and M. Leysen, "*Flooded forest mapping at regional scale in the Central Africa Congo River basin: First thematic results derived by ERS-1 and JERS-1 radar mosaics*", *Proceedings of Second International Workshop on "Retrieval of Bio- and Geo-physical Parameters from SAR Data"*, ESA ESTEC, 21–23 October 1998.
5. G.F. De Grandi, P. Mayaux, M. Simard, and S. Saatchi, "*Fusion of the L-band GRFM and C-band CAMP wide area Central Africa radar mosaics: A new data set with unprecedented potential for regional scale vegetation mapping*", *Proceedings of the IEEE IGARSS'00, CD-ROM paper MO01_05*, Honolulu, Hawaii, USA, 2000.
6. G.F. De Grandi, P. Mayaux, M. Massart, A. Baraldi, and M. Sgrenzaroli, "*A vegetation map of the Central Congo Basin derived from microwave and optical remote sensing data using a variable resolution classification approach*", *Proceeding of the IEEE IGARSS'01, CD-ROM paper FR06_02*, Sydney, Australia, 2001.

7. G.F. De Grandi, P. Mayaux, J.P. Malingreau, A. Baraldi, M. Simard, and S. Saatchi, "*Cornerstones and epilogue of the GRFM Africa project: A gallery of regional scale vegetation maps*", *Proceeding of the IEEE IGARSS'02, CD-ROM paper WE01_326*, Toronto, Canada, 2002.
8. Lebrun, J., and Gilbert, G. 1954. *Une Classification Écologique des Forêts du CONGO*, I.N.E.A.C. série scientifique n°63, 89 pp. Institut National pour l'Etude Agronomique du Congo Belge, Bruxelles.
9. Evrard, C. (1968) *Recherches Écologiques sur le Peuplement Forestier des sols Hydromorphes de la Cuvette Centrale Congolaise, ONRD*, p. 295. Institut National pour l'Etude Agronomique du Congo Belge, Bruxelles.
10. Junk, W.J. (editor), 1997, *The Central Amazon Floodplain, Springer Series on Ecological Studies*, vol. 126, (Berlin: Springer Verlag).
11. M. Simard, G.F. De Grandi, S. Saatchi and P. Mayaux, "Mapping tropical coastal vegetation using JERS-1 and ERS-1 radar data with a decision tree classifier", *International Journal of Remote Sensing*, vol. 23, no. 7, April 2002.
12. M. Simard, S. Saatchi, and G.F. De Grandi, "*Comparison of a Decision Tree and Maximum Likelihood Classifiers: Application to SAR Images of Tropical Forest*", *Proceedings of the IEEE IGARSS'00, CD-ROM paper INT28_14*, Honolulu, Hawaii, USA, 2000.
13. G. Patane, and M. Russo, "ELBG implementation", *International Journal of Knowledge-based Intelligent Engineering Systems*, vol. 4, pp. 94–109, Apr. 2001.
14. G. Patane, and M. Russo, "The enhanced-LBG algorithm," submitted to *IEEE Transactions on Knowledge and Data Engineering*, 14, 1219–1237, 2001.
15. G. Patane, and M. Russo, "Fully automatic clustering system," submitted to *IEEE Transactions on Neural Networks*, 13, 1285–1298, 2001.
16. A. Baraldi, P. Blonda, F. Parmiggiani, and G. Satalino, "Contextual clustering for image segmentation" *Optical Engineering*, vol. 39, no. 4, pp. 1–17, April 2000.
17. I.R. Joughin, D.B. Percival, and D.P. Winebrenner, "Maximum likelihood estimator of K distribution parameters for SAR data", *IEEE Transactions on Geoscience and Remote Sensing*, vol. 31, no. 5, pp. 989–999, September 1993.
18. M.L. Williams, "*Influence of canopy shape on SAR speckle distribution over woodland*", *Proceedings of the International Geoscience and Remote Sensing Symposium*. (Singapore, 3–8 August), IEEE Piscataway NJ, pp. 755–757, 1997.
19. Boulvert, Y. (1986) *Carte phytogéographique de la République Centrafricaine à 1:1 000 000, Notice Explicative No104*, Editions de l'ORSTOM, Paris.
20. CENADEFOR, 1985, Carte Écologique du Couvert Végétal du Cameroun, Yaoundé.
21. Laudet D. (1967), *Congo. Phytogéographie*, Editions de l'ORSTOM, Paris.
22. SPIAF, Carte Forestiere de Synthese de la Republique du Zaire, Service Permanent d'Inventaire et d'Amenagement Forestiers, Kinshasa, D.R. Congo, 1995.
23. Belward, A.S., Estes, J.E., and Kleine, K.D. (1999) The IGBP-DIS global 1 km land cover data set discover: A project overview. *Photogrammetric Engineering and Remote Sensing*, vol. 65, 1013–1020.
24. Mayaux, P., Janodet, E., Blair-Myers, C.M., and Legeay-Janvier, P. (1997) Vegetation map of Central Africa, Scale 1/5M, p. 32 + 1 map, *TREES Publications Series D*, N°1, EUR 17322 EN, European Commission, Luxembourg.
25. Laporte, N.T., Goetz, S.J., Justice, C.O., and Heinicke, M., 1998. A new land cover map of central Africa derived from multi-resolution, multi-temporal AVHRR data, *International Journal of Remote Sensing*, vol. 19, 3537–3550.

4 Evolution of the Species

The ALOS PALSAR Africa Mosaic

4.1 INTRODUCTION

At the time of the GRFM initiative, awareness of the dramatic problems related to climate change prompted the international community, under the umbrella of the United Nations Framework Convention on Climate Change (UNFCCC), to set the stage for an international treaty aimed at reducing emissions of greenhouse gasses [1]. The Protocol was adopted by Parties to the UNFCCC in 1997 and came into force in 2005.

In the same time frame, the radar remote sensing community responded to the observational needs raised by climate change studies with two major initiatives: GRFM in 1997, and the Kyoto and Carbon (K&C) project in 2007.

The Kyoto and Carbon (K&C) project [2,3] was an initiative of the Japanese Aerospace Exploration Agency (JAXA). The goal was to support terrestrial carbon science and related international treaties through thematic information derived by systematic – across time and space – observations of selected earth ecosystems from orbital instruments. The main data source of the project was the JAXA Phased Array L-band Synthetic Aperture Radar (PALSAR) aboard the Advanced Land Observing Satellite (ALOS) spacecraft [4,5].

The K&C project was designed to deliver thematic information satisfying criteria of spatial consistency over very large areas (regional/continental scale), and at the same time, of high spatial resolution (order of 100 m). To this purpose, large mosaics composed of radar images acquired during hundreds of orbits were assembled.

A collaborative effort among several partners of the K&C project was set up for the generation of these datasets. In this section we expound on one such collaborative effort (hereafter K&C Africa mosaic), conducted by the JRC in Europe and JAXA in Japan, that was aimed at developing a wide-area PALSAR mosaic over the whole African continent. A snapshot of the prototype mosaic is shown in Figure 4.1. The mosaic is a dual-band (HH–HV) dataset referenced to a latitude–longitude coordinate system with pixel spacing of 8.333 10^{-3} degrees (roughly 100 m at the Equator).

The CAMP and GRFM projects were indeed the precursors of the K&C Africa mosaic, and these projects set the stage for building the technical know-how required for the development and analysis of wide-area high-resolution radar mosaics.

However, the K&C Africa mosaic's development is to be positioned one notch higher in the scale of performance and complexity with respect to its precursor, because of:

1. Availability of the cross-polarized backscattering coefficient (HV) on top of the co-polarized backscatter (HH).
2. Availability of a high-quality digital elevation model (SRTM [6]) with comparable spatial resolution over the extent of the whole mosaic.

FIGURE 4.1 The K&C continental-scale Africa mosaic. The mosaic was assembled from PALSAR fine beam dual pol (FBD) images which were geocoded into a geodetic (latitude, longitude) coordinate system with a pixel spacing of 0.8333 millidegrees (roughly 100 m at the equator). Courtesy of JAXA and EC JRC.

3. A new type of SAR image product developed by JAXA (here dubbed "path images"), based on focusing the raw data along extended segments of the sensor's orbit, and thus more convenient for large area coverage.
4. Improved sensor performance with respect to radiometric (stability) and geometric accuracy.
5. Optimized data acquisition plan, this guaranteeing full coverage of the area of interest (continent) in the shortest time compatible with the sensor's orbital cycle (2 months in the case of Africa).

The availability of the cross-polarized channel is very important – for example, in thematic applications related to forest mapping – since cross-polarization arises from volume scattering, and therefore it is proportional to the woody biomass in the forest crown, and less sensitive to direct ground contribution.

The availability of a high-quality digital elevation model paves the way for the utilization of precision geocoding, of radiometric corrections for effects induced by topography, and of terrain morphology measures as additional training sets in a classification procedure.

These latter two aspects are of paramount importance for the generation of seamless mosaics, which in turn is a precondition for the successful thematic exploitation of the datasets.

The sheer size and improved characteristics of the K&C radar mosaic called for new approaches in data processing and analysis. This effort was balanced by the benefit of enabling new and improved thematic applications on the regional or continental scale in terms of spatial resolution or estimation accuracy of quantities of interest [7]. Cases in point are, for example, as far as the Africa mosaic is concerned, woodlands biomass estimation and deforestation rate assessments.

Next, we shall describe topics concerning the generation of the K&C Africa mosaic, an assessment of the dataset quality, and thematic observations that document the richness in information content, and the potential for future thematic applications. The reader will find additional information in [8].

4.2 THE MOSAIC PROCESSING CHAIN

The K&C PALSAR raw data were focused on a routine basis by JAXA during the fall of 2007, using a proprietary SAR processor [9] (SigmaSAR). The processor generates path images in slant range, amplitude data at two polarizations (HH, HV), with 16 Looks in azimuth and 2 Looks in range, a pixel spacing of 37.47 m in range, 19.2 m in azimuth, and image size corresponding, in ground range, to approximately 70 km in range, and up to 2000–3000 km in cross-range.

Features of the K&C path datasets (i.e. data volume, non-linear geometry and radiometric anomalies) called for the development of bespoke processing chains for the assemblage of wide area mosaics with suitable geometric and radiometric quality.

For the purpose, a number of algorithms for the geometric and radiometric revision and the path images assembly were studied and implemented.

Functionalities implemented in the K&C Africa mosaic software included:

Housekeeping routines to handle the ingestion and the file structure of the JAXA path image datasets.

Adaptive calibration revision of the original slant range datasets. This module checks automatically for the presence of radiometric anomalies and calibrates the data accordingly.

Extraction of subsets corresponding to the geographical extent of each projected strip image from a continent-wide digital elevation model (DEM) over Africa. The DEM was derived from Shuttle Radar Topography Mission (SRTM) data [6].

Geocoding into a geographic reference coordinate system (e.g. unprojected latitude–longitude) using the solution of the range-Doppler equations. This step also produced auxiliary data holding the effective local incidence angles for each pixel of the backscatter amplitude image.

Compression/decompression of the geocoded imagery to optimize data volume and processing time.

Assemblage of the geocoded strips within a geographic bounding box. The module used an inter-strips amplitude-blending algorithm to avoid edge effects.

Mosaic's radiometric revision to correct for seasonality effects and residual calibration errors, based on the estimation of backscatter differences along overlapping borders of neighboring strips. The bespoke modules were interfaced through a batch processor to the underpinning functions (e.g. geocoding) provided by the commercially available software SARscape by Sarmap SA [10]. In this way, large batches of data could be processed in the background, while the operator's intervention was limited to some critical steps, such as the mosaic's calibration revision. Indeed, the optimization of processing resources (memory, disc space) and processing time was one of the challenges posed by the construction of a high resolution continental scale dataset.

In the following we shall expand on the main issues that had to be resolved in the design and implementation of the processing chain in connection with the characteristics of the Africa mosaic's radar imagery.

4.3 CORRECTION OF RANGE DEPENDENT RADIOMETRIC BIAS IN PATH IMAGES

By block averaging in the along path direction over a stationary area (e.g. dense lowland forest) a profile is obtained that encapsulates the dependence of the backscattered power on the slant range coordinate. One such range profile, giving the average amplitude as a function of the incidence angle $\sqrt{I} = f(\vartheta)$, is shown in Figure 4.2 (black line).

Two features can readily be noted; an anomalous abrupt power loss at the end of the swath, and a linear decreasing trend from near toward far range. Barring the drop at far range, the power ratio from near to far range is of –2 dB for the test case considered. This level of power loss cannot be justified either by the dependence on effective scattering area, nor by the dependence of the radar cross-section on incidence angle (in the case of scattering from a dense vegetation layer). In the first case a drop of –0.04 dB may be expected for the range of incidence angles within the path data. In the second case, wave scattering models predict a change of the order of –0.13 dB.

This analysis reveals the presence of an anomalous bias on top of the intensity variation as a result of the aforementioned phenomena. The problem is compounded by the fact that the radiometric anomaly is not observed systematically and consistently throughout the datasets used in the compilation of the mosaics.

This scenario strongly suggests that the radiometric correction for the bias in the K&C strip data may not be based on theoretical global correction functions, even if parameterized by free variables, but should perforce rely on an adaptive algorithm driven by local estimation of the radiometric trends. The basic assumptions underlying the algorithm are as follows.

The negative linear trend in range is system induced (e.g. processing or SAR electronics) and can be accounted for by multiplicative intensity gain function that is linear with the range coordinate:

$$g(r) = g(0) + m\,r$$

where r is the range coordinate (pixel index in the column direction of the slant range data) and m and $g(0)$ are the parameters of the linear fit of averaged range profiles in homogeneous regions.

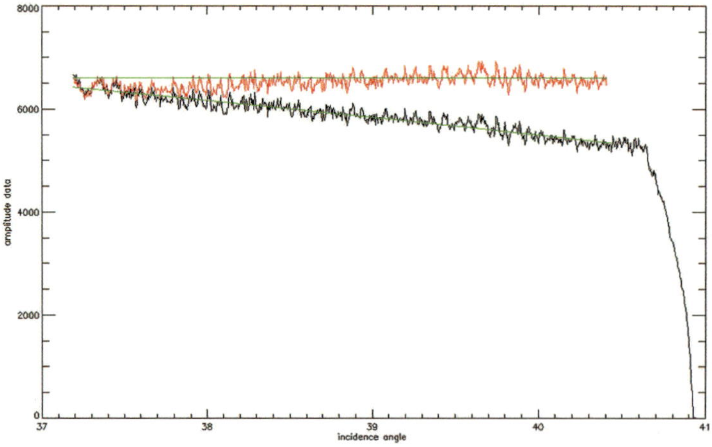

FIGURE 4.2 Range profile as a function of incidence angle showing the power loss at the end of the swath and the linear negative trend within the swath (black line in graph). The red line shows the corrected profile after the adaptive calibration procedure, which is based on linear fitting of the most homogeneous block-averaged (in along-path direction) range profiles (linear fitting function in green).

The best candidate region for the range profile estimation is searched by dividing the strip-image into along-track segments and selecting the one where two conditions are verified: minimum global backscatter variance and normalized difference between mean backscatter values at near and far range of less than 10%.

A correction gain function is derived from the fit parameters:

$$c(r) = 1 - \frac{m\,r}{g(0)}$$

In the same dataset, the rate of decay of the linear trend (if present) can be characterized by two functions, corresponding (in a loose sense) to areas dominated by volume scattering (e.g. forests and woodlands) and surface scattering (e.g. bare soil, sparse vegetation), respectively. These areas are identified by segmenting the range of backscatter values in the dataset into two classes: high and low backscatter. Specific threshold values for identifying the two classes are chosen as a function of the polarization. The two-class segments are a disjoint along the path direction. Therefore the region in between is characterized by a continuous smooth functional of the two functions $F[g_1(r), g_2(r)]$.

An example of the adaptive radiometric correction is reported in Figure 4.2, where the original range profile (green line) is graphed together with the corrected one (red line). Notice the linear intensity bias with negative slope and the sudden power drop at end of the swath in the original profile.

4.4 CORRECTION FOR ADDITIVE THERMAL NOISE IN HV STRIP IMAGES

The noise equivalent σ^0(NES) for PALSAR dual-pol imagery is rated at −29 dB. In some desert areas of Northern Africa, the response at cross-polarization falls below such a figure. Therefore the contribution of thermal noise becomes noticeable in areas of low backscatter (e.g. HV surface scattering) and is dependent on the incidence angle. The theoretical noise floor is a nonlinear function of the antenna pattern, the range distance, and the incidence angle [16]. In our approach, the additive noise component was estimated empirically by fitting a second-degree polynomial to a range intensity profile measured over a homogeneous (from near to far range) area with a backscatter coefficient lower than the NES and assumed constant throughout the swath. The estimated additive noise component was then used to compensate the trend in range of all HV datasets. A typical range profile corresponding to a low backscatter area before and after the correction is shown in Figure 4.3.

4.5 RADIOMETRIC INTER-STRIP MOSAIC BALANCING

Analysis of the mosaicked images compiled from the geocoded and range calibrated strips revealed the presence of inter-strip radiometric discontinuities. In particular, some strips present remarkable differences in backscatter values (order of 2 dB) with respect to adjacent ones (see Figure 4.4).

One of the reasons underlying this effect may be traced to the difference in seasonality and weather conditions between the acquisitions of the adjacent strip images. Such an effect is particularly evident in the west of Africa, where the acquisition dates of PALSAR (June to August 2007) spanned both dry and wet seasons, with the transition period being around mid-June. This difference in time can account for difference in radar reflectivity resulting from evolution in the target's properties (e.g. change in vegetation cover, change in soil and vegetation moisture).

In particular, a study was conducted to investigate the correlation between inter-strip intensity changes and soil and vegetation property changes. In line with what is suggested in [22] (albeit in a

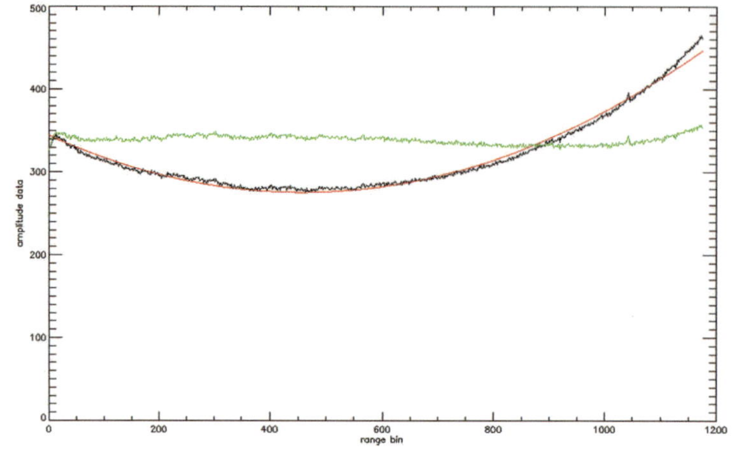

FIGURE 4.3 Range profile of a (slant range) HV dataset estimated in a block corresponding to a homogeneous (along the full swath) target featuring an average backscattering coefficient σ^0 is on the order of the noise equivalent backscattering coefficient (−29 dB). The impact of system thermal noise becomes relevant and results in a nonlinear increasing trend in intensity data (black line in the graph). An empirical additive correction function is derived by polynomial fitting of this range profile (red line), and it is applied to calibrate all HV datasets. The green line shows the same range profile after the correction for thermal noise.

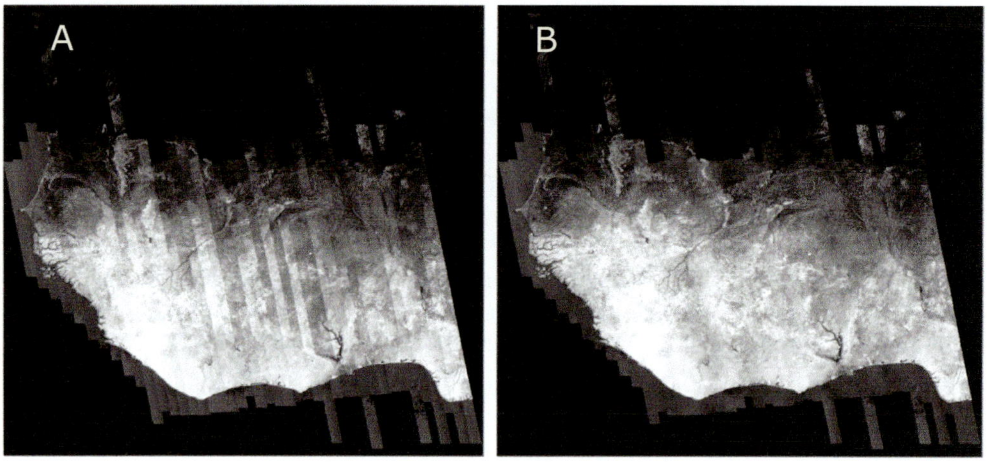

FIGURE 4.4 The West Africa part of the mosaic. Relevant radiometric differences (order of 2 dB) are clearly visible in the left-hand frame (A) between adjacent strips in the original (uncorrected) mosaic.

different thematic context), Advanced Microwave Scanning Radiometer—EOS (AMSR-E) brightness polarization ratio data, overlapping the ALOS acquisition dates, were used in this analysis. The rationale is that at 6.6 GHz, the microwave polarization difference index (MPDI) contains information on both the canopy optical depth and soil dielectric properties through emission. Therefore, a trend in a yearly time series of MPDI can be related to changes in vegetation development and water content, while the residual response, which is more random in time, can be linked to variations in soil moisture.

The anomaly signature (proportional to soil moisture) and the seasonal trend (an indicator of vegetation water content) were mapped into the area covered by the data of each strip. From these map and acquisition dates, regions were identified where the vegetation water content was low with respect to the annual mean $\Re_{low}$ and where the soil moisture was high with respect to annual mean $\Re_{high}$.

Assuming a correlation between vegetation water content, soil moisture and strip images' radiometric imbalance, the $\Re_{low}$ areas should correspond to strip images featuring a lower backscattering coefficient, and requiring a positive correction, while the $\Re_{high}$ areas should correspond to strip images requiring a negative radiometric correction. Comparison with the actual radiometric correction map suggests the following conclusions.

Matching between the areas affected by water content and radiometric corrections can be found only in the western part of the mosaic. Here at higher latitudes (where there is the presence of more open vegetation cover), the radiometric discrepancies between adjacent strips seem to be caused mainly by the effect of vegetation water content ($\Re_{high}$ patterns, and high correction values), except when this effect is compensated by higher than usual soil moisture ($\Re_{low}$ patterns). At lower latitudes, the radiometric imbalance could not be explained by this type of analysis. Other causes must be accounted for, as is explained next.

One more effect to consider is related to incidence angle differences in the overlap area between two adjacent strips. Here, the same target on the ground is seen by the radar with a different incident angle (far and near range). Assuming a cosine law for the radar cross-section dependence on incidence angle, the difference in backscatter because of near–far range incidence angle difference is roughly 0.78 dB, hence of the order of PALSAR radiometric accuracy. However, this figure in dB corresponds to a power ratio of 1.196, which would still give visible edge effects in the imagery. Other minor inter-strip unbalancing is produced by residual errors in range calibration.

The first approximation mosaic therefore needed to be corrected to arrive at a more radiometrically homogeneous version, which could be suitable for successive automatic interpretation, generation of land-cover maps, or for the estimation of bio-physical parameters, such as forest biomass.

This correction was achieved by an inter-strip balancing algorithm that relied on information derived by along-track profiles of the backscatter values estimated within the area of overlap between one strip and all the surrounding ones.

Since the radiometric discrepancies depend on the position within the strip (land cover and conditions), it was not possible to define a global strip-wise gain factor. Therefore, a gain function was defined for each strip in the geocoded geometry.

A strip frame is defined as the set S_j of the available values in the geocoded image. Each set S_j presents an overlap region O_{j-1} with the previous S_{j-1} (with ordering is Easting), and an overlap region O_{j+1} with the successive S_{j+1}.

Within each O_j regions, by averaging in the Easting direction, two intensity profiles, P_j and P_{j+1}, are obtained, e.g. one for the S_j set and one for the S_{j+1} set (see Figure 4.5). These profiles portray the trend of the radiometric discrepancy between two strips in the along-track direction. The radiometric differences are not constant along track, which suggests the use of a variable gain function.

Thus, calling $P_k^{left}(i)$ and $P_k^{right}(i)$ estimates of the profile values along the left and right borders at position i along track of strip k, and $P_{k-1}^{right}(i)$, $P_{k+1}^{left}(i)$ estimates of the corresponding profiles of adjacent strips at left and right of strip k, discrepancy measures are defined as:

$$D^{left}(i) = \frac{P_{k-1}^{right}(i),}{P_k^{left}(i)}$$

$$D^{right}(i) = \frac{P_{k+1}^{left}(i)}{P_k^{right}(i)}$$

These discrepancy measures form the basis for the mosaic correction algorithm.

The gain functions are computed by a piecewise linear approximation in the along-track of the discrepancy measures D^{right} and D^{left} (see Figure 4.5).

In fact, the D_k functions, being ratios of two samples from a speckle noise process, would introduce artifacts if used as multiplicative factors of the strip amplitude data. Therefore the D_k functions are first low-pass filtered and down-sampled by a factor of 64 to obtain a slow-varying trend.

This trend function is split iteratively into an increasing number of consecutive segments. At each step of the iteration the segments are linearly fitted, and the global root mean squared (RMS) differences computed. The procedure is stopped when the RMS differences between two steps increase or after a maximum number of divisions. The piecewise linear trend is then up-sampled to the original resolution by spline interpolation.

Once the linear trends are estimated, the adjustment of two neighboring strips proceeds using gain functions with a "moving half way up and half way down" strategy to avoid error propagation throughout the mosaic. Thus:

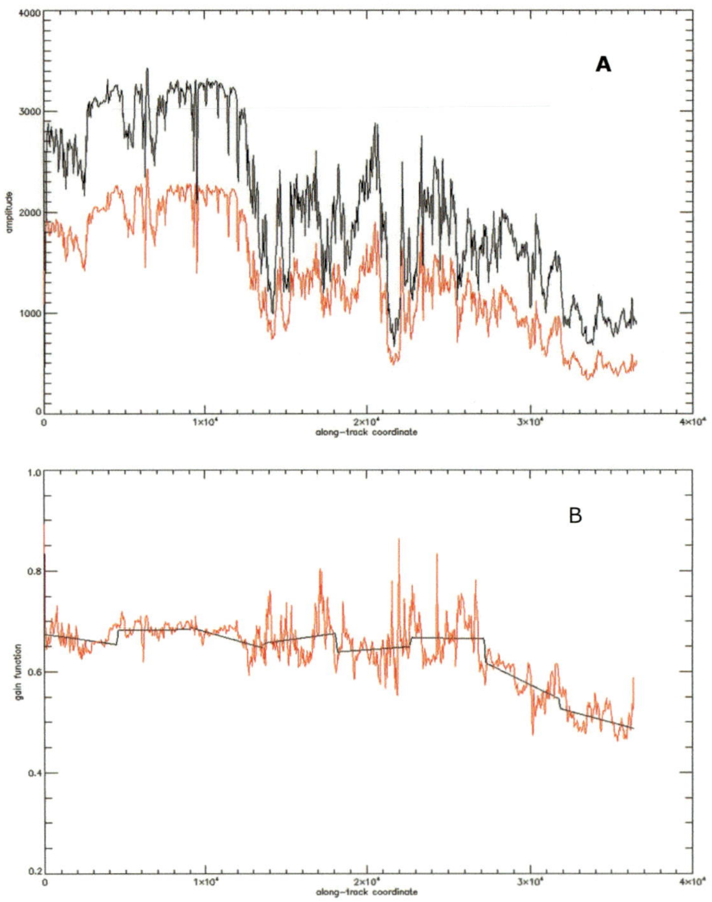

FIGURE 4.5 Profiles in the along-track direction of mean DN (digital number) values within the overlap area at the margin between two adjacent strips (the black line is the first strip profile, and the red line the overlapping second strip profile in graph A). The radiometric differences are not constant along-track, and suggest the use of a variable gain function. This gain function is estimated by a piecewise linear fitting of a smoothed version of the ratio between the two along-track profiles (red line in graph B). The segments in black are the gain function resulting from the piecewise linear fitting algorithm.

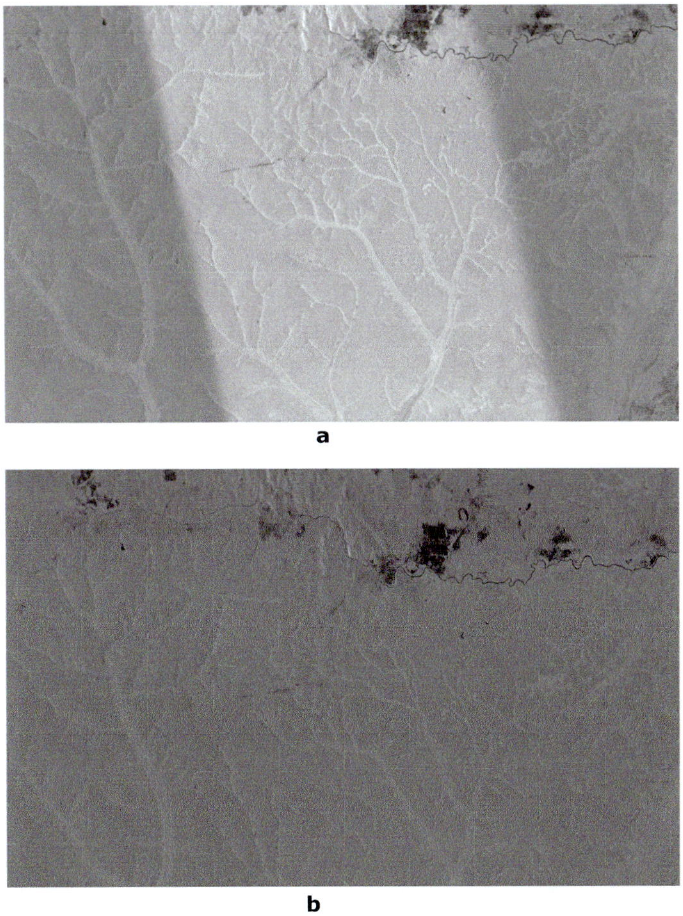

FIGURE 4.6 Result of the strip balancing algorithm (b) when applied to a set of strips with one radiometric anomaly, namely a higher backscattering coefficient, possibly because of vegetation water content and soil moisture change (a).

$$g_k^{left}(i) = \frac{1}{2}\mathcal{L}\left[D_k^{left}(i)\right] + \frac{1}{2}$$

$$g_k^{right}(i) = \frac{1}{2}\mathcal{L}\left[D_k^{right}(i)\right] + \frac{1}{2}$$

Where $\mathcal{L}$ is the piecewise linear trend operator.

An example of the effects obtained by the strip balancing algorithm is shown in Figure 4.6.

4.6 GEOCODING

Terrain geocoding of the slant range path datasets was performed using a module of the SARscape software [10]. The process relies on a continent-wide digital elevation model (DEM) of Africa derived from the Shuttle Radar Topography Mission (SRTM) data [6]. This DEM is provided in a geographic projection (Lat/Lon coordinates) using the WGS-84 horizontal datum, with a pixel size of 3 arc seconds (i.e. 0.8333 millidegree, approximately 90 m at the equator).

The geocoded PALSAR strip images were generated in the same projection and spatial resolution of the DEM. This choice assures the best geometric and radiometric accuracy. Moreover, past experience with thematic applications based on GRFM datasets indicates that the adopted pixel spacing is suitable for regional scale vegetation mapping studies. Higher spatial resolution is only needed for particular tasks, such as selective logging monitoring, which are conducted on a local scale.

SARscape geocoding is based on the classical range-Doppler approach. Note that this is the only appropriate way to obtain precision geocoding of SAR data. In fact, SAR systems cause nonlinear distortions (in particular in the presence of topography), and thus SAR datasets cannot be corrected using polynomials as in the case of optical images, where (in the case of flat Earth) an affine transformation is sufficient for conversion into a cartographic reference system.

Mapping from the slant range SAR geometry into a cartographic projection is obtained by considering the sensors and the image focusing parameters. Therefore the range-Doppler approach calls for the solution of the following relations:

$$\overrightarrow{R_s} = \vec{S} - \vec{P} \tag{4.1}$$

$$f_D = \frac{2\left(\overrightarrow{v_p} - \overrightarrow{v_s}\right)}{\lambda \left|R_s\right|} \tag{4.2}$$

where $\vec{S}$ and $\vec{P}$ are the sensor and target position vectors, $\overrightarrow{v_p} - \overrightarrow{v_s}$ is the velocity of the target relative to the sensor, and λ is the wavelength of the carrier.

The geocoding is implemented by a backward solution of the range-Doppler equations. Staring from each point of the DEM reference system, the corresponding point in the slant range radar geometry is found using the range-Doppler equations. Elements from the SAR geometry frame are then re-sampled into the earth projection (DEM) coordinate system.

Notice that the JAXA K&C strip products are not processed at zero-Doppler. Therefore in the solution of Equations (4.1) and (4.2) the proper value of the Doppler centroid had to be used. These values are supplied by JAXA in an auxiliary file for each line (along-track direction) of the slant range image as a polynomial approximation of the Doppler dependence on slant range distance.

Sources of geometric errors in SAR geocoding propagate from satellite orbit, range time, Doppler frequency and DEM accuracies [11]. Assessment of the geocoding accuracy was conducted a posteriori by establishing common features with the same DEM used in the geocoding procedure. It is therefore a relative accuracy measure. Difference vectors between control points in the SAR image and in the reference DEM image were defined by visual inspection. The maximum difference vector magnitude was found to be one pixel.

4.7 RADIOMETRIC NORMALIZATION FOR TOPOGRAPHIC EFFECTS

SAR radiometry is affected by topography because of changes of the ground scattering area, and changes of the target radar cross-section, these being affected by local incidence angle variation. Correction of these effects is important, to make the dataset suitable for automatic classification [12,13].

On the other hand, application of these corrections tend to smooth features related to terrain morphology, and therefore render the dataset less amenable to visual interpretation by the expert, at least for certain thematic applications. Therefore it was decided to apply radiometric normalizations for topography a posteriori in a second installment of the mosaics.

4.7.1 Correction of Effective Scattering Area

The PALSAR intensity products are calibrated in such a way that the backscattering coefficient σ^0 is obtained for each pixel by applying the following formula:

$$\sigma^0\left(dB\right) = 10\log_{10} DN^2 - 83$$

where *DN* is the pixel value (digital number) of the original data.

This backscattering coefficient represents the normalized radar cross-section for flat terrain; that is, the radar cross-section σ divided by the ground scattering area when the terrain is flat, noted A_{flat}.

This area corresponds to the pixel area projected from the slant range plane to the horizontal flat ground plane, and is therefore given by the following expression:

$$A_{slope} = \frac{r_a r_s}{\sin \vartheta_{loc}}$$

where r_a and r_s are the azimuth and slant range pixel spacing, and ϑ_{loc} is the SAR incident angle, i.e the angle between the incident electromagnetic wave vector $\vec{d}$ and the vertical at the terrain surface $\vec{n}$ (see Figure 4.7).

The incident electromagnetic wave vector $\vec{d}$ depends on two angles related to the SAR observation configuration: the incident angle θ and the viewing azimuth angle φ.

The normal at the terrain surface $\vec{n}$ depends on two other angles related to local topography: the local slope α and the aspect angle of the slope β.

The geometry is summarized schematically in Figure 4.7.

The theoretical expression of ϑ_{loc} is:

$$\cos \vartheta_{loc} = \frac{-\vec{d} \cdot \vec{n}}{\left|\vec{d}\right|\left|\vec{n}\right|} = \cos\alpha \cos\vartheta + \sin\alpha \sin\vartheta \cos\left(\beta - \varphi\right)$$

Here, $\vec{n}$ is directly derived from the DEM at each pixel (x,y) by calculating the cross-product of the two three-dimensional vectors formed by the adjacent pixels in longitude and latitude, respectively:

$$\vec{n} = \upsilon_1 \wedge \upsilon_2$$

These two vectors are given by:

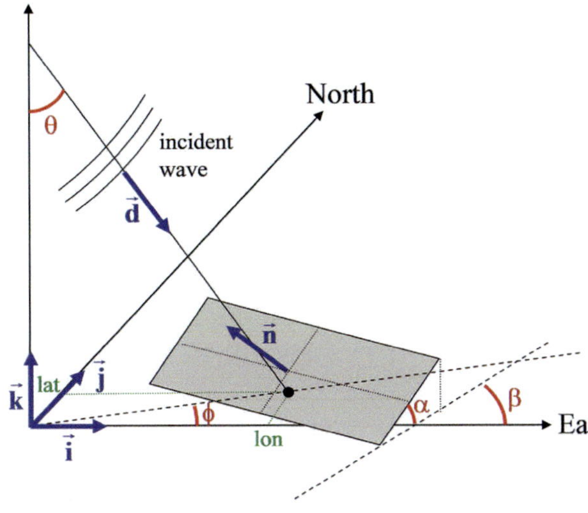

FIGURE 4.7 Configuration of the observation and scattering geometry over hilly terrain.

$$\vec{\upsilon}_1 = \begin{pmatrix} 2p_m \\ 0 \\ H_{x+1,y} - H_{x-1,y} \end{pmatrix}$$

$$\vec{\upsilon}_2 = \begin{pmatrix} 0 \\ 2p_m \\ H_{x,y-1} - H_{x,y+1} \end{pmatrix}$$

where p_m represents the pixel size in meters, and $H_{x,\,y}$ the altitude of the terrain at pixel (x,y) as given by the DEM.

The pixel size p_m is related to the pixel size in degrees p_d by the relation $p_m = R_e p_d$ where R_e is the Earth radius, given at latitude λ by the WGS-84 ellipsoidal model:

$$R_e = a\sqrt{\frac{(\cos\lambda)^2 + \left(\frac{b}{a}\right)^4 (\sin\lambda)^2}{(\cos\lambda)^2 + \left(\frac{b}{a}\right)^2 (\sin\lambda)^2}}$$

with a = 6378.137 km and b = 6356.752 km.

The correction of the effective scattering area can therefore be achieved by multiplying the DN^2 term by a correction factor C_{area}:

$$C_{area}\left(\vartheta_{loc}\right) = \frac{\sin\vartheta_{loc}}{\sin\vartheta_{nom}}$$

where ϑ_{nom} is the nominal incidence angle for flat terrain.

4.7.2 Correction for the Dependence of the Backscattering Coefficient on Incidence Angle

The effect of incidence angle on the backscattering coefficient depends on the nature of the target as well as on the polarization. Two possible avenues can be taken for the correction of these effects: (a) postpone the correction to the classification phase, given that a priori information on the target type at the scale of the whole mosaic will be available from an auxiliary dataset (e.g. MODerate Resolution Imaging Spectrometer (MODIS optical data); (b) tailor the correction to one specific thematic application, accepted to be suboptimal for a general land cover mapping task. In view of applications of interest regarding tropical forest and woodland mapping, it was decided to adopt a correction model for vegetation backscattering, i.e. dominated by volume scattering.

The angular dependence of the forest backscattering coefficient, when volume scattering dominates, is well represented by a cosine function, as proposed by the water could model in [14].

$$\sigma^0(\vartheta) = B m_v \cos\vartheta$$

Where B is a constant and m_v is the cloud water content.

Therefore the related correction factor is:

$$f(\vartheta) = \frac{\cos\vartheta}{\cos\vartheta_{nom}}$$

Where ϑ_{nom} is the nominal incidence angle.

Each intensity element with local incidence angle ϑ is corrected by:

$$I_{cor}(i,j,\vartheta) = I(i,j,\vartheta)\frac{1}{f(\vartheta)}$$

4.7.3 Assessment of the Radiometric Correction for Topography

The backscatter coefficient corrected for the overall effect of topography is:

$$\sigma^0_{cor}(dB) = 20\log_{10}(DN) + 10\log_{10}\left(\frac{\tan\vartheta_{loc}}{\tan\vartheta_{nom}}\right) - 83\,dB$$

The efficacy of the topographic corrections is documented by the following cases.

Figure 4.8 shows the effect of the topographic correction for a subset of the mosaic at the border between Congo and Gabon. The original dataset is in (a) and the corrected dataset in (b). Textural

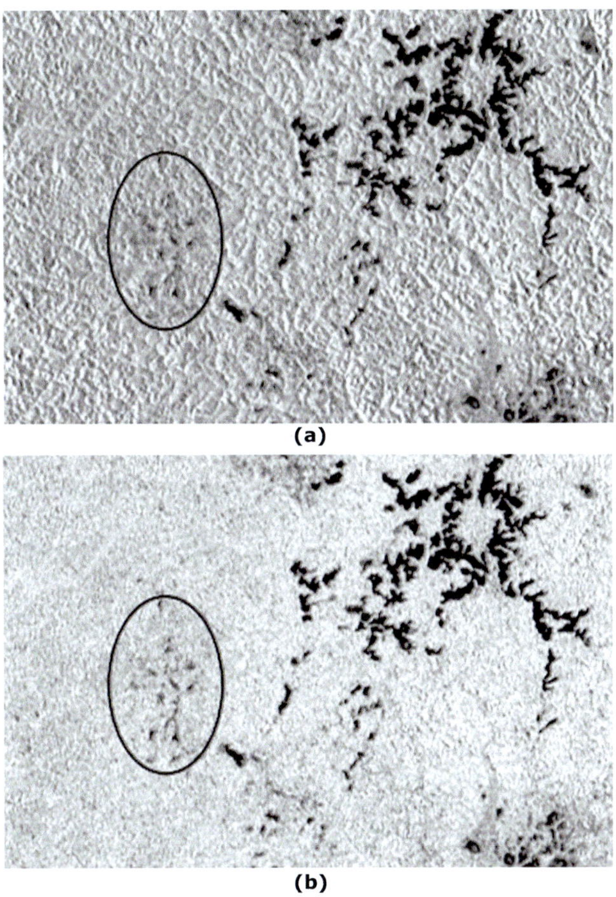

FIGURE 4.8 Subset of the mosaic located in an area between Congo and Gabon. The original dataset is in (a) and the corrected dataset in (b). Textural information related to topography is evident in (a), while it is smoothed in (b). However, non-forested areas (highlighted) appear much better resolved in the corrected dataset.

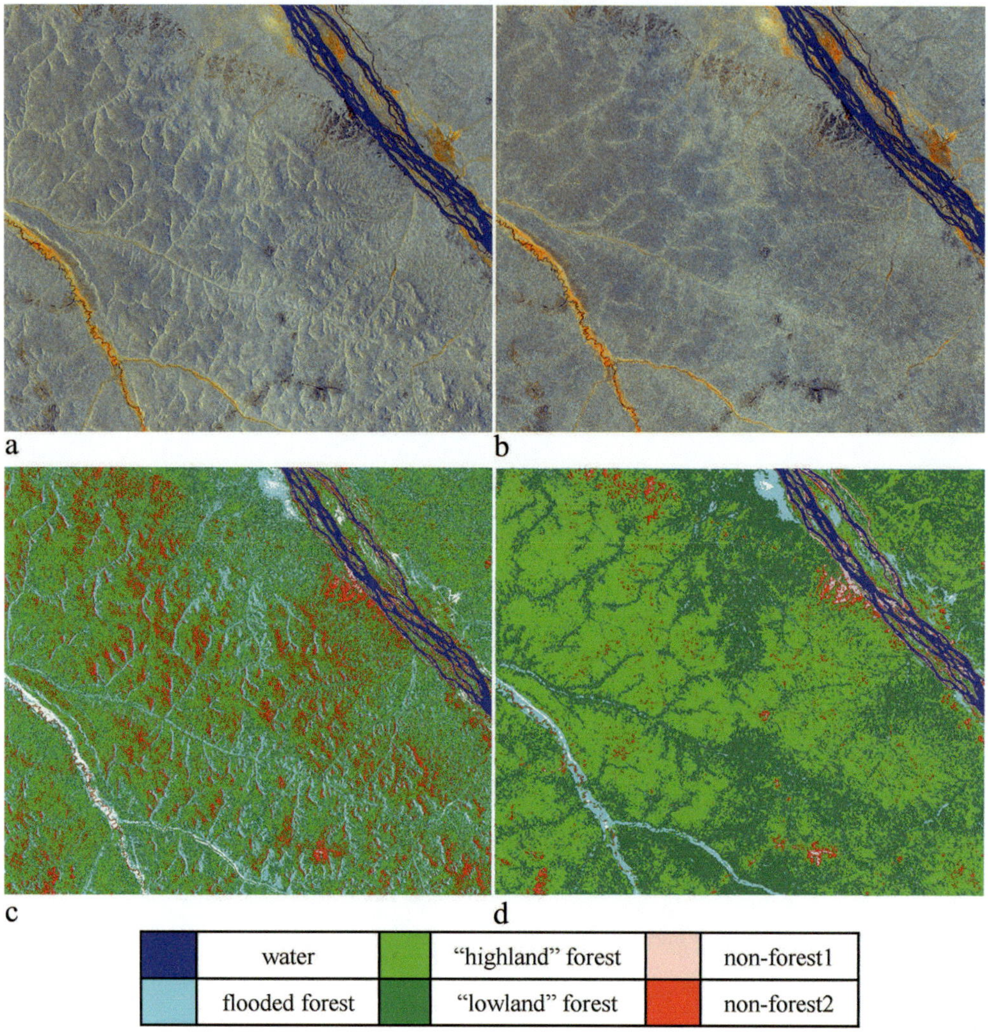

FIGURE 4.9 A subset of the mosaic (RGB channels HH, HV, HV/HH) portraying an area in the Democratic Republic of Congo (a) before, and (b) after topographic correction. K-means unsupervised clustering and supervised class labeling was applied to this dataset to test the impact of the radiometric corrections on classification accuracy. Classification from original data in (c) and from the corrected data in (d).

information related to topography is evident in (a), while it is smoothed in (b). However, non-forested areas (highlighted) appear much better resolved in the corrected dataset.

Figure 4.9 shows a subset of the mosaic (RGB channels HH, HV, HV/HH) portraying an area in Democratic Republic of Congo. The image in (a) is from the original dataset, while the image in (b) is obtained after topographic correction. The original and corrected images (HH and HV) have been used in an unsupervised clustering procedure (K-means) with supervised class labeling to test the classification improvement. The classification obtained from the corrected dataset (d) looks much more consistent than the uncorrected one (c), where terrain slope introduces errors in the classification.

Finally, we mention that polarization diversity may also be exploited to mitigate the influence of topography, on top of the aforementioned procedures. For example, the HV/HH ratio decreases the effect of terrain slope because the same effective area is illuminated in both channels, and the

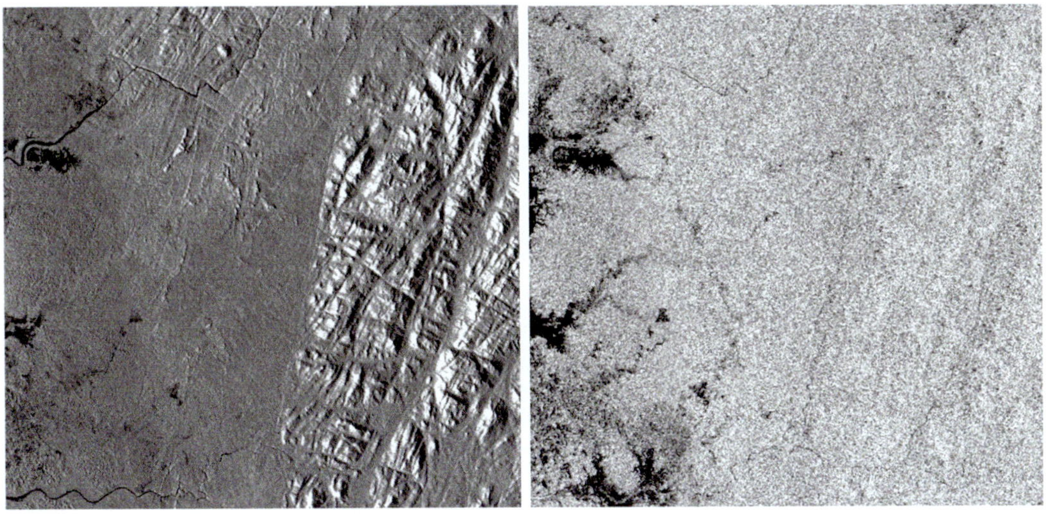

FIGURE 4.10 Capability of the HV/HH polarization ratio in reducing topographic effects. HH intensity dataset (left) and HV/HH dataset (right).

difference in radar cross-section sensitivity to local incidence angle is low between the HH and HV channels. The effectiveness of the HV/HH ratio in reducing topographic effects is documented visually in Figure 4.10, where the HH intensity dataset (left) is compared with the HV/HH dataset (right).

4.8 OVERVIEW OF THE THEMATIC INFORMATION CONTENT

In this section some thematic observations are reported, mainly concerning vegetation in the Central and Western parts of the continent. These observations show evidence that the Africa mosaic offers a rich source of thematic information, especially related to the humid forest to savanna interface, to deforestation and logging patterns, and to the characterization of the dry forest and woodlands at the margin with Saharan Africa.

4.8.1 Comparison with the GRFM Africa Dataset

The availability of the Global Rain Forest Mapping (GRFM) radar mosaic (see Chapter 2), forms the basis of investigating ecosystem changes in Central and West Africa over a time span of 10 years. Automatic change detection between the two datasets will pose some challenges, since the two sets were acquired by opposite satellite viewpoints (ascending orbit for PALSAR, descending for JERS-1). However, at least in flat or gently undulating areas, some comparisons are possible even by a simple combination of the imagery. A few examples are given below.

In Figure 4.11 (a) a subset of the PALSAR mosaic (color composite HH-HV-HH in the RGB channels) around a region near the town of Pokola, Republic of Congo, is shown. Thin linear features corresponding to logging roads are strikingly visually clear in the image. The corresponding GRFM dataset is shown in Figure 4.11 (b). Here the features related to logging roads are not detectable.

This case might well indicate that the PALSAR dataset offers better performance in small but genuine features detection, because of better radiometric resolution with respect to JERS-1. However, further analysis using a third information source will be needed to rule out the possibility of features' true changes between the two dates.

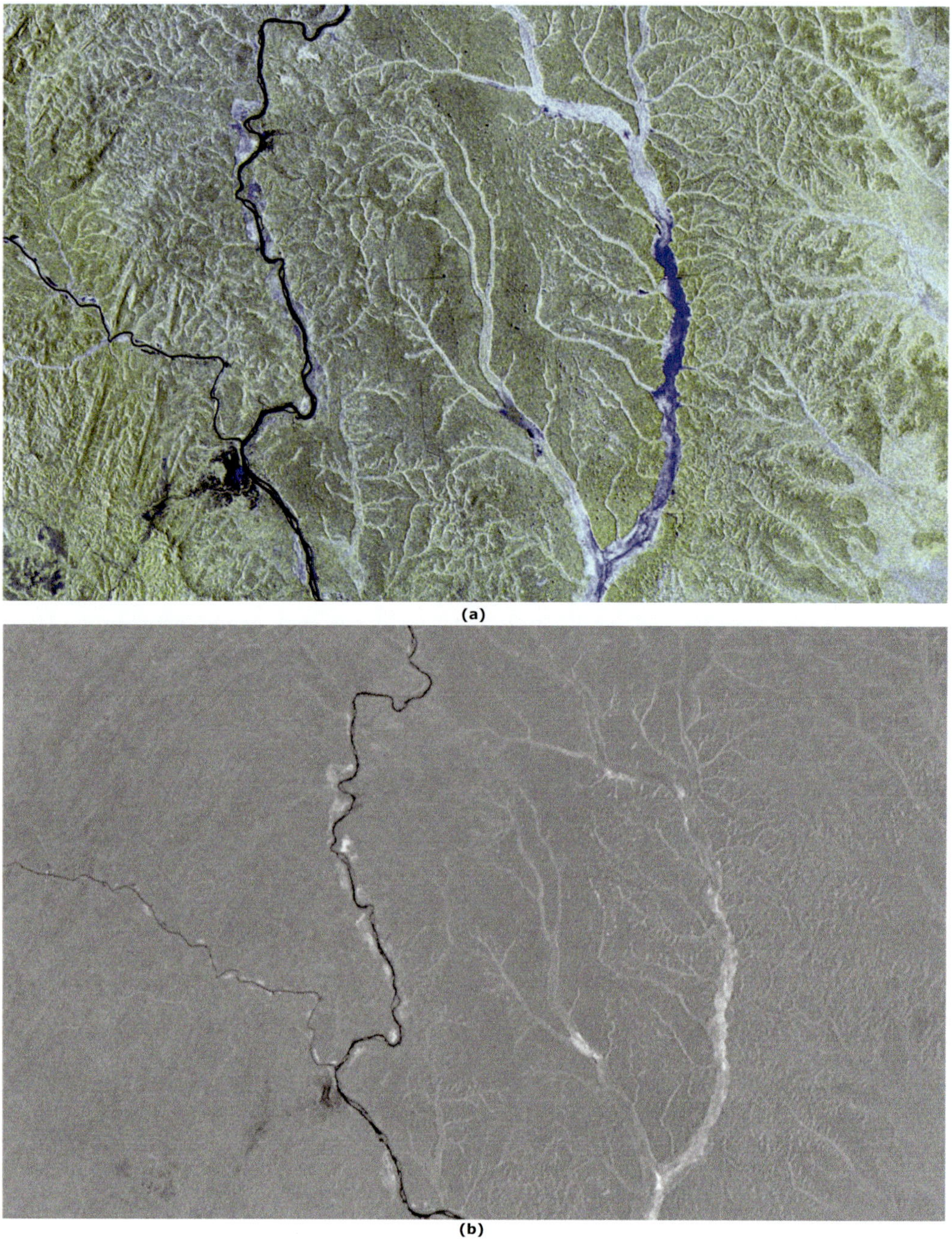

(a)

(b)

FIGURE 4.11 PALSAR Africa mosaic's subset (a) of an area near the town of Pokola in the Republic of Congo (1.868°N, 16.252°E) obtained in 2007, and corresponding image (b) obtained by JERS-1 in the context of the GRFM project in 1997.

In Figure 4.12 an area of the PALSAR mosaic corresponding to the interface between the rain forest and savanna around latitude 4.4 N is shown.

Visual comparison with the corresponding area in the GRFM mosaic highlights that the encroaching of the savanna into the rain forest domain (a result of anthropogenic impact) proceeds at a fast pace. This enhances the need for frequent monitoring of these areas, as can be done with PALSAR.

(a)

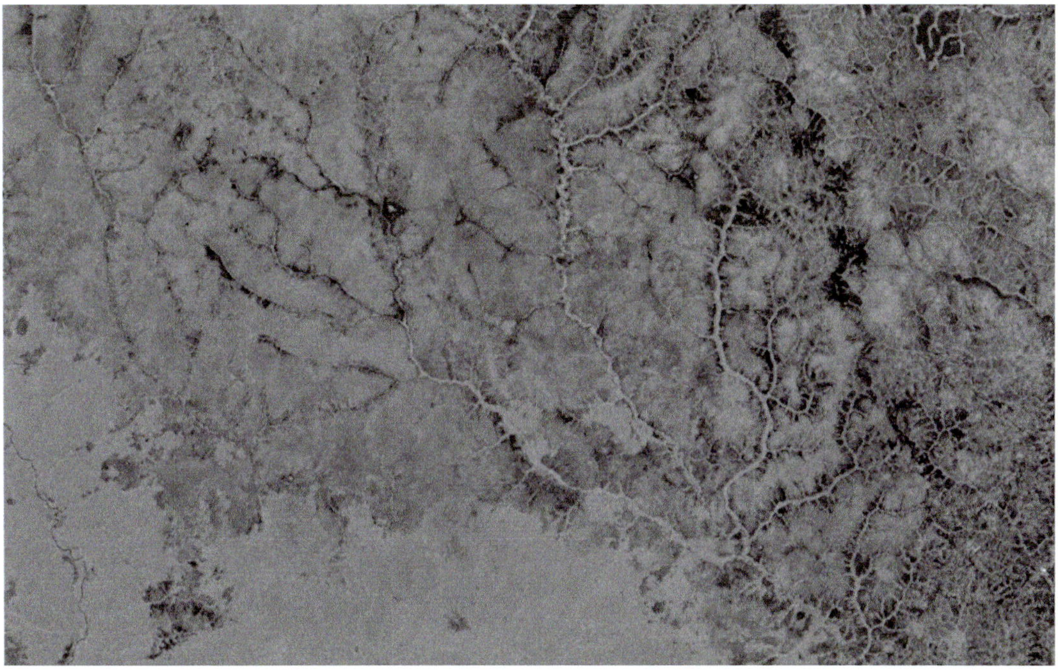

(b)

FIGURE 4.12 PALSAR mosaic's subset of an area at the margin between rain forest and savanna in Central Africa (a). The corresponding JERS-1 image from the 1997 GRFM mosaic is shown in (b).

It is also to be noted that the ability to distinguish between savanna and forest is increased compared to JERS-1, thanks to the presence of the HV channel. Over other areas, the cross-polarized band seems to offer the possibility to discriminate woody (shrub or tree) savanna from grass savanna (see next section).

4.8.2 Grass and Woody Savannas

A mosaic of savanna patches intertwined within the primary forest in the Congo Basin, as seen in a composite PALSAR image, is shown in Figure 4.13.

The RGB channels are assigned respectively to HH, HV and the HV/HH ratio (amplitude). For reference, a Google Earth image of the same area is shown in the frame at right of the same figure. By visual inspection and matching the patterns in the two images, one can conclude that the PALSAR image allows for the separation of class forest (in bright azure color), woody savanna (in orange) and grassland (in brown).

4.8.3 Flooded Forest

In L-band backscatter data, the HH scattering coefficient increases when standing water is present under the canopy through double-bounce interaction between the surface and trunks, while the HV backscattering coefficient, mainly generated by volume scattering from the canopy, will not be modified, or increases to a lesser extent, if the crown– surface interaction term is not negligible. As a consequence, the HV/HH ratio decreases in flooded forests.

This scattering mechanism can be exploited to monitor flooding conditions even under the thick canopy of the tropical forest. A case in point is illustrated by the PALSAR image of a swamp forest along the Congo River near Brazzaville (Figure 4.14, with a colour composite image HH, HV and HV/HH amplitude data). A Google Earth Landsat (land satellite) image is also shown for comparison.

The different orange shades in the swamp forest at the center of the image indicate the state of the soil: from very wet to flooded. Logging roads are also visible in the primary forest area (bluish colour) on the left.

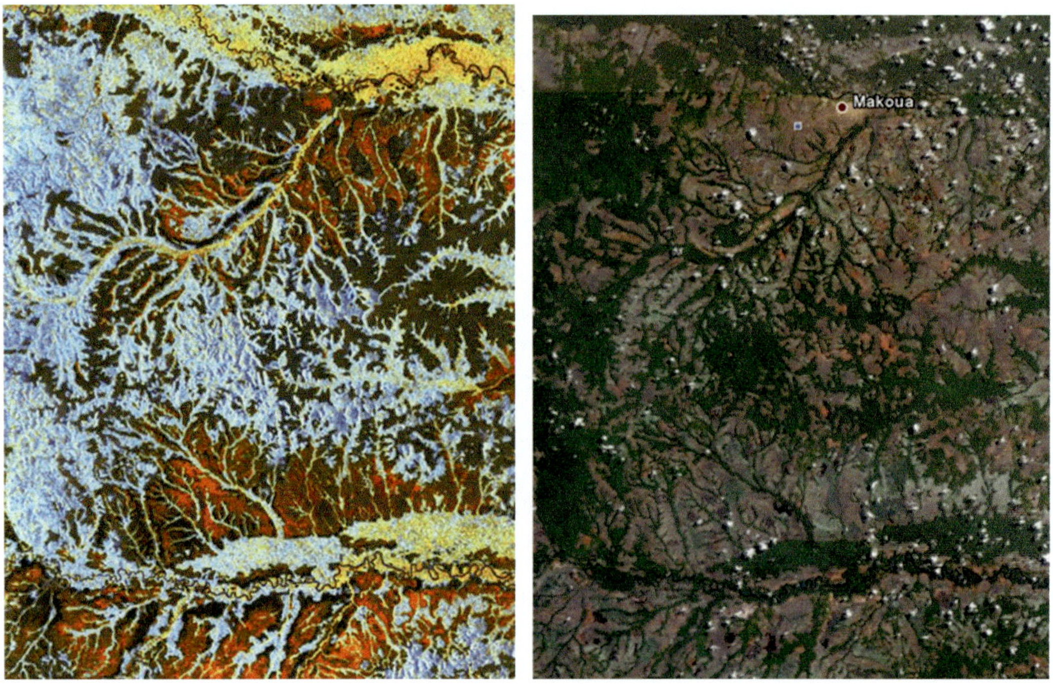

(a) (b)

FIGURE 4.13 A mosaic of savanna patches intertwined with the rain forest as portrayed by a PALSAR image (a) (color composite RGB with HH, HV, HV/HH) and by a Google Earth optical image (b).

FIGURE 4.14 PALSAR image (color composite RGB are HH, HV, HV/HH) of the swamp forest along the Congo River near Brazzaville (a). The different orange shades in the swamp forest at the center of the image indicate the state of the soil: from very wet to flooded. An optical image from Google Earth is shown for comparison (b).

A second example is reported in Figure 4.15. The scene in this case portrays the coastline with mangrove forests near the Nigeria–Cameroon border. Radar backscatter distinguishes two types of mangrove (flooded in orange shades and non-flooded in blue shades). Comparison with a Google Earth Landsat image indicates that this distinction is not possible using optical data.

The following classes were considered in a supervised analysis: rainforest, mangrove, flooded mangrove, swamp forest, flooded swamp forest. Estimates of the backscattering coefficients (and their variance) of these classes are reported in Figure 4.16.

4.8.4 Plantations

A subset of the PALSAR mosaic (color composite HH-HV-HV/HH in the RGB channels) portraying a plantation area near Kribi, Cameroon is shown in Figure 4.17 (a). The same area as seen by Landsat ETM (color composite band 4-band 5-band 6 in the RGB channels) is shown in the same figure (b).

In the PALSAR data, the bright orange patches represent oil palm plantations, which are not clearly detectable in the Landsat image. On the other hand, the bright areas at the bottom left of the Landsat image correspond to hevea plantations, which are not discriminated from the surrounding primary forest in the PALSAR mosaic. These two figures illustrate the complementarity of the two sensors for tree plantation monitoring.

4.8.5 Secondary Forest

Delineation of secondary forest formations poses a well-known problem in forest mapping by L-band radar observations [15]. Single polarization data cannot detect this thematic class unambiguously,

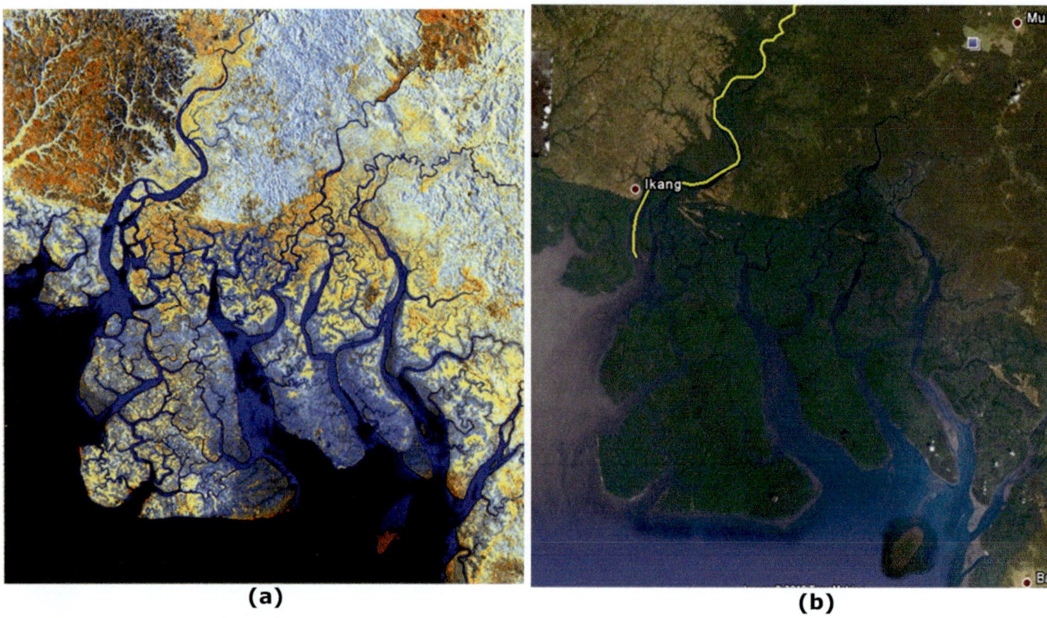

FIGURE 4.15 The coastline with mangrove forests near the Nigeria–Cameroon border as portrayed by a PALSAR image (a) and by a Landsat image (b). Radar backscatter distinguishes two types of mangrove (flooded in orange shades and non-flooded in blue shades). Comparison with the Google Earth Landsat image indicates that this distinction is not possible using optical data.

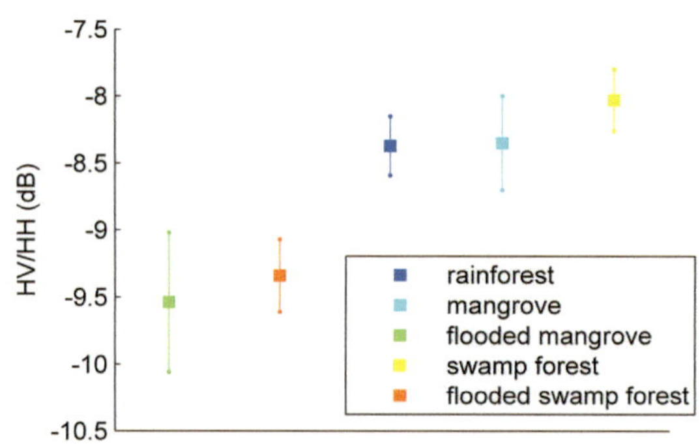

FIGURE 4.16 Supervised analysis of the thematic classes' separability provided by the HV/HH ratio for the dataset in Figure 4.15.

because of confusion with competing vegetation classes with similar backscatter coefficient. However, we present here a case that could indicate the potential of the HV/HH ratio for the purpose. Of course, the case should be validated by consistent tests, repeated in different areas and circumstances.

The images in Figure 4.18 shows a scene over a region in South East Cameroon as portrayed both by the PALSAR HV/HH amplitude ratio and a Landsat image. A rural complex emerges clearly from the primary surrounding forest in the Google Landsat image (b). The same pattern is detected by the HV/HH polarization ratio (a). A precise and automatic delineation of the pattern in the PALSAR image seems, however, difficult so far.

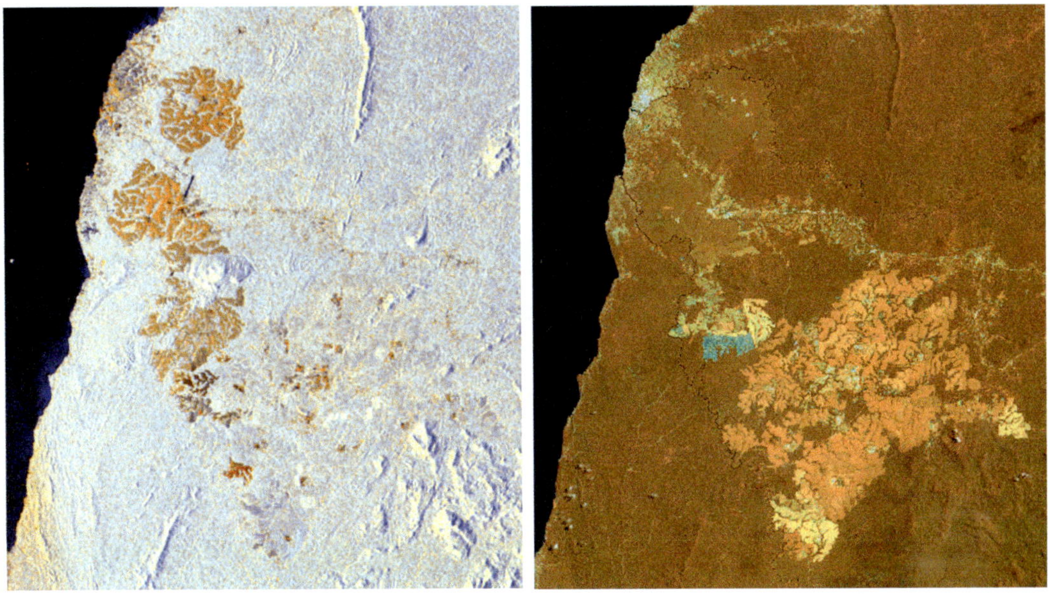

(a) (b)

FIGURE 4.17 A plantation area near Kribi, Cameroon, as seen in a PALSAR color composite image (a) The same area is shown in the Landsat color composite image (band 4-band 5-band 6 in the RGB channels) (b). Bright orange patches represent oil palm plantations in the PALSAR image, but they cannot be detected in the optical image. On the other hand, hevea plantations appear in the Landsat image (orange patches) but cannot be seen in the PALSAR image.

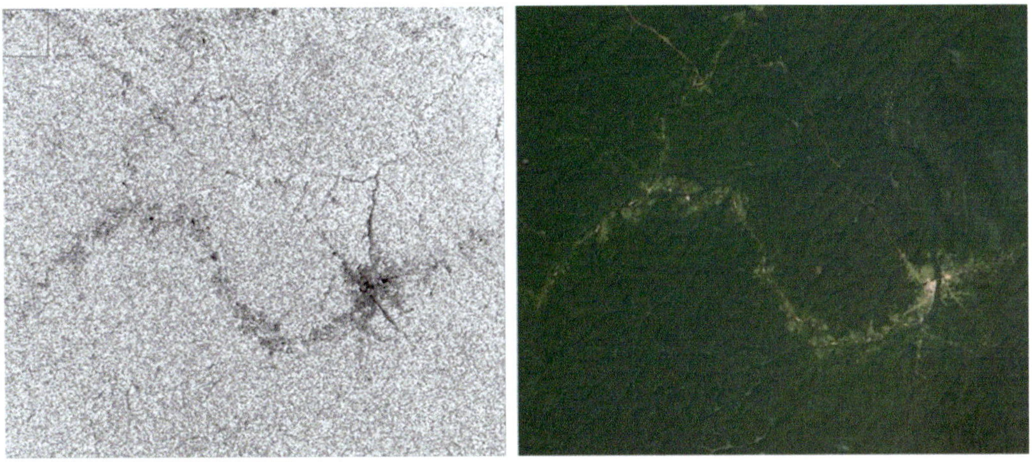

(a) (b)

FIGURE 4.18 An area in South East Cameroon where secondary forest (a rural complex) emerges from the primary forest as seen by a PALSAR polarization ratio image (a), and from a Landsat image (b).

REFERENCES

1. Kyoto Protocol Reference Manual, ISBN 92-9219-055-5, Information Services of the UNFCCC secretariat, 2008.
2. A. Rosenqvist, R.M. Lucas, B. Chapman, Lowry, L. Hess, and P. Paillou. "The ALOS Kyoto & Carbon Initiative, Science Plan v.3.1", Report, *JAXA Earth Observation Research Center*, Tsukuba, Japan, 2008.

3. A. Rosenqvist, M. Shimada, and A.K. Milne, "*The ALOS kyoto & carbon initiative*", *Geoscience and Remote Sensing Symposium. IGARSS'07*, pp. 3614–3617, 2007.
4. A. Rosenqvist, M. Shimada, N. Ito, and M. Watanabe "ALOS PALSAR: A pathfinder mission for global-scale monitoring of the environment", *IEEE Transactions on Geoscience and Remote Sensing*, vol. 45, no. 11, Part 1, 3307–3316, 2007.
5. M. Shimada, O. Isoguchi, and A. Rosenqvist, "*PALSAR CALVAL and generation of the continent scale mosaic products for Kyoto and Carbon Project*", *IEEE Geoscience and Remote Sensing Symposium, IGARSS 2008*, vol. 1, pp. I-17–I-20, 2008.
6. Jarvis, A., H.I. Reuter, A. Nelson, and E. Guevara, "Hole-filled SRTM for the globe", Version 4, 2008, available from the CGIAR-CSI SRTM 90m Database: http://srtm.csi.cgiar.org.
7. A. Bouvet, and G. De Grandi, "*A land cover map of Africa at 100 m resolution using a mosaic of ALOS PALSAR dual-polarization data: Preliminary developments*", *Proceeding IEEE Igarss*, Vancouver, Canada, 2011.
8. G. De Grandi, A. Bouvet, R.M. Lucas, M. Shimada, S. Monaco, and A. Rosenqvist, "The K&C PALSAR mosaic of the African Continent: Processing issues and first thematic results", *IEEE Transactions on Geoscience and Remote Sensing*, vol. 49, no. 10, October 2011.
9. M. Shimada, *Imaging from Spaceborne and Airborne SARs, Calibration, and Applications*, CRC Press, 2018.
10. SARscape ©2002–2020, available at: http://www.sarmap.ch/pdf/SARscapeTechnical.pdf.
11. G. Schreier, J. Raggam, and D. Strobl, "*Parameters for Geometric Fidelity of Geocoded SAR Products*", *Proceeding IEEE Geoscience and Remote Sensing Symposium, IGARSS '90*, pp. 305–308, 1990.
12. J.J. Van Zyl, "The effect of topography on radar scattering from vegetated areas", *IEEE Transactions on Geoscience and Remote Sensing*, vol. 31, no. 1, pp. 153–160, 1993.
13. F. Holecz, U. Wegmüller, E. Rignot, and Y. Wang, "*Observed radar backscatter from forested areas with terrain variations*", *Proceedings IEEE International Geoscience and Remote Sensing Symposium*, Firenze, vol. 1, pp. 613–615, 1995.
14. E.P. Attema, and F.T. Ulaby, "Vegetation modeled as a water cloud", *Radio Science*, vol. 13, 357–364, 1978.
15. P. Mayaux, G.F. De Grandi, and J.P. Malingreau, "Central Africa forest cover revisited: A multi-satellite analysis", *Remote Sensing and Environment*, vol. 71, pp. 183–196, 2000.

Part II

Measures of SAR Random Fields in the Scale–Space–Time Domain

5 The Stuff Backscatter Random Fields Are Made Of

5.1 INTRODUCTION

In the coming chapters, we shall talk about SAR backscatter spatial random fields. A spatial random field is a generalization of a random process, where the index of each point is no longer a single real number (e.g. time) but is vector spanning a bi-dimensional manifold. They will be our playground, where we shall try to devise statistical means to capture spatial patterns that will give us hints on the structure of the portrayed underlying natural reality. So the question naturally arises: What stuff are these fields, a mathematical abstraction, made of? Moreover, how are they created in the first place?

The beginning of the story unfolds from the sequence of electromagnetic wave pulses transmitted by the radar aboard the satellite and reflected back from terrain. With a slew of hand (aka signal processing), from the returned energy an "image" is created, that our eyes, accustomed to a different view of reality provided by light, can still interpret and understand.

We are not going to delve here into the details of how the SAR image is formed (on the topic see [1–4]). We assume somebody is able to produce an image, a field where the values are proportional locally to the energy per unit time that is scattered from the target back to our radar. The question is: Given a certain natural target, for example, a forest, how much energy should we expect to measure? Does this energy carry information on physical properties of the target, such as biomass, or its structural properties, such as canopy distribution? Is it dependent on environmental conditions external to the target? Is it dependent on the viewing geometry? Can we establish an invertible one-to-one relationship between the back-scattered power and some bio-physical parameter of the target?

Qualitative answers to these questions are a prerequisite for the correct interpretation of our radar scene, which in turn would lead to good thematic mapping. Quantitative answers set the ground for deriving from the radar backscatter field the spatial distribution of physical quantities related to the target of interest.

Wave scattering models provide a fundamental tool for answering these questions. We want to take advantage of this technique here to illustrate a simple study case related to radar backscatter from a dense primary tropical forest. This exercise will shed light on a number of phenomena that underpin our interpretation of the backscatter field, and allow us to build measures of the spatial relations among values. Therefore it is a good prelude to the material we shall expose in the following sections.

The area of wave scattering modeling in the microwave domain and for vegetation is a huge area of active research and development. Many different approaches have been pursued, following different avenues for the description of the medium and description of the wave propagation and scattering.

Regarding the former, we cite media descriptions at stand level, single-tree level, continuous isotropic or discontinuous random cloud of scattering elements, aggregation of elementary scatterers into vegetation components by tree growth models.

For the solution of the wave scattering problem, analytical theory based on Maxwell equations and transport theory dealing with energy balance were used. We want to look from the model's user's point of view, and show how one such modeling approach can lead to useful insight into our case of interest.

Our modeling tool of choice is a wave scattering model based on radiative transfer theory developed by the University of Texas at Arlington (UTA) Wave Scattering Research Center [5,6].

Before delving into the core of the topic, let us review the underpinning theory of radiative transfer and some of the basic workings of the model.

5.2 TRANSPORT THEORY

The UTA model is based on radiative transfer theory [7–11]. The basic notion in this approach is the energy associated with a wave traveling in a random medium, whereas other properties associated with fields, such as phase and cross correlations between fields are not taken into account. As a result, diffraction and interference effects are not included in the theory. The net effect on a wave interacting with a random media is then described by the sum in power of a number of elementary interactions.

To describe local properties in space and time of the energy associated with a wave, traveling and interacting with matter, and therefore varying in time, in amplitude and direction, we need a suitable differential descriptor. This descriptor is called *specific intensity* (the opticians call it brightness) and is defined as pictured in Figure 5.1.

For a given direction, it gives the average power flux within a unit frequency band and within a unit solid angle. It is thus measured in Watt m^{-2} sr^{-2} Hz^{-1}, where sr is the unit solid angle. If we take an elementary area da, as shown in the figure, oriented along a unit vector $\vec{s}_0$ at an angle ϑ with respect to $\vec{s}$, then the amount of power dP flowing through this area, an infinitesimal solid angle $d\omega$ and in a frequency interval ν, $\nu + d\nu$ is:

$$dP = I\left(r,\vec{s}\right)\cos\left(\vartheta\right) da\, d\omega\, d\nu \tag{5.1}$$

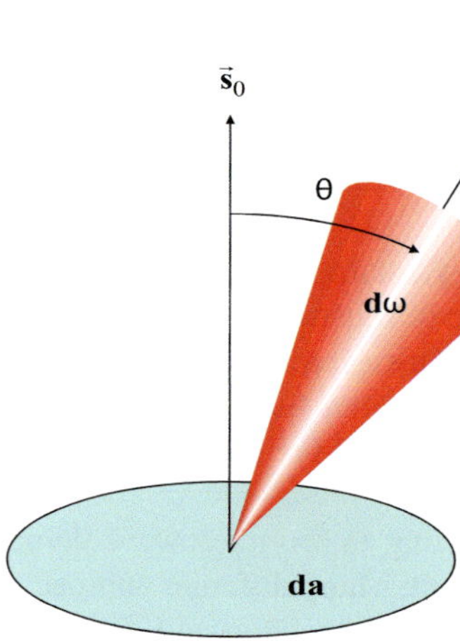

FIGURE 5.1 Specific intensity.

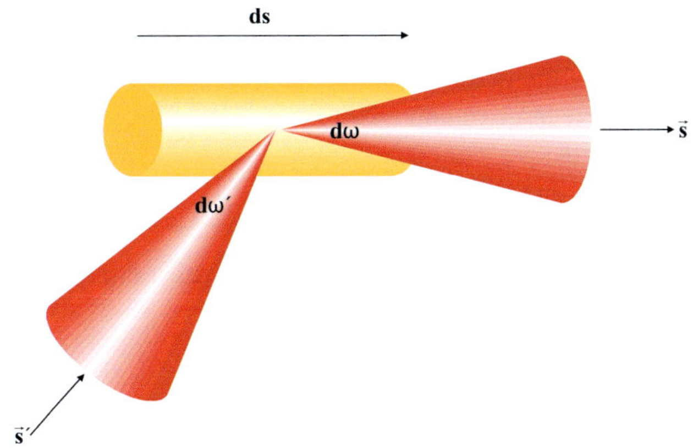

FIGURE 5.2 Scattering of specific intensity incident on volume of length *ds* from direction $\vec{s}'$ into specific intensity along direction $\vec{s}$.

Let us now consider a medium composed of a large number of particles, whose characteristics in terms of spatial distribution (geometry) and electrical properties are described in terms of statistical quantities. This is what we call a random medium. Initially, let us consider an infinitesimal cylindrical element of unit cross-section and length *ds* (see Figure 5.2), and see what happens to a pencil of energy (described by a specific intensity) impinging on the cylinder along the unit vector $\vec{s}$. Suppose the cylinder contains $\rho\, ds$ particles, where ρ is the number of particles per unit volume. The specific intensity will be reduced because of absorption (electromagnetic energy converted in other forms of energy) and by scattering outside the volume.

A convenient way of characterizing the energy–matter interaction for a single particle is through the concept of cross-section. In case of scattering, it gives the virtual area of a particle that would produce by isotropic scattering the observed total power at all angles surrounding the particle, when a unitary power flux density is incident upon the particle. The scattering cross-section is a relative of the geometric cross-section of a particle, and is equally measured in units of area, e.g. square meters.

The geometric cross-section can tell us something about the wave-particle interaction when the size of the particle is much greater than the wavelength. In other words, when we may picture the event using rays and geometrical optics. The scattering cross-section plays a similar role, when the wavy nature of the incoming radiation comes into effect because of the comparable size between wavelength and particle. Indeed, for particles whose size is much greater than the wavelength, the scattering cross-section can be approximated by twice the geometrical section.

Thus the particles in our cylinder will absorb and scatter power $\sigma_T I$ and the specific intensity will decrease by:

$$dI\left(\vec{r},\vec{s}\right) = -\rho ds \sigma_T I \tag{5.2}$$

At the same time, energy will be pumped into the cylinder, when pencils of energy impinging on the cylinder from another direction $\vec{s}_1$ are scattered into direction $\vec{s}$. To account for this contribution, we have to bring into play another character: the scattering amplitude $f\left(\vec{s}_1,\vec{s}_2\right)$. This function characterizes, for a single particle, the angular dependency of the scattering phenomenon. More precisely, *f* represents the phase and amplitude of the scattered field in the far range (where it can be approximated by a spherical wave) in the direction $\vec{s}_1$ when a plane wave of unit amplitude illuminates the particle into direction $\vec{s}_2$.

$$E_s\left(r\right) = f\left(\vec{s}_1,\vec{s}_2\right)\frac{e^{-ikr}}{r} = Ae^{i\phi}\,\frac{e^{-ikr}}{r} \tag{5.3}$$

The modulus squared of the scattering amplitude relates the incident specific intensity in direction $\vec{s}_2$ to the scattered specific intensity in direction $\vec{s}_1$ for a single particle. In our case (see Figure 5.2) the scattered specific intensity in the direction $\vec{s}$ due to the incident power flux density (amount of energy flowing per unit time through unit area) in the direction $\vec{s}$ is:

$$I_s(\vec{r},\vec{s}) = \left|f(\vec{s},\vec{s}')\right|^2 I_i(\vec{r},\vec{s}') \tag{5.4}$$

The specific intensity scattered in the direction $\vec{s}$ by $\rho\, ds$ particles in the volume ds is obtained by summing the incident fluxes from all directions $\vec{s}'$:

$$\int_{4\pi} \rho\, ds \left|f(\vec{s},\vec{s}')\right|^2 I_i(\vec{r},\vec{s}')\, d\omega' \tag{5.5}$$

The specific intensity within volume ds can also increase because of emission within the volume. In the microwave domain of our interest, however, this term is negligible and here is disregarded.

We recollect that the flow of energy in our problem, where we consider the local properties in an infinitesimal volume, is governed by two processes: (i) absorption and scattering outside the volume, which contributes to the decrease of the incident specific intensity; and (ii) scattering from other directions into the volume, which contributes to the increase of the specific intensity.

Now, adding the two contributions, we obtain an integro-differential equation that describes mathematically the physics of our wave-matter interaction.

$$\frac{dI(\vec{r},\vec{s})}{ds} = -\rho\sigma_T I(\vec{r},\vec{s}) + \int_{4\pi} \rho\, ds \left|f(\vec{s},\vec{s}')\right|^2 I_i(\vec{r},\vec{s}')\, d\omega' \tag{5.6}$$

This is the "equation of transfer" and it constitutes the cornerstone of a method that has very wide applicability, from astrophysics to optical and microwave remote sensing, to nuclear physics. Indeed, the same formalism can also describe neutron transport.

5.2.1 An Illustrative Case: Propagation Through A Plane Parallel Medium

Before delving into the main topic of this chapter, which is to characterize qualitatively radar backscatter for a tropical forest using wave scattering modeling, let us go through a simpler illustrative example, which will clarify how the radiative transfer approach works in practice for the numerical solution of a physical problem. The problem is to study wave propagation through a medium bounded by infinite parallel planes. This case is not only useful as a tutorial on radiative transfer technique, but is representative of practical applications of radiative transfer in studies concerning media that can be approximated by a plane parallel geometry, such as planetary atmospheres, ocean waters and clouds. Our presentation is derived from material presented in [7].

The geometry of our test case is depicted in Figure 5.3. We are concerned with a collimated plane wave of flux F_0, impinging normally on a plane-parallel medium of width $z = d$. The medium is surrounded by free space, and consists of free space filled uniformly with a density ρ of equal particles, each with total cross section σ_T. We are interested in computing the specific intensity $I(z,\vec{s})$ at each point z of the slab, and along direction $\vec{s}$ Since the medium is unbounded in the x and y directions, and because the incident radiation is a plane wave, the specific intensity will only vary along the z axis, and therefore the intensity will be a function of the angle ϑ with respect to the z axis, and the z coordinate. The variable $\mu = \cos\vartheta$ is introduced for convenience in the solution of the transfer equation.

It is also convenient to encapsulate the properties of the particles (cross-section and density) into a non-dimensional variable related to distance into the medium. This variable is called

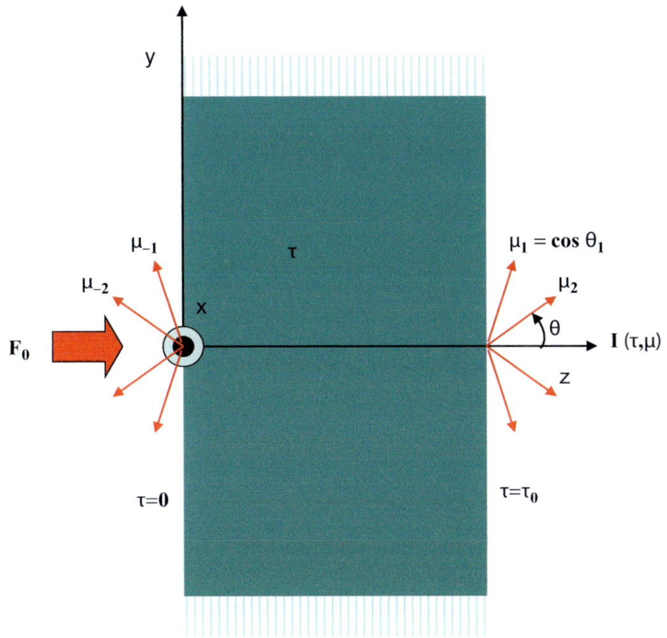

FIGURE 5.3 Geometry and quantities of interest for the plane parallel medium illuminated by a plane wave.

optical distance, and is defined in the general case where the cross-section and density vary in space, as:

$$\tau = \int \rho\sigma_T ds \tag{5.7}$$

in our case, since we assume cross-section and density to be constant, $\tau = \rho\sigma_T z$. Looking back at Equation (5.2), we see that τ is the parameter governing the rate of the exponential decay of intensity because of scattering and absorption. The same attenuation can be obtained by a trade-off between particle density, cross-section and path length. For example, when $\tau = 1$,the intensity is down of $e^{-1} = 0.368$.

One more definition is at hand. In radiative transfer problems, it is often convenient to split the intensity conceptually into two parts: one part corresponding to the incident intensity that is attenuated, as it progresses through the medium, by the process of scattering and absorption governed by Equation (5.2). This is called the reduced incident intensity I_{ri}.

The second part is the intensity that is created within the medium by scattering, and is called the diffuse intensity I_d. This conceptual splitting allows us to define boundary conditions at the surface of the volume containing the scattering elements. Indeed, since the diffuse component is created by definition only within the medium, it cannot enter the medium, and therefore we require that the diffuse component at the surface be zero for directions pointing inward into the volume:

$$I_d\left(\tau = 0, \mu > 0\right) = 0 \tag{5.8}$$

$$I_d\left(\tau = \tau_0, \mu < 0\right) = 0$$

In our case, the reduced intensity for an incident collimated plane wave is given by:

$$I_{ri}\left(\tau, \mu\right) = F_0 e^{-\tau}\delta\left(\mu - 1\right)\delta\left(\phi\right) \tag{5.9}$$

where δ is the Dirac distribution.

The directional effects in scattering from a single particle are often described in terms of the so-called *phase function*, which is derived from the scattering amplitude:

$$p\left(\vec{s}_1,\vec{s}_2\right)=\frac{4\pi}{\sigma_t}\left|f\left(\vec{s}_1,\vec{s}_2\right)\right|^2 \tag{5.10}$$

Therefore, the phase function also includes the notion of the strength of scattering and absorption through the cross-section σ_t. The term *phase* is not related to the phase of the wave, but rather is related to the angular information, since it derives historically from astronomy, where it refers to lunar phases.

For isotropic scattering, the phase function is simply:

$$p\left(\mu_1,\phi_1;\mu_2,\phi_2\right)=p\left(\mu_1;\mu_2\right)=\frac{\left|f\left(\mu_1;\mu_2\right)\right|^2}{\sigma_t}=\frac{\sigma_s}{\sigma_t}=W_0$$

The quantity W_0 is called *albedo* and measures how much of the incident radiation is scattered, being captured by the particle through absorption and scattering.

For Rayleigh scattering (lossless particle with dimensions that are small compared with the wavelength) the phase function is (in the azimuthally isotropic case):

$$p\left(\mu_1,\mu_2\right)=\frac{3}{4}\left[1+\mu_1^2\mu_2^2+\frac{1}{2}\left(1-\mu_1^2\right)\left(1-\mu_2^2\right)\right]$$

The radiative transfer Equation (5.6), expressed in terms of the phase function, becomes:

$$\frac{dI\left(\vec{r},\vec{s}\right)}{ds}=-\rho\sigma_T I\left(\vec{r},\vec{s}\right)+\frac{\rho\sigma_t}{4\pi}\int_{4\pi}p\left(\vec{s},\vec{s}'\right)I_i\left(\vec{r},\vec{s}'\right)d\omega' \tag{5.11}$$

Substituting I_{ri} from Equation (5.9) into Equation (5.11) we arrive at the following integro-differential equation for the diffuse intensity:

$$\mu\frac{dI_d\left(\tau,\mu\right)}{d\tau}+I_d\left(\tau,\mu\right)=\frac{1}{2}\int_{-1}^{+1}d\mu' p\left(\mu,\mu'\right)I_d\left(\tau,\mu'\right)+\frac{1}{4\pi}p\left(\mu,1\right)F_0e^{-\tau} \tag{5.12}$$

The integral in Equation (5.12) can be evaluated numerically by Gauss quadrature. In a nutshell, Gauss quadrature approximates a Riemann integral by a weighted sum of the function at points x_j, which are selected so as to minimize the approximation error.

$$\int_a^b f\left(x\right)dx=\sum_{j=1}^{m}a_jf\left(x_j\right) \tag{5.13}$$

$$\varepsilon=\int_a^b f\left(x\right)dx-\sum_{j=1}^{m}a_jf\left(x_j\right) \tag{5.14}$$

The full derivation of the Gauss quadrature formula can be found, for example, in [7,8]. In general, a function $f(x)$ can be interpolated by a polynomial of degree m $F_m = \prod_j(x - x_j)$ by setting $F_m(x_j) = f(x_j)$ if:

$$f(x) = \sum_{j=1}^{m} f(x_j) \frac{F(x)}{(x - x_j) F'(x_j)} \tag{5.15}$$

This is the so-called Legendre interpolation. Substituting Equation (5.15) into Equation (5.13) gives the following condition, which assures zero error in case the function $f(x)$ that we wish to integrate is also a polynomial of degree m:

$$a_j = \frac{1}{F_m'(x_j)} \int_a^b \frac{F_m(x)}{x - x_j} dx$$

Gauss quadrature assures zero error even when the function $f(x)$ is not itself a polynomial by exploiting the remaining degrees of freedom, namely the choice of the points x_j and the form of the polynomial F_m. It turns out in general that the condition for zero error is:

$$\int_a^b F(x) x^l dx = 0 \quad l = 0, \ldots, m-1 \tag{5.16}$$

In our case, the position $\mu = \cos\vartheta$ allows us to change the range of integration to $[-1, 1]$. The condition Equation (5.16) is satisfied if we choose $F_m$to be a Legendre polynomial P_m (which is by definition orthogonal to x^l) and we choose x_j to be zeros of P_m, namely: $P_m(x_j) = 0$. Then the coefficients in the series Equation (5.13) are given by:

$$a_j = \frac{1}{P_m'(x_j)} \int_{-1}^{1} \frac{P_m(x)}{x - x_j} dx \tag{5.17}$$

The integral in Equation (5.12) is finally computed choosing an even order $m = 2N$ and dividing the range $[-1, 1]$ of $\mu = \cos(\vartheta)$ in N points for $\mu < 0$ and N points for $\mu > 0$:

$$\int_{-1}^{1} f(\mu) d\mu = \sum_{j=-N}^{N} a_j f(\mu_j) \tag{5.18}$$

The μ_j are $2N$ zeros of the Legendre polynomial of order $2N$, and are labeled $\{\mu_{-N}, \mu_{-N+1}, \ldots \mu_{-1}, \mu_1 \ldots, \mu_{N-1}, \mu_N\}$.

For example, for $N = 4$ the zeros of P_4 are: μ_j={-0.861136, -0.339981, 0.339981, 0.861136}. They correspond to ϑ= -49.3395, -19.4795, 19.4795, 49.3395.

Using the Gauss quadrature formula, the integral in Equation (5.12) becomes:

$$\frac{1}{2} \int_{-1}^{+1} d\mu' p(\mu, \mu') I_d(\tau, \mu') = \frac{1}{2} \sum_{j=-N}^{N} a_j p(\mu_i, \mu_j) I_d(\tau, \mu_j) \tag{5.19}$$

Therefore, the integro-differential Equation (5.12) has been now reduced to a system of linear differential equations, by discretizing the angular dependence in μ. The system in matrix form can be written as:

$$\frac{d}{d\tau} I_j(\tau) + \bar{S}_{ij} I_j(\tau) = \bar{B}_j e^{-\tau} \tag{5.20}$$

where the unknown vector I_j contains solutions at each discrete orientation μ_j. The matrix $\vec{s}$ is given by:

$$S_{ij} = \frac{1}{\mu_i} - \frac{a_i p(\mu_i, \mu_i)}{2\mu_i} \quad when\, i = j \tag{5.21}$$

$$S_{ij} = -\frac{a_i p(\mu_i, \mu_j)}{2\mu_i} \quad when\, i \neq j$$

$$B_i = -\frac{p(\mu_i, 1)}{4\pi\mu_i} F_0$$

The system of linear differential Equations (5.20) can be solved numerically; for example, using the Runge Kutta algorithm, and the following boundary conditions (see Equation 5.8):

$$I_d(\tau = 0, \mu_i) = 0 \quad i = 1, 2, \ldots, N$$

$$I_d(\tau = \tau_0, \mu_i) = 0 \quad i = -1, -2, \ldots, -N \tag{5.22}$$

A few test runs of this numerical solver are illustrated next. Notice at the outset how the properties of the medium in Equation (5.20) are condensed in the optical length τ, as far as attenuation and particle density are concerned, and in the phase function p in relationship to the scattering mechanism.

For our test case, let us suppose considering an atmospheric layer with hail, which, at a wavelength of 1 cm, can be modeled as spheres and by Rayleigh scattering. For hail with water content

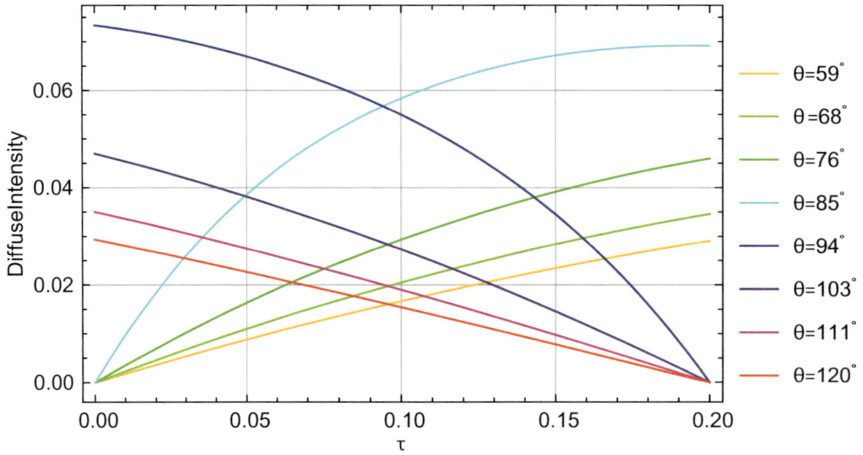

FIGURE 5.4 Diffuse intensity computed for a plane-parallel layer of 1 km containing spherical particles with attenuation $\rho\sigma_t = 0.2\ m^{-3}$ at $\lambda = 1$ cm. The Rayleigh scattering regime is assumed. The legend of the figure gives the angle ϑ in degrees between the propagation direction and the z-axis (see geometry of the problem in Figure 5.3) and the corresponding colors of the family of curves in the graph.

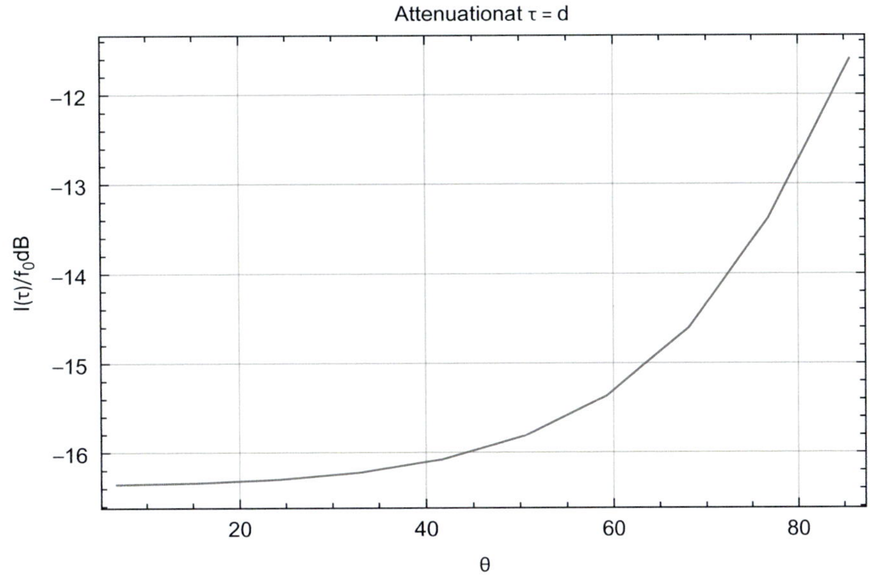

FIGURE 5.5 Intensity attenuation (in dB) at $\tau = 0.2$ (emerging radiation at end of slab) as a function of angle ϑ.

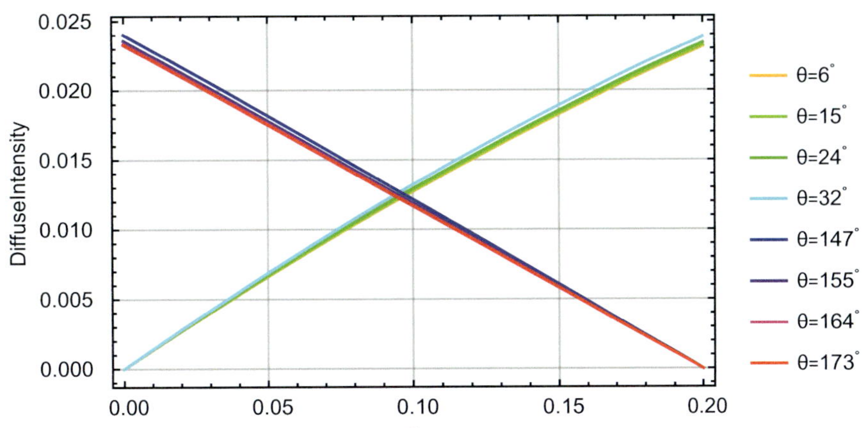

FIGURE 5.6 Diffused intensity for the same plane-parallel problem but for small angles ϑ.

of 0.3 $g/m^3$and number density 10 m^{-3} the attenuation at $\lambda = 1\ cm$, as reported in the literature [7], is: $\rho\sigma_t = 0.2\ (km^{-1})$. A layer of 1 km would therefore correspond to $\tau = 0.2$.

The numerical solver of Equation (5.20) for this test case was run using 20 steps for the Gauss quadrature, and assuming an incident collimated plane wave with flux $F_0 = 1$ Watt m^{-2} Hz^{-1}. Results for large forward angles with respect to the z axis are presented in Figure 5.4. Results for small forward angles are shown in Figure 5.6. In the direction of propagation ($\vartheta < 90°$) the diffuse intensity along a given direction builds up within the medium, because of the scattering of the surrounding particles in this direction. Energy of the incoming beam, which was collimated in the $\vartheta = 0$ direction, becomes spread out in several directions as it propagates through the medium, as a result of scattering. Notice also how the boundary conditions are fulfilled. The intensity attenuation, at the end of the slab as a function of direction, is graphed in Figure 5.5.

The angular dependencies of diffuse intensity at the boundaries of the medium are shown in Figure 5.7. At $\tau = 0.2$ (end of the slab) there is no diffuse intensity entering the medium.

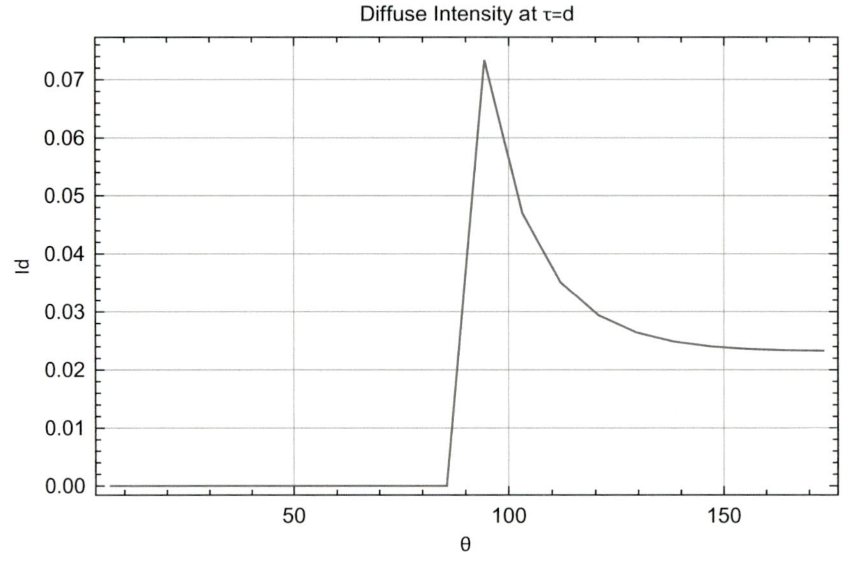

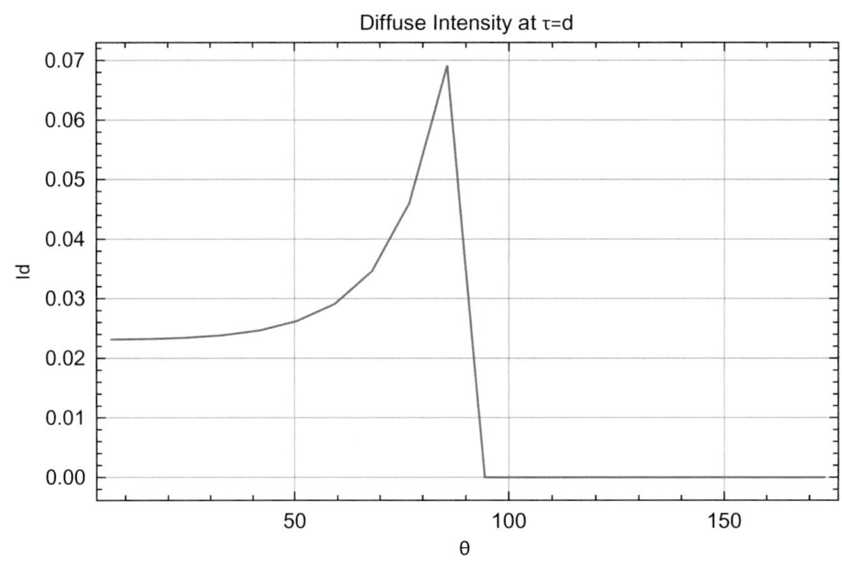

FIGURE 5.7 Angular dependency of the diffuse intensity emerging from the slab at $z = 1$ km and at $z = 0$.

5.3 THE UTA WAVE SCATTERING MODEL FOR LAYERED VEGETATION

Let us now see how the radiative transfer theory is used to construct the UTA model. We shall sketch here only the principles underlying the model, leaving the reader to access details of the paper in [1].

At a first level of simplification, we assume that the problem can be reduced to a single semi-infinite random layer of particles, bordered on one side with the atmosphere, from where the incident radiation comes, and on the other side with a dielectric surface (see Figure 5.8). With respect to the plane parallel problem illustrated above, we introduce therefore one more level of complexity, given by the fact that one boundary of the slab is composed of a different medium.

For the sake of illustration of the main principles, one more simplification in the following exposition concerns the wave polarization. We assume we are dealing with a fully polarized wave, and that the medium does not introduce any depolarization effects. Indeed, to take polarimetric effects into account (as carried out by the UTA model) the scalar quantities related to specific intensities

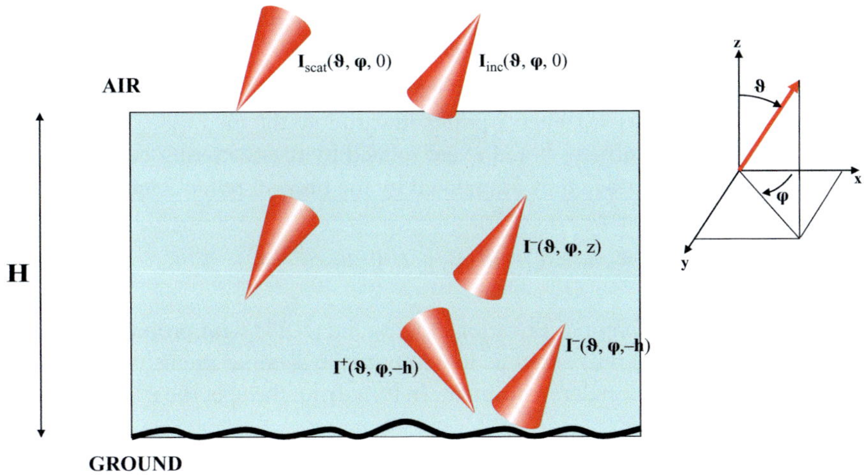

FIGURE 5.8 Geometry of the plane-parallel problem. The up-welling and down-welling specific intensities and their boundary conditions are indicated.

need to be substituted by vector quantities – the Stokes vector – that fully describe the polarimetric phenomena.

Within our medium, we split conceptually the traveling radiation into a down-welling part and an up-welling part. Also, the geometry of the problem suggests the use of a coordinate system where ϑ (the incidence angle), ϕ (the azimuth angle) are defined as indicated in Figure 5.8, with respect to a Cartesian coordinate system where the plane-parallel medium lies in the (z, x) plane. Correspondingly, we define two specific intensities

$$I^+\left(\mu_s, \varphi_s, z\right) \text{ and } I^-\left(-\mu_s, \varphi_s, z\right)$$

where $\mu_s = cos\ (\vartheta_s)$ and $-\mu_s = cos\ (\pi - \vartheta_s)$. In this setting, Equation (5.6) becomes:

$$\mu_s \frac{dI^+\left(\mu_s, \phi_s, z\right)}{dz} = -\kappa I^+\left(\mu_s, \phi_s, z\right) + S^+\left(\mu_s, \phi_s, z\right) \tag{5.23}$$

$$-\mu_s \frac{dI^-\left(-\mu_s, \phi_s, z\right)}{dz} = -\kappa I^-\left(-\mu_s, \phi_s, z\right) + S^-\left(-\mu_s, \phi_s, z\right)$$

where $S^+(\mu_s, \phi_s, z)$ is a source term given by:

$$\begin{aligned} S^+\left(\mu_s, \phi_s, z\right) = \rho \int_0^{\pi} d\phi_T \int^{\frac{\pi}{2}} \sin \vartheta_t d\vartheta_t \Big\{ \left| f\left(\mu_s, \phi_s \mu_t, -\phi_t\right) \right|^2 I^+\left(\mu_t, \phi_t, z\right) \\ + \left| f\left(\mu_s, \phi_s - \mu_t, -\phi_t\right) \right|^2 I^-(-\mu_t, \phi_t, z) \Big\} \end{aligned} \tag{5.24}$$

An analogous expression holds for $S^-(\mu_s, \phi_s, z)$.

Proper boundary conditions are imposed on the specific intensities at the interfaces of the layer. In our case (one layer) there are two interfaces: one with the atmosphere and the other with the ground surface. At the boundary with the atmosphere, assuming that there is no scattering in the atmosphere, the down-welling specific intensity is matched with a collimated beam of radiation in a selected $\vec{s}$ direction. Mathematically this is described by means of a Dirac distribution as in:

$$I^{-}\left(-\mu_s,\phi_s,0\right)=I_0\delta\left(\mu_s-\mu_i\right)\delta\left(\phi_s-\phi_i\right) \tag{5.25}$$

The up-welling specific intensity must only leave the medium, and therefore at the surface the condition is imposed that $I^{+}\left(\vec{s}\right)=0$ when $\vec{s}$ points inward into the medium.

At the boundary with the surface, I^{+} and I^{-} are related by the scattering behavior of the surface. For example, for a specular surface they are related by the Fresnel reflection coefficients:

$$I^{+}\left(\mu_s,\phi_s,-d\right)=\left|R\left(\mu_s\right)\right|^2 I^{-}\left(-\mu_s,\phi_s,-d\right) \tag{5.26}$$

Solution of the integro-differential Equations (5.23) and (5.24) with proper boundary conditions (e.g. Equations 5.25 and 5.26) will allow us to compute the specific intensity that is backscattered by the random medium into our radar instrument. In turn, from the specific intensities we can derive the backscattering cross-section:

$$\sigma^0=4\pi\frac{I^s}{I^i} \tag{5.27}$$

However, a closed form solution of these equations is not available. Therefore, one has to resort to numerical solutions. In a nutshell (a detailed description is given in [11]) the numerical approach calls for an iterative procedure that comprises the following steps: (i) the integro-differential Equations (5.23) are integrated and transformed to integral equations; (ii) at zero order of the iterative process, a solution is sought considering only scattering from the ground and extinction from the volume, i.e. the source elements $S^{+}$and $S^{-}$are set to zero; (iii) at first order, the intensities derived by solutions at zero order are used to compute the source functions $S^{+}$and S^{-}; and (iv) higher order solutions are obtained using the same mechanism, namely correcting the source functions at order n with specific intensities that are solutions of the integral equations at order n–1.

Having established the theoretical background, or the nuts and bolts of radiative transfer theory, let us see how these nuts and bolts are used in the UTA model to build the higher-level machinery.

Vegetation is modeled as an ensemble of objects that are made "visible" by the radiation in the microwave domain, distributed stochastically in a volume that is layered in one dimension. By "visible" we mean bodies of dielectric material that give rise to scattering and absorption processes in the microwave domain. These elementary scattering centers mimic the response of corresponding vegetation elements, such as stems, branches, leaves and needles. The objects occupy the volume as a cloud of particles whose relative positions are not relevant for the description of the medium. They are divided into categories or subsets.

Each population of objects within a category is described by a set of statistical parameters, either in terms of average values (e.g. stem diameter and length) or in terms of probability density functions (e.g. angular distributions). It is important to notice that this picture is suitable to portray cases where vegetation behaves as a homogeneous, non-fluctuating target within the radar resolution element. In other words, a situation where the radar cross-section is constant in space and time. The vegetation itself can be highly non-homogeneous from the botanic point of view, namely it can be composed of several different woody or leafy constituents. On the other hand, cases where the medium is discontinuous, such as an open fragmented canopy, cannot be described adequately by this model.

The model's view of vegetation is thus something like the one pictured by a computer rendition in Figure 5.9.

Since our case study is concerned with a tropical forest, from here on the term vegetation will refer to a dense population of trees. Vegetation components in our canopy are modeled by elementary scattering elements: dielectric cylinders for branches, stems and trunks; dielectric discs for leaves. Electrical behavior of the single scattering elements is described by a scattering amplitude $f\left(\vec{s},\vec{s}'\right)$, which is a function of their shape, size, dielectric constant and angular distribution. These amplitude functions enter in the source terms of the radiative transfer equations.

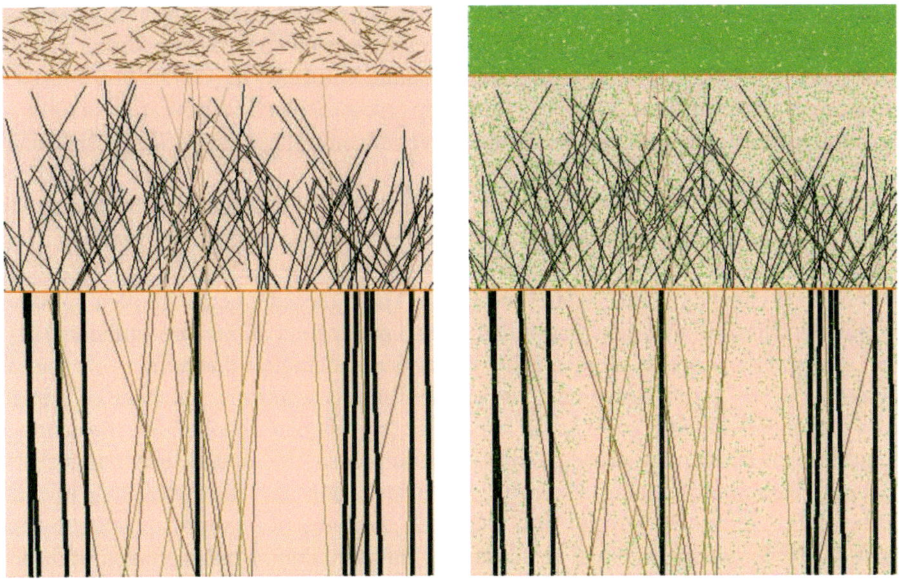

FIGURE 5.9 The model's view of the forest in terms of layered random clouds of elementary dielectric objects. Branches and stems on the left, represented by cylinders, and leaves on the right, represented by thin discs.

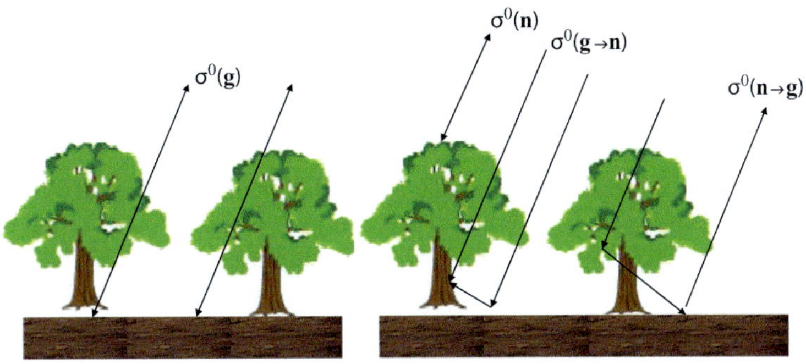

FIGURE 5.10 Scattering mechanisms coming into play in the zero-order solution (left) and first order solution (right).

The total contribution of all scattering elements is computed, adding the specific intensities. This is the trademark of the radiative transfer approach: specific intensities can be added. This possibility paves the way to break down, virtually, the contribution of different scattering mechanisms, whose sum in energy contributes to the total radar cross-section.

In particular, at zero order attenuation by the canopy and scattering from the ground are considered (see Figure 5.10, left diagram). At first order, direct back-scattering from the canopy volume, direct back-scattering from the ground, attenuation by the canopy and indirect scattering (interaction) from canopy volume and ground are considered (see right diagram in Figure 5.10).

Notice that direct back-scattering from the trunks is disregarded, because vertical cylinders, which are large compared to the wavelength of interest, scatter mainly in the forward direction. This is something to keep in mind when talking about woody biomass sensing. As we shall see next, the main contribution from the large vertical stems results from indirect scattering, which in turn is sensitive to the soil conditions.

As we mentioned, the model is driven by a description of the elementary scatterers in the layered clouds (cylinders and discs), in terms of statistical parameters related to geometric and dielectric properties.

How do we go from a botanic view of the forest to this dielectric view of the forest? Indeed, proper description of the medium is a crucial issue for a successful exploitation of wave-scattering models.

The problem is compounded for several reasons. Nature is not easy to portray. In a forest a rich variety of forms are displayed that are driven by competing processes, such as the search for light, in an evolutionary manner. Drastic simplifications are therefore at hand. These simplifications' validity hinges on the sensitivity of the model to specific input parameters, and on the specific modeling scenario. Obviously, a managed homogeneous plantation is easier to model than a natural fragmented forest.

Currently, information on the plant constituents comes from forest inventories prepared by national forest management authorities, or from measurements performed by ground surveys, which are carried out by ad hoc field campaigns for research purposes. These measurements are then compounded by analytical relationships (allometric equations) that map parameters of plant constituents [12] (such as tree height to diameter at breast height), and by statistical techniques that extrapolate measurements to a population (stands, ecological zone). The problem in this respect is that some of the parameters required by the models are difficult to gauge in practice (e.g. angular distributions, dielectric constant).

Difficulties in the description of the medium have somehow hindered the use of this class of models for quantitative remote sensing. However, wave-scattering modeling is still an invaluable tool as a qualitative probe into the physics that underpins the radar observations. Modeling results are thus a prerequisite for a correct visual or automatic interpretation of radar images, and for deriving simulations or thought experiments on new ways of processing the radar signal (see Chapter 6), or to design new instruments. It is with this perspective that we present a case study on the simulation of a dense, humid tropical forest.

5.4 BACKSCATTER SIMULATION FOR A DENSE TROPICAL PRIMARY RAIN FOREST

In our case, the description of the medium was derived from the expert knowledge of ecologists, and from material published in the literature [13–15].

The rain forest is modeled using three layers, measuring 5 m, 15 m and 20 m from top to bottom. Three leaf categories (one for each layer), and eight branch categories are defined, subdivided into two branch categories in the top layer, three branch categories in the middle layer and three stem categories in the bottom layer. Details on the geometric and dielectric parameters for leaves and branches are given in Table 5.1. In a first batch of simulations, the ground is modeled as a humid

TABLE 5.1

Geometric and Dielectric Parameters of the Scattering Elements

Layer	Type	D (cm)	H (cm)	N (n/m^3)	gv
1	Leaves	3.0	0.1	280	0.7
2	Leaves	2.5	0.1	17	0.7
3	Leaves	3.5	0.1	4	0.7
1	Branches	1.0	100.0	2	0.65
1	Branches	2.0	200.0	0.1	0.65
2	Branches	5.0	800.0	0.03	0.65
2	Branches	10.0	1500.0	0.015	0.65
2	Branches	30.0	1500.0	0.0025	0.55
3	Stems	20.0	2000.0	0.015	0.55
3	Stems	50.0	2000.0	0.0025	0.55
3	Stems	200.0	2000.0	0.00005	0.55

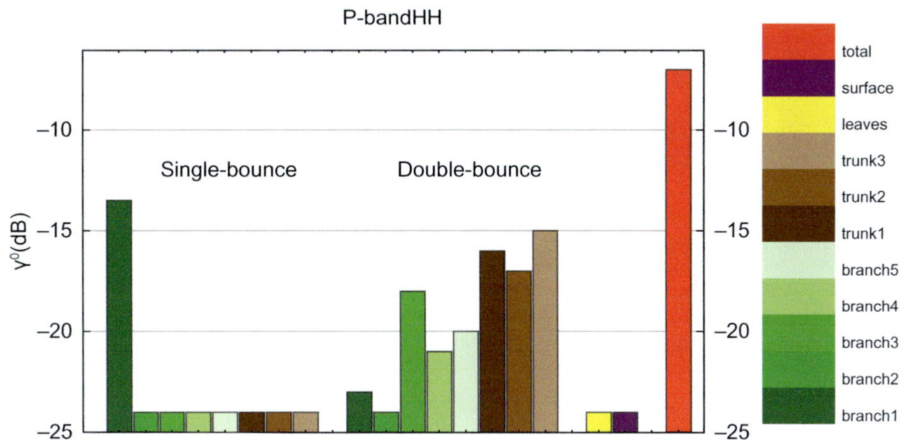

FIGURE 5.11 P-band HH backscattering coefficient. The total backscatter is broken down into the contributions of the single scattering components (branches, trunks, leaves, surface), that are color coded as indicated in the legend. Moreover, the contributions are split between single-bounce and double-bounce scattering.

sandy soil with rms height of 2 cm, correlation length of 15 cm, and gravimetric moisture content of 0.35.

Simulations are run at three frequencies (0.46 GHz λ = 65cm P-band, 1.20 GHz λ = 23.6cm L-band, 5.3 GHz λ = 5.7cm C-band) and incidence angle $\vartheta = 34^0$.

Results are discussed next.

First, let us examine the backscatter coefficient, reported in dB, for each center frequency and HH polarization (Figures 5.11, 5.12, 5.13).

The model also provides a breakdown of the total return into contributions of the single scattering elements categories and into different scattering mechanisms (single-bounce and double-bounce) that come into play in the first order solution. In the next three bar plots, the five branch categories in the upper and middle layers are depicted in green, the vertical stems in the lower layer in brown, the total return from leaves in yellow, the surface in crimson, the total backscattering coefficients in red (see legend in Figure 5.11). Also, direct scattering from branches and stems, and indirect scattering (branch-surface) are reported for each category.

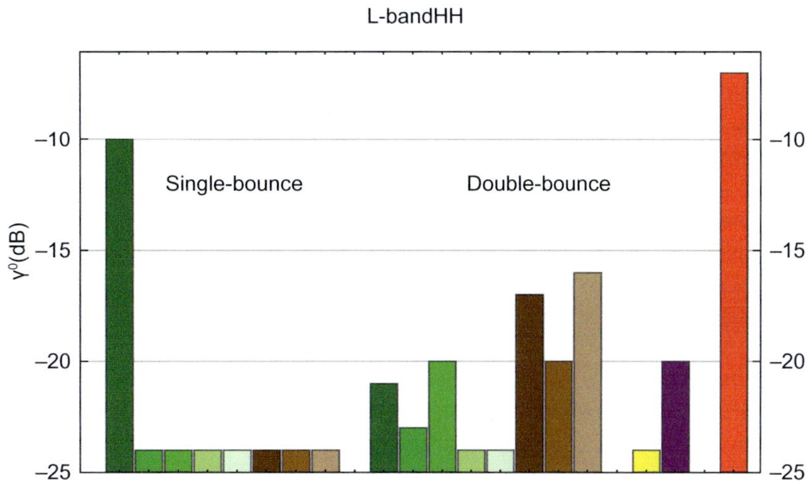

FIGURE 5.12 L-band HH backscattering coefficient.

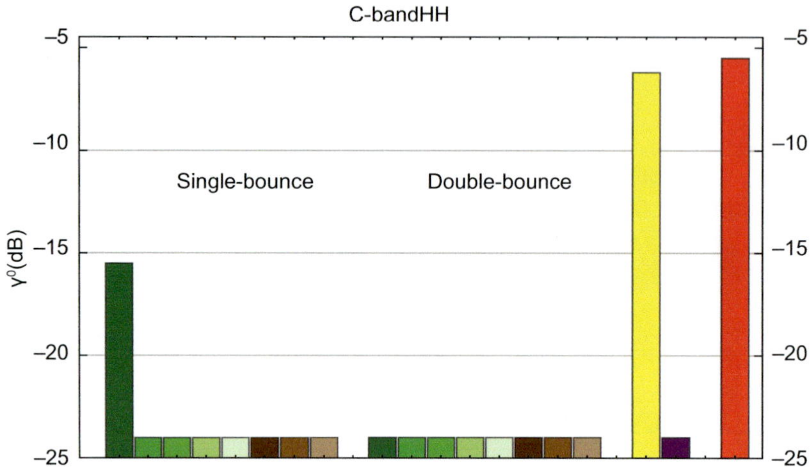

FIGURE 5.13 Backscattering coefficient C-band HH.

The following observations can be made. The total backscatter does not vary significantly with frequency, the difference between P- and C-bands being roughly 2 dB. However, the underpinning mechanisms that come into play at the three frequency bands, are quite different.

At P-band (Figure 5.11) only the upper small branches give meaningful direct scattering. This is because of the high density of branches in this category with respect to the others. As mentioned previously, vertical stems do not contribute to direct scattering.

An entirely different situation arises in connection with indirect branch–surface scattering. The vertical stems now supply a contribution similar to the branch direct scattering. Also, category 3 of long branches in the middle layer weighs in. The factor playing a role here is the length of the branches.

Leaves are transparent at P-band (too small with respect to wavelength). Also, direct scattering from the surface is not significant for this type of soil.

In a nutshell, we can say that radiation at P-band sees through the dense canopy rather well, reaching the soil and sensing through direct or indirect scattering most of the woody biomass. The fact that, at this wavelength, we can sense, through indirect scattering, the soil conditions under the canopy is important thematically to detect flooding conditions in the case of the swamp forests. We shall come back and document this effect quantitatively later.

Results at L-band are reported in Figure 5.12. The trends are similar to P-band, but the following differences may be noted. Direct scattering from the small upper branches is higher. This is a "size" effect, since those branches are becoming bigger with respect to wavelength and therefore less transparent. The opposite is true for indirect scattering. The effect is still there, but mainly dominated by the vertical stems. Also, direct scattering from the surface is increased (because of increased roughness relative to wavelength).

At C-band the situation is changing dramatically (Figure 5.13). Now the dominating contributions come all from direct scattering by small elements in the upper canopy, namely leaves and small branches. Attenuation prevents elements in the lower layers from playing a role. The C-band radiation does not "see through" the canopy, but, in a complementary way to P- and L-bands, senses the properties of the upper canopy stratum. Diagrams of the loss coefficients will reinforce this analysis (see below).

Difference in the scattering mechanisms at L- and C-bands may be exploited successfully to differentiate the primary rain forest from the swamp forest, and to monitor their flooding conditions.

Let us now have a look at the role of polarimetry. Results at P-band are reported in Figure 5.14 (HV) and Figure 5.15 (VV).

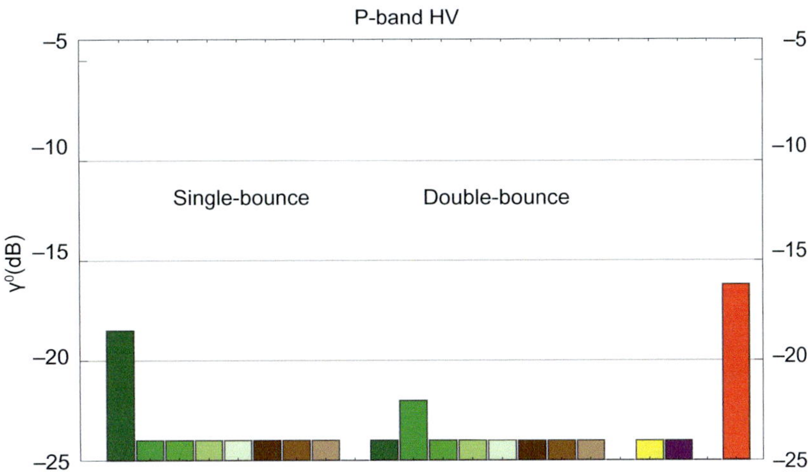

FIGURE 5.14 Backscattering coefficient P-band HV.

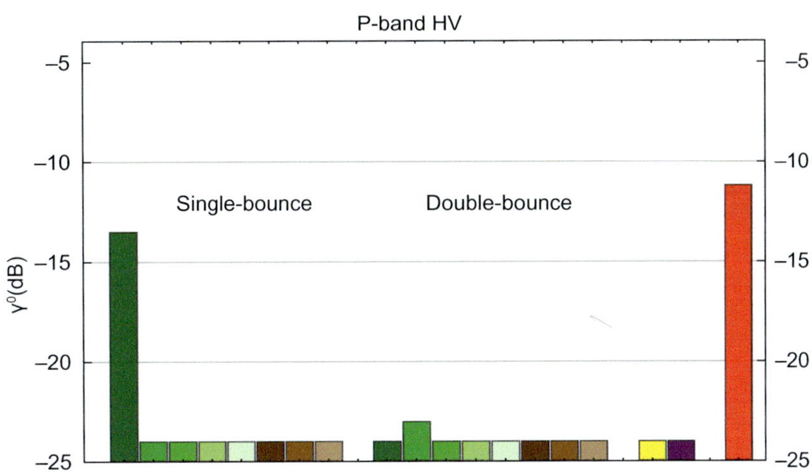

FIGURE 5.15 Backscattering coefficient P-band VV.

The HV contribution is always significantly lower (5 to 9 dB) with respect to the co-polarized states, and stems mainly from volume scattering from the small upper branches. Notice, therefore, that the ratio cross-polarized to co-polarized backscatter can be used as an indicator of the relative weight between direct volume and volume–surface backscatter.

The most striking difference appears between the co-polarized channels and indirect volume-surface scattering. Here, HH polarization is the only one that responds significantly through this mechanism (see Figures 5.11 and 5.12). This polarization is therefore the most suitable to sense the soil conditions under a dense canopy. The difference in indirect backscattering is responsible for the total backscatter difference between the co-polarized channels of 4 dB.

Similar considerations apply in the case of L-band (Figures 5.16 and 5.17). However, here the relative contribution of indirect scattering is smaller than at P-band, and therefore the unbalance between the co-polarized channels is less pronounced (only 2 dB).

Again, there is a totally different story for C-band (Figures 5.18 and 5.19). Here, the only contributions stem from direct volume scattering in the co-polarized channels from small elements in the upper canopy. Total responses at HH and VV are similar, and therefore no significant polarimetric diversity appears at C-band.

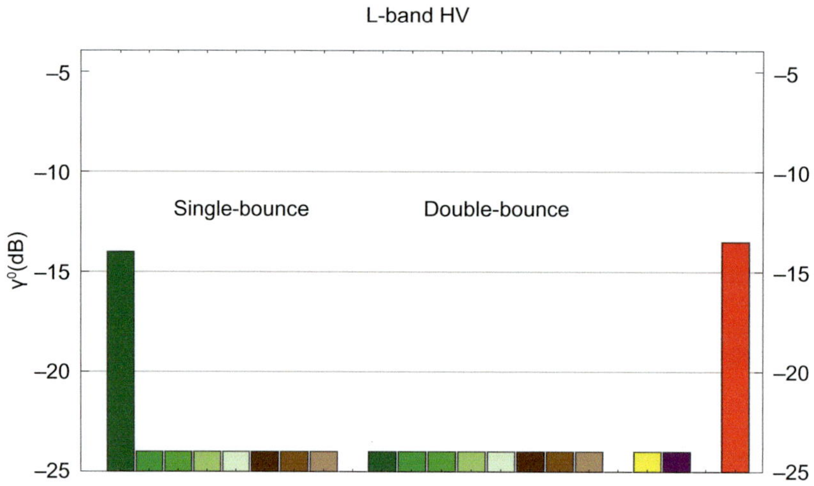

FIGURE 5.16 Backscattering coefficient L-band HV.

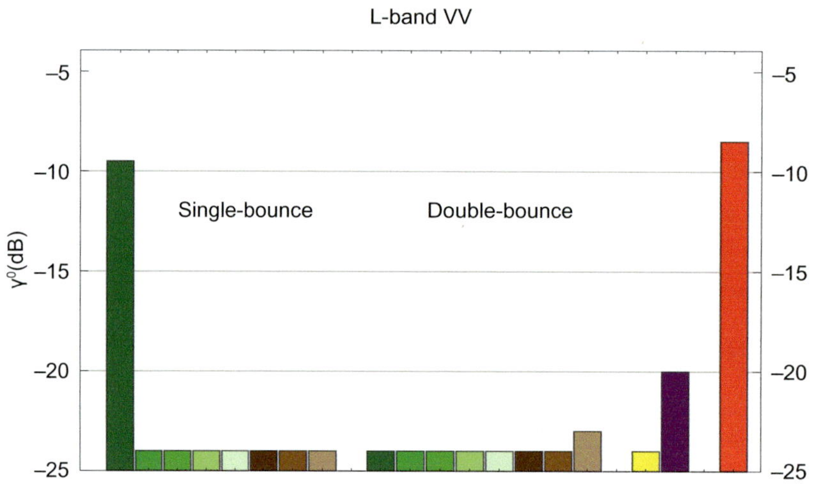

FIGURE 5.17 Backscattering coefficient L-band VV.

Finally, let us have a look at the influence of extinction on the overall radiative transfer process. In the graph in Figure 5.20, the one-way transmission coefficients for P-, L-, C-bands, H and V polarization and the three medium layers are reported. The transmission coefficient (inverse of loss coefficient) gives the fraction of power that seeps through a layer:

$$L_H = e^{-\mathrm{K}h/\cos\vartheta}$$

where K is the extinction coefficient, h the height of the layer, and ϑ the incidence angle.

The bars in the graph are arranged in three groups of six bars each. The first group refers to the upper vegetation layer, the second to the middle, and the third to the lower layer. Within each group, the first three bars refer to V polarization, the second three bars to H polarization, and within each group the bars are ordered for P-, L- and C- bands.

In the first layer, attenuations are comparable at H and V polarization. Attenuation increases from P-band (60% gets through) to L-band (50% gets through), and C-band is already highly attenuated (30% gets through) because of the large amount of attenuation caused by leaves.

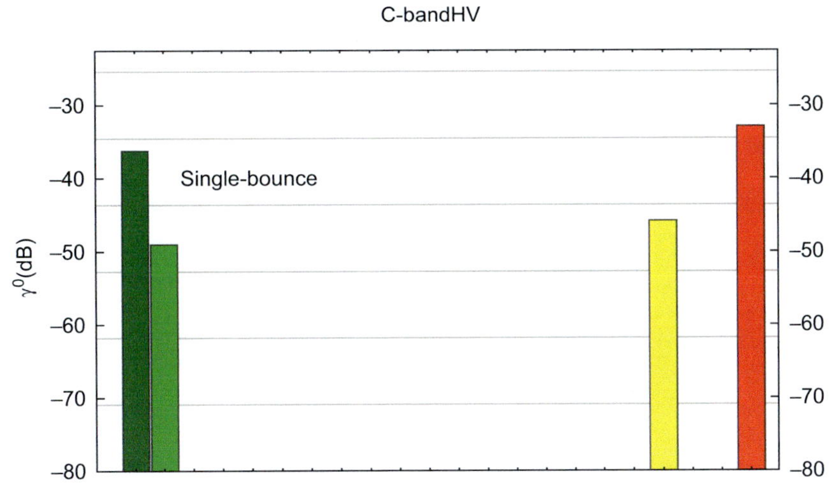

FIGURE 5.18 Backscattering coefficient C-band HV.

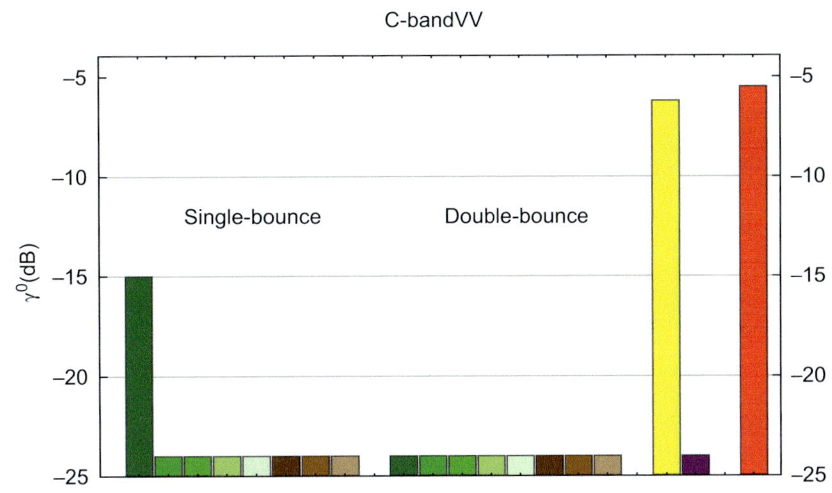

FIGURE 5.19 Backscattering coefficient C-band VV.

In the middle layer, C-band radiation is again drastically reduced (25% gets through), whereas the role of P- and L-bands is reversed. P-band extinction is higher as a result of higher interaction with branch categories 3-4-5 in the middle layer.

The lower layer is transparent to both P- and L-band radiation. The transmission coefficient for C-band is not available because no appreciable radiation reaches the bottom layer.

As we have seen, radiation at P- and L-bands reaches the soil even through a dense canopy, such as the one modeled in our case study. Let us therefore investigate what happens when the properties of the underlying surface are changed. Let us consider the extreme situation, when, instead of a rough, humid surface, we substitute a very smooth, wet surface, mimicking, for example, standing water under the canopy in a flooded forest.

Total backscatter at L-band HH for the flooded condition is shown in Figure 5.21. We can observe a significant increase of all contributions related to indirect branch–surface scattering (around 3 dB with respect to the non-flooded conditions – see Figure 5.12). This phenomenon is obviously caused by the increased specular reflectivity of the surface.

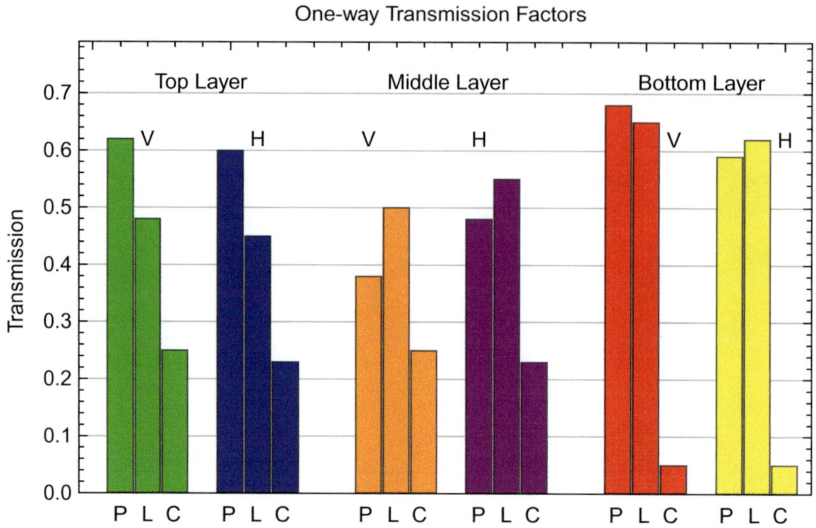

FIGURE 5.20 Transmission coefficients of the top, middle and lower layers, for the H polarized and the V polarized wave and for P-, L- and C-bands.

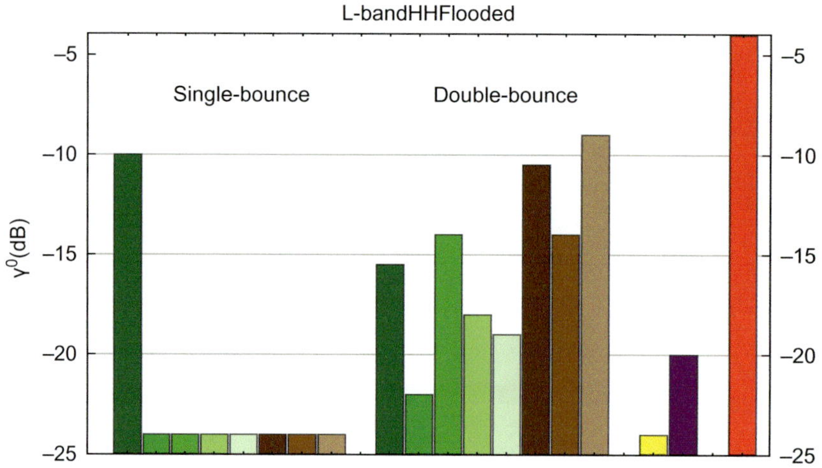

FIGURE 5.21 Backscattering coefficient L-band HH for the flooded forest condition.

Concerning dependency on polarization, we can notice that also VV and HV returns are enhanced by indirect scattering from the specular surface (see Figures 5.16 and 5.17). In particular, major contributions for VV scattering are arising from the large stems in the bottom layer. The HV return is significantly increased by indirect scattering from category three branches in the middle layer. Overall, VV and HV backscattering coefficients are boosted by 1 dB.

REFERENCES

1 I.G. Cumming, and F.H. Wong, *Digital Processing of Synthetic Aperture Radar*, Artech House, Norwood, MA, USA, 2005. ISBN 1-58053-058-3.
2 J.P. Fitch, *Synthetic Aperture Radar*, Springer-Verlag, 1988, ISBN 0-387-96665-X.
3 J.C. Curlander, and R.N. McDonough, *Synthetic Aperture Radar – Systems and Signal Processing*, John Wiley & Sons, 1991.

4 M. Shimada, *Imaging from Spaceborne and Airborne SARs, Calibration and Applications*, 2018, CRC Press.
5 M.A. Karam, A.K. Fung, R.H. Lang, and N.S. Chuahan, "A microwave scattering model for layered vegetation", *IEEE Transaction on Geoscience and Remote Sensing*, vol. 30, no. 4, pp. 767–784, 1992.
6 M.A. Karam, F. Amar, and A.K. Fung, "Electromagnetic wave scattering from a forest or vegetation canopy: Ongoing research at the University of Texas at Arlington", *IEEE Antenna and Propagation Magazine*, vol. 35, no. 2, pp. 18–26, 1993.
7 A. Ishimaru, *Wave Propagation and Scattering in Random Media*, IEEE Press and Oxford University Press, NewYork, 1997, ISBN 0-7803-4717-X.
8 S. Chandrasekhar, *Radiative Transfer*, Oxford University Press, New York, 1960.
9 V. Kourganoff, *Basic methods in transfer problems, International Series of Monographs in Physics*, Oxford University Press, London, 1952.
10 F.T. Ulaby, R.K. Moore, and A.K. Fung, *Microwave Remote Sensing*, Artech House, Inc., Norwood, MA, USA, 1986, vol. III, pp.1085–1202.
11 A.K. Fung, *Microwave Scattering and Emission Models and their Applications*, Artech House, Inc., Norwood, MA, USA, 1994, ch. 2, pp. 49–122.
12 E.C. De Grandi, "Understanding the Relationship between Radar Microwave Backscatter and Vegetation Biomass through Allometric Equations and Wave Scattering Models: A Case Study for Pinus Sylvestris", BSc dissertation, University of Wales, Aberystwyth –IGES.
13 C. Evrard, *Recherches ècologiques sur le peuplement forestier des sols hydromorphes de la Cuvette centrale congolaise,* O.N.R.D and I.N.E.A.C, sèrie scientifique no. 110, 1968.
14 M. Quinones Fernandez, *Polarimetric Data for Tropical Forest Monitoring – Studies at the Colombian Amazon*, Tropenbos International, The Netherlands, 2002, ISBN 90-5113-061-9.
15 J.J. van der Sanden, *Radar Remote Sensing to Support Tropical Forest Management*, Tropenbos-Guyana Programme, Guyana, 1997, ISBN 90-5485-77801.

6 Statistical Measures of SAR Random Spatial Fields

Fingerprints of the Forest Structure

6.1 INTRODUCTION

In this chapter we begin a journey through the main theme of the book: spatial analysis techniques. At first, we shall investigate the question of how, from the random field generated by radar back-scattered radiation, one can derive information over the spatial arrangement of the forest canopy. Next, in the following chapters, we shall encounter discontinuities (edges, point targets) and their characterization through concepts taken from mathematical analysis. We will next consider ways of exploiting the spatial operators to produce better or cleaner imagery. Finally, we shall encounter objects, set up measures of their sizes and shapes, and introduce the notion of temporal change, aimed at characterizing the forest's dynamic evolution.

As the entrance door to the subject, we would like to invite the reader to climb with us to the top of a mountain, and from this vantage point take a look at the woodland that covers the neighboring valley.

The reader will possibly have a view similar to that in the picture in Figure 6.1. He will be able to resolve the tops of the trees' canopies, though they somehow intersect each other, he will discern

FIGURE 6.1 A woodland in a valley as seen from a mountain top. Laying a blanket on top of the canopy would suggest an analogy of the way we propose to establish textural measures of the forest's spatial distribution.

bumps and troughs, being created by shadows from the nearby trees, and therefore gain an overall impression of the woodland's spatial arrangement over the hillside.

Now imagine laying down a blanket over the top of the canopies. If the blanket is made from a rigid material, such as plexiglass, then all bumps and troughs will be concealed, the reader will not be able to resolve the single crowns, and s/he will lose information about the trees' spatial arrangement. The only visual notion left will be of the hillside slope.

Yet if the blanket is made of a flexible material, such as a thick cloth or a carpet, then it will in some way follow the canopies' bumps and troughs, provided it is laid out properly, with the correct tension to avoid missing too many canopies.

Now, the main tenet of our spatial analysis techniques is that radar can lay out a radiation blanket over the woodland in the same way as our imaginary material blanket can do. The radar blanket can be manufactured using different cloths. It can be provided by backscattered radiation, or by surface (top of canopy) height estimates through the correlation of two beams in an interferometer.

The radar blanket will result in a spatial random field (SRF, a two-dimensional set) that will encapsulate information on the spatial arrangement of the vegetation, given a number of conditions. Following the metaphors of the rigid and flexible blankets, the radar blanket will also need to match some constraints. It will need to have the proper spatial resolution to be able to follow the bumps and troughs of the canopies (or at least of a meaningful cluster) without smearing out the variations. It should have the right penetration into the vegetation volume to feel the development in space of the top of the canopy without too much influence from the underlying terrain.

The rest of our journey will be devoted to quantifying the physics of those boundary conditions, and to find appropriate yardsticks to measure the layout of the radar blanket. The metaphor of the blanket immediately suggests that we shall have to look for statistical means of detecting the bumps and valleys of the blanket, these being the features correlated with the forest structure. The yardsticks will therefore have to be built out of second order one-point or two-point statistics of the random fields – in other words, out of texture measures.

6.2 RANDOM FIELDS FROM BACKSCATTER OBSERVATIONS

Let us begin by reviewing some key points of SAR backscatter one-point and two-point statistics. These concepts will allow us gradually to set the ground for understanding the connection between the layout in space of the forest structure and the statistical properties of the random field provided by backscatter.

Let us consider surface scattering. In this case we assume that the wave penetration into the target is small (only a few wavelengths), in such a way that the radiated field is only a result of the currents induced at the surface boundary between the air and the material.

The scattered complex field from a rough surface, illuminated by a monochromatic coherent radiation, is the result of many dephased components generated by the ensemble of scattering centers within the resolution element. The notion of scattering centers can be visualized, for example, as facets, or scattering areas, along a height profile of the surface, each one contributing a component to the overall resulting field. Thus the received field is the sum of many elementary waves that are represented mathematically by a phasor (a point in the complex plane that encapsulates the amplitude and absolute phase of the wave).

Assuming a number of statistical properties of the individual phasors, we are then able to derive the probability density function (PDF), for example, of the resulting field complex amplitude, by performing the complex addition of the N phasors:

$$\vec{E}(x,y) = \frac{1}{\sqrt{N}} \sum_{k=1}^{N} |e_k| e^{i\varphi_k} \tag{6.1}$$

This PDF will characterize the distribution of repeated outcomes, these being either the result of repeated experiments or the realization of a random field composed by many resolution elements, i.e. a speckle pattern.

The statistical assumptions for the individual phasors include: (i) A large number of discrete scattering centers within each resolution cell (allowing for the application of the Central limit theorem); (ii) The phasors $\vec{E}(x,y)$ are independent (amplitude and phase) of one another and have a stationary distribution with the same PDF in all cells; and (iii) The surface is rough compared to the wavelength, resulting in many turns of the phase within the interval $-\pi$, $+\pi$ and therefore in a uniform distribution of the phases φ_k.

Computing the first two moments of the complex field $\langle\vec{E}\rangle,\langle\vec{E}^2\rangle$ by the summation in Equation (6.1), and keeping track of the initial statistical assumptions, we arrive at the conclusion that the complex field is asymptotically ($N \to \infty$) circular Gaussian [1,2]. By detection, the field intensity obeys a negative exponential distribution, while the phase obeys a uniform distribution:

$$p_I(I) = \frac{1}{2\sigma^2} e^{-\left(\frac{I}{2\sigma^2}\right)}$$

$$p_\theta(\theta) = \frac{1}{2\pi} \quad -\pi < \theta < \pi$$

Of importance for our scope are the moments of intensity, these being:

$$\langle I^n \rangle = n!\left(2\sigma^2\right)^n = n!\langle I \rangle^n$$

In particular:

$$VAR(I) = \langle I^2 \rangle - \langle I \rangle^2 = \langle I \rangle^2$$

Since our aim is to find ways of capturing variations of the radar field that could reflect variation of the imaged target, we realize here signs of a problem: areas of different backscatter intensity but macroscopically similar (i.e. with constant radar cross-section) would have different intensity variance, because they are microscopically different. We shall see later how this problem affects our textural yardsticks, and how it can be sidetracked.

So far, we have considered variations of the intensity of a single point in the random field. Another important statistical property is the correlation among points in the field. The speckle pattern resulting from coherent scattering from a surface presents a coarse granular structure in space. Therefore, to account for this property we must look at a two-point statistic of the field, such as the autocorrelation function of intensity. Correlations in a speckle pattern arise from the size and form of the illumination beam and from the radiation collection capability of the instrument.

The full theory can be found in [1], for example. We summarize here only the main outcome. In the first case (free space geometry, shape of the illuminating beam), the autocorrelation function of the fields $A(x,y)$ turns out to be, under some simplifications, the Fourier transform of the intensity distribution of the illuminating spot:

$$\left\langle A(x_1,y_1)A^*(x_2,y_2)\right\rangle = \frac{\kappa}{\lambda^2 z^2}\iint \left|P(\xi,\eta)\right|^2 e^{i\frac{2\pi}{\lambda z}\left[\varsigma(x_1-x_2)+\eta(y_1-y_2)\right]} d\zeta\, d\eta$$

where $|P(\xi,\eta)|^2$ is the intensity distribution of the illuminating beam on the scattering spot.

For example, for a square scattering spot of width L, the ACF of intensity will be a sync function:

$$R_I\left(\Delta x,\Delta y\right)=\langle I\rangle^2\left[1+sinc^2\frac{L\Delta x}{\lambda z}sinc^2\frac{L\Delta y}{\lambda z}\right]$$

When the imaging instrument collection property is taken into account (the pupil lens in optics), then the spectral density of the speckle intensity field results in a Dirac δ at zero lag, plus an extended spatial frequency component with the shape of the normalized autocorrelation function (ACF) of the instrument intensity transmittance.

The second lesson we learn here is that when we set up two-point statistical measures of the radar field we shall always have to take into account an instrumental correlation component that will be superimposed on the component related to the underlying surface reflectivity.

Will this instrumental correlation component be the same over the whole backscatter image? The answer is no. In the slant-range geometry there will be differences in range and cross-range direction caused by differences in the illumination beam and the processing. Moreover, in ground range geometry, the ACF will change with range because the ground-range resolution will increase with the range coordinate (i.e. will degrade at a small off-nadir angle).

Let us now move to volume scattering, which is the case, for example, of forest imaging. We realize immediately that the statistical assumptions of the former case no longer hold true. Indeed, the probability distribution of the elementary scatterers within the resolution cell will not be identical from one cell to the next, and will not be stationary. Take the radiative transfer model described in Chapter 5. Here, the total contribution to the radar cross-section was modeled by the distribution of a cloud of dielectric cylinders within a layered infinite half space. This distribution will change from one cell to the next, reflecting the heterogeneity of the forest structure (leaves, branches, trunks). Moreover, the reflectance of the scattering elements will not be distributed uniformly. There will be a few elements with strong reflectance, such as trunks, and many elements with weaker reflectance, such as branches. The scattering elements will be in a finite number and they will not be independent (second order scattering).

Clearly, a different statistic (a non-Rayleigh statistic) should be called into play to model the random field related to this scattering process. Different approaches were undertaken historically to arrive at a model of the one-point statistic.

In the discrete setting, it is assumed that the number of scatterers within the illuminated cell fluctuate randomly about some mean value, with a negative binomial distribution of the number of scatterers. In the continuous case, it is assumed that the cross-section fluctuates according to a Gamma PDF. In both cases, a K distribution of the intensity is obtained [3,4]:

$$p\left(\sigma\right)=\frac{1}{\langle\sigma\rangle}\left(\frac{\sigma}{\langle\sigma\rangle}\right)^{\upsilon-1}\frac{1}{\Gamma\left(\upsilon\right)}e^{\left(-\frac{\sigma}{\langle\sigma\rangle}\right)}$$

$$p\left(I\right)=\frac{2}{\langle I\rangle}\left(\frac{I}{\langle I\rangle}\right)^{(\upsilon-1)/2}\frac{1}{\Gamma\left(\upsilon\right)}K_{\upsilon-1}\left[2\left(\frac{I}{\langle I\rangle}\right)^{1/2}\right]$$

where $\Gamma(\upsilon)$ is the gamma function of order υ and $K_{\upsilon-1}$ is the K distribution (modified Bessel function) or order $\upsilon - 1$.

The order parameter ν of the Gamma distribution controls the relative variance:

$$\frac{1}{\nu}=VAR\left(\sigma\right)/\langle\sigma\rangle^2$$

Whereas in the case of a flat cross-section the relative variance did not provide any information, being always equal to unity, now it is a textural parameter that captures the variation in intensity of resolution cells in the illuminated field. It is a first indication that textural measures can provide information on the forest structure.

What about the two-point statistic? Seminal work on modeling a correlated non-Rayleigh statistic of radar clutter was performed in the 1980s [3–5]. Here, analytical models were derived, including the effect of a correlated surface, of the influence of the illumination beam and instrument radiation collection. Again, a random process describing the underlying cross-section is the key mechanism to derive the intensity statistic, including the correlation properties. In [4] the surface cross-section process is assumed to have a Lorentzian power spectrum and a Gamma PDF. As a consequence, the normalized ACF of the surface is:

$$ACF(\sigma) = \frac{\langle \sigma(x)\sigma(x+kd) \rangle}{\langle \sigma^2 \rangle} = 1 + \frac{1}{2\upsilon} e^{(-\Gamma d|k|)} \left[1 + \cos(\omega dk)\right] \tag{6.2}$$

where an array of scatterers is assumed, each having length d, and the array is indexed by k.

Therefore this model accounts for both the surface correlation length and the periodicity. The ACF of intensity results in three terms: a constant corresponding to the uncorrelated background, a Gaussian term representing the coherent interference of many random scattering elements within the resolution cell, and a third term encapsulating the properties of the surface random process Equation (6.2). This term is a rather complicated function of the parameters that characterize the surface process (order parameter, correlation length, surface period) and the observing instrument (illuminated length, wavefront curvature)

For our purposes, it is sufficient to notice that the intensity ACF predicted by the model takes the form depicted in Figure 6.2. The main difference with respect to volume scattering models that consider independent resolution elements is the presence of periodic correlations (and anti-correlations) that appear at delays τ longer than the correlation length. It is precisely this statistical property that we shall attempt to exploit to capture the forest structure.

It has been proved experimentally that the K distributed correlated model matches well the backscatter statistic in radar observations of woodlands. The model was derived from purely statistical

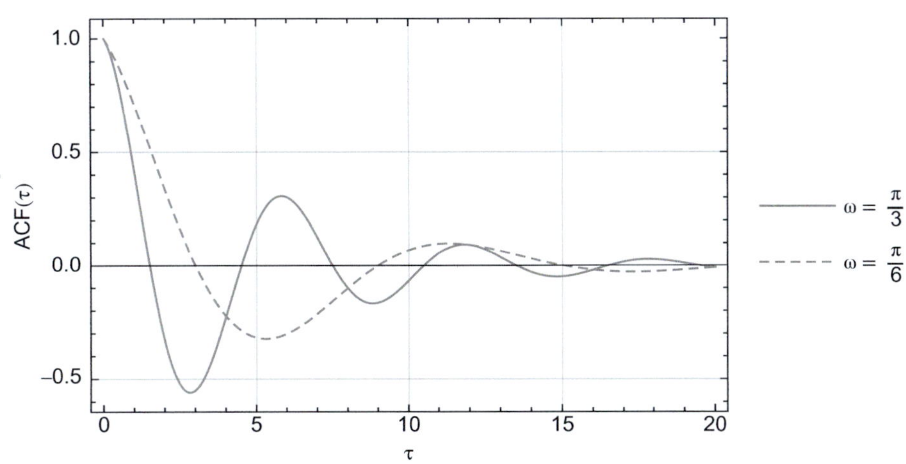

FIGURE 6.2 Proposed exponential cosine autocorrelation function (ACF). Examples with the cosine frequency $\omega = \frac{\pi}{3}(rad/\sec)$ and $\omega = \frac{\pi}{6}(rad/\sec)$.

assumptions on the distribution of elementary scatterers within a resolution cell, and the correlation among neighboring cells (the Gamma Lorentzian surface). In this representation, a description of the structure and physical properties of the illuminated natural target (e.g. the forest) is missing. In other words, the connection between the signal and forest structure is missing. There is another way of looking at things that takes a closer view of the interaction between the electromagnetic wave and the forest structure [6].

For illumination at high frequency (C-, X-band) and for a dense forest, the main backscattering contribution comes from leaves in the upper canopy layer (see radiative model in Chapter 5). As suggested in [6], since the radiation penetration length is short, we can model the wave–canopy interaction by a thin layer whose upper surface (air–vegetation interface) follows the shape of the canopy. The situation is illustrated by the schematic in Figure 6.3. The reader will surely now notice the connection with our metaphor of the blanket on top of the woodlands.

We propose here a toy model developed along this idea. The model is discretized by slicing through the vegetation volume in one spatial direction. For each slice, the scattering problem consists of a layer of Rayleigh scatterers whose boundary geometry is governed by the shape of the canopy. Thus the local incidence angle is modulated by the direction of the top canopy surface's derivative at the point of intersection with the slice. Notice that according to the vertical structure of the canopy, regions of shadowing or layover can also occur.

It is a toy model because it makes a number of simplifying assumptions, such as the independence of the slices. Moreover, Rayleigh scattering by small particles is certainly not the best way to model the scatterers in the upper canopy layer of a tropical forest. The model, therefore, is meant to be only a proof of concept.

In more detail, the canopy geometry is fed into the model by a known canopy height profile (CHM). In the test case illustrated in the following, the CHM is provided by an airborne LiDAR measurement.

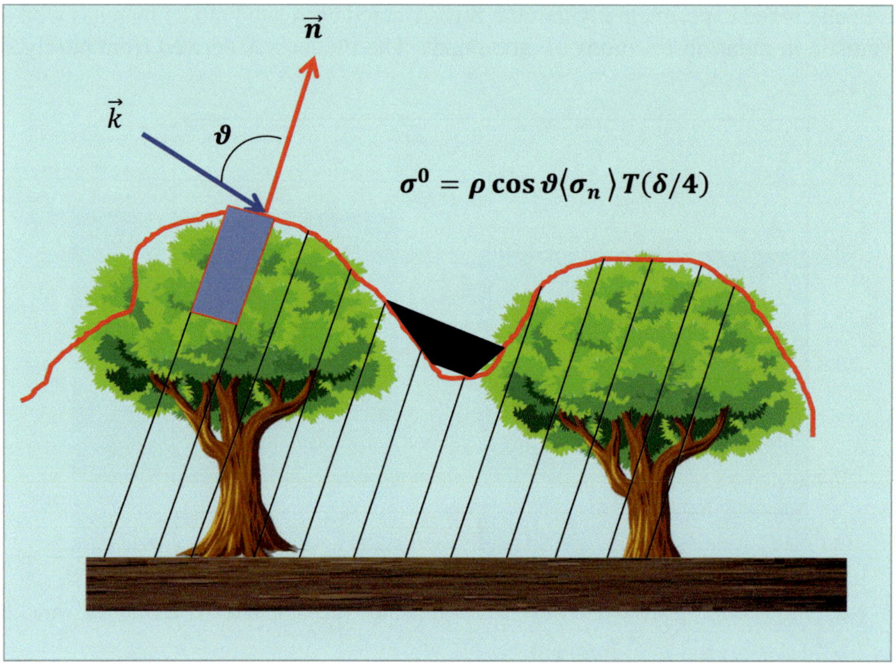

FIGURE 6.3 A high frequency model accounting for the influence of the canopy shape on backscatter.

The CHM profile is low-pass filtered and subsampled to match the radar resolution. The CHM samples $h(x_k)$ are fitted by a trigonometric series to obtain a continuous function of the space coordinate x:

$$CHM(k) = h(x_k) \Rightarrow CHM(x) = FIT\left[h(x_k), \sum_{k=1}^{N} \sin k\pi x/L\right]$$

The local slope of each slice is obtained by the derivative of the CHM function at the sampling point x_i:

$$\phi_i = \tan^{-1}\left[\frac{dCHM(x)}{dx}\right]_{x=xi}$$

The local incidence angle ϑ_i of the tilted layer with respect to the radar incidence angle ϑ is:

$$\vartheta_i = \vartheta - \phi_i \tag{6.3}$$

For each tilted layer, the low dielectric Rayleigh layer model proposed in [7] is used. The model is characterized by three parameters: absorption coefficient K_a, albedo α, and optical depth τ. K_a depends on the complex dielectric constant and volume fraction of the Rayleigh scatterers, and the dielectric constant of the medium.

As an illustrative example, we report here the results of a model run intended to simulate the situation that would typically occur with a high resolution (5 m) X-band ($f = 9.6\ GHz$) radar, such as TanDEM-X.

The LiDAR CHM (resolution 1 m) was captured by an airborne LiDAR over the lowland tropical forest of Sungai Wain, Indonesia (see Section 6.9). It is shown in Figure 6.4 together with the smoothed and subsampled version at 5 m resolution, and the local incidence angle ϑ_i computed as in Equation (6.3). Values $\vartheta_i \cong 0°$ and $\vartheta_i \cong 90°$ correspond to shadow and layover situations.

The autocorrelation function ACF of the CHM profile at 5 m resolution is shown in the graph of Figure 6.5. Three regions with different decorrelation trends can be observed. The first region has

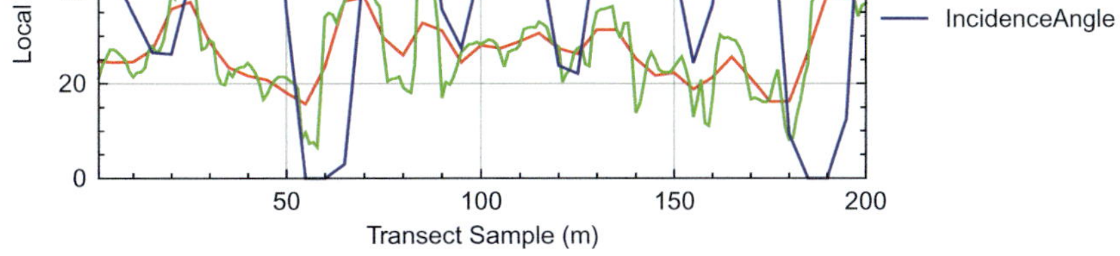

FIGURE 6.4 Canopy height profile provided by a LiDAR measurement (red line), the subsampled version to match the radar resolution (4 m) (green line) and the resulting local incidence angle for each slice of the model (blue line).

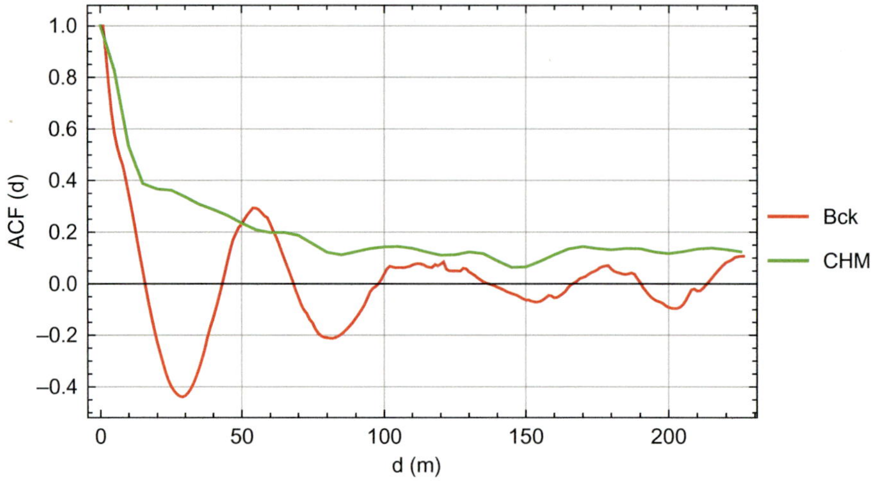

FIGURE 6.5 Autocorrelation function of the LiDAR CHM profile (green line). The ACF of the HH backscatter from the model is overlaid (red line).

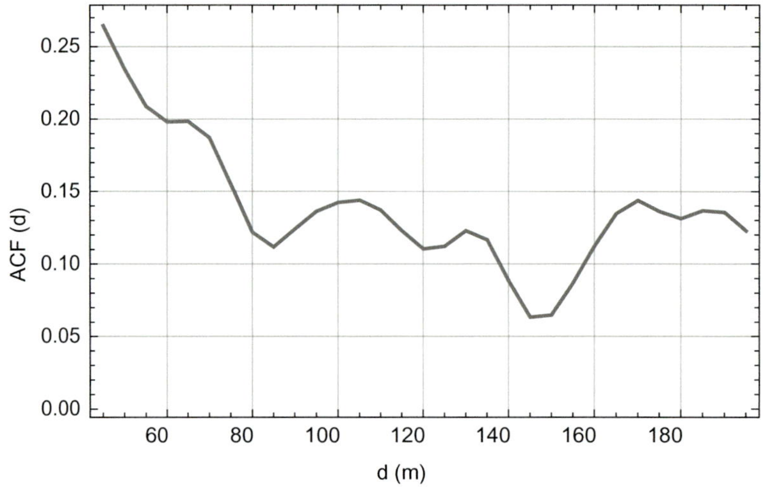

FIGURE 6.6 Detail of the CHM autocorrelation function for distances from 60 m to 180 m.

rapid decorrelation down to 40% and a distance of approximately 20 m. This region corresponds to rapid variation of backscatter return from the heterogeneous canopy structure.

A second decorrelation region appears from 20 m to 50 m, this possibly corresponding to larger structures such as single tree crowns or the merging of several crowns. At greater distances there are still persistent correlations (long memory process), as can be noted in the ACF segments from 60 m to 180 m reported in Figure 6.6. These segments show a periodicity of some 50 m.

The ACF of the HH scattering coefficient is shown overlaid to the CHM ACF in Figure 6.5. At short lags the two functions decay with the same derivative. Therefore the backscatter is affected in the same way by the canopy variations at short distances. The second region in the CHM ACF corresponds to strong anti-correlation in the radar ACF, signaling the onset of shadow and layover. Beyond the second region (d >50 m) the radar ACF presents persistent periodic correlation patterns.

All in all, both points of view, the one based on purely statistical considerations and the other based on some empirical positions and microwave scattering theory, suggest the presence of periodic correlation patterns in the radar random fields. This correlation periodicity corresponds to patterns in the spatial distribution of the forest canopy, when some form of stationarity (homogeneity) is assured by a dense canopy cover. We shall see that correlation in random fields will be one mechanism we will try to exploit to characterize the forest structure.

6.3 RANDOM FIELDS FROM INSAR COHERENCE OBSERVATIONS

An InSAR digital surface model (DSM) may also supply a field holding shadows of the forest structure. To clarify this claim let us consider the complex coherence in a volume scattering layer, such as the canopy of a dense forest.

While for backscatter the main effect was caused by variation of the local incidence angle with respect to a thin layer lying on top of the canopy, now the full extension of the vegetation volume, this being the source of interferometric decorrelation, needs to be considered.

Within the volume, the main driver affecting the coherence is the distribution of the scattering elements and their scattering and extinction coefficients along the vertical coordinate (vertical structure function). Suppose the forest structure is intrinsically spatially heterogeneous, or it is locally changed by a disturbance, then the structure function will also be spatially variant, and, in turn, the complex coherence will be a function of the space coordinates. This is the basic mechanism whereby we expect to see forest structural patterns reflected on to InSAR DSMs.

A more quantitative insight can be gained by a simple scattering model, consisting of a uniform layer and an exponential profile for the scattered power along the vertical coordinate [8]:

$$\gamma = f\left(h,\sigma_e,\beta_z\right) = \frac{p\left(e^{(p+i\beta)h}-1\right)}{\left(p+i\beta\right)\left(e^{ph}-1\right)}\quad p = 2\sigma sec\vartheta \tag{6.4}$$

where γ is the complex coherence, h is the volume height, σ is the extinction coefficient, ϑ is the incidence angle, $\beta_z = \dfrac{4\pi B_\perp}{\lambda R\sin\vartheta}$ is the vertical wavenumber, a parameter that summarizes the interferometer's sensitivity.

What this model says is that the complex coherence will be the result of a coherent sum of all the returns from the volume weighted by the vertical structure function.

The InSAR DSM over the forest reports a height h_{dsm} that will be the sum of the topographic height and the position of the complex coherence phase center within the vegetation volume. The fractional phase center height (P_c), i.e. the position of the phase center with respect to the vegetation volume height, is dictated by the amount of volume decorrelation, and is defined in [8] as:

$$P_c = \frac{\phi}{h_v\beta_z} \tag{6.5}$$

where is ϕ the interferometric phase; h_v is the height of the volume and β_z is the vertical wavenumber. Therefore, P_c is the ratio of the coherence phase (radians) resulting from decorrelation induced by volume scattering to the volume height measured in unit of the vertical wave number.

From the model Equations (6.4) and (6.5), we can evaluate P_c as a function of the sensor's parameters (e.g. wavelength, incidence angle and baseline), and the extinction coefficient of the vegetation volume. As an illustrative example, we have computed the distribution of P_c values determined by the variation of the canopy height along a transect measured experimentally by a LiDAR across a dense tropical forest. Results are reported in Figure 6.7. The three frequency distributions

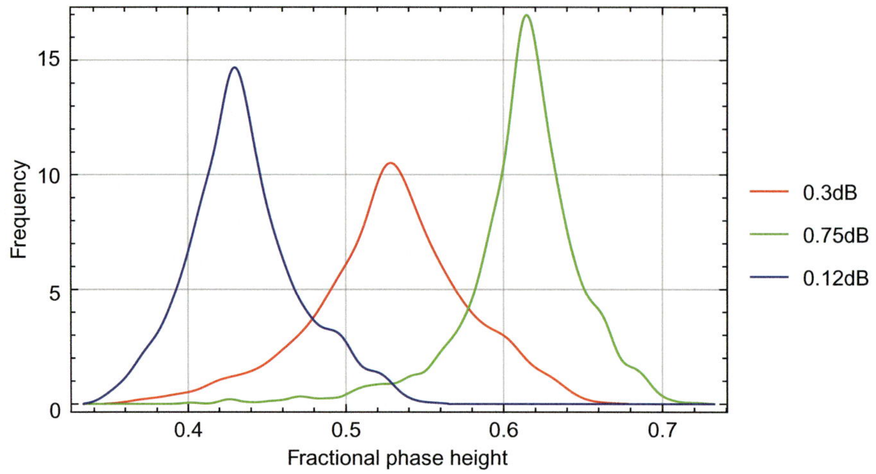

FIGURE 6.7 Frequency distributions of phase center fractional height within a vegetation volume with variable canopy height measured along a transect in a dense tropical forest. The distributions are parametrized by different extinction coefficient values.

correspond to extinction coefficients of 0.12 dB, 0.3 dB and 0.75 dB. Note how the phase center is always located within the vegetation volume, and increasingly with extinction towards the top of canopy.

6.4 WAVELET BASED TEXTURAL MEASURES OF RANDOM FIELDS

The arguments expounded in the previous section suggest that patterns of the forest spatial distribution are reflected on to the random fields of SAR observations. It was pointed out that one possible way to characterize this connection was by means of a two-point statistic, the autocorrelation function. So far, we have assumed stationarity of the random field, and accordingly the ACF was only a function of the relative distance among resolution elements.

We now wish to generalize this situation. On the one hand, we would like to account for a change of the correlation properties from place to place. This would happen, for example, when crossing the border between two different forest types, e.g. the lowland and the swamp forest. On the other hand, we would like to account for different generating random processes, including fractal noises and intermittency. For convenience, let us summarize with a unique name all the realizations of different spatial random processes: textures.

In the same way that perhaps the farmer would set up yardsticks in his cornfield for measuring the acreage, we need now to find and set up some flexible yardsticks in our random fields for measuring the patterns of interest. Our choice for the yardsticks is wavelets. For alternative yardsticks and different viewpoints of the texture mystery as proposed by the competition please refer to [9–16].

So, why wavelets? One good reason is that wavelets are used extensively for time series analysis and random process characterization in geophysics and earth system sciences [17–19].

However, stepping out of this general statement, we shall try to put our exposition in the context of our goal, finding good textural measures. We shall first offer an intuitive justification for using wavelets, and move into a more formal terrain later (see also the Appendix).

Wavelets are functions that oscillate and decay rapidly (Figure 6.8). They can be used to decompose a signal (field) in a complete way (without losing information) into more elementary building blocks (or oscillatory waveforms) [20]. A 'mother' wavelet is the main function used to decompose the signal by dilation and translation (Equation 6.6).

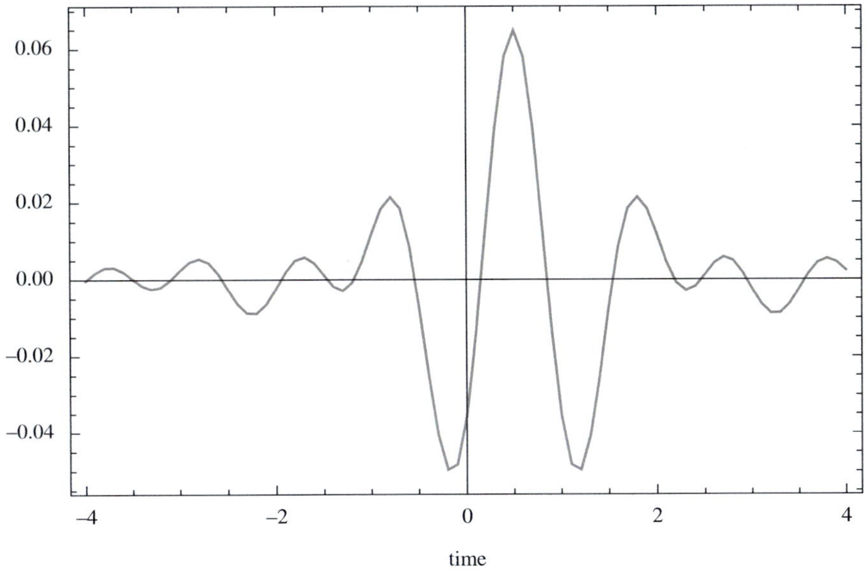

FIGURE 6.8 A wavelet is an oscillating function that decays rapidly in time (or space).

There are many other ways of building a basis aside from wavelets. For example, sine waves (functions that oscillate at a specific frequency) are widely used, as in the well-known Fourier transform. The main advantage of the wavelet basis is to adapt the space–frequency resolutions in such a way that their product is constant [20]. Frequency and space resolution cannot be made independently and arbitrarily better. For example, the basic function of the Fourier transform (sine wave) presents zero spread in frequency (pure tone), while infinite spread in space (it oscillates everywhere). What would be desirable is to have more resolution in space when dealing with smaller-scale (higher frequency) analysis, and lower resolution in space when dealing with processes that are evolving slowly.

Wavelets do exactly that, by adapting the space–frequency resolution. The capability is of importance when the dynamics of the signal change at specific points in space, or, in our context, when we want to detect different textures in the field. A rigorous mathematical formulation of the space–frequency resolution (the Heisenberg box) can be found in the Appendix.

We shall now proceed by considering a one-dimensional problem, with a continuous signal $f(x)$ (instead of a field) defined on the space of real numbers $\Re$. The representation in the wavelet basis is obtained by applying translations and dilations of the wavelet to the signal to be decomposed [20,21]:

$$W_f(s,b)\Big|_{1/2} = s^{-\frac{1}{2}} \int_{-\infty}^{+\infty} f(x)\psi\left(\frac{x-b}{s}\right) dx \tag{6.6}$$

where $W_f(s,b)|_{1/2}$ are the wavelet coefficients at point b in space and scale s.

The wavelet $\psi(x,s)$ must be designed subject to a mathematical constraint (that is why it is so difficult to design wavelets), called the admissibility condition, which assures that the transform in Equation (6.6) is invertible, i.e. the signal can be fully reconstructed from its representation on the wavelet basis:

$$f(x) = \frac{1}{C_\psi} \iint W_f(s,b)\Big|_{1/2} s^{-\frac{1}{2}} \psi\left(\frac{x-b}{s}\right) db \frac{ds}{s^2}$$

where C_ψ is a normalization factor that encapsulates the admissibility condition.

The possibility of inverting the transform is not of paramount importance for our texture measures, but will be exploited in algorithms for filtering radar fields, which will be described in later chapters.

To gain an insight into what Equation (6.6) is telling us, let us consider a wavelet at scale s=1 (the shortest one), and imagine moving the wavelet along a signal of interest. At each point in space, the wavelet is multiplied by the signal $f(x)$. Given that the wavelet has short support (i.e. decays fast in space) it will capture the similarities (correlation) with the signal in a small neighborhood of $f(x)$ [17]. Since the wavelet oscillates fast around this point, it will be sensitive to details of the signal that are also changing rapidly. The first wavelet coefficient will therefore capture information about rapidly changing features at each point in space.

Now, using a dilated version of the wavelet (e.g. s = 2) more slowly evolving features will be captured, given that these will resemble more the dilated wavelet. Thus the wavelet coefficients will provide indication of where (in space) the signal presents dynamical features and how "fast" they are (in term of spatial frequency).

We have here a first intuitive indication of how a wavelet representation can be useful for analyzing patterns. The transform sorts out fast-varying and slow-varying features and, importantly, places where they are.

Let us now move to the discrete setting, which, of course, is the one that fits the way the radar fields in remote sensing are represented. In this case, the transform is obtained by discretizing the s, b parameters in $W_f(s,b)|_{1/2}$ to obtain the discrete wavelet transform (DWT) [20,21]:

$$Wf\left(m,n\right) = a_0^{-m/2} \int f\left(x\right)\psi\left(a_0^{-m}x - n\,b_0\right) \tag{6.7}$$

where $a = a_0^m$ with $m \in Z$ is the discretization of the scale parameter, n is an integer number, $n\,b_0$ with $b_0 > 0$ is the discretization of the translation parameter.

However, the passage from WT to DWT does not happen easily. The difficulty is to assure that the discrete basis is complete (e.g. it can be inverted to reconstruct the signal) and stable (it does not produce infinities for small variations of the signal). The mathematical conditions to achieve this goal give rise to a particular basis, composed of functions that are not orthogonal, and introduce redundancy in the representations [21,22]. These functions are called frames.

We then definitely need a frame, if we want to use a discrete wavelet transform. Yet, which frame? Let us go back for a second to the illustrative description of the wavelet transform. We saw that the transform is sensitive to how fast a signal wiggles. This is certainly good if we want to catch the pitch of a piece of music. But, in our case, for pattern analysis, what is more important is how much the signal differs in magnitude from point to point. We conclude that we must search for a differentiator.

Luckily, a frame that acts as a multiscale differentiator was designed by Mallat [20] (see the Appendix on details of how the frame is constructed). When discretized, the wavelet consists just of two points (-0.5, 0.5), as it is depicted in Figure 6.9, where the corresponding continuous function is also overlaid.

Let us remember that our goal is to recognize patters in random fields. Obviously, differencing and looking at single values of a random sequence would not provide much information. We need some statistics of the wavelet coefficients. Since the first moment of the coefficients is zero, let us consider the second moment or the wavelet variance. Indeed, the wavelet variance is a widely used statistic in many techniques in geophysics for the characterization of random processes [17].

We start to explore the capabilities afforded by the wavelet variance from a very simple case: a discrete stationary random signal $f(j)$. When the non-dilated wavelet frame (Figure 6.9) is applied to the signal by a convolution, as in Equation (6.7), successive shifts of the wavelet will detect differences in the signal spaced by one pixel. The wavelet variance is computed by squaring the

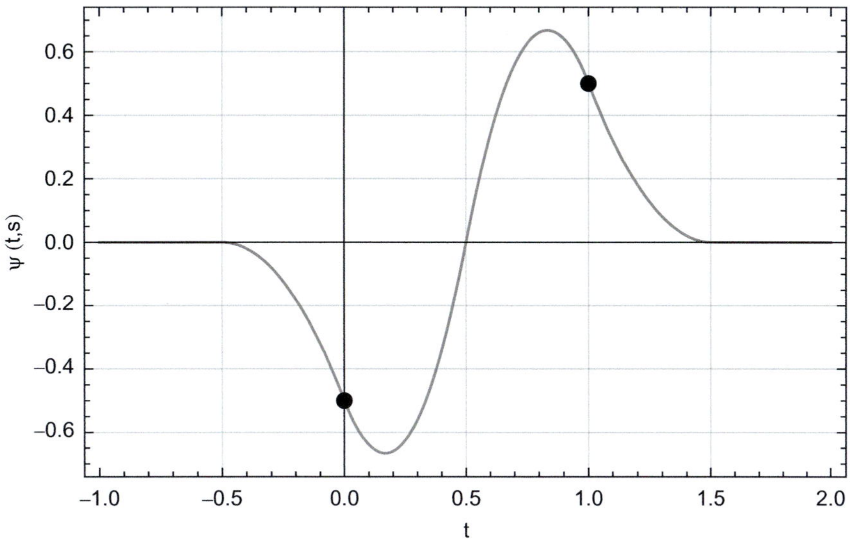

FIGURE 6.9 Mallat's wavelet frame.

coefficients and taking a local average in the neighborhood of a point. Then, assuming stationarity, i.e. $\langle f^2(j)\rangle = \langle f^2(j+n)\rangle$:

$$\langle W_{s=1}^2\rangle = \left\langle \left(f(j+1)-f(j)\right)^2\right\rangle = 2\langle f^2\rangle - 2\langle f(j+1)f(j)\rangle$$

where s is the scale and j is the index of a sample in the signal.

This equation shows that the variance of the wavelet coefficients at scale 1 captures the signal second moment (first term) and the autocorrelation of the signal at lag 1 pixel (second term), when estimated in the neighborhood of point j.

When the wavelet is dilated by 2^m the resulting wavelet variance will be:

$$\langle W_{s=2^m}^2\rangle = \left\langle \left(f(j+2^m)-f(j)\right)^2\right\rangle = 2\langle f^2\rangle - 2\langle f(j+2^m)f(j)\rangle$$

In this case, the wavelet variance will be proportional to the autocorrelation of the signal at lag 2^m pixels when estimated in the neighborhood of point j.

Therefore we obtain the following general relationship between the wavelet variance as a function of scale and the discrete time autocorrelation function $\mathcal{R}(k)$ as a function of the lag k between two points:

$$\langle W_{s=k}^2\rangle = 2\langle f^2\rangle - 2\mathcal{R}(k)$$

The same relation holds for the continuous wavelet transform and the continuous time autocorrelation function $(\langle f(x)f(x+\tau)\rangle$, which we shall discuss later in this section.

The limiting condition of (W_s^2) for small and large scales are:

$$\lim_{s\to 0} W_s^2 = 0$$

$$\lim_{s\to\infty}\langle W_s^2\rangle = \langle f^2\rangle - \langle f\rangle\langle f\rangle = \langle f^2\rangle - \langle f\rangle^2 = VAR(f)$$

Therefore, the wavelet variance of a stationary signal for large scales approaches the variance of the signal asymptotically. Moreover, if the wavelet variance is normalized by an estimate of the signal mean value at a large scale, then it approaches the square of the coefficient of variation, this being a one-point textural measure:

$$\lim_{s\to\infty}\frac{\langle W_s^2\rangle}{\langle x\rangle^2}=\frac{VAR(x)}{\langle x\rangle^2}$$

Therefore, the wavelet variance encapsulates both a two-point and a one-point statistic that provide textural measures.

At this point of the discussion, students always ask: why, then, not use the classical Fourier analysis instead of looking for trouble with the wavelets? Apart from issues on computational complexity and estimator errors (these are dealt with later), the main reason for using wavelets stems from their better Heisenberg box, and their better localization in the space–frequency domain. This capability allows us to deal with non-stationary processes, which indeed, as we mentioned, are of importance in our thematic context. We shall illustrate this point with a toy signal.

Suppose one constructs a signal made of two consecutive segments, each consisting of Gamma correlated random deviates, but with different correlation properties. The test signal was generated by the Monte Carlo method and is shown in Figure 6.10. The one-point statistics of the two segments, x_1, x_2 are equal, these being $\langle x_1\rangle = \langle x_2\rangle = 1.3$, $VAR(x) = 0.8$, order parameter $\upsilon = 2$. However, the exponential cosine autocorrelation functions $ACF(x_1)$, $ACF(x_2)$ have oscillator period $T = 8$ and, importantly, a different correlation length: $\delta(x1) = 0.1$ $\delta(x2) = 1.0$ (see graph in Figure 6.11).

The ACFs estimated taking samples within each segment are shown in Figure 6.11. Now, suppose you want to measure the different correlation properties and identify in which parts of the signal they occur by means of the statistic afforded by the Fourier bases. The ACF estimated by all samples of the signal is overlaid (blue line) in the graph of Figure 6.11. It is evidently mostly sensitive to the long correlations of the first segment, while it misses completely the short correlations of the first one. With the Fourier basis one cannot detect *where*, only *what* is happening.

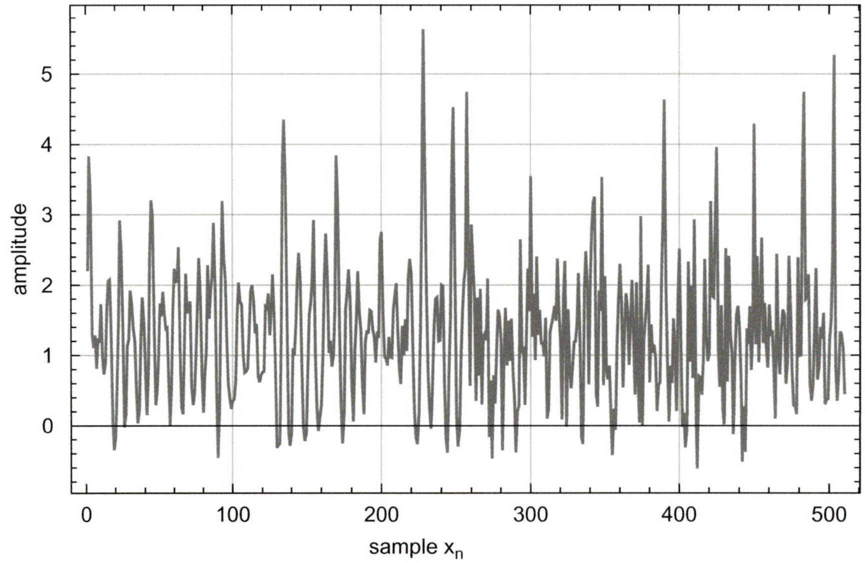

FIGURE 6.10 A test signal composed of two segments with Gamma distributed random deviates and exponential cosine ACFs with different correlation lengths.

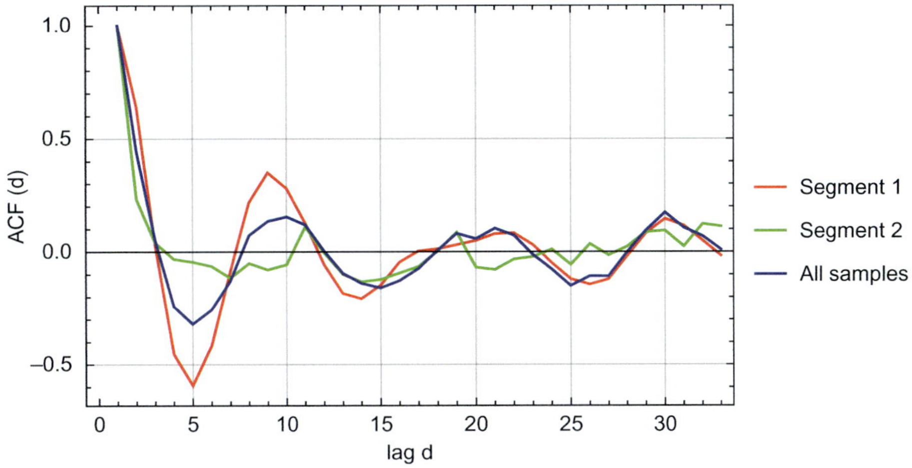

FIGURE 6.11 Auto-correlation functions estimated by samples within each segment $ACF(x1)$ and $ACF(x2)$ (red and green lines). The overlaid (blue line) is the ACF that would be estimated by samples over the whole signal.

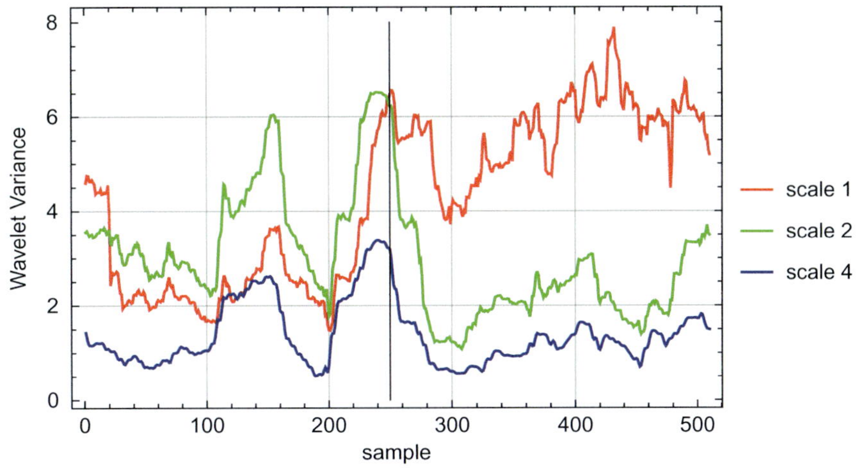

FIGURE 6.12 Wavelet variance analysis of the test signal. Dependence of the wavelet variance on scale permits the identification of differences of texture between the two segments.

Let us see what happens with wavelets, and apply the DWT based on the frame in Figure 6.9. Local estimates of the wavelet variance at 3 dyadic scales, obtained by a convolution with a low-pass filter, are shown in Figure 6.12. At scale 1 (short scale) the wavelet variance (red), being sensitive to high frequency phenomena, increases in the passage from segment x_1 to segment x_2 (see vertical bar delimiting the two segments), since x_2 is characterized by short correlation length noise. While the variance at larger scales (green, blue) detects patterns associated with long persistent correlations induced by the oscillator, responds more strongly within the first segment, and decays in the second. Moreover, the trajectory in scale of the wavelet variance is a further discriminator between the two processes. Within the first segment, the variance trajectory in non-monotonic – it increases at short scales and then decreases at all sampling points, while in the second segment the variance steadily decreases at all sampling points.

This illustrative example suggests that the dependence of wavelet variance on scale can provide measures of the correlation properties of signals (fields), these being adaptive to changes in space of these properties. These observations lead to the concept of wavelet scaling signatures and spectra, that we shall meet in a moment.

We next explore further the connection between two characterizations of random processes, a space-scale wavelet statistic, such as the wavelet variance and covariance, and the Fourier domain power spectrum.

Before starting on this subject, we note that one more way of interpreting the statistic of a wavelet that acts as a differential operator is to realize that the second moment of the wavelet coefficients provides a proxy of a two-point statistic known in geoscience as "structure function":

$$\left\langle c_j^2[x]\right\rangle \approx \left\langle \left(f(x)-f(x+\tau)\right)^2\right\rangle$$

where scale τ in the structure function plays the role of the dilation parameter 2^j in the wavelet transform, position x in the structure function is substituted for the translation parameter of the wavelet, and $\langle \odot \rangle$ is a local spatial average.

This connection is developed, for example, in [18], where it is shown how structure functions (and their wavelet counterparts) can be used to qualify and quantify non-stationarity in geophysical processes, such as non-stationary mono- and multi-fractals. Wavelet variance analysis therefore provides ways to deal with a wide gamut of random processes.

6.5 CONNECTION BETWEEN WAVELET SPACE–SCALE ANALYSIS AND FOURIER SPECTRAL ANALYSIS

For this discussion, we switch to the domain of continuous real functions and use the wavelet transform in Equation (6.6). The normalization $s^{-1/2}$ preserves the L^2 norms of the dilated wavelets.

The wavelet variance is:

$$\left\langle W_f^2\right\rangle(s) = \int_{\varepsilon}\left[W_f(s,b)\Big|_{1/2}\right]^2 db \tag{6.8}$$

where ε is the estimation interval.

For this analysis we adopt the mother wavelet in Figure 6.9, which is the derivative of a box spline of order 3, whose Fourier transform is (see [20] and the Appendix):

$$\hat{w}(\omega) = -i\frac{\omega}{4}\left(\frac{\sin\left(\frac{\omega}{4}\right)}{\frac{\omega}{4}}\right)^4 e^{-i\frac{\omega}{2}}$$

The wavelet operator is linear. Since the transform Equation (6.6) is a convolution of the signal with the mother wavelet dilated by s, in the Fourier domain the transform at scale s is equivalent to:

$$\hat{W}_f(\omega) = s^{-1/2}\,\hat{w}_s(\omega)\Im_f(\omega)$$

where $\hat{w}_s(\omega) = s\,\hat{w}(s\omega)$ is the Fourier transform of the dilated wavelet at scale s.

Given an input noise process with known spectral properties $S_f(\omega)$, the spectrum of the signal at the output of a linear system with transfer function $H(\omega)$ is:

$$S_{out}(\omega) = \left|\mathrm{H}(\omega)\right|^2 S_f(\omega)$$

Substituting the wavelet spectrum:

$$S_{\text{out}}(\omega,s) = \left| s^{-1/2}\, \hat{w}_s(\omega) \right|^2 S_f(\omega)$$

The variance of the wavelet coefficients can be computed from the ACF $\mathcal{R}(\tau)$ of the output signal at lag $\tau = 0$, this being given for a wide-sense stationary (WSS) process by the inverse Fourier transform of the power spectrum $S_{out}(\omega, s)$:

$$\left\langle w(s)^2 \right\rangle = \mathcal{R}(\tau = 0) = \mathfrak{I}^{-1}\left(\mathrm{S}_{out}(\omega,s)\right)\Big|_{\tau=0} = \frac{1}{2\pi}\int \mathrm{S}_{out}(\omega,s)\,d\omega \tag{6.9}$$

where $S(\omega) = \hat{W}_f\, \hat{W}_f^{\,*}$, and $\mathcal{R}(\tau)$ is the ACF of the stationary process.

Regarding the wavelet covariance, we can derive it from the cross-correlation function (CCF) at lag 0:

$$\left\langle \left\langle w_X w_Y \right\rangle (\mathrm{s}) \right\rangle = \mathrm{CCF}\left(w_X w_Y, \tau = 0\right)$$

Let us consider the wavelet transforms w_X, w_y of two signals x, y. The cross-correlation function $CCF(w_X w_Y)$ of the wavelet transform coefficients w_X, w_Y is linked to the cross-correlation of the input signals x, y through the dilated wavelet frequency response. Indeed, the CCF($w_X w_Y$, τ) in the time domain can be seen as the convolution of the first signal $w_x(t)$ with a time reversed version of the second signal $w_y(-t)$. In the frequency domain:

$$\mathfrak{I}\left(CCF\left(w_X w_Y,\tau\right)\right) = \mathfrak{I}\left(w_X \otimes w_Y\right) = \mathfrak{I}\left(w_X\right)\mathfrak{I}^*\left(w_Y\right) = s^{-1}\left|\hat{W}\right|_s^2 \mathfrak{I}_X(X)\mathfrak{I}_Y^*(Y)$$

$$= s^{-1}\left|\hat{W}\right|_s^2 \mathfrak{I}(X \otimes Y) = s^{-1}\left|\hat{W}\right|_s^2 \mathfrak{I}\left(CCF(x,y,\tau)\right)$$

This equation shows that the CCF of the wavelet transform coefficients w_X, w_Y of two signals x, y is linked to the CCF of the two signals.

Finally, the wavelet covariance $\langle w_X w_Y \rangle(\mathrm{s})$ is computed from the $CCF(w_X w_Y, \tau)$ at lag = 0:

$$\left\langle w_X w_Y \right\rangle(\mathrm{s}) = \mathfrak{I}^{-1}{}_{\tau=0}\left(s^{-1}\left|\hat{W}\right|_s^2 \mathfrak{I}_X(X)\mathfrak{I}_Y^*(Y)\right) \tag{6.10}$$

These formulations will be used next to study the wavelet variance scaling response in a number of test cases consisting of input random processes with different statistical properties. Symbolic and numerical computations have been carried out using Mathematica.[1] We shall see how the functional dependence of the wavelet variance with scale provides a measure of the underlying random process two-point statistic. This fact suggests calling this dependence a signature of the observed process. The following numerical analysis based on some theoretical random process models will provide guidelines for the interpretation of experimental wavelet signatures in experiments with SAR data that will be described in later sections.

6.5.1 White Noise

An example of white noise is δ correlated speckle statistics with constant radar reflectivity, with autocorrelation function $\mathcal{R}(\tau)$ and power spectrum S(ω) given by:

$$\mathcal{R}(\tau) = \langle I \rangle^2 \delta(\tau)$$

$$S(\omega) = \langle I \rangle^2$$

The wavelet variance by Equation (6.9) is:

$$\left\langle w(s)^2 \right\rangle = \frac{2\langle I \rangle^2}{3}$$

The wavelet variance $\langle w(s)^2 \rangle$ is scale independent.

6.5.2 1/f Process

Examples of 1/f processes (e.g. fractal surfaces) can be found in [23–25].

An 1/f process has a power spectrum obeying a power law of the form:

$$S(\omega) = \frac{\langle I \rangle^2}{|\omega|^\beta}$$

The corresponding ACF is: $\mathcal{R}(\tau)$

$$\frac{1}{\pi}\left(\langle I \rangle^2 |\tau|^{\beta-1} \Gamma(1-\beta) \sin\frac{\pi\beta}{2}\right)$$

By integrating the output spectrum (Equation 6.9), for instance setting $\beta = 1.5$, the wavelet variance is:

$$\left\langle w(s)^2 \right\rangle = const \langle I \rangle^2 s^{3/2}$$

The dependence of the wavelet variance on scale is linear in log-log scale, with the first derivative proportional to the spectral exponent.

6.5.3 Correlated Surface (Gamma Distributed RCS) with Exponential ACF (Lorentzian Spectrum)

The ACF $\mathcal{R}(\tau)$ and the power spectrum $S(\omega)$ are:

$$\mathcal{R}(\tau) = \langle I \rangle^2 + \frac{\langle I \rangle^2}{\upsilon} e^{-\beta|x|}$$

where I is the scattered intensity, ν is the order parameter of the K distribution, and β is the inverse correlation length.

$$S(\omega) = \frac{\langle I \rangle^2 \beta}{\beta^2 + \omega^2} + 2\pi \langle I \rangle^2 \delta(\omega)$$

where $\delta(x)$ is the Dirac distribution.

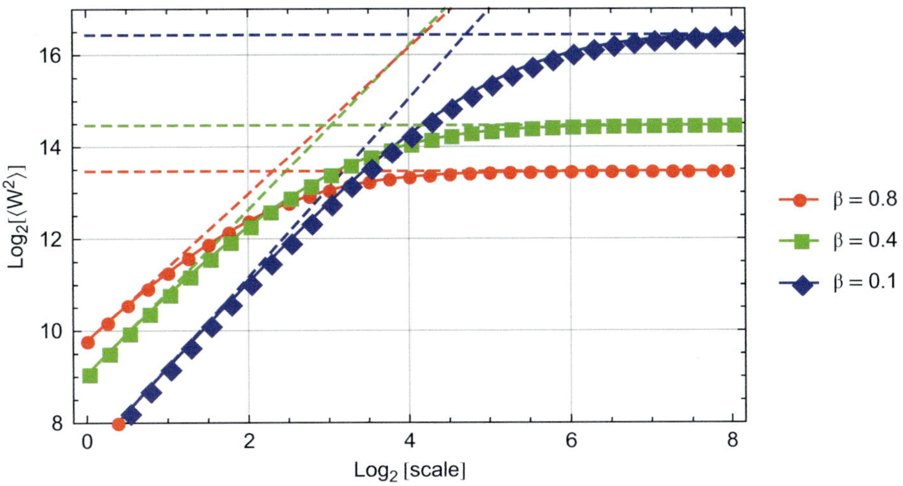

FIGURE 6.13 Wavelet variance dependence on scale for correlated Gamma deviates with exponential ACF.

The integral Equation (6.9) (wavelet variance) was evaluated numerically as a function of scale and for parameters $\upsilon = 0.375$ and $\beta = 0.1, 0.4, 0.8$. The resulting curves are shown in the log-log graph in Figure 6.13.

The red (circles), the green (squares) and the blue (diamond) solid lines correspond respectively to the inverse correlation length values $\beta = 0.1, 0.4, 0.8$. The dotted lines mark the tangent of the curve at scale s=1, and the asymptotic values of the wavelet variance.

Let us consider the derivative of the wavelet variance at scale s=1. Taking the Fourier series of the sampled wavelet at scale 1, the wavelet spectrum at scale s can be approximated by:

$$S(\omega,s)=\left|-\frac{1}{2}+\frac{1}{2}e^{-is\omega}\right|^2$$

The derivative of the wavelet variance at scale 1, this corresponding to $\Re(0)'$, can be then derived from the derivative under the integral sign in Equation (6.9). Retaining only first order terms in ω:

$$\left\langle w(1)^2\right\rangle' \simeq \frac{\langle I\rangle^2}{\upsilon}\tan^{-1}\frac{2\pi}{\beta}$$

For $\beta \leq 1$ the dependence of the signature slope at $s = 1$ is linear in β and proportional to the signal intensity normalized second moment $\frac{1}{\nu}$. Therefore $\langle w(1)^2\rangle'$ *and* $\langle w(1)^2\rangle$ supply information on the RCS distribution order parameter, and on the ACF inverse correlation length.

The asymptotic value of the wavelet variance can be derived by considering that at large scales the input signal becomes uncorrelated with spectrum $G(\omega) = \langle I^2\rangle - \langle I\rangle^2$. Taking the limit under the integral sign in Equation (6.9):

$$\lim_{s\to\infty}\int\langle I\rangle^2\left|s^{-1/2}\hat{w}_s(\omega)\right|^2 d\omega=\frac{2}{3}\pi\left(\langle I^2\rangle-\langle I\rangle^2\right)$$

Therefore, the asymptotic value of the wavelet variance is proportional to the variance of the random process.

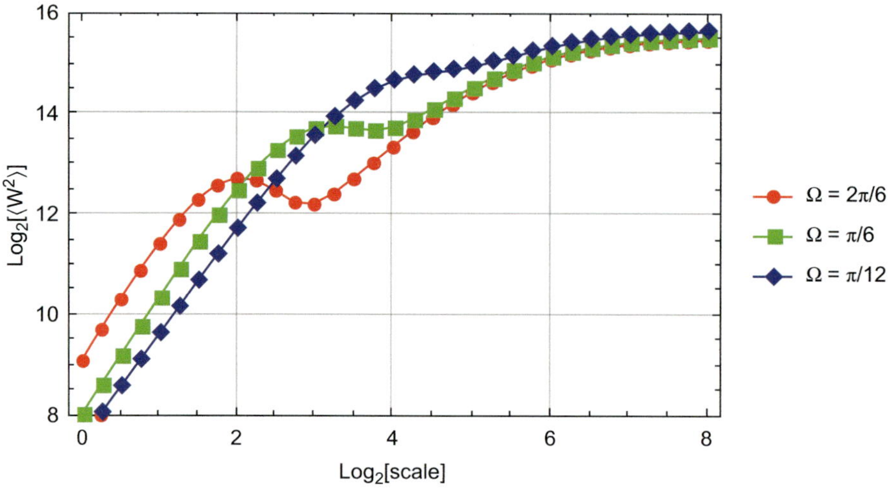

FIGURE 6.14 Wavelet variance as a function of scale (log-log plot) for the exponential cosine ACF. The three curves correspond to different values of the cosine term frequency, these being $\Omega = 2\pi/6$, $\Omega = \pi/6$, $\Omega = \pi/12$. Shorter period oscillation of the ACF affects the wavelet variance at a corresponding shorter scale.

6.5.4 Correlated Surface with Exponential Cosine ACF

A model of SAR correlated clutter based on this statistic is discussed in [3,26].

$$\mathcal{R}(\tau) = \langle I \rangle^2 + \frac{\langle I \rangle^2}{\upsilon} e^{-\beta|x|}\left(1 + \cos \Omega x\right)$$

The wavelet variance Equation (6.9) was evaluated numerically as a function of scale and the following ACF parameters: $\langle I \rangle^2 = 100$, $\nu = 0.735$, $\beta = 0.1$, $\Omega = \frac{\pi}{3}, \frac{\pi}{6}, \frac{\pi}{12}$. The corresponding period in scale units is: $T = 6, 12, 24$. The results are shown in Figure 6.14.

As for the case of the exponential ACF, similarly the wavelet variance derivative at scale 1 $\langle w(1)^2 \rangle'$ is proportional to the correlation length β in all three test cases. At large scales, where the process becomes uncorrelated, the wavelet variance tends asymptotically to the process variance. In the intermediate range of scales, the correlation variation induced by the oscillatory term is captured by the wavelet variance at the scale that resonates with the oscillator period.

6.5.5 Effects from Coherent Imaging and Illumination Beam Size

In the previous analysis the connection was established between the wavelet variance scaling and the correlation properties of the illuminated surface (i.e. the radar cross-section – RCS). In other words, we have assumed a situation where perfect reconstruction of RCS had been somehow accomplished before estimating the wavelet statistics. We consider now the case where the coherent interaction of the scattering elements and the illumination beam finite size are included in the model of the backscattered intensity ACF, such as in the correlated K-distributed model in [26] Equation (5.21), and in [4] Equation (6.2):

$$\mathcal{R}(\tau) = \langle I \rangle^2 + \langle I \rangle^2 e^{-\frac{\tau^2}{\sigma^2}} + \frac{\langle I \rangle^2}{\upsilon}\left(e^{-\beta\tau^2} \otimes \left|e^{-\frac{\tau^2}{\sigma^2}}\right|^2\right) \tag{6.11}$$

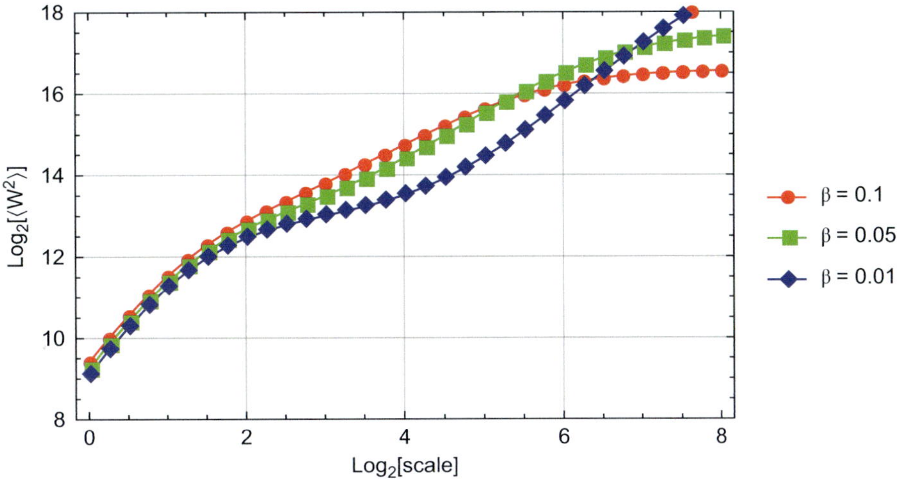

FIGURE 6.15 Wavelet variance scaling signature in the case of a correlated surface and taking into account the imaging geometry (illumination beam with a Gaussian profile). The curves are parametrized by correlation length $\beta = 0.1, 0.05, 0.01$.

where the illumination beam is assumed to have a Gaussian profile with the standard deviation σ being the 1/e width of the profile, ν is the order parameter of the K distribution, and β is the inverse correlation length.

Numerical evaluation of the integral Equation (6.9) using the intensity ACF Equation (6.11) yields the wavelet variance signatures represented in Figure 6.15. Parameters were: $2\sigma = 3$, $\upsilon = 0.375$ and $\beta = 0.01, 0.05, 0.1$, these corresponding to the lines with diamond markers, squares and circles, respectively.

Information about the underlying RCS at short scales$[2^0, 2^1]$, these being within the instrument resolution, is shadowed by the impulse response function of the instrument, and is therefore unobtainable. The RCS correlation length can still be estimated from the onset of the asymptotic value, albeit only in the region of scales outside the influence of the instrument point spread function (PSF).

6.5.6 Cross-correlation between Two Stationary Processes with a Gaussian CCF

Test cases were developed by numerical evaluation of integral Equation (6.9) and assuming two stationary random input processes with a Gaussian CCF. Moreover, the Schwartz inequality condition is imposed on the second moments of intensity:

$$\left|\langle I_1 I_2 \rangle\right|^2 \leq \langle I_1^2 \rangle \langle I_2^2 \rangle$$

We have considered, for example, processes with a given intensity one-point statistic $\langle I_1 \rangle$, $\langle I_2 \rangle$, $\langle I_1 I_2 \rangle$ and Gaussian CCFs with different variance, these being:

$$\sigma = (1.0,\ 4.0,\ 8.0)$$

Results are reported in Figure 6.16; the lines with the round, square and diamond markers correspond to the spread (correlation length) of each of the CCFs listed above.

Note that the asymptotic point of the wavelet covariance is reached proportionally to the amount of correlation between the two input processes. In other words, the wavelet covariance provides a measure of the spatial similarity (diversity) between the two processes.

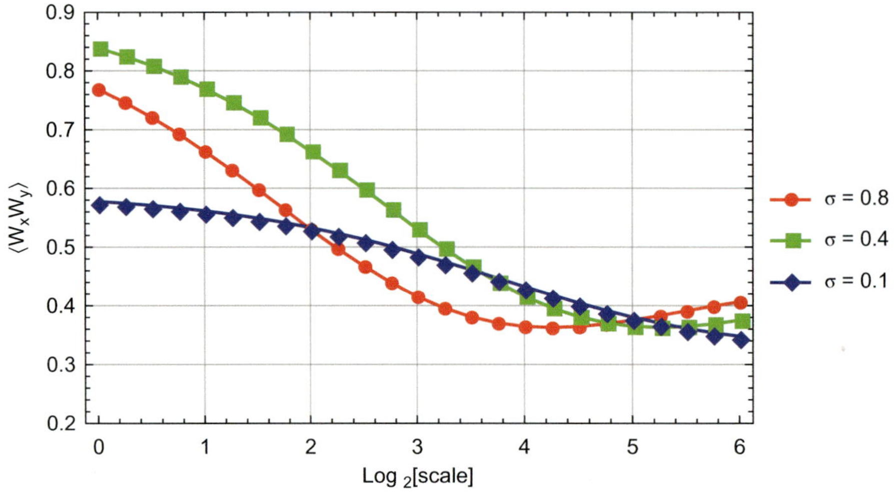

FIGURE 6.16 Wavelet normalized covariance. Three test cases are studied, where two random input processes are assumed to have different auto- and cross-correlation functions.

6.6 ACCURACY OF WAVELET VARIANCE ESTIMATORS

Having decided, following so many advertisements, to buy wavelets, we now need to read the user's manual. This section is about the tool's specifications, such as operating limits and performance figures. In later sections of this chapter, we will talk about the assembly of the parts to build the desired measurement tools, and finally we shall show how to use them, how to perform measures of real SAR data.

First, we turn our attention to a statistical analysis that is specific to the discrete wavelet frame representation of SAR random fields. We recall that our choice of the machinery for setting up yardsticks that would measure patterns in SAR random fields falls on a wavelet designed by Mallat [20] that acts as a differential operator. The reader will find details of the mathematical derivations and the computational schemes based on the à trous algorithm in the Appendix. We recap here the essentials we shall need for the ensuing derivations.

The frame is derived from an interpolating function that is a shifted spline of order 2 with Fourier transform:

$$\hat{\varphi}(\omega) = \left(\frac{\sin\left(\frac{\omega}{2}\right)}{\frac{\omega}{2}} \right)^3 \exp\left(-i\frac{\omega}{2}\right)$$

The Fourier transform of the low-pass filter associated with this interpolating function, after renormalization by a factor $1/\sqrt{2}$ to make it energy conserving, is:

$$\hat{g}(\omega) = \left(\cos\frac{\omega}{2}\right)^3 \exp\left(-i\frac{\omega}{2}\right)$$

The high-pass filter normalized by the same factor $1/\sqrt{2}$ is:

$$\hat{h}(\omega) = -isin\frac{\omega}{2}\exp\left(-i\frac{\omega}{2}\right)$$

The filter coefficients computed by inverse DTFT (Discrete Time Fourier Transform) are:

$$g[n] = \sqrt{2}\{0.125,\ 0.375,\ 0.375,\ 0.125\} \tag{6.12}$$

$$h[n] = \{0.0, -0.5,\ 0.5,\ 0.0\} \tag{6.13}$$

6.6.1 Prelude: Probability Density Function of the Wavelet Coefficients of a Speckle Pattern

For uncorrelated 1-Look intensity speckle, the probability density function (PDF) of the wavelet coefficients at scale 2°can be calculated as the difference between two independent exponentially distributed random variables.

Indeed, the high pass filter $\hat{h}(\omega)$ performs the following operations on two samples of the speckle pattern:

$$w = 0.5(x_1 - x_2) \tag{6.14}$$

The speckle samples x are exponentially distributed, say with:

$$f_x(x) = ae^{-a\,x}U(x)$$

where $U(x)$ is the Heaviside step function.

Let us consider first the transformation $z = x - y$. The PDF of the sum of two independent random variables is equal to the convolution integral of their PDFs [27]:

$$z = x + y$$

$$f_z(v) = \int_{-\infty}^{+\infty} f_x(x) f_y(v - x)\,dx$$

By the change of variables $z = x + (-y)$, since $f_{-y}(v - x) = f_y(x - v)$:

$$f_z(v) = \int_{-\infty}^{+\infty} f_y(x - v) f_x(x)\,du = \int_{-\infty}^{+\infty} a^2 e^{-a(x-v)} U(x - v) e^{-a\,x} U(x)\,dx \tag{6.15}$$

For positive values of v the integral Equation (6.15) can be evaluated as:

$$f_y(v) = \int_{v}^{+\infty} a^2 e^{-a(u-v)} e^{-au}\,du = \frac{a}{2} e^{-av}$$

For negative v the integral Equation (6.15) can be evaluated as:

$$f_y(v) = \int_{0}^{+\infty} a^2 e^{-a(u-v)} e^{-au}\,du = \frac{a}{2} e^{av}$$

Joining the two results, the PDF of the difference of two exponentially distributed random variables results in a Laplace distribution:

$$f_y(v) = \frac{a}{2} e^{-a|v|}$$

The variance of the Laplace distribution is $VAR(v) = 2\left(\frac{1}{a}\right)^2$.

The variable w in Equation (6.14) (wavelet coefficients) is also scaled by a factor of 2; therefore, the resulting PDF of the wavelet coefficients is given by:

$$f_w(w) = 2 f_z(2v) = a e^{-2a|w|} \tag{6.16}$$

Since the variance of an exponentially distributed speckle process with mean intensity $\langle I \rangle$ is:

$$\sigma^2_{speckle} = \frac{1}{a^2} = \langle I \rangle^2$$

The Laplace distribution in Equation (6.16) can be expressed as:

$$f_w(v) = \frac{1}{\sigma_{speckle}} e^{\frac{-2|v|}{\sigma_{speckle}}}$$

The corresponding variance of the wavelet coefficients at scale 1 is (Figure 6.17):

$$\langle w_1^2 \rangle = 2\int_0^{\infty} \frac{1}{\sigma_{speckle}} e^{-\frac{2x}{\sigma_{speckle}}} dx = \frac{\sigma_{speckle}{}^2}{2} = \frac{\langle I \rangle^2}{2} \tag{6.17}$$

$$\langle w_1^2 \rangle = \frac{\sigma^2_{speckle}}{2} = \frac{\langle I \rangle^2}{2}$$

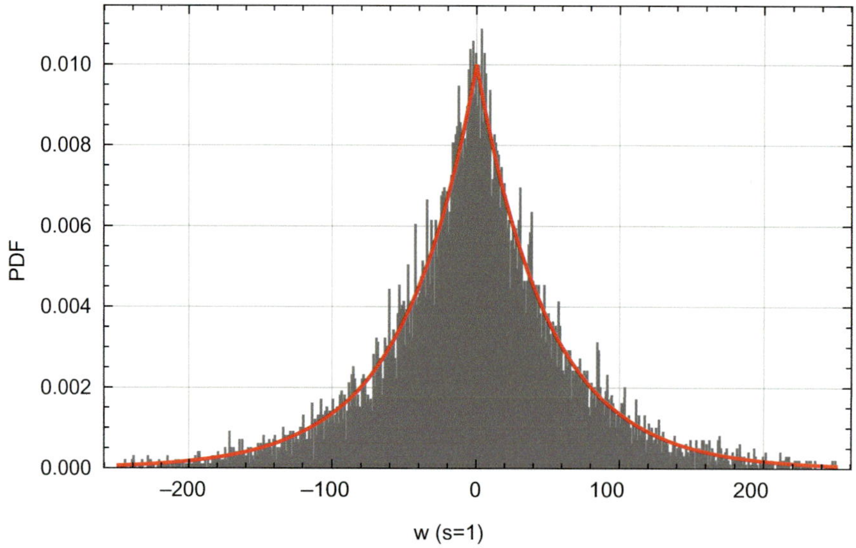

FIGURE 6.17 Experimental histogram of the wavelet coefficients at scale 1 corresponding to 1-Look intensity speckle with $\langle I \rangle = 100$. The theoretical PDF is overlaid in red.

At scale 2 the high pass à trous filter again extracts and makes the difference between two samples of the smooth signal, but now the samples are Gamma distributed because they are taken at the output of the low pass filter that performs a weighted average of a number of samples. Neighboring samples of the smooth signal are indeed now correlated (the over-sampled version of the discrete wavelet transform is used here) but also the high pass filter is dilated by two samples with zero insertion. Therefore correlation effects can be disregarded. As in the preceding analysis, the wavelet PDF can be estimated by a convolution integral of the Gamma PDFs that characterize the smooth signal at scale 2.

Let us denote for convenience the Gamma PDF of the variable x as:

$$f_x(x) = \frac{\lambda^N x^{N-1}}{\Gamma(N)} e^{-\lambda x}$$

in terms of the speckle variance σ^2 and intensity I, $\lambda = \dfrac{N}{\sigma} = \dfrac{N}{\langle I \rangle}$

By the change of variables $z = x + (-y)$, and considering the case with $z < 0$ the convolution integral becomes:

$$f_z(z) = \int_0^{\infty} f_x(x) f_y(x - z) dx = \frac{\lambda^{2N}}{\Gamma^2(N)} e^{\lambda z} \int_0^{\infty} e^{-2\lambda x} x^{N-1} (x - z)^{N-1} dx \tag{6.18}$$

The equivalent number of Looks of the speckle patterns in the smooth signal can be estimated in the Fourier domain from the impulse response function of the filter Equation (6.12):

$$H(\omega) = \left(\cos\frac{\omega}{2} \right)^3 e^{-i\omega/2}$$

Assuming a speckle pattern with an exponential ACF with $\langle I \rangle = 1$ and $VAR(I) = 1$ and correlation length of 1 pixel we have:

$$\langle I^2 \rangle' = \int |H(\omega)|^2 \mathcal{F}[ACF(x)](\omega) d\omega = 1.38$$

$$Var'(I) \cong \frac{1}{3} Var(I)$$

Therefore, the variance reduction corresponds approximately to N=3 equivalent number of Looks.

The integral $\mathcal{M}(z) = \int_0^{\infty} f(z,\lambda,N) dz$ in Equation (6.18) for N=3 and $z < 0$ is:

$$\mathcal{M}(z) = \frac{(z\lambda - 3)(z\lambda + 3)}{4\lambda^5} \Rightarrow z < 0$$

By symmetry considerations, since $f_z(z) = f_z(-z)$:

Finally, the PDF of the wavelet coefficients at scale 2, since $f_w(w) = 2 f_x(2x)$ and by symmetry considerations, since $f_z(z) = f_z(-z)$:

$$f_w(w) = \frac{2\lambda^6 e^{2w\lambda}}{\Gamma^2(N)} \frac{(2w\lambda - 3)(2w\lambda + 3)}{4\lambda^5} \Rightarrow z < 0 \tag{6.19}$$

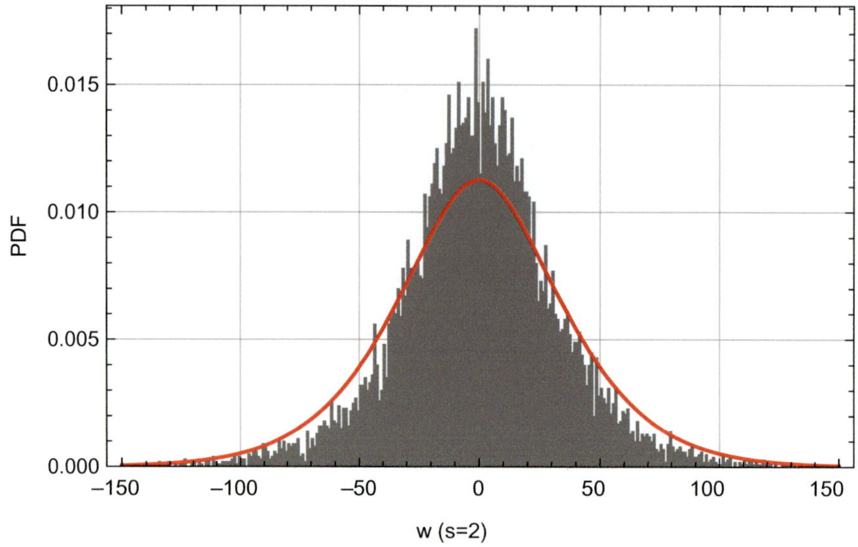

FIGURE 6.18 Experimental histogram of the wavelet coefficients at scale 2 corresponding to 1-Look intensity speckle with $\langle I \rangle = 100$. The theoretical PDF is overlaid in red.

$$f_w(w) = \frac{2\lambda^6 e^{-2w\lambda}}{\Gamma^2(N)} \frac{(-2w\lambda - 3)(-2w\lambda + 3)}{4\lambda^5} \Rightarrow z > 0 \tag{6.20}$$

From the PDF $f_w(w)$in Equations (6.19) and (6.20):

$$\langle w_2^2 \rangle = \frac{3}{2\lambda^2} = \frac{\langle I \rangle^2}{6} = \frac{\sigma^2_{speckle}}{6}$$

The histogram of the wavelet coefficients at scale 2 computed with 1-Look intensity experimental data is shown in Figure 6.18. The theoretical PDF in Equations (6.19) and (6.20) is overlaid with a dashed red line.

Variance of the wavelet coefficients computed from the theoretical PDFs is in good agreement with experimental data. For example, for a speckle process with $\langle I \rangle = 100$ the RRMSE (root mean square error relative to the mean error) at scale 1 and 2 are, respectively, 1% and 5%.

It is also interesting to consider the statistics of $z = w^2$ because it will give us information on the higher moments of w that will be useful to derive the variance of the estimator $\langle w^2 \rangle$ as implemented by convolution with a cubic B-spline (see below).

The PDF of the transformation of a random variable by a square law $z = x^2$ is given by:

$$f_z(v) = \frac{U(v)}{2\sqrt{v}} \left[f_x(\sqrt{v}) + f_x(-\sqrt{v}) \right] \tag{6.21}$$

From Equations (6.16) and (6.21), at scale 1 the PDF of w^2 is:

$$f_{w^2}(v) = a \frac{e^{-2a\sqrt{v}}}{\sqrt{v}} U(v)$$

Mean value and second moment by integration are:

$$\left\langle w_1^2 \right\rangle = \int_0^{+\infty} av \frac{e^{-2a\sqrt{v}}}{\sqrt{v}} dv = \frac{1}{2a^2} = \frac{1}{2}\sigma_{speckle}^2$$

$$\left\langle w_1^4 \right\rangle = \int_0^{+\infty} av^2 \frac{e^{-2a\sqrt{v}}}{\sqrt{v}} dv = \frac{3}{2a^4} = \frac{3}{2}\sigma_{speckle}^4$$

The theoretical variance of the random variable w^2 at scale 1 (assuming we had an infinite number of samples at our disposal) is therefore:

$$Var\left(w_1^2\right) = \frac{5}{4}\sigma^4 \tag{6.22}$$

At scale 2 by the same procedure but using the PDF in Equation (6.20) we have:

$$\left\langle w_2^2 \right\rangle = \frac{\sigma_{speckle}^2}{8}$$

$$\left\langle w_2^4 \right\rangle = \frac{15}{256}\sigma_{speckle}^4$$

$$VAR\left(w_2^2\right) = \frac{11}{256}\sigma_{speckle}^4$$

$$Var\left(w_2^2\right) = \frac{1}{18}\sigma^4 - \frac{1}{12}\sigma^2 \tag{6.23}$$

6.6.2 Expected Value and Variance of the Wavelet Variance Estimator

The results of the previous section will enable us to characterize statistically the wavelet variance estimators in terms of their expected value and variance. We begin by considering the case of uncorrelated speckle. In the following derivations, S_j is the smooth signal at scale 2^j, $s = 2^j$ is the scale, and $Y_{in}^j(\omega)$ is the Fourier spectrum of the input process at scale j of the filter bank.

6.6.2.1 Uncorrelated Speckle Pattern

It is important to notice that the theoretical wavelet variance corresponding to the speckle process Equations (6.22) and (6.23) is proportional to the mean speckle intensity squared $\langle I \rangle^2$. This is highly undesirable in our texture analysis context, because it would result in a measure of the speckle strength instead of texture. In other words, constant reflectivity areas without texture would give wavelet variance values that are dependent on the mean backscatter coefficient.

One possibility for counteracting this effect, originally proposed in [28,29], consists in normalizing the wavelet coefficients with the smooth signal at the same scale j:

$$\hat{W}_j = \frac{W_j}{S_j}$$

Notice that the smooth signal S_j at scale 2^j, when computed by the à trous algorithm (see Section 6.7), must be normalized by $\frac{1}{2^{j/2}}$ to preserve the mean value, since the à trous filter does not have unitary norm.

Let us analyze the consequences of this position on the estimator that is implemented in our algorithm by local averages using a finite number of samples:

$$\left\langle \left(\frac{\tilde{w}}{\tilde{s}} \right)^2 \right\rangle = \left\langle \sum_{i=1}^{N} \left(\frac{w_i}{s_i} \right)^2 \right\rangle$$

It is thus important to evaluate the expected value and variance of such an estimator. Analysis will be carried out at decomposition scale 1. For this purpose, we consider a Taylor expansion of the function $\frac{x}{y}$ around the mean value [30]:

$$\frac{x}{y} = \left.\frac{x}{y}\right|_{\bar{x},\bar{y}} + \left(x-\bar{x}\right)\left.\frac{\partial}{\partial x}\left(\frac{x}{y}\right)\right|_{\bar{x},\bar{y}} + \left(y-\bar{y}\right)\left.\frac{\partial}{\partial y}\left(\frac{x}{y}\right)\right|_{\bar{x},\bar{y}} +$$

$$+\frac{1}{2}\left(x-\bar{x}\right)^2\left.\frac{\partial^2}{\partial x^2}\left(\frac{x}{y}\right)\right|_{\bar{x},\bar{y}} + \frac{1}{2}\left(y-\bar{y}\right)^2\left.\frac{\partial^2}{\partial y^2}\left(\frac{x}{y}\right)\right|_{\bar{x},\bar{y}} +$$

$$+\left(x-\bar{x}\right)\left(y-\bar{y}\right)\left.\frac{\partial^2}{\partial x \partial y}\left(\frac{x}{y}\right)\right|_{\bar{x},\bar{y}} + O_3 \tag{6.24}$$

Applying the expected value operator $E\{\}$on each term, an approximation of the expected value of the ratio is given by:

$$E\left\{\left\langle\frac{x}{y}\right\rangle\right\} \cong \frac{\bar{x}}{\bar{y}} + \operatorname{var}\left(y\right)\frac{\bar{x}}{\bar{y}^3} - \frac{\operatorname{cov}\left(x,y\right)}{\bar{y}^2} \tag{6.25}$$

where $\bar{x},\bar{y}$ are theoretical mean values of the random variables, *var*(y) is the theoretical variance of y and *cov*(x, y) is the covariance between x and y.

Setting $x = w^2$, $y = s^2$, we can evaluate Equation (6.24) as a function of the mean, variance and covariance of these random variables. Regarding s^2, we know that, at scale 1, the smooth signal follows a Gamma PDF with an equivalent number of 4 Looks. Therefore the PDF of s^2 is given by the transformation of the Gamma PDF according to Equation (6.18). By integration we have:

$$\left\langle s^2 \right\rangle = \frac{4}{3}\left\langle I \right\rangle^2$$

$$\left\langle s^4 \right\rangle = \frac{40}{9}\left\langle I \right\rangle^4$$

$$\left\langle s^2 \right\rangle = \frac{8}{3}\sigma^2 \tag{6.26}$$

$$\left\langle s^4 \right\rangle = \frac{80}{9}\sigma^4 \tag{6.27}$$

$$Var\left(s^2\right) = \frac{16}{9}\sigma^4 \tag{6.28}$$

Regarding w^2 from Equation (6.17):

$$\langle w^2 \rangle = \frac{1}{2}\sigma^2 \tag{6.29}$$

Substituting Equations (6.26), (6.28) and (6.29) in Equation (6.25) gives:

$$E\left[\left(\frac{w}{s}\right)^2\right] = \frac{\langle w^2 \rangle}{\langle s^2 \rangle} + Var\left(s^2\right)\frac{\langle w^2 \rangle}{\langle s^2 \rangle^3} - \frac{\mathrm{cov}\left(w^2, s^2\right)}{\langle s^2 \rangle^2} = \frac{15}{64} - \frac{9}{64\langle I \rangle^4}\mathrm{cov}\left(\langle w^2 \rangle, \langle s^2 \rangle\right) \tag{6.30}$$

The term involving the covariance of the variables does not depend on the mean intensity, being proportional to $\langle I \rangle^4$. This point can be proved as follows. At scale 1, w and s are signals at the output of the à trous high-pass and low-pass filters $g(k)$ and $h(k)$ when the input is the speckle process, say $x(t)$. Therefore, setting $z = s^2$ and $y = w^2$ the following relations hold for the cross-correlation functions between the input–output of the filters:

$$R_{zy}(\tau) = E\left[z(t)y(t+\tau)\right] = E\left[z(t)\int \tilde{g}(\xi)x^2(t+\tau-\xi)d\xi\right] =$$

$$= \int E\left[z(t)x^2(t+\tau-\xi)\right]\tilde{g}(\xi)d\xi$$

where $\tilde{g}$ is the equivalent filter that would generate $y = w^2$ variates from x^2.

$$E\left[z(t)x^2(t+\tau-\xi)\right] = R_{zx^2}(\tau) = R_{x^2x^2}(\tau)\otimes\tilde{h}(-\tau)$$

from the relationship linking the cross-correlation function between input–output $y = L(x)$ and the autocorrelation function of input in a linear system [27]:

$$R_{xy}(\tau) = R_{xx}(t)\otimes h(t)$$

$\tilde{h}$ is an equivalent filter that would generate $z = s^2$ variates from x^2; $\otimes$ is the convolution operator.

$$R_{zy}(\tau) = R_{x^2x^2}(\tau)\otimes\tilde{h}(-\tau)\otimes\tilde{g}(\tau)$$

Since the speckle is uncorrelated:

$$R_{x^2x^2}(\tau) = \langle I \rangle^4 \delta_\tau$$

Therefore, finally:

$$R_{w^2s^2}(0) = \mathrm{cov}\left(w^2, s^2\right) + \langle w^2 \rangle\langle s^2 \rangle = \langle I \rangle^4 \, const$$

Numerical evaluation using experimental data confirms this position. Therefore we can conclude that the estimator of the square of the normalized wavelet coefficients is, as desired, independent of the speckle strength $\sigma = \langle I \rangle$. The estimator is biased, though, as can be appreciated from Equation (6.25).

Let us now consider the variance of the estimator. From the Taylor series expansion of the function $\frac{x}{y}$ we have:

$$\operatorname{var}\left(\left\langle\frac{x}{y}\right\rangle\right)=E\left\{\left(\left\langle\frac{x}{y}\right\rangle-\frac{\overline{x}}{\overline{y}}\right)^2\right\}\approx\frac{1}{N}\left(\frac{\operatorname{var}(x)}{\overline{y}^2}+\frac{\overline{x}^2\operatorname{var}(y)}{\overline{y}^4}-\frac{2\overline{x}\operatorname{cov}(x,y)}{\overline{y}^3}\right) \tag{6.31}$$

Setting $x = w^2$, $y = s^2$, we can evaluate Equation (6.31) substituting, as before, values from Equations (6.26)–(6.29):

$$\left\langle Var\left(\frac{w^2}{s^2}\right)\right\rangle\cong\frac{1}{N}\left(0.18-\frac{1}{20}\frac{\operatorname{cov}\left(w^2,s^2\right)}{\langle I\rangle^4}\right) \tag{6.32}$$

Therefore, to second-order Taylor expansion, the variance of the estimator tends to 0 for an infinite number of samples, and the estimator is consistent.

Standard error of the estimator (standard deviation over the mean) as a function of the number of samples in the average is shown in Figure 6.19 (dashed line). The error is computed from Equation (6.30) and (6.32), setting $\frac{c}{\langle I\rangle^4}=1.8$ as derived from experimental data.

A second possibility to cancel out the dependency of the wavelet variance from the mean speckle intensity would be to normalize the variance by the squared mean value of the smooth signal at the same scale:

$$\overline{\left\langle W_j^2\right\rangle}=\frac{\left\langle W_j^2\right\rangle}{\left\langle S_j\right\rangle^2}$$

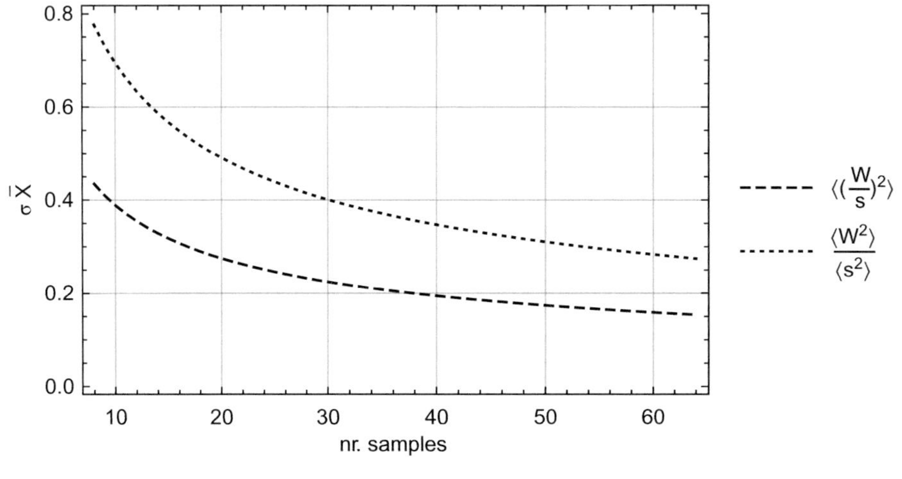

FIGURE 6.19 Standard error $\sigma_{\overline{X}}$ of the estimator $\left\langle\left(\frac{W}{S}\right)\right\rangle^2$ as a function of the number of samples computed from Equations (6.30) and (6.32) (dashed line). Standard error of the estimator $\frac{\langle W^2\rangle}{\langle S\rangle^2}$ as a function of the number of samples computed from Equations (6.33) and (6.34) (dotted line).

To evaluate the expected value and variance of this estimator we proceed as in the previous case but consider this time a Taylor expansion of the function x/y^2. At scale 1 and setting $x = w^2$, $y = s$ we have:

$$E\left\{\frac{\langle x\rangle}{\langle y\rangle^2}\right\} \approx \frac{\overline{x}}{\overline{y}^2} + \frac{1}{2}\operatorname{var}\left(\langle y\rangle\right)\frac{6\overline{x}}{\overline{y}^4} - \frac{2\operatorname{cov}\left(\langle x\rangle,\langle y\rangle\right)}{\overline{y}^3} \approx$$

$$\approx \frac{\overline{x}}{\overline{y}^2} + \frac{1}{n}\left(\operatorname{var}(y)\frac{3\overline{x}}{\overline{y}^4} - \frac{2\operatorname{cov}(x,y)}{\overline{y}^3}\right)$$

This estimator is therefore asymptotically unbiased. Substituting values from Equations (6.26)–(6.29) we have:

$$E\left\{\frac{\langle w^2\rangle}{\langle s\rangle^2}\right\} = \frac{1}{2} + \frac{1}{n}\left(\frac{5}{2} - \frac{2\operatorname{cov}\left(w^2, s\right)}{\langle I^3\rangle}\right) \tag{6.33}$$

From arguments similar to the ones expounded above it turns out also that

$$\operatorname{cov}(x,y) = \langle I\rangle^3 \, const$$

Therefore this second estimator also achieves the intended goal, namely to cancel the dependency on the mean speckle intensity.

Let us now consider the variance of the estimator:

$$\operatorname{var}\left(\frac{\langle x\rangle}{\langle y\rangle^2}\right) = E\left\{\left(\frac{\langle x\rangle}{\langle y\rangle^2} - \frac{\overline{x}}{\overline{y}^2}\right)^2\right\} \approx \left(\frac{\operatorname{var}(\overline{x})}{\overline{y}^4} + \frac{4\overline{x}^2\operatorname{var}(\overline{y})}{\overline{y}^6} - \frac{2\overline{x}\operatorname{cov}(\overline{x},\overline{y})}{\overline{y}^5}\right) \approx$$

$$\approx \frac{1}{N}\left(\frac{\operatorname{var}(x)}{\overline{y}^4} + \frac{4\overline{x}^2\operatorname{var}(y)}{\overline{y}^6} - \frac{2\overline{x}\operatorname{cov}(x,y)}{\overline{y}^5}\right)$$

Substituting the theoretical expected values from Equations (6.25)–(6.28) we have:

$$Var\left(\frac{\langle w^2\rangle}{\langle s\rangle^2}\right) = \frac{1}{n}\left(\frac{35}{12} - \frac{\operatorname{cov}\left(w^2, s\right)}{\langle I^3\rangle}\right) \tag{6.34}$$

Also in this case the variance decreases to zero for an infinite number of samples, and therefore the estimator is consistent. Results of a numerical evaluation of the standard error of the estimator are graphed in Figure 6.19 (dotted line).

This second estimator is asymptotically unbiased but results in a higher standard error with respect to the first. Since a bias in the expected value is not critical in segmentation problems where relative distances are considered, we conclude that the first form of wavelet normalization is advantageous with respect to the behavior in the presence of speckle noise.

6.6.2.2 Correlated Speckle

For this analysis, an approximation to the expected value $E\left(\left\langle \frac{w^2}{s^2} \right\rangle\right)$ and the variance of the estimator $VAR\left(\left\langle \frac{w^2}{s^2} \right\rangle\right)$ at dyadic scales2^jcan be obtained by Taylor expansions Equations (6.25) and (6.31), disregarding the covariance terms.

To compute the expected value, Equation (6.25) setting $x = w^2$ and $y = s^2$ we need estimates of the second and fourth moments: $\langle w^2\rangle\langle w^4\rangle\langle s^2\rangle\langle s^4\rangle$.

The second moments can be computed numerically in the Fourier domain considering the à trous algorithm as a filter bank. Thus, calling $h(\omega)$ and $g(\omega)$ the low-pass (à trous) filter and the high pass filter in the bank, we have at successive decomposition scales 2^j:

$$H_j(\omega) = h(\omega) \quad j = 0 \tag{6.35}$$

$$H_j(\omega) = h(2^j\omega)\prod_{k=0}^{j-1} g(2^k\omega) \quad j \geq 1$$

As it was pointed out in Section 6.5 regarding Equation (6.9), the wavelet variance can be derived from the spectrum at the output of the filter bank:

$$\langle w(s)^2 \rangle = \mathcal{R}(\tau = 0) = \mathfrak{I}^{-1}(Y(\omega,s))\Big|_{\tau=0} = \frac{1}{2\pi}\int Y^j_{out}(\omega,s)\,d\omega \tag{6.36}$$

with:

$$S^{2j}_{out}(\omega) = |H_{2^j}(\omega)|^2 S_{in}(\omega)$$

Assuming as auto-correlation function for the input speckle process:

$$R(x) = \langle I \rangle^2\left[1 + \sin c^2\left(\frac{x}{d}\right)\right]$$

where $\langle I \rangle$is the mean speckle intensity, d is the position of the first null of the sin c function and is an indication of the average width of a speckle grain.

The sin c function is defined as:

$$\sin c(x) = \frac{\sin(\pi x)}{\pi x}$$

Indeed, for single Look data:

$$VAR(I) = \langle I \rangle^2$$

Therefore:

$$\langle I^2 \rangle = 2\langle I \rangle^2 = R(0)$$

The corresponding power spectrum is:

$$S_{in}(\omega) = \langle I \rangle^2 \left[\delta_\omega + d\Lambda\left(d\frac{\omega}{2\pi} \right) \right]$$

where δ_ω is the Dirac distribution, and $\Lambda(x)$ is the triangle function defined as:

$$\Lambda(x) = 1 - |x| \quad for\,|x| \le 1$$

The wavelet variance Equation (6.36) is:

$$\langle w_{2^j}^2 \rangle = \frac{1}{2\pi} \int \langle I^2 \rangle \left[2\pi\delta + d\Lambda\left(d\frac{\omega}{2\pi} \right) \right] |H_j(\omega)|^2 \, d\omega$$

By the sifting property of δ and the definition of Λ the integral reduces to:

$$\langle w_{2^j}^2 \rangle = \frac{\langle I^2 \rangle}{\pi} \int \left[d\left(1 - d\frac{\omega}{2\pi} \right) \right] |H_j(\omega)|^2 \, d\omega \tag{6.37}$$

$$\langle W_{2^j}^2(x) \rangle = \langle I \rangle^2 \frac{1}{\pi} \int_0^{\frac{2\pi}{d}} \left(d\left(1 - d\frac{\omega}{2\pi} \right) \right) |H_j(\omega)|^2 \, d\omega$$

Results of the numerical evaluation of the integral in Equation (6.37) for 4 dyadic scales, and $\langle I \rangle = 1$ are reported in Figure 6.20. The family of curves in the graph are parametrized by four values of the speckle correlation length (d = 1, 2, 4, 8) and marked respectively with symbols circle, square, diamond and triangle. The following observations can be made.

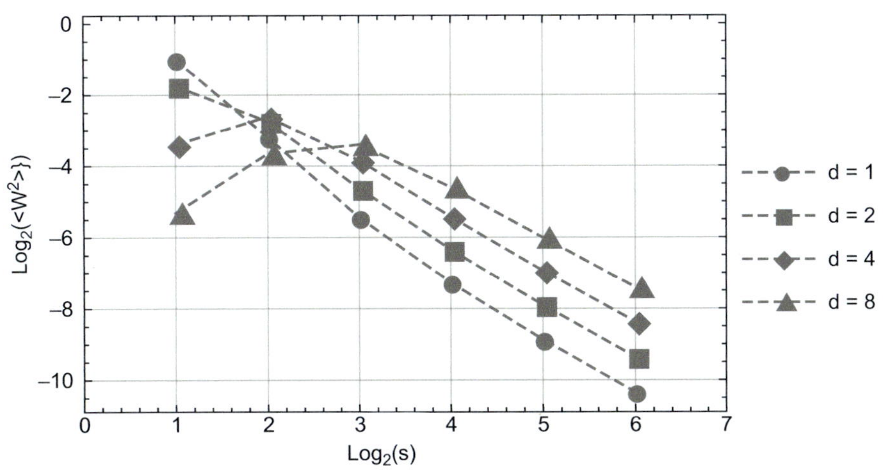

FIGURE 6.20 Wavelet variance for correlated 1-Look speckle as a function of four dyadic scales (log-log plot). The family of curves corresponds to different speckle correlation length d (d=1 circle, d=2 square, d=4 diamond, d=8 triangle).

If the scale is greater than the speckle correlation length, then the wavelet variance decreases with scale. If the scale is less than the correlation length, then the wavelet variance increases. The turning point occurs at a scale equal to the correlation length. This fact indicates that the wavelet variance is a measure of the auto-correlation properties of the input random process, therefore care must be taken in choosing the proper scale where speckle will not dominate over genuine textural features.

For correlation parameter $d = 1$ the wavelet variance decays by only 40% from scale 2 to scale 4. For $d = 2$ (a common case in SAR imagery), the wavelet variance at scale 4 decays by 72% with respect to scale 2 but it is higher than the corresponding variance for $d = 1$. Therefore the first two dyadic scales in the decomposition are still heavily influenced by speckle.

For multilook correlated speckle the autocorrelation function is:

$$R(x) = \langle I \rangle^2 \left[1 + \frac{1}{nl} \sin c^2 \left(\frac{x}{d} \right) \right]$$

where nl is the number of Looks.

The wavelet variance is:

$$\left\langle w_{2^j}^2 \right\rangle = \frac{\langle I \rangle^2}{\pi} \int_0^{\frac{2\pi}{d}} \frac{d}{nl} \left(1 - d \frac{\omega}{2\pi} \right) \left| H_j(\omega) \right|^2 d\omega \tag{6.38}$$

Therefore, the results are the same as in the case of single-Look speckle, but are scaled by the number of Looks.

Let us now go on to consider the smooth signal terms $\langle S^2 \rangle$ and $\langle S^4 \rangle$ that are needed in the expansion Equations (6.25) and (6.31).

$\langle s^2 \rangle$ may be computed by integration as in Equation (6.35), but using a cascade of low pass filters in the filter bank.

$$\left\langle s_{2^j}^2(x) \right\rangle = \frac{1}{2\pi} \int \left(\langle I \rangle^2 \left[2\pi\delta_\omega + d\Lambda \left(d \frac{\omega}{2\pi} \right) \right] \right) \left| Q_{2^j}(\omega) \right|^2 d\omega$$

$$Q_{2^j} = \prod_{i=1,j} G_i$$

An approximation to $\langle s^4 \rangle$ is obtained by assuming that the distribution of s tends to normality. In this case the fourth raw moment is:

$$\left\langle s^4 \right\rangle = 12 \left(\left\langle s^2 \right\rangle - \left\langle s \right\rangle^2 \right)$$

The mean value $\langle w^2 \rangle$ is obtained numerically by the integral Equation (6.37).

An approximation to $\langle w^4 \rangle$ is obtained assuming that it is the fourth raw moment of a variable w that is distributed according to a Laplacian PDF with known raw second moment: $\langle w^4 \rangle = 5\langle w^2 \rangle^2$.

Given these positions and assuming correlated 16-Looks speckle, numerical evaluation gives the results reported in Figures 6.21 and 6.22. In Figure 6.21 the ratio of the estimator variance to the estimator expected value for 32 samples in the averages is graphed as a function of scale. The family of curves in the graph are parametrized by four values of the speckle correlation length (d = 1, 2, 4, 8) (see legend in figure). It is to be noted how the variance of the estimator for long correlation lengths increases at first with scale, and features a maximum at the scale that best matches the correlation length.

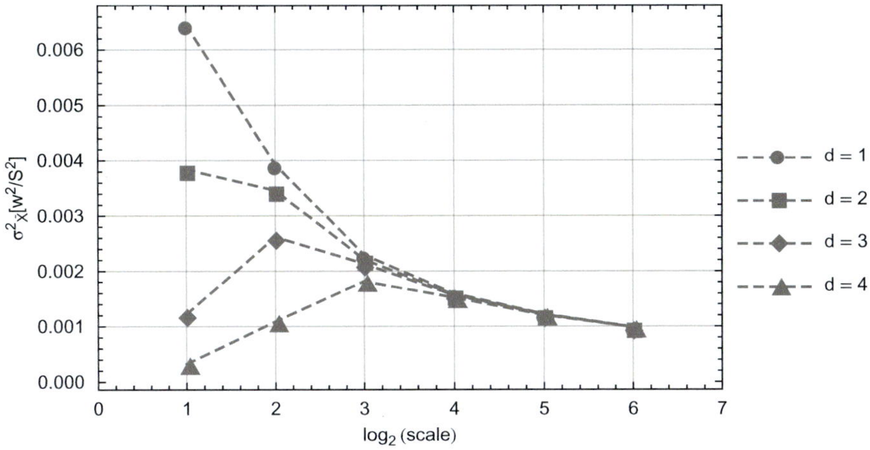

FIGURE 6.21 Normalized variance $\sigma_{\bar{X}}^2$ of the estimator $\langle W^2/S^2 \rangle$ for 16 -Looks correlated speckle as a function of scale, and for 32 samples in the estimator. The family of curves corresponds to different speckle correlation lengths *d* (see legend).

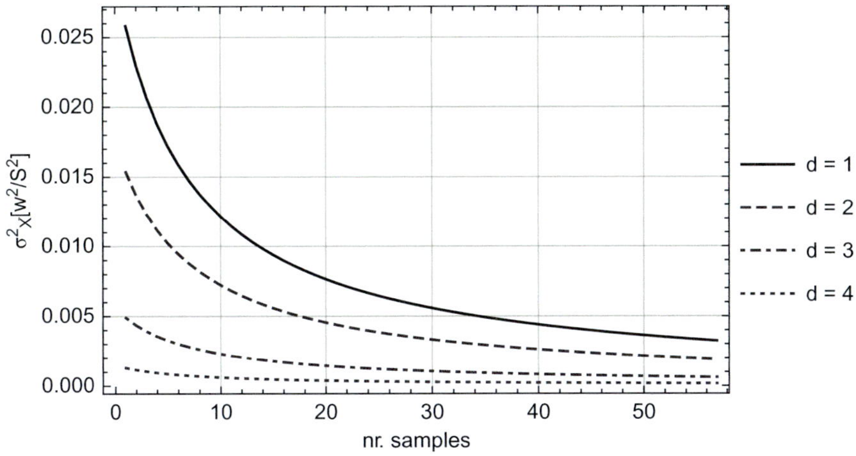

FIGURE 6.22 Normalized variance of the estimator $\langle W^2/S^2 \rangle$ at scale 2 for 16-Looks correlated speckle as a function of the number of samples used in the estimator. The family of curves is parametrized by the speckle correlation length (*d*=1 solid line, *d*=2 dashed, *d*=3 dot dashed, *d*=4 dotted).

In Figure 6.22 the normalized variance at scale 2^1 is graphed against the number of samples used in the estimator. The curves are parametrized by the speckle correlation length (see Figure caption).

6.7 TOOLS FOR TEXTURAL ANALYSIS OF SAR RANDOM FIELDS

Enough of the theory. Let us now move on to practice and set up tools that will enable us to perform textural measurements of SAR random fields.

6.7.1 A Multi-Voice Discrete Wavelet Transform

The basic machinery is built around a discrete wavelet transform (DWT) that maps the SAR random field (space domain) into the scale–space (frequency–space) domain of the wavelet coefficients.

From the statistic of these coefficients we shall then derive our textural measures. The DWT is implemented with an à trous algorithm, with oversampling of the translation parameter and with a multi-voice scheme. We recap first the main features of this implementation (see also the Appendix).

One way to look at the discrete signal representation in the wavelet frame basis by the à trous algorithm is to realize that an à trous filter is an interpolator (see [31]). Consider the continuous wavelet transform where the variables scale $s = 2^m$ and translation b are discretized in the domain of natural numbers ($m, b \in Z$). At a given scale $s = 2^m$, the wavelet coefficient are obtained by a discrete convolution in the translation space b of the signal with a dilated version of the wavelet frame at scale 2^m:

$$\psi_{m,n}\left(x\right)=2^{-\frac{m}{2}}\psi\left(2^{-m}x-nb\right)$$

The dilated frame is obtained by inserting a suitable number of zeros between the wavelet samples, and then interpolating the missing values using the à trous filter.

The decomposition by the à trous algorithm can also be seen as an iterated filter bank scheme (see [31]):

$$\begin{aligned} W_1 &= S_0 \otimes g_0 \\ S_1 &= S_0 \otimes h_0 \end{aligned}$$

$$\left.\begin{aligned} W_{j+1} &= S_j \otimes \dddot{g}_j \\ S_{j+1} &= S_j \otimes \dddot{h}_j \end{aligned}\right. \; j = 1 \cdots n \tag{6.39}$$

where, S_0 is the input signal, g_0 is the high pass filter associated with the wavelet frame, h_0 is the low-pass à trous filter, $\dddot{g}_j$ is a dilated filter where 2^{j-1} zeros are inserted between the samples of the g_0 filter, $\dddot{h}_j$ is a dilated filter where 2^{j-1} zeros are inserted between the samples of the h_0 filter, W_{j+1} are the wavelet coefficients at scale 2^j, and S_{j+1} is the smooth (low pass filtered) signal at scale 2^j.

When dealing with two-dimensional (2D) regions within a field, an extension of the decomposition in Equation (6.39) is adopted to implement a 2D discrete wavelet transform. Two wavelet frames are defined for two directions (x, y) of the region (rows and columns in a matrix representation): $w_x(i,j,s)$, $w_y(i,j,s)$ [20,32]. The impulse response of the associated high pass filter is separable: $h(\omega) = h_x(\omega)\, h_y(\omega)$. Therefore, the two-dimensional convolutions required by the 2D filter bank are realized by successive 1D convolutions, the first along the x disrection and the second along the y direction:

$$\begin{aligned} W_{1x} &= S_0 \otimes_x g_0 \\ W_{1y} &= S_0 \otimes_y g_0 \\ S_1 &= S_0 \otimes_x \otimes_y h_0 \end{aligned}$$

$$\begin{aligned} W_{j+1,x} &= S_j \otimes_x \dddot{g}_j \\ W_{j+1,y} &= S_j \otimes_y \dddot{g}_j \; j = 1 \cdots n \\ S_{j+1} &= S_j \otimes \dddot{h}_j \end{aligned}$$

where $\otimes_x$ is the convolution by row operator, and $\otimes_y$ is the convolution by column operator.

We have implemented a variant of the à trous algorithm with 4 voices per octave. The design of the multi-voice scheme entails the following steps. For each voice in the first octave, a fractionally dilated wavelet is computed in the frequency domain from the Fourier transform of the wavelet:

$$\hat{\Psi}(\omega) = -i\frac{\omega}{4}\left(\frac{\sin\left(\frac{\omega}{4}\right)}{\frac{\omega}{4}}\right)^4 e^{-i\frac{\omega}{2}}$$

$$\Psi_{\text{dil}}(\omega) = \frac{1}{\sqrt{2^d}} 2^d \Psi\left(2^d\omega\right) e^{-i\omega(\frac{1}{2}\left(2^d-1\right)}$$

where the fractional dilation factor is $d = 0.25\,k$, $k = 0\cdots3$. The dilated wavelet is normalized by $\frac{1}{\sqrt{2^d}}$ to preserve the norms. It is shifted by $e^{-i\omega(\frac{1}{2}\left(2^d-1\right)}$ in such a way as to match the zero crossings of the original mother wavelet.

Importantly, the multi-voice scheme using the à trous algorithm can only be implemented with $s^{-1/2}$ normalization to assure equal norms among the mother wavelets that sample the frequency lattice within one octave. In this way, wavelet coefficients corresponding to all voices and octaves carry comparable energy, and can be used in forming measures, e.g. two-point statistics of the signal.

The approximation of the continuous space wavelet ϕ by means of an interpolator filter is expressed in the time and frequency domain by:

$$\frac{1}{\sqrt{2}}\psi\left(\frac{t}{2}\right) = \sum_n h[n]\phi(t-n) \tag{6.40}$$

$$\frac{2}{\sqrt{2}}\Psi_{\text{dil}}(2\omega) = H_j(\omega)\Phi(\omega) \tag{6.41}$$

The high pass filter coefficients for each fractionally dilated wavelet are computed from Equation (6.41) by inverse discrete Fourier transform (DFT):

$$h_j[n] = \frac{1}{2\pi}\int_{-\pi}^{+\pi} H_j(\omega)e^{i\omega n}d\omega$$

where j is the voice index.

The dilated high pass filter h_j for each voice is finally used in a recursive à trous decomposition scheme to generate wavelet coefficients at the corresponding dyadic scales 2^{d+j} $j = 1..\,n$:

$$c(s,n) = h_s(n)\otimes f(n)$$
$$h_s = g\otimes\left(Dh_{s-1}\right)$$

where $c(s, n)$ are wavelet coefficients at level (scale) s and position n, $\otimes$ is the convolution operator, $f(n)$ is the input signal, h_s is high pass filter at level s, obtained by interpolation of the filter at level s–1 up-sampled by the dilation operator D.

6.7.2 Wavelet Signatures

As we have documented from the theoretical standpoint, the functional dependence of a wavelet statistic on scale (e.g. wavelet variance) encapsulates the statistical properties of the underlying random process. In our case, the interest is then to associate this dependence with the natural target structure that is portrayed in turn by the radar field.

In general, a wavelet statistic is defined in a scale–space domain, i.e. it depends on the scale of decomposition and the spatial coordinates (i, j) of the field:

$$\wp DWT = f\left(s,i,j\right)$$

If the estimation of the wavelet statistic is performed by averaging over the coordinates of a region of interest $\Re$, this being defined by supervised inspection of the data field, then we single out the dependence on scale. In this case, we call the analysis tool a wavelet scaling signature, in short a WASS [33]. These signatures are graphed in a log_2-log_2 plot, therefore the dependence is on the scale index m.

We shall consider the following wavelet statistics:

1. Wavelet variance

$$\left\langle c_x\left(j,2^m\right)^2\right\rangle_{\mathcal{R}} = f\left(2^m\right) \tag{6.42}$$

$$\left\langle c_y\left(j,2^m\right)^2\right\rangle_{\mathcal{R}} = f\left(2^m\right) \tag{6.43}$$

$$\left\langle c_x\left(i,j,2^m\right)^2 + c_y(i,j,2^m)^2\right\rangle_{\mathcal{R}} = f\left(2^m\right) \tag{6.44}$$

$$\left\langle \tan^{-1}(c_x\left(i,j,2^m\right),c_y(i,j,2^m))\right\rangle_{\mathcal{R}} = f\left(2^m\right) \tag{6.45}$$

The variance in Equations (6.42) and (6.43) provides a measure along one direction of the field, this being, for example, range or azimuth in a slant range geometry, or latitude or longitude if the dataset was geocoded. If the field is anisotropic these signatures will be different and will afford information on this spatial property.

The variance in Equation (6.44) is tantamount to an estimate of the field gradient modulus squared. The statistic in Equation (6.45) provides an estimate of the gradient angle. It is used to detect the presence of particular directions along which strong field variations do occur.

2. Wavelet covariance

$$\left\langle c_{x1}\left(i,j,2^m\right)c_{x2}\left(i,j,2^m\right)\right\rangle = f\left(2^m\right) \tag{6.46}$$

$$\frac{\left\langle c_{x1}\left(i,j,2^m\right)c_{x2}\left(i,j,2^m\right)\right\rangle}{\sqrt{\left\langle c_{x1}\left(i,j,2^m\right)\right\rangle^2\left\langle c_{x2}\left(i,j,2^m\right)\right\rangle^2}} = f\left(s\right) \tag{6.47}$$

where x_1, x_2 are two processes. The normalized version in Equation (6.47) is also called the wavelet correlation [17].

This signature is used to assess the spatial similarity between two random fields, or how the two fields are related on a scale-by-scale basis.

3. Wavelet normalized 4th moment (kurtosis)

$$\frac{\left\langle c_x\left(j,2^m\right)^4\right\rangle}{\left\langle c_x\left(j,2^m\right)^2\right\rangle^2} = f\left(2^m\right) \tag{6.48}$$

This statistic, also called the flatness factor, is a measure of intermittency in the random field. Intermittency (the presence of short but intense bursts within the otherwise stationary process) generates large isolated values with high derivatives that will populate the tails of the wavelet coefficients' probability distribution. The fourth moment, being proportional to the flatness of the distribution's tails, is therefore taken as an indicator of the presence of intermittency.

Regarding the wavelet variance, interpretation of the scaling signature can be guided by considerations of the statistical properties of the input random process, such as the autocorrelation function and the one-point moments. This way of proceeding is supported by the analysis expounded in Section 6.5. However, there is an alternative way of interpreting the signature, based on the concept of the partition of the input process energy with scale (instead of correlation properties).

To illustrate this approach, let us consider again the link between the wavelet variance and the spectral density of the input process:

$$\left\langle w(s)^2\right\rangle = \mathcal{R}(\tau=0) = \frac{1}{2\pi}\int S_{out}(\omega,s)\,d\omega = \frac{1}{2\pi}\int\left|s^{-1/2}\,\hat{w}_s(\omega)\right|^2 S_f(\omega)$$

The wavelet high pass filter $\left|s^{-1/2}\,\hat{w}_s(\omega)\right|^2$ slices the spectral density of the input process within a range of frequencies that is inversely proportional to the wavelet dilation s (see Figure 6.23). Therefore, the higher the value of the signature at a given scale, the higher the energy content of the input process within a range of frequencies that is inversely proportional to the scale. This criterion is useful when comparing the signatures of two processes, because it points out the relative strength in terms of variance of the processes.

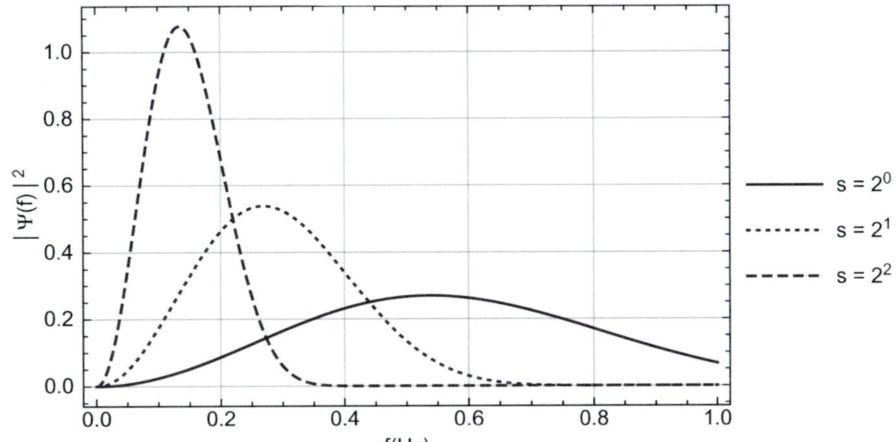

FIGURE 6.23 Wavelet frequency response $H(f)$ for dilation scale 2^0, 2^1, 2^2.

6.7.3 Wavelet Spectra

The WASS analysis, where the expected values of the wavelet statistic (e.g. variance) are obtained by averaging over a region of interest, characterizes the scaling property of the random field within a homogeneous (stationary) area at a given location. To take into account variations in space of the random field properties (e.g. caused by forest fragmentation or degradation), we perform local (in space) point-estimates of the wavelet statistics. These estimates are possible because of the good wavelet space-frequency resolution. These wavelet statistics in the space–scale domain are called wavelet spectra. Spectra are obtained by convolution of the wavelet coefficients with a smoothing spline filter. Thus the wavelet variance spectrum (in one dimension and two dimensions) is:

$$Ws\left(j,2^m\right) = c_x\left(j,2^m\right)^2 \otimes \beta\left(\mathrm{j}\right) = f\left(2^m,\mathrm{j}\right)$$

$$Ws\left(i,j,2^m\right) = \left(c_x\left(i,j,2^m\right)^2 + c_y(i,j,2^m)^2\right) \otimes \beta\left(\mathrm{i}\right) \otimes \beta\left(\mathrm{j}\right) = f\left(2^m,\mathrm{i,j}\right)$$

where ⊗ is the convolution operator, and β is a spline smoothing filter.

An alternative representation can be built by considering the wavelet spectrum as a function. In this case, each point in the discrete spectrum field will be a continuous function $F(s)$ of the scale variable:

$$Ws\left(i,j,s\right) = F\left(f_{i,j}\left(s\right)\right)$$

with $f_{i,\,j}(s)$ being interpolating functions of the local wavelet variance estimates. WASS analysis suggests that a third-degree polynomial provides a good estimate of the wavelet variance trajectory in scale. Thus:

$$f_{i,j}\left(s\right) = a_0 + a_1\,s + a_2\,s^2 + +a_3\,s^3$$

The functions $f_{i,\,j}(s)$ are represented in the discrete setting by the four a_i polynomial coefficients, while the functions $f_{i,\,j}(s)$ are reconstructed when manipulating the spectrum by sampling linearly the scale variable by an interval δs in the $\log_2 s$ domain.

Spectra, being functions of space and scale, pave the way, as we shall see later, for the implementation of textural classifiers.

6.8 WASS Analysis of SAR Backscatter Fields

We now move on to the application of the concepts presented in the previous sections, and in particular show how WASS analysis is applied to SAR backscatter fields.

6.8.1 Lowland Rainforest and Swamp Forest Signatures in ERS-1 Data

We start with a test case that highlights neatly how a textural measure can provide added value with respect to a measure based only on backscatter intensity.

Let us consider two thematic classes of importance in the tropical ecosystems of Central Africa: the lowland rainforest, and the swamp forest. These forest types have already been described in Chapter 3, dealing with the GRFM mapping experiment of Central Africa. More details from the ecological standpoint can be found in [34]. The main characteristics of interest for us here are the following.

FIGURE 6.24 The upper canopy of the lowland rainforest (right) and the swamp forest (left) as seen from above, showing a notable difference in the spatial structure. (Courtesy of the Nouabalé-Ndoki project in Congo-Brazzaville.)

The lowland rainforest is the climax ecosystem in the Central Congo basin. The upper stratum (35–45 m) of this formation is composed of a few evergreen or briefly deciduous species, whose age distribution is quite large. Age distribution makes the upper layer quite irregular, and the canopy, observed from high above, appears discontinuous and rolling.

Swamp forest formations are influenced by the presence of frequent floods and soils with poor drainage. The upper canopy layer (up to 45 m in some cases), composed of a small number of species, is quite homogenous compared to the lowland rainforest (Figure 6.24).

Let us now resume the analogy proposed at the start of this chapter, and imagine laying a virtual blanket on top of the upper canopy of the two formations. In the case of the swamp forest, we shall obtain some regular and homogeneous wavy patterns, while in the case of the lowland rainforest there will be a bumpy layout because of the steep and intermittent variations in canopy height of the different species.

Our backscatter model (see Section 6.2) will then suggest that these different spatial patterns will also be reflected in the backscatter field, if the field is generated by short wavelength and sufficiently high spatial resolution observations.

Let us verify that the WASS signature is sensitive to these different spatial patterns, by providing measures of the different spatial statistics. For the purpose, we selected an ERS-1 Precision Image Product (PRI) dataset (ground range amplitude VV data, 12.5 m resolution, 3 Looks, incidence angle 23°at mid-range) covering a site in the Congo River floodplain in Central Africa 0° 20′ 13″ *S*; 20°06′ 48″ *E*. The data set was acquired in 1997 in the context of the joint European Space Agency–Joint Research Centre (ESA–JRC) Central Africa Mosaic Project (see Chapter 1).

The backscatter image down-sampled to 200 m is shown in Figure 6.25. Two regions of interest were selected using expert knowledge corresponding to areas of the swamp (SF) and lowland forest (LF). The estimated backscatter mean values relative to the two areas of interest are SF $\sigma^\circ = -5.55$ dB and LF $\sigma^\circ = -5.49$ dB. Being the difference of the two values within the speckle noise variance, it would be impossible to discriminate between the two classes based only on non-contextual information, i.e. spatial statistic, such as the one afforded by the wavelet variance.

Let us now see how the WASS analysis performs in this case. The signature for the two classes and variance in the range and cross-range directions is reported in Figure 6.26. In the plot, the variance samples at discrete fractional scales are marked with round and square symbols. Moreover, the continuous lines corresponds to a 3rd degree polynomial fitting of the samples.

Observe that the functional dependence of the wavelet variance on scale is similar for both classes, and is characteristic of a stationary random process with an exponential autocorrelation function (see Section 6.5). Yet the LF signature is consistently higher than the SF one, indicating that at all scales the energy of the LF random process is stronger than the SF one. In other words, the two classes are texturally well separated, at least for a global measure over an area of interest.

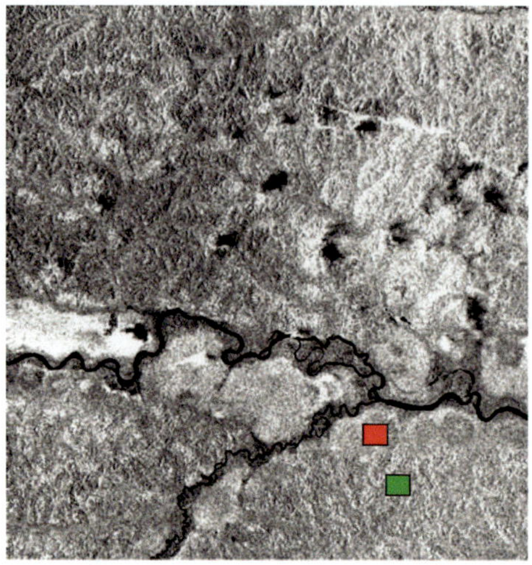

FIGURE 6.25 ERS-1 image acquired over the Congo River Central Floodplain (ESA–JRC Central Africa Mosaic project). Two regions of interest are marked, corresponding to the swamp forest (red box) and the lowland forest (green box).

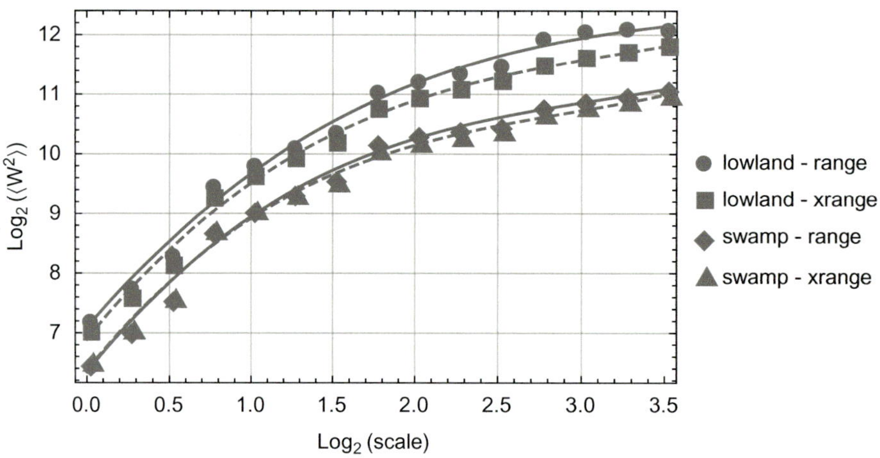

FIGURE 6.26 WASS variance signature computed by averaging over the two regions of the ERS-1 backscatter data set corresponding to the swamp and lowland rainforest. The wavelet variance in the range and cross-range direction are compared in the graph.

We shall see later that, by introducing the concept of wavelet spectra, i.e. local textural measures, we shall also be able to discriminate between the two classes down to the resolution of a few data elements and in an unsupervised way.

Finally, notice that the difference between the variance in range and cross-range directions is small for the swamp class, and increases for the lowland class. We conclude that the SF field is more homogeneous but also more isotropic than the LF one.

Let us next have a look at the WASS flatness factor ($FF(s)$) signature. We remind the reader that for a wide sense stationary WSS (homogeneous) Gaussian random process (i.e. no intermittency) $FF(s) = 3.0$ independent of scale. The SF signature indicates therefore a stationary (homogeneous) process up to scale 2.5. Estimation of $FF(s)$beyond this scale is not possible because of the size of the regions of interest (ROIs). Whereas, the LF process shows a slightly higher $FF(s)$, a sign of the presence of small-scale non-homogeneities within the field (Figure 6.27).

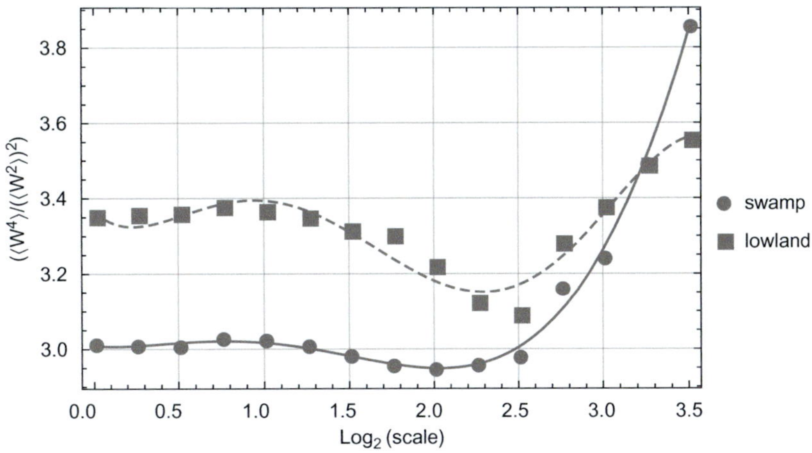

FIGURE 6.27 WASS flatness factor signatures for the two regions of interest (ROIs) corresponding to the swamp and lowland rainforest classes.

6.8.2 TanDEM-X Signatures in the same Thematic Context

In a second WASS experiment with the same thematic flavor we use a TanDEM-X dataset acquired over a site in the Democratic Republic of Congo, centered around the Lulanga River, Basankusu (1°13′20″N, 19°48′10″E). Data were provided by DLR in the framework of the announcement of opportunity AO2010, project VEGE0330. The X-band data were acquired with incidence angle $\vartheta = 23.3°$, baseline 150.9 m, and were processed to achieve a a ground range pixel size of 4.6 m. Therefore in this experiment we use SAR data still with short wavelength (X -band) but higher resolution and dual- polarization mode (HH, HV).

The HH TanDEM-X backscatter image is shown in Figure 6.28. North of the river Lulanga, a large area of intact dense and homogeneous riverine forest can be seen. The town of Basankusu can be picked out south of the river because of the bright spots, corresponding to buildings, and the low backscatter runway of the airport. South and to the left of the town there is a large area that was converted from primary forest for agricultural practice. The area consists of a mosaic of cultivated fields, bare soil and patches of forest regrowth. Remnants of the primary forest can be spotted in different parts of the image.

Four regions of interest were selected for the WASS analysis (see markers over image). These correspond to areas of the swamp forest (SF), intact forest (IF), degraded forest (DF), and a cassava nursery (NU). Cassava is one of the most important crops in the tropics, being rich in calories and used for human food. The plants are cultivated by nursery transplanting. The nursery is maintained in a corner of a larger field in a raised bed where the plants are arranged in a regular and close pattern. In the radar image the nursery appears as a low backscatter clear cut, with a regular array of stronger returns resulting from the single plants.

We start with a number of cases using data at HH polarization, and will later discuss the difference between WASS signatures using HV data. The variance signatures for the swamp forest and the intact lowland forest IF regions are shown in Figure 6.29. The functional dependence on scale is similar to the one observed in the case of the ERS-1 data, i.e. characteristic of stationary random process with exponential ACF. The correlation length (point of asymptotic reach) is slightly shorter for SF (scale=2^3 or 37 m) than for IF (scale=2^4 or 73 m). The correlation lengths are in agreement with the crown sizes, these being smaller for the SF. As in the case of ERS data, the LF signature is consistently higher than that of the SF, pointing to the relatively higher smoothness of the field associated with SF.

Comparison between the variance signatures of the IF and DF is reported in Figure 6.30. The DF is one step higher in the ladder of variance (or process energy) at all scales with respect to SF

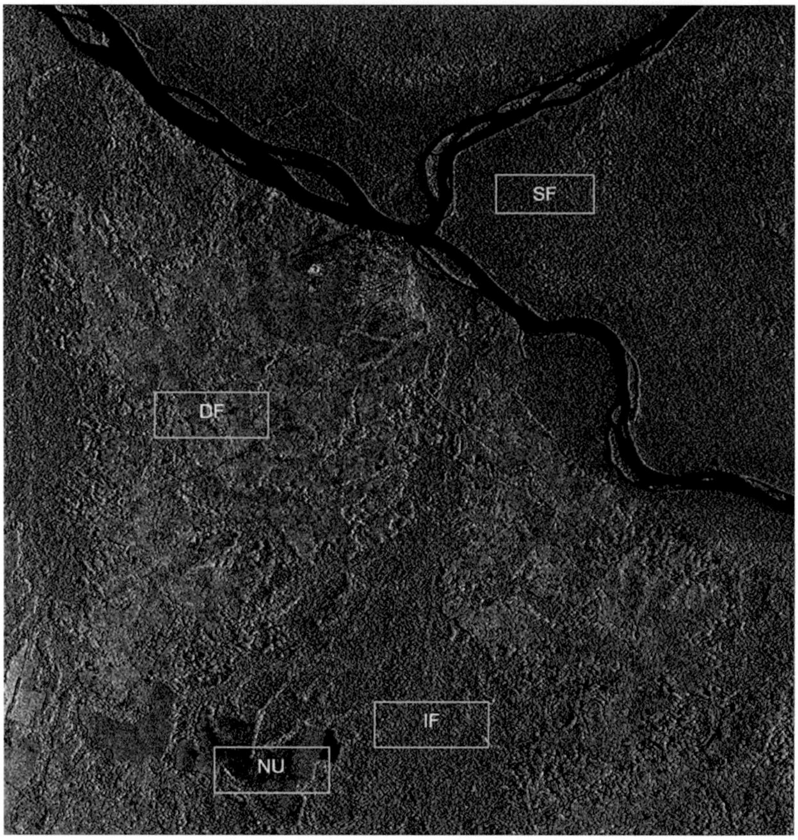

FIGURE 6.28 TanDEM-X HH backscatter image acquired in the Democratic Republic of Congo, near the town of Basankusu. Four regions of interest were selected for the WASS analysis (see markers over image). These correspond to areas of the riverine forest (SF), intact forest (IF), degraded forest (DG), and a cassava nursery (NU). (Courtesy DLR AO project.)

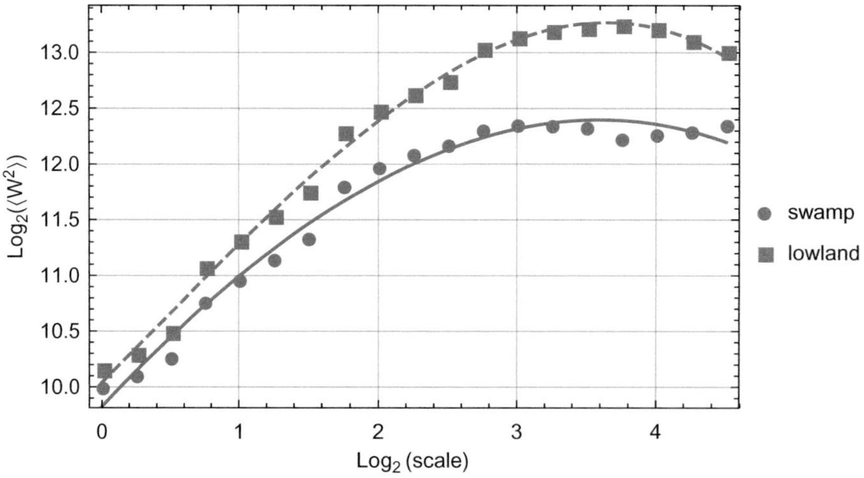

FIGURE 6.29 WASS variance signatures for the SF and IF regions (HH polarization).

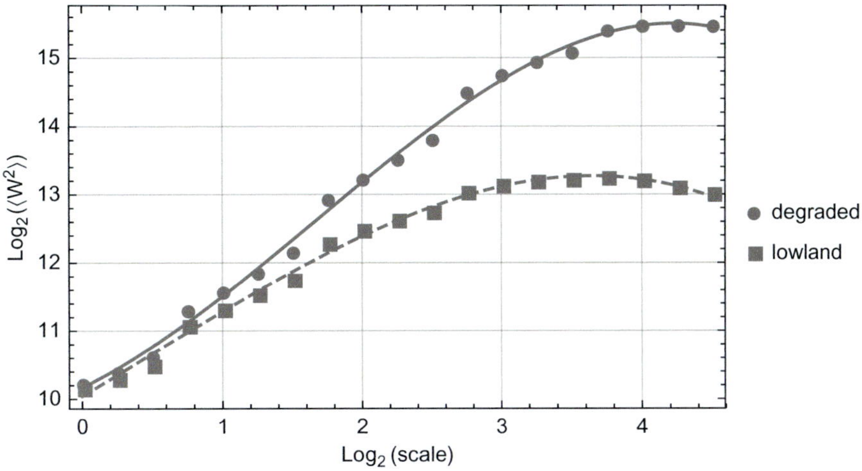

FIGURE 6.30 WASS variance signatures for the IF and DF regions (HH polarization).

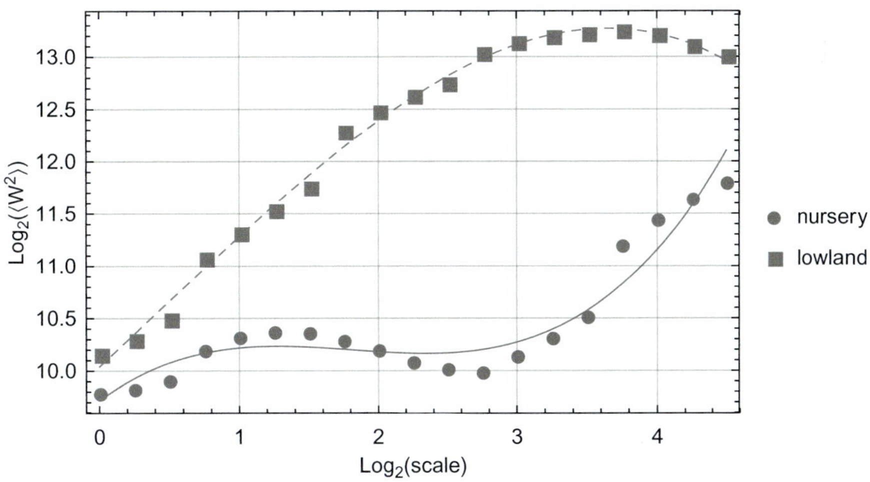

FIGURE 6.31 WASS variance signatures for the IF and NU regions.

and IF. Moreover, the difference between the two signatures increases with scale, up to the point of asymptotic reach, this happening on a slightly longer scale for the DF.

Comparison between the NU and the IF WASS is shown in Figure 6.31. Approximately up to scale 2^3 the NU signature is almost flat with respect to scale, i.e. the dependence expected for white noise. Therefore, at short scales the field appears as a smooth noisy process, as expected for the return from bare soil. After scale 2^3 correlations start to appear, signaling the onset of the backscattered signal returned from the plants at regular intervals. At all scales the variance associated with IF is higher than the NU. All in all, regions of the mosaic with cultivated fields appear much less textured than regions within the forest.

We next examine the eventual textural differences between HH and HV fields. For the purpose, we exploit the wavelet covariance. As explained in Section 6.7.2, this statistic provides a measure of the textural similarity between two fields. The covariance WASS signatures for the four regions of interest (SF, IF, DF, NU) is shown in Figure 6.32.

The trend with scale of the covariance signatures for all classes is similar to that of thevariance for the respective classes. In other words, the two processes (HH, HV) behave in the same way in

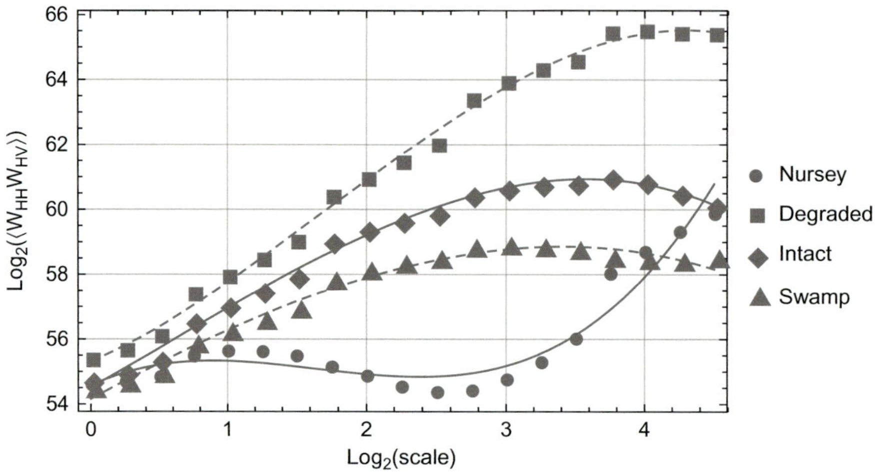

FIGURE 6.32 Wavelet covariance between HH and HV fields for the four regions of interest (SF, IF, DF, NU).

the scale domain. Yet there are notable differences in energy (variance) of the processes at all scales and depending on the class.

In particular, the NU presents lower covariance in the range of scales where surface scattering from bare soil dominates ($s = 2°, 2^3$); while, the covariance increases rapidly when scattering from the plant canopy weighs in $s > 2^3$. In this range of scale the HV and HH fields are largely spatially dissimilar, because the dominating scattering mechanisms at HV and HH are different (volume scattering, and double bounce stalk-terrain).

Of the forest classes, the SF presents the least covariance, while the covariance is larger for IF and even more so for DF. Understandably, passing from SF to DF means that the forest cover becomes increasingly more heterogeneous. As a consequence, the micro-topography effects from reflections off the top layers of the crowns tend to dominate at HH backscatter over variations of volume scattering at HV backscatter.

6.8.3 Intact and Degraded Forest Detection by Functional Analysis of WASS Signatures

A case study for characterizing the degraded forest in Cameroon by SAR backscatter WASS was conducted using ENVISAT ASAR and ALOS PALSAR data. The experiment is described in detail in [35]. We take a leaf from this study to discuss here a way of extracting information on the underlying random field spatial statistic by functional analysis of the WASS. It has been proven that differential properties of a WASS fitting function provide a more robust condition to discriminate two thematic classes (e.g. intact and degraded forest) rather than a point-wise distance measure between the signature points in the scale domain.

We consider here four WASS signatures estimated on regions of interest (ROIs) defined by supervised inspection within an ENVISAT ASAR amplitude dataset. The dataset was geocoded to WGS84 UTM projection (zone 33N) at 15-m pixel spacing, set using a 90-m SRTM DEM. Importantly, the dataset was speckle filtered using a multi-temporal filter based on the principles proposed in [36] (for a similar technique, see also Chapter 9). In this way, the influence of speckle on the WASS error budget was minimized.

The four ROIs (40 x 40 pixels each) correspond to the thematic classes intact forest (IF), degraded forest (DF), savanna and grassland (FS), secondary forest and agriculture mosaic (FAM). The WASS signatures are shown in Figure 6.33 together with the corresponding fitting functions (a 3rd degree polynomial). Clearly, discrimination of the classes based, for example, on a distance measure in the scale domain between the signatures would be feasible only by regrouping the classes into a forest

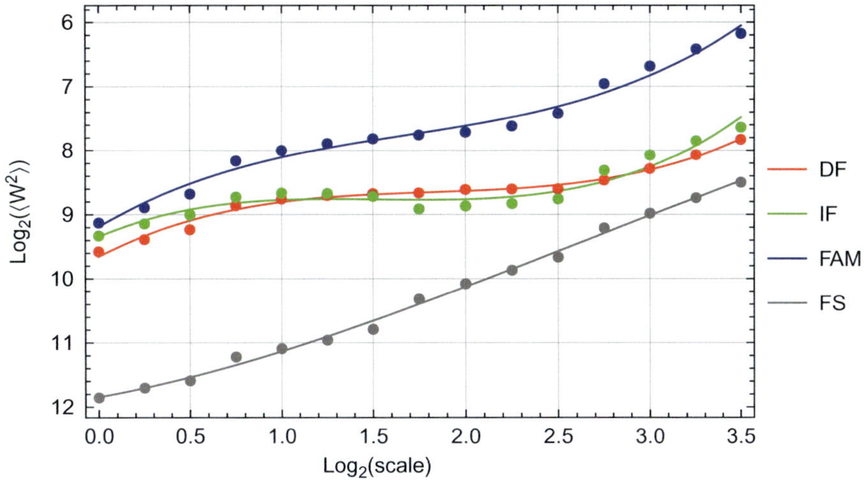

FIGURE 6.33 The WASS signatures for the thematic classes IF, DF, FAM, FS. The fitting functions (3rd degree polynomial) are overlaid.

and non-forest problem, while it would be problematic for handling the important case of degraded/intact forest detection.

Let us suppose, for example, to assume as distance measure the integral of the fitting function:

$$d(i) = \int_0^{3.5} f_i(s)\,ds$$

where $f_i(s)$ is the fitting polynomial of class i.

The $d(i)$ values for the four classes are listed in Table 6.1. Assuming a mean standard estimation error of 0.7 in the $\log_2$ scale, we obtain figures of the class distance by the "effect size" statistic (a measure of the difference between two mean values; see Chapter 10), this being defined as:

$$Ef(i,j) = \frac{|m_i - m_j|}{\sigma}$$

where m_i, m_j are mean values and σ is the pooled standard deviation of the populations.

This simple derivation reinforces the assertion about the difficulty of discriminating the intact and degraded forest classes by a distance measured between signatures.

TABLE 6.1
Distance Measures d(i) and their separability in terms of the "Effect size" Statistic

	DF	IF	FS	FAM
d(i)	−30.39	−30.28	−36.10	−26.92
DF		0.16	8.16	4.95
IF	0.16		8.32	4.79
FS	8.16	8.32		13.11
FA	4.95	4.79	13.11	

Let us now see what can be done by looking closely at the geometry of the signatures, in particular at some differential properties. The two signatures of interest are shown separately for clarity in Figure 6.34. The first derivative of the IF and DF fitting functions are shown in Figure 6.35, while the second derivative is graphed in Figure 6.36.

First, we notice that at scale 2° the DF first derivative is higher than the IF. As discussed in Section 6.5.3, the WASS first derivative at scale 2° is proportional to the correlation length and to the radar cross-section Gamma distribution order parameter υ for the *K*-distributed, exponential cosine random process. This is an indication that the DF field is more textured than the IF.

Second, the minima of the first derivative, which indicate the onset of the sill in the signature, are shifted significantly in scale between the DF and the IF, with the DF happening at longer scale. This feature tells us that the DF process presents more persistent correlations.

The same condition is captured by the zero crossing of the second derivative, with the DF always occurring at longer scales.

The combination of these functional parameters provides a consistent condition to discriminate between the two classes. This proposition was proven by a statistical hypothesis test of the difference of the parameters' mean values using 20 samples for each class of interest (H_0: xDEF – xIF = 0).

For both parameters the H_0 hypothesis was rejected, indicating that there is a significant difference between the means at a 0.05 significance level.

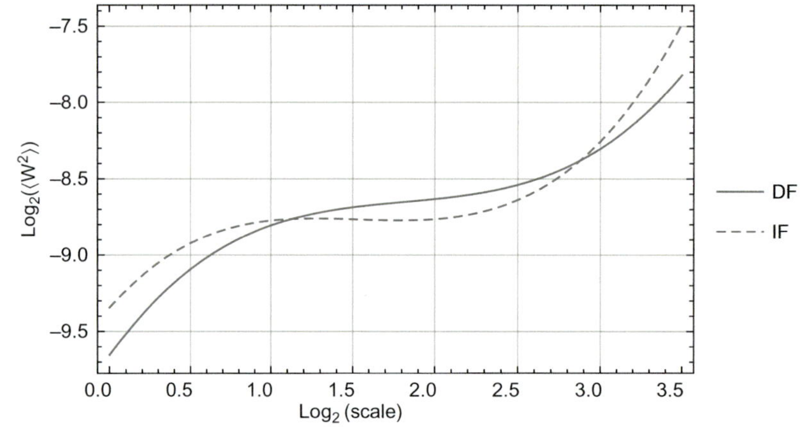

FIGURE 6.34 The IF and DF WASS signatures.

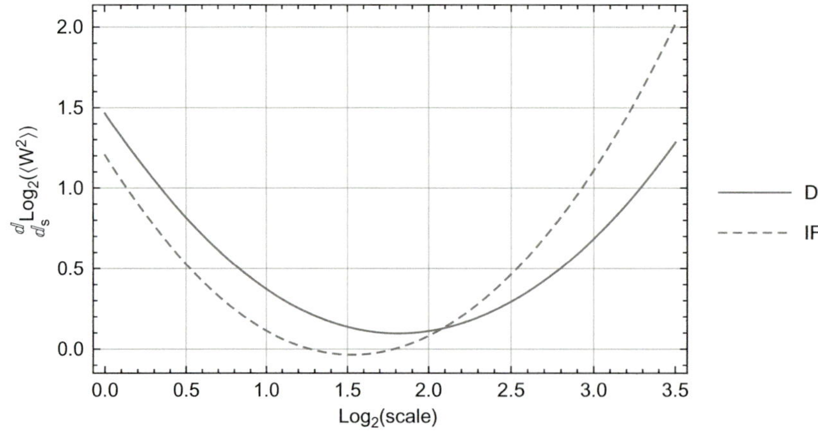

FIGURE 6.35 First derivative of the IF and DF WASS fitting functions.

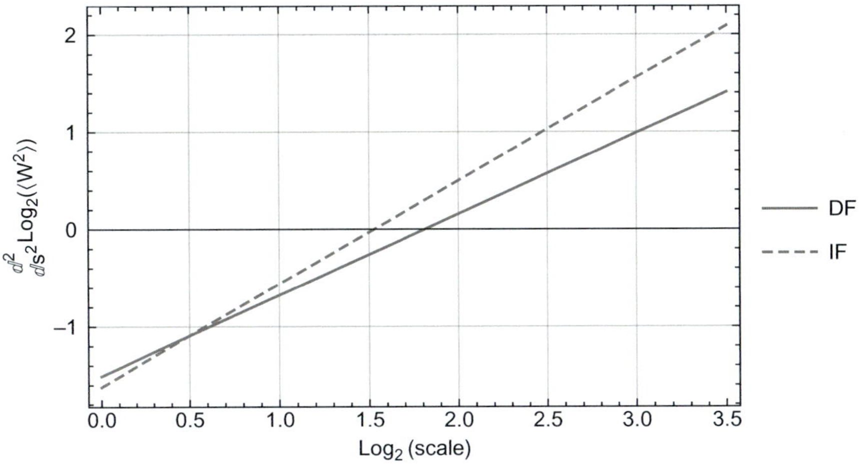

FIGURE 6.36 Second derivative of the IF and DF WASS fitting functions.

6.9 WASS ANALYSIS OF INSAR AND LIDAR DIGITAL SURFACE MODELS

We shall deal in this section with wavelet textural measures of a different kind of fields. Whereas, in the previous discussion, measures were applied to radiation (backscatter coefficient) fields, now fields holding geometric information, elevation, come into play. This change is made possible, in the case of InSAR observations, by an intermediate step, whereby the coherence between two radiation fields is converted into surface height. As anticipated in Section 6.3, when a vegetation volume lies on the ground, we expect that surface height variation will carry information on the forest structure.

For dealing with this matter, we shall lean on a research work that we undertook in the framework of the German Aerospace (DLR) Announcement of Opportunity AO VEGE6702. The work aimed, from the remote sensing methodological standpoint, at investigating the capability of capturing information on the upper forest canopy structure through wavelet space–scale analysis of digital surface models derived from InSAR observations, while the thematic focus was on mapping secondary and degraded forests. Secondary forests are the end-point of a regrowth process enacted by deforestation (clearing of the intact primary forest). Degraded forests are also the result of disturbance events, such as fires, pests and logging roads. Estimates of the total carbon flux to/from the atmosphere associated with these forest dynamic processes are still marked by considerable uncertainty. Hence the importance of this theme.

In particular, our research addressed the discrimination between a primary lowland dipterocarp forest located in the Sungai Wain Protection Forest (East Kalimantan, Indonesian Borneo), and a secondary forest that developed as a consequence of fire disturbance.

Details of the project can be found in [37,38]. We summarize here only some aspects that are relevant for the discussion on the WASS and intermittency analysis, these being the subjects of this section.

TanDEM-X data for the AO project were acquired on 11 December 2014 in StripMap bistatic mode (single polarization HH) at 41° incidence angle. The scene covers 32 km × 25 km (UL Geo: 116°46′14.08″E, 0°59′57.91″S). The data were processed using SARScape 5.0 [39] to generate a digital surface model (DSM) geocoded at 4.6 m resolution to WGS84 datum, Zone 50 South.

LiDAR data were made available by Prof. Dirk Hoekman of Wageningen University, the Netherlands, co-investigator of the AO project. Data were acquired on 14 November 2014 using the Optech ORION M300 system (frequency: 55 Hz). The airborne LiDAR instrument was flown at an altitude of 700 m above ground level, with a scan angle of ±18°, and swath width of 455 m with an overlap of 227 m. The resulting point density was 5.75 points per m^2 (excluding overlap). The vertical accuracy for all LiDAR returns on clear ground was 0.08 m (standard error at 1 sigma). The data covers 9.7 × 0.7 km^2(Site A) and 4.5 × 0.8 km^2 (Site B). The LiDAR products were geocoded in

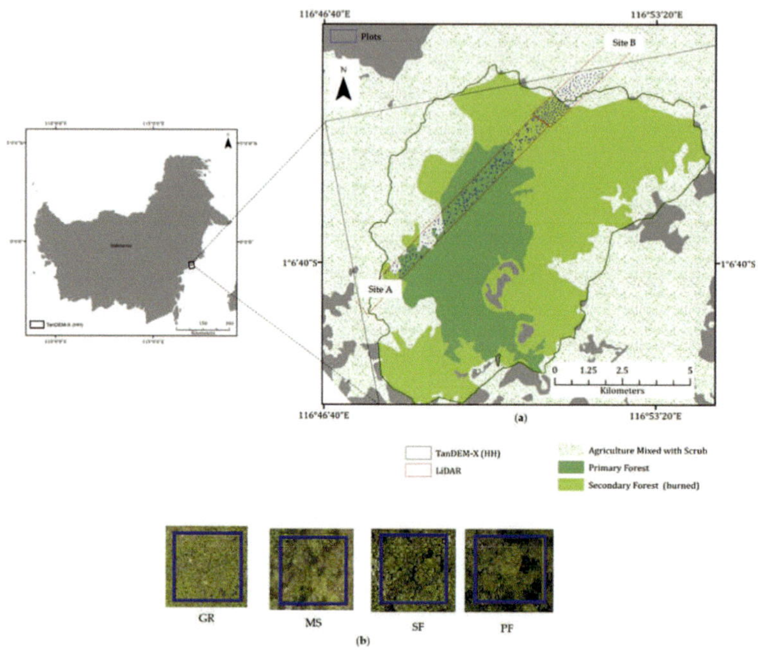

FIGURE 6.37 Location of the digital surface model (DSM) derived from LiDAR data. The data cover 9.7 × 0.7 km² (Site A) and 4.5 × 0.8 km² (Site B). The primary and secondary forest are mapped in site A, while site B includes an area with grassland and mixed shrubs at the margin of the secondary forest.

a UTM projection, Zone 50 South, and to WGS84 datum. Lidar2dems open source software developed by Applied GeoSolutions was used to process the LiDAR data. Three products were derived: a digital surface model (DSM) at 1 m spatial sampling, this product mapping the top forest canopy; a digital terrain model (DTM), this mapping the underlying ground; a canopy height model (CHM), obtained by subtracting the DTM from the DSM. From the thematic point of view, the primary and the secondary forest are present in site A, while site B includes grassland and mixed shrubs at the margin of the secondary forest (see Figure 6.37).

As discussed in Section 6.3, the InSAR DSM over the forest reports a height h_{dsm} that will be the sum of the topographic height and the position of the complex coherence phase center within the vegetation volume. We are interested in variations of the DSM that are associated with spatial patterns of the forest structure. These variations are caused by shifts of the phase center within the vegetation volume. We wonder, therefore, how topographic variations, brought on to the DSM by the phase return from terrain, may influence, or worse obscure, our textural features of interest.

To answer this question, we selected a 6.95 km-long transect in primary and disturbed forest crossing the coverage of the LiDAR CHM in test Site A. A section of the TanDEM-X DSM is shown in Figure 6.38, with the corresponding LiDAR DTM section overlaid on the same graph.

Clearly, the TanDEM-X DSM appears to be affected by the terrain elevation. To investigate the matter further, we bring into play WASS analysis of the TanDEM DSM and the LiDAR DTM, DSM and CHM. The respective variance signatures are reported in Figure 6.39.

Since the LiDAR measures split the components resulting from ground and top of canopy, the WASS analysis allows us to characterize these components in terms of spatial frequencies, or, in other words, how spatial patterns are modulated by the ground and canopy variations.

Let us then begin considering the LiDAR surface models. In the range of scales from 1 m to 13 m the DSM and the CHM signatures overlap. This means that in this range the DSM is not affected by topography, but it follows the canopy spatial patterns. On the other hand, for scales $s > 13\ m$, the DSM and CHM signatures depart from one another. While the CHM decreases, indicating that less spectral energy is supplied by the canopy variations, the DSM increases and follows the same

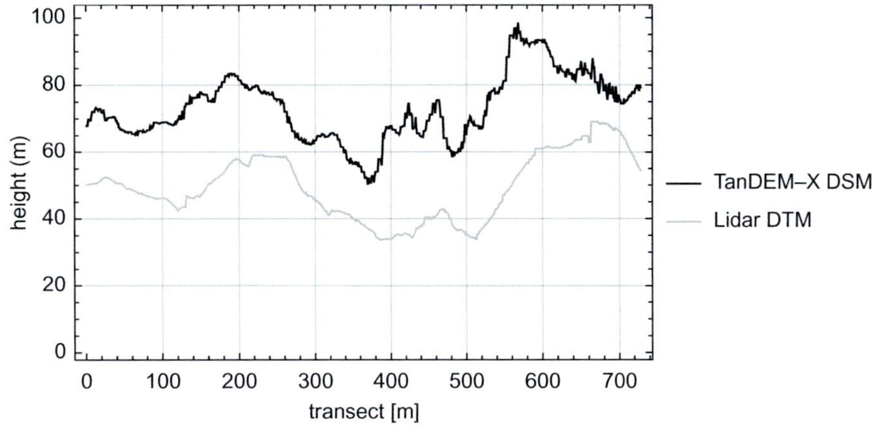

FIGURE 6.38 A section of the TanDEM-X DSM belonging to a transect crossing site A. The corresponding LiDAR DTM is overlaid in the same graph. The TanDEM-X DSM is the result of the phase return from ground and the phase distribution within the vegetation volume. Therefore, it encapsulates both the terrain elevation variations and patterns associated with the forest structure.

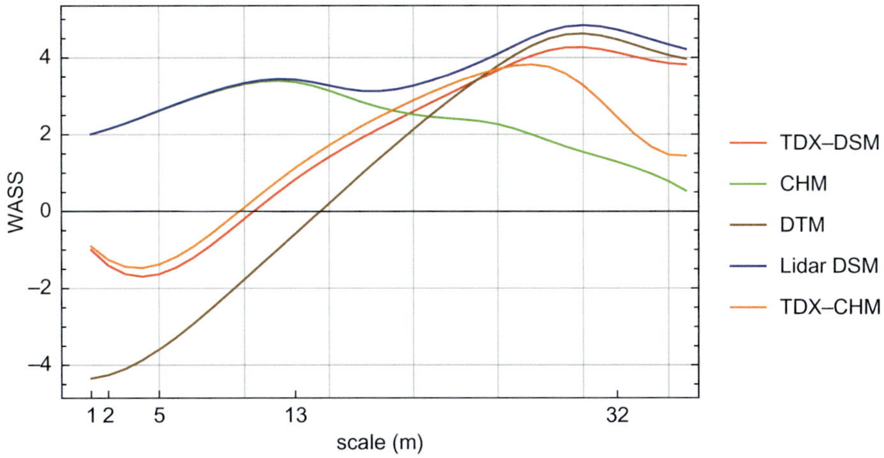

FIGURE 6.39 WASS variance signatures of the TanDEM-X DSM, and the LiDAR DTM, DSM and CHM.

dependence on scale as does the DTM. We conclude that in this range of scale the DSM texture is mainly affected by the terrain elevation variations.

Importantly, these observations suggest that, at least for the LiDAR measures, there is a spectral frequency range where patterns associated with canopy are reflected into the DSM without being contaminated by the terrain elevation variance.

Let us see if a similar frequency splitting can work for the radar DSM. The TDX-DSM signature is higher than the DTM signature in the range of scale $s < 32$ m. This means that within this range the radar DSM carries more information than the topography variations. For longer scales, all signatures merge, indicating the frequency zone where only textural information on the ground variations can be obtained.

This statement raises two further questions: if there is more textural information than just topography in the range $s < 32$, how is this information related to that captured by the LiDAR CHM? And second, how could we extract this information of interest?

We propose a solution that starts from the last question. We can indeed extract from the radar DSM a signal that contains only the scales in the range $s < 32$ m by wavelet thresholding and the inverse wavelet transform.

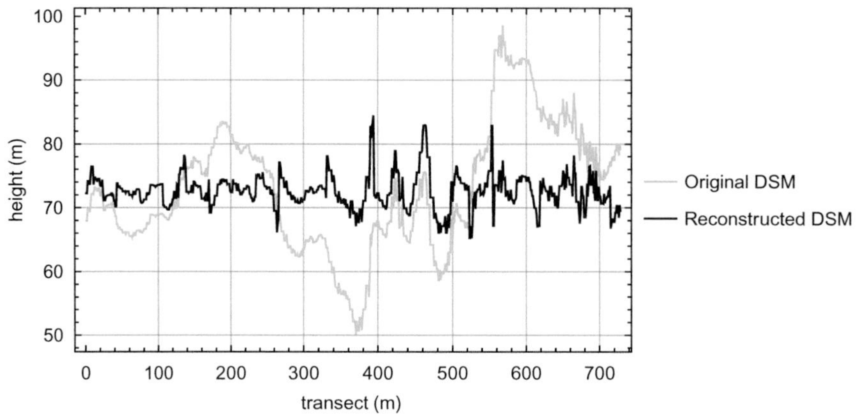

FIGURE 6.40 The original radar DSM signal and the reconstructed signal (TDX-Canopy), where the spatial frequencies belonging to terrain variation were removed by wavelet thresholding.

The discrete wavelet transform decomposes the DSM signal in 13 levels. All the wavelet coefficients up to scale $2^5$are "hard" thresholded, and the modified signal is reconstructed by the inverse wavelet transform:

$$\left[c(i,s),A(i,s)_{s=1\to 13}\right] = DWT\left(h_{DSM}(i)\right)$$

$$c(i,s)_{s=6\to 13} = 0$$

$$h_{canopy}(i) = IDWT(\left[cth(i,s),A(i,s)_{s=1\to 13}\right]$$

A section of the original radar DSM signal $h_{DSM}(i)$ and the reconstructed signal $h_{canopy}(i)$, where the spatial frequency belonging to terrain variation were removed, is shown in Figure 6.40. The section of 720 m is the same as the one in Figure 6.38.

Clearly, the largest height excursions related to topography were cancelled in the reconstructed signal, while the short-scale variance is still present, and the trend line has a zero slope.

Still, we need to check what kind of statistic is left in the reconstructed signal by purging the long scale variations. The WASS signature of $h_{canopy}(i)$ is overlaid to the signatures of LiDAR-DSM, TDX-DSM, LiDAR-CHM and LiDAR-DTM in the graph of Figure 6.41. The h_{canopy} signature (orange) now departs significantly from the TDX-DSM one. The signature follows a functional dependence with scale similar to the LiDAR-CHM, with strong energy in the scale range $1 < s < 13\ m$ and decaying thereafter.

This observation verifies that the reconstructed signal carries textural information in the given spectral range that is comparable to what the LiDAR observes, and therefore it captures a forest structural pattern. Of course, the absolute values of the radar signature are lower than those of the LiDAR; that is, the spectral energy in the radar signal is lower, because the physics underpinning the respective measurements is different. The LiDAR optical reflectivity at 1 m spacing obviously captures the canopy variations at a much finer vertical and horizontal resolution with respect to the SAR height estimates, with these depending on the phase center position within the vegetation volume.

For scale $s > 20$ m the SAR signature now flattens out, in the same way as the LiDAR-CHM does. Only for scale $s > 30$ m is there a slight influence of topography on the SAR signature, while the LiDAR-CHM keeps losing energy as the scale increases.

All in all, this analysis shows that, by suitable signal processing, information on the forest canopy spatial patterns, at least in a range of scale within some 20 m, can be retrieved from the InSAR DSM even in the absence of a DTM.

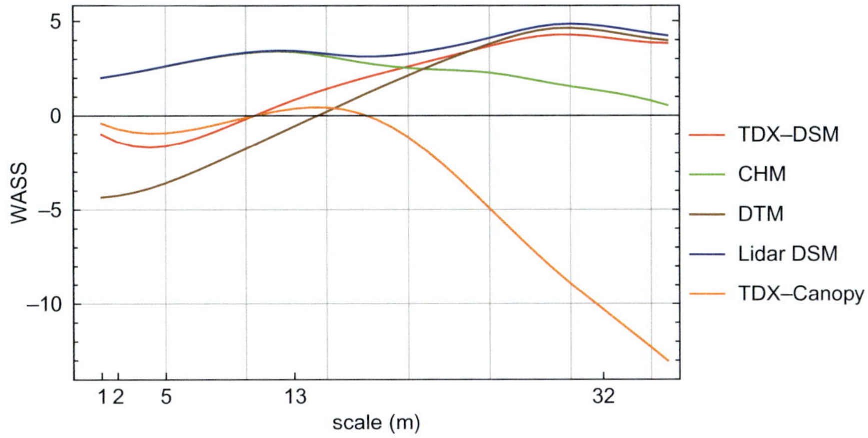

FIGURE 6.41 WASS signature of the reconstructed signal $h_{canopy}(i)$ overlaid to the signatures of LiDAR-DSM, TDX-DSM, LiDAR-CHM and LiDAR-DTM. The graph should be compared with the one in Figure 6.39.

Additional information on the spatial statistic of the LiDAR and SAR surface models may be inferred from the wavelet coefficients' flatness factor, using a scaling signature here, for short called WFLATS.

The signatures for the LiDAR CHM, DSM and DTM, and for the SAR DSM, are reported in Figure 6.42. The signatures were computed using a DWT with 10 dyadic scales and 4 voices.

At short scales ($s < 13$ m) all signatures reveal a strong departure from normality of the wavelet coefficients' distribution (WFLATS > 3), or the presence of strong intermittency in the signals. For the CHM and DSMs this is because of the heterogeneity of the forest, which is captured at a scale size comparable with the presence of a single (or a group of) emerging crowns. The DTM presents the highest values, these being given by the sharp signal transitions that occur with repetition at large scale.

For larger scales (s > 13) all signatures tend to level off around the values indicating normality in the coefficients' distribution (WFLATS = 3). Albeit, there is a characteristic oscillation in all DSMs and the CHM, with the one of the SAR DSM occurring at larger scale with respect to the LiDAR ones. We postulate that these patterns, being present in both the DSMs, could be a reflection of the intermittency of the DTM. Yet, surprisingly, the patterns is also present in the CHM, which should

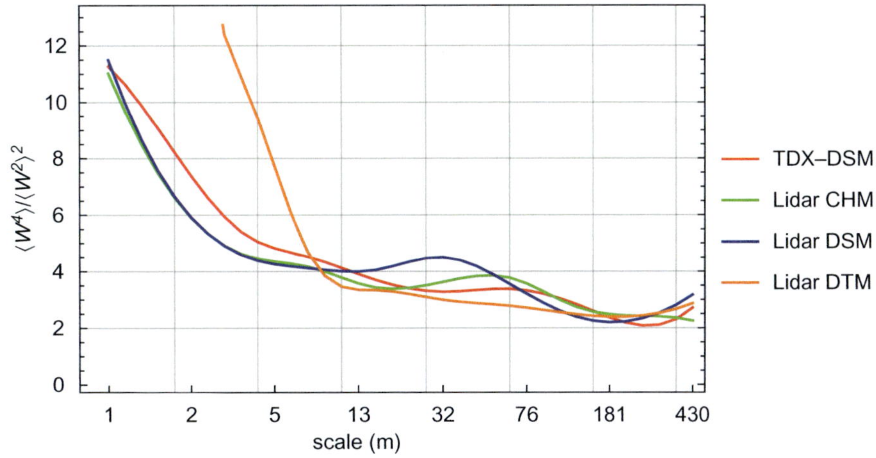

FIGURE 6.42 The wavelet flatness factor signatures for the LiDAR and the SAR surface models.

not be sensitive to terrain variation. We surmise that this coupling could confirm the fact that forest in valleys is higher than forest on ridges. This is because of several factors, such as: (a) higher soil fertility and moisture availability in valleys; and (b) forests in valleys need to grow taller to improve exposure to sunlight.

So far, we have integrated the wavelet coefficients over a spatial coordinate. We now seek to obtain information on the spatial distribution of the forest cover by local estimates of the wavelet variance obtained along a transect. We move, therefore, to an analytical tool that provides the dependence of the wavelet variance on the combined scale and space, i.e. the wavelet spectrum.

The spectrum of the LiDAR-CHM was computed over the same transect as in the previous analysis. As mentioned, and as depicted in Figure 6.37, the LiDAR observations over test site A cover the transition from the intact primary dipterocarp forest to the fire-damaged secondary forest. We want to investigate whether the structural differences between the two thematic classes may be captured by the wavelet spectrum.

The spectra at scales $s = 1\ m$, $s = 512\ m$, $s = 1024\ m$ are plotted in Figure 6.43. The large-scale information (s=512 and 1024 m) shows patterns that are split into two spatial segments. The spectrum decays strongly from small to large scale in the first segment, this being up to 3000 m. After that point, the decay with scale becomes less important, and the large-scale variance tends to acquire energy and grow. The two patterns in the scaling behavior of the wavelet variance mark the transition from the primary to the secondary forest. Indeed, the secondary forest structure is much patchier and heterogeneous; therefore, at large scales the wavelet variance becomes sensitive to these discontinuities.

This analysis can be appreciated by a different graphical rendition in Figure 6.44. Here the full spectrum $Wsp(s_j, x_i)$, including the space and scale dependence, is shown as a density plot. Large-scale components of the spectrum appear after the space coordinate $x_i = 3000\ m$ signaling a point of discontinuity in the spatial distribution of the forest cover.

Interestingly, the same functional dependence of the wavelet variance spectrum can be observed in the reconstructed SAR-DSM (TDX-canopy) (see Figure 6.45). In the scale domain, the variance decays from small to large scale. In the space domain, the large-scale components increase after space coordinate x_i = 3000 m. This analysis reinforces the evidence that the reconstructed SAR-DSM carries equivalent information on the forest structure as the LiDAR-CHM.

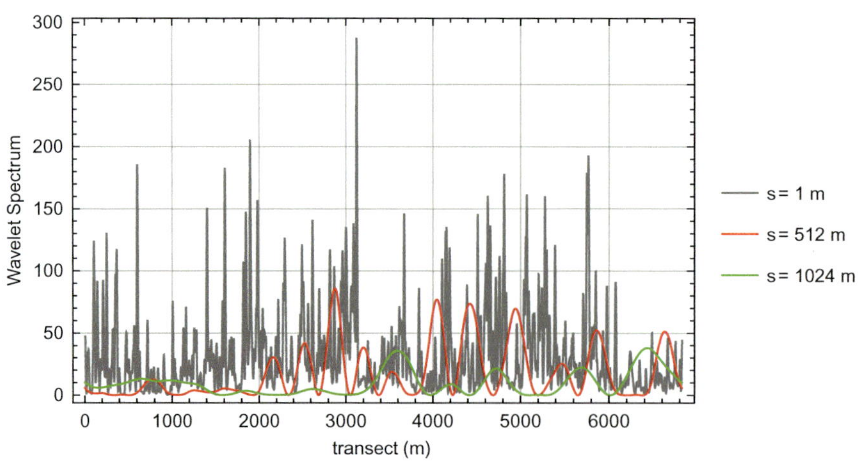

FIGURE 6.43 Wavelet spectra computed over the LiDAR-CHM transect and at scales $s = 1$ m, $s = 512$ m and $s = 1024$ m. The large-scale information shows patterns that are split into two spatial segments. The two patterns in the scaling behavior of the wavelet variance mark the transition from the primary to the secondary forest.

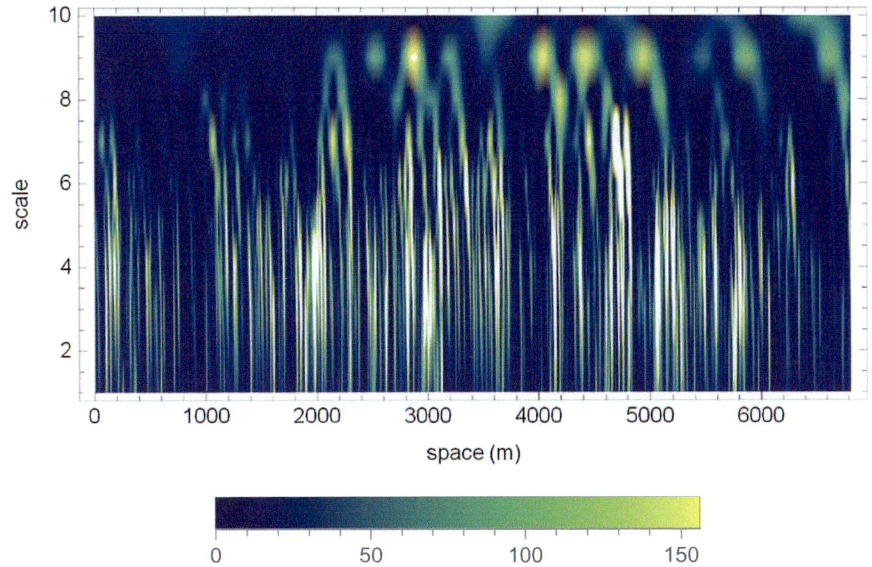

FIGURE 6.44 The wavelet variance spectrum of the LiDAR-CHM transect as a function of scale and space in a density plot, where the intensity is color coded.

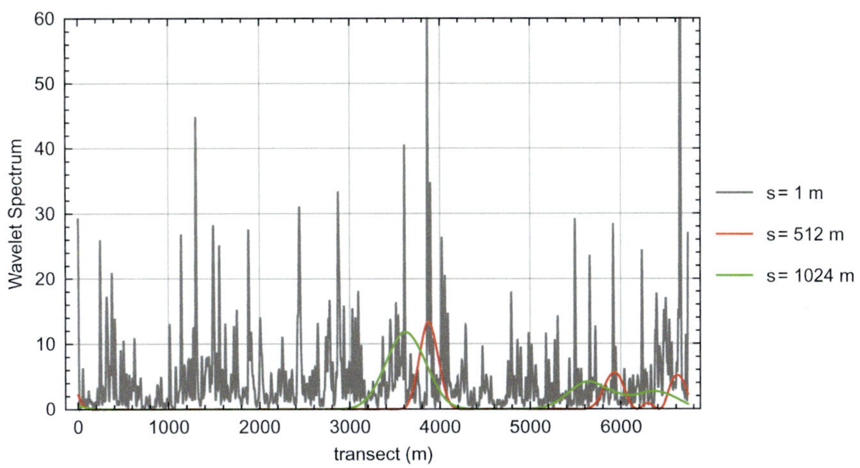

FIGURE 6.45 Wavelet variance spectra at scales $s = 1$ m, 512 m, 1024 m of the reconstructed signal TDX-canopy. The spectra show the same scaling and space dependence as the LiDAR-CHM spectra, reinforcing evidence that the SAR reconstructed signal carries equivalent information on the forest structure to the LiDAR-CHM.

For a more complete analysis of the two thematic classes' textural separability (primary and secondary forest) provided by the wavelet variance statistic, the reader is referred to [37].

6.10 2D WAVELET VARIANCE SPECTRA OF BACKSCATTER FIELDS: TOWARD A TEXTURAL CLASSIFIER

Wavelet variance estimates of the SAR backscatter field (2D wavelet spectra) can pave the way to the implementation of unsupervised textural classifiers, as we demonstrate in this section.

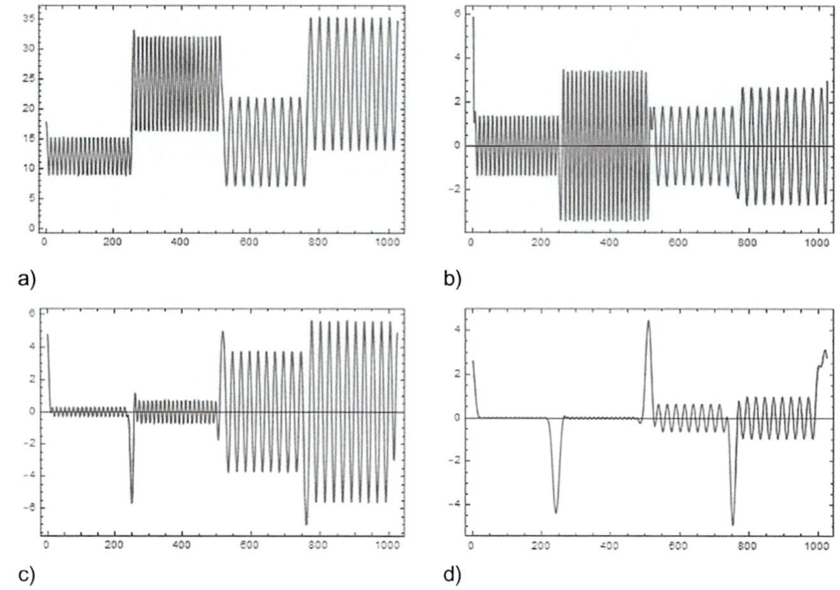

FIGURE 6.46 A toy signal to demonstrate the principle underpinning the wavelet textural segmentation method.

The basic problem is to find ways for aggregating the multi-dimensional information provided by the wavelet variance at several scales and at each point of the field into separable clusters, these reflecting different textural patterns of the field. To begin to gain insight into the problem, let us consider a simple 1D case, a toy signal (see Figure 6.46).

Our toy signal is composed of four segments with different edge patterns (see Figure 6.46(a)). They mimic profiles of surfaces with different roughness. In the first two segments transitions occur at short scale; moreover, the swing of the edges in the second segment is greater with respect to the one in the first segment. In the third and fourth segments transitions occur at longer scale and again the swing in the last segment is higher with respect to the former.

In Figure 6.46(b)–(d) the wavelet coefficients at scales 2, 8, 16 are shown. We can notice that the wavelet coefficients' swing and density varies locally following the patterns in the four segments. Also, we observe that there are characteristic scales where certain patterns are emphasized or damped. For example, the coefficients related to the higher frequency transitions in segments 1 and 2 dominate at short scales, whereas they disappear at coarse scale. The opposite is true for the coefficients associated with the slower transitions in segments 3 and 4.

Graphs of the wavelet variance estimated by a sliding window of length $l = 2\tau_1$, where $\tau_1$is the period of the patterns in segments 1 and 2, are shown in Figure 6.47(a) and (b). From graph (a) we can conclude that indeed the local variance of the wavelet at scale 2 estimated by a window at scale $l = 2\tau_1$ is stable within the segments and is sensitive as expected to the different roughness of the four patterns. This is a very desirable property in view of segmentation of the signal into homogeneous textured regions. We can say that the within-region variance is much smaller than the between-regions variance.

However, the wavelet variance associated with segment 3 is quite low, and could fall beneath the noise threshold if our signal was contaminated by noise. Moreover, the variance associated with segment 4 is quite near the one associated with segment 1, and therefore discrimination between the two textures becomes difficult. These facts indicate that scale 2 is not suitable for characterizing textures with slower variation, namely the ones of segments 3 and 4.

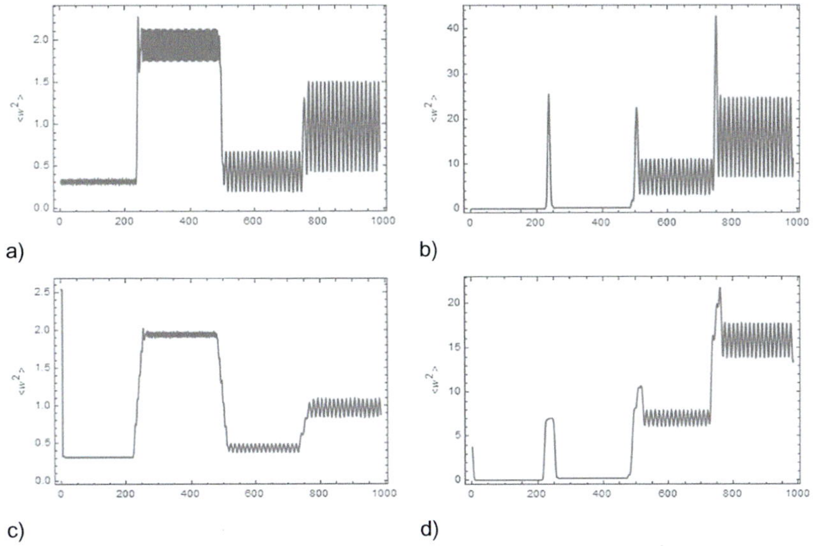

FIGURE 6.47 Local estimates of the wavelet variance. Short estimation window and scale 2^0 (a) and scale 2^4 (b). Long estimation window and scale 2^0 (c) and scale 2^4 (d).

Scale 16, on the contrary, captures quite well the slow textures associated with segments 3 and 4, while the fast variations of segments 1 and 2 have been smoothed out and disappear completely. Yet the wavelet variance estimates at this scale are quite noisy, or in other words the ratio of within-region to between-region variance is not as good as in the case of scale 2. To obtain stable estimates at scale 16 we have to widen the estimation window to $l = 8\tau_1$. This case is shown in Figure 6.47(c) and (d).

The lesson learnt from this simple test case is that successful texture discrimination (and hence segmentation) can only be achieved by the aggregate of measures of wavelet variances taken at different scales and by different local estimation neighborhoods. In the two-dimensional case (backscatter fields) there is in principle one more parameter to account for, namely the distribution of directions of the gradient field. We assume here we are dealing with spatially isotropic textures, and therefore information on the gradient angle is not useful and is therefore dropped from the feature vector.

This aggregation is achieved by a method of feature reduction suggested by M. Unser in [40–42]. The reader is referred to these papers for a rigorous mathematical treatment. In a nutshell, the unsupervised feature reduction method rests on an approximate solution of the multi-discriminant transform (MDT) that does not require a priori knowledge of the statistics of the feature vectors within and between the regions. Estimation of these statistics would require knowledge of the partitions of the textured areas, and would therefore not be suitable for unsupervised segmentation. MDT is the optimal transform that maximizes the ratio of the between-class and the within-class scatter matrices associated with the feature vectors. The approximate solution to the MDT problem takes advantage of the availability of multi-scale feature planes and requires only knowledge of a global statistics associated with the feature vector.

It is composed of the following sequence of operations. First, the original feature vector is smoothed by an appropriate low pass filter at scale m and rotated into its principal components by a Karhunen-Loève transform [22,27,43]. The rotated components are then normalized by the square root of the corresponding eigenvalue. Smoothing is then performed a second time at scale $2m$. The approximate MDT solution is then given by the KLT transform of the smoothed vector. The components of the resulting feature vector are ranked in order of discriminatory power.

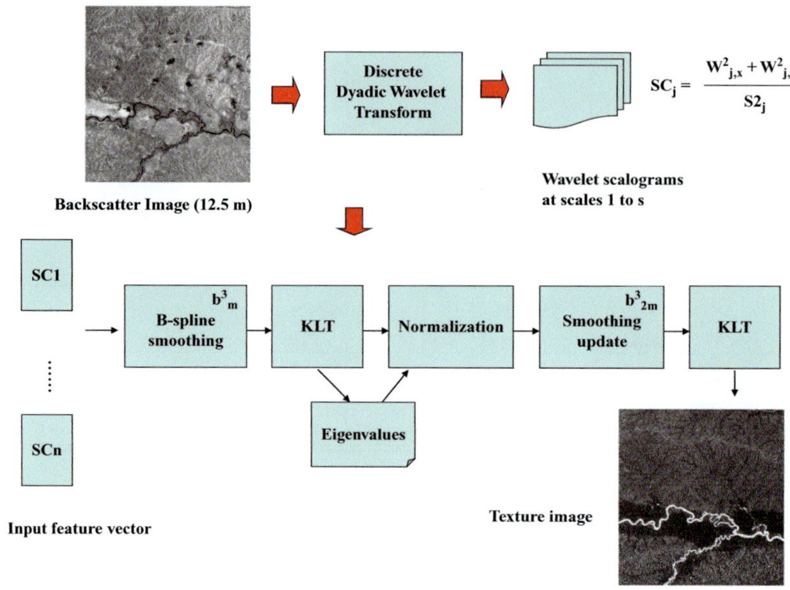

FIGURE 6.48 Flow chart of the unsupervised textural classifier method.

We are now in a position to summarize the operations of the unsupervised texture segmentation approach (see Figure 6.48).

The original backscatter image is decomposed by the discrete dyadic wavelet transform into n scales ($2^{\circ} \cdots 2^{n-1}$). For each scale, a feature vector is built by summing the squares of the wavelet coefficients proportional to the x and y (rows and columns) components of the gradient field and by normalizing by the square of the smooth signal at the same scale.

$$SC_j = \frac{W_{x,j}^2 + W_{y,j}^2}{S_j^2}$$

where $W_{x,j}^2 W_{y,j}^2$ are the wavelet coefficients at scale 2^j, and S_j is the smooth signal at scale2^j.

Normalization by the smooth signal is necessary to equalize the speckle multiplicative noise contribution to the wavelet coefficients (see Section 6.6.2). This contribution is higher for regions of higher backscatter, and would therefore produce different texture measures even for textureless homogeneous areas with different mean backscatter values.

The feature vector contains normalized gradient coefficients up to scale $n = 2m$. This feature vector is smoothed at scale m by convolution with a cubic B-spline kernel dilated by a factor m. The coefficients of this filter are computed by the Z-transform representation of the dilated cubic B-spline with upsampling index m:

$$B_m^3(z) = \frac{z^{2(m-1)}}{m^3} \left(\frac{1 - z^{-m}}{1 - z^{-1}} \right)^4 B_1^3(z)$$

$$B_1^3(z) = \frac{z + 4 + z^{-1}}{6}$$

A Karhunen-Loève transform (KLT) is applied for the first time at this stage. The resulting eigenvector components are normalized by the square-root of the corresponding eigenvalues. Smoothing is then performed at scale $2m$ by convolution with the B-spline kernel B_{2m}^3.

Finally, the KLT transform is applied a second time. Inspection of the eigenvalues associated with the resulting eigenvector allow us to choose the N most discriminatory components that will constitute our reduced feature vector to be used as input to a segmentation algorithm, such as the classical K-Means or ISODATA [43].

6.10.1 A Test Case: Texture-Based Forest Mapping in the Congo Floodplain by ERS-1 Data

As an illustration of the method's performance, we present here a test case related to the thematic focus on swamp forest mapping in the Congo floodplain (i.e. the same case handled in Section 6.8.1). The ESA ERs-1 backscatter image of the test area in the Congo floodplain is shown in Figure 6.25. This is the input product for the wavelet texture classifier. It is a 3-Looks image (therefore heavily contaminated by speckle) with a pixel spacing of 12.5 m. The image in Figure 6.25 was resized by pixel aggregation to 200 m pixel spacing.

The maximum eigenvalue component of the feature vector after the KLT at the second smoothing stage is shown in Figure 6.49 (left). Notice how in the backscatter image the differences in smoothness between the two types of forest are barely visible, whereas the local estimates of the wavelet variance capture these textural differences (see the dark ribbon along the river, and the fern-like patterns spreading from the river bed into the forest domain).

The ratio of between regions to within regions variances in the texture image is high enough to allow us to use this image as input product for the segmentation stage. In our experiment, a K-Means algorithm with ten clusters produces the result shown in Figure 6.49 (right). The clusters were labeled as four thematic classes: lowland forest, swamp forest, secondary forest, river.

The area covered by the swamp forest (corresponding to low texture) is visible along the river (dark green). Also, the thin swamp forest ribbons along the drainage pattern are well delineated. The river is mapped into a high-texture class. Moreover, a thin structure with high texture appears at the top of the image (red). This structure corresponds to a ribbon of secondary forest that delineates the division of waters between two catchments.

As an important point, we emphasize the fact that the wavelet texture segmentation preserves the pixel spacing of the original image, though the resolution is degraded to 2 m, where m is the scale parameter that corresponds to the width of the window used to estimate the wavelet variance locally. This is a definite advantage with respect to the technique originally employed in the generation of the GRFM mosaics. This technique, as explained in Chapter 3, is based on the estimate of the K

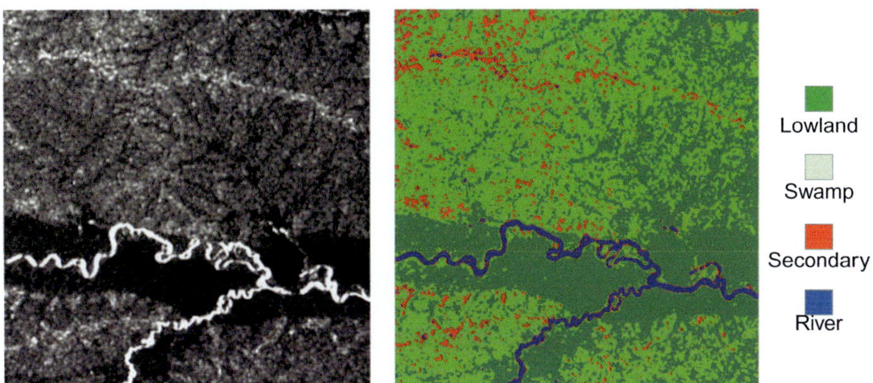

FIGURE 6.49 The feature vector at the output of the multi-discriminant transform (a). The textural classification map derived by K-Means clustering (4 clusters).

distribution order parameter υ using a block window. As such, it degrades the pixel spacing proportionally to the estimation window size.

6.10.2 Floodplain Mapping Revisited by Sentinel-1 data

The launch of the Sentinel-1 mission in 2014 opened up novel possibilities for SAR remote sensing of tropical forests, as a result of the improved data availability with increased sampling in space and time. It was therefore of interest to revisit the mapping experiment reported in the previous section, keeping the same thematic context and using Sentinel-1 data.

A study site (center coordinates: 1° 25.024' N, 16° 41.547' E) was selected in the Republic of Congo overlapping thematically the site used for the ERS-1 experiment.

Sentinel-1 Interferometric Wide Swath mode imagery (IW, 250 km swath width, incidence angle range 29.1°, 46.0°) acquired on November 4, 2019 over the Republic of Congo was downloaded from the Copernicus Open Access Hub. The Sentinel-1 Single Look Complex (SLC) imagery was acquired in descending mode, dual polarization.

Pre-processing was performed using SNAP Toolbox (v6.0). We performed the following processing steps: data import, thermal noise removal, calibration, deburst and merge, multi-Looking (2 Looks in range and in azimuth) and slant to ground range conversion. Range Doppler terrain corrected geocoding was performed using the Shuttle Radar Topography Mission (SRTM) 3 sec DEM (bi-linearly interpolated). The projected coordinate system (WGS84, UTM 33N) was adopted with a resulting pixel spacing of 15 m. No speckle filtering was performed.

Only a subset of the full Sentinel-1 imagery was used for the purpose of the analysis, since it covered fully the study area selected.

The same unsupervised texture classifier described in this section was applied to the backscatter intensity image (see Figure 6.50(a)). The KLT transform (first eigenvalue) of the wavelet variance dataset is shown in Figure 6.50(b). A vegetation map for the study area adapted from work in [44] is reported in Figure 6.50(c) for visual comparison.

Areas that appear darker in the KLT of the wavelet variance dataset correspond to hardwood swamp (shown in yellow on the vegetation map in (c)),while, areas that appear as a lighter grey correspond to terra firme forest (depicted in green on the vegetation map in (c)). The wavelet variance

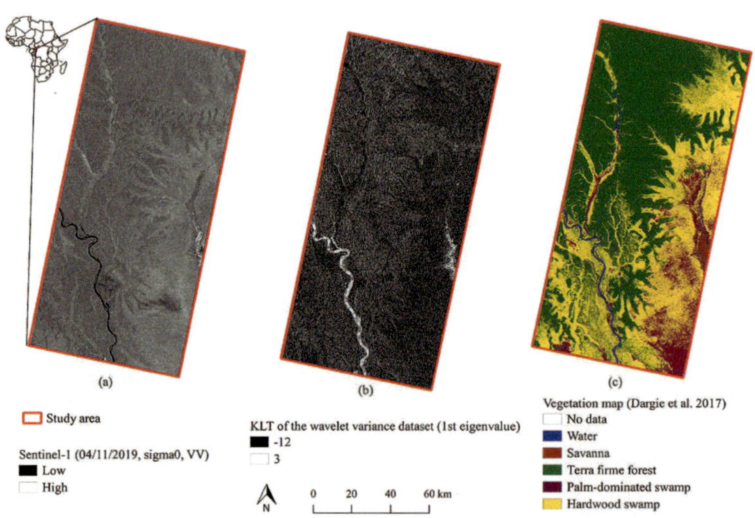

FIGURE 6.50 (a) Sentinel-1 imagery acquired on November 4, 2019 (VV polarization, 15 m); (b) Karhunen–Loève (KLT) transform of the wavelet variance dataset (1st eigenvalue); and (c) Vegetation map for the study area adapted from [44].

data also highlights swamp forest areas around the river (dark grey), the river is well delineated (white) in (b). Note that the correspondence between Sentinel-1 intensity and the map in (c) is not as evident.

Importantly, this experiment, because of the availability of Sentinel-1 time-series, paves the way to adding a time dimension to the wavelet representation in the scale–space domain, enabling the extension to multi-temporal analysis of textural patterns.

6.10.3 An (Experimental) Wavelet Spectrum Functional Classifier

The abstraction of the wavelet spectrum as a functional leads naturally to considering the realization of a functional classifier, i.e. a classifier where the features are functions $f(s)$of a variable s in $\Re$ (the space of real numbers).

We are experimenting with a prototype functional classifier in a test case where we use the wavelet spectrum of a digital surface model (DSM) derived from a TanDEM-X dataset. The data set was obtained over the Sungai Wain Protection Forest, within the project described in Section 6.9.

The functional classifier is based on a multilayer perceptron (MLP) [45]. The MLP architecture is composed of an input linear layer with 50 nodes (each node provides a weighted linear combination of the inputs and a hyperbolic tangent as activation function), and a linear hidden layer with a scalar output.

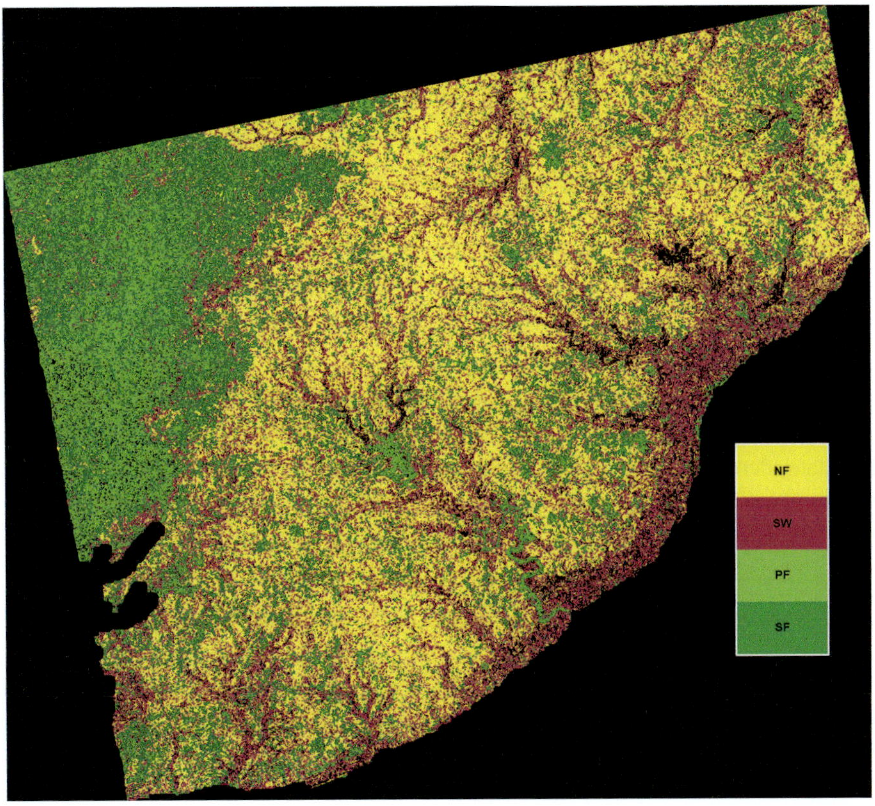

FIGURE 6.51 Classification map of the Sungai Wain test site derived by the experimental functional classifier. The supervised classification exercise detects four thematic classes (see legend) from the textural information provided by the functional representation of the wavelet spectrum.

The MLP is trained by wavelet variance trajectories $f(s)$ computed from the average of the spectrum polynomial coefficients within a region of interest (ROI). ROI selection was assisted by WASS analysis (see Section 6.9) and the link to thematic classes was found by inspection of ancillary data. Four classes are defined for this experiment: primary forest (PF), secondary disturbed forest (SF), swamp forest (SW), and non-forest (NF) including crop fields, mixed scrub and built-up areas. The trained MLP is then applied to each term of the spectrum functional $F(f_{ij}(s))$ for the prediction step.

The resulting classification map is shown in Figure 6.51. Qualitative evidence of the textural classification accuracy was inferred by comparison with spatial measures provided by LiDAR observations and other ancillary data (see [37]). At the time of writing, this experiment can therefore only be considered as a proof of concept.

6.11 EXTENSION TO POLARIMETRY

The key question we shall now address is whether, and how, a change in the wave polarization basis will affect the two-point statistic of the SAR backscatter random field. To introduce the topic and supply an intuitive justification of the matter, let us consider a thought experiment.

Consider a scene with a timber plantation, where the trees are arranged in a regular pattern with equal spacing between the trunks. If the penetration depth allows for the wave reaching the surface, there will be a strong dihedral retro-reflection between trunks and ground, mainly affecting the HH component. If the sensor's spatial resolution is higher than the trunk spacing, we would expect in the observed HH field a spatial frequency component that matches the periodicity of the trees' spacing. The HH component spatial statistic (as measured, for example, by the wavelet variance scaling) will then show a strong correlation around the scale corresponding to the trunk spacing.

On the other hand, the HV component, which is mainly given by multiple scattering among the elementary scattering centers (leaves, branches) in the canopy, will yield a spatially uncorrelated field, if the upper canopy layer is a continuum, and the number density of the scatterers varies spatially in a random way within a canopy and between neighboring canopies.

In this simple scenario, we have surmised that two different polarization configurations (HH, HV) can yield backscatter fields with completely different spatial correlation properties – the first with strong autocorrelation at the frequency resonating with the target periodicity, and the second with weak auto-correlation. We would like to generalize this idea for taking into account the dependence of the observed field spatial statistic on different and more complete configurations of the polarization space.

Polarimetric signatures, introduced by van Zyl [46,47], are graphic representations that encapsulate in a 3D plot the dependence of the co- and cross-polarization backscattered power on the polarization state identified by the orientation and ellipticity angles of the polarization ellipsis. We take inspiration from this formalism and propose a representation of the wavelet variance integral spectrum on a polarization state. As the van Zyl signature depicts the dependence of the one-point first moment backscatter statistic on wave polarization, we now address the dependence of a two-point backscatter statistic on wave polarization.

This texture signature is dubbed a WASP (Wavelet Polarimetric Signature). The WASP (a new polarimetric insect) is an extension of the wavelet scaling signature (WASS) aimed at including the dependence on a polarization state. It is defined as the wavelet variance estimated over a region $\mathcal{R}$ of a backscatter intensity field as a function of scale $sc = 2^m$, with the intensity field being synthesized at polarization state $\wp_{xcpol}(\phi, \tau)$:

$$\left\langle \mathrm{W}_I\left(\vec{P},\mathrm{sc}\right)^2 \right\rangle_{\vec{P}\in\mathcal{R}} = f\left(\mathrm{sc}, \wp\left(\phi,\tau\right)\right)$$

where W_I are the wavelet coefficients of the intensity field I that is synthesized in the cross-polarized or co-polarized state $\wp_{xcpol}(\phi, \tau)$, and ϕ, τ are the orientation and ellipticity angle.

For a 3D graphic representation of the WASP, the domain of the independent variables must be reduced to two variables, i.e. scale and an indicator of the polarization state. This indicator is obtained by considering the following power synthesis cases:

Cross-polar/co-polar power in a linear rotated basis:

$$\left\langle W_I\left(\vec{P},\mathrm{sc}\right)^2\right\rangle_{\vec{P}\in\mathcal{R}} = f\left(\mathrm{sc},\phi,\tau=0\right)$$

Cross-polar/co-polar power in an elliptical basis:

$$\left\langle W_I\left(\vec{P},\mathrm{sc}\right)^2\right\rangle_{\vec{P}\in\mathcal{R}} = f\left(\mathrm{sc},\tau,\phi=0\right)$$

To gain an insight into the possible dependence of the wavelet variance on the polarization state, we shall next present a numerical model that, starting from a stationary random intensity field measured on the (H, V) linear basis with given two-point statistic, derives the wavelet variance of the backscatter intensity when synthesized in a different polarization basis.

6.11.1 The WASP of Correlated Backscatter Textures: A Numerical Model

The scope of the model is to compute the wavelet variance of an intensity field resulting from power synthesis in any polarization basis (linear or elliptical), and assuming two-point statistics (autocorrelation functions) of the covariance matrix elements in the original orthogonal (H, V) basis.

For fully polarimetric SAR observations, a wide sense stationary random field in the linear (H, V) basis can be characterized by an ensemble of target feature vectors:

$$\underline{\Omega} = \left[S_{hh}\ \sqrt{2}\,S_{hv}\ S_{vv}\right]$$

The $\sqrt{2}$ term makes the norm of the vector equal to the span of the corresponding scattering matrix.

The components of the feature vector Ω in the orthogonal (H, V) basis are transformed in any orthogonal elliptical basis defined by the polarization ratio ρ by the linear transformation $T(\rho)$:

$$\rho = \frac{\cos 2\tau \sin 2\phi + j \sin\tau}{1+\cos 2\tau \cos 2\phi}$$

$$T(\rho) = \frac{1}{\left(1+\rho\rho^*\right)}\begin{bmatrix} 1 & \sqrt{2}\rho & \rho^2 \\ -\sqrt{2}\rho & \left(1-\rho\rho^*\right) & \sqrt{2}\rho \\ \rho^{*2} & -\sqrt{2}\rho^* & 1 \end{bmatrix}$$

$$\underline{\Omega}' = \underline{T}(\rho)\cdot\underline{\Omega}$$

Let us first consider the transformation into a rotated linear basis ($\tau = 0$). In this case:

$$\rho = \frac{\sin 2\phi}{1+\cos 2\phi}$$

$$T(\rho)=\begin{vmatrix} \cos\phi^2 & \frac{\sqrt{2}}{2}\sin 2\phi & \sin\phi^2 \\ -\frac{\sqrt{2}}{2}\sin 2\phi & \cos 2\phi & \frac{\sqrt{2}}{2}\sin 2\phi \\ \sin\phi^2 & -\frac{\sqrt{2}}{2}\sin 2\phi & \cos\phi^2 \end{vmatrix}$$

The cross-polar power element in the averaged rotated vector is:

$$\begin{aligned} P^{\phi} &= \langle \Omega_2'\Omega_2' \rangle^* = 2|\mathrm{Shv}|^2\cos[2\phi]^2 + 2\left(|\mathrm{Shh}|^2+|\mathrm{Svv}|^2\right)\cos[\phi]^2\sin[\phi]^2 \\ &-2\langle \mathrm{Shh}\,\mathrm{Shv}^*\rangle\cos[2\phi]\sin[2\phi]+2\langle \mathrm{Shv}\,\mathrm{Svv}^*\rangle\cos[2\phi]\sin[2\phi]-\langle \mathrm{Shh}\,\mathrm{Svv}^*\rangle\sin[2\phi]^2 \end{aligned}$$

To establish a mapping between the two-point statistic of the covariance matrix elements in the original orthogonal linear (H, V) basis and the wavelet variance statistic of the transformed power P^{ϕ} we cast the problem in the Fourier domain.

The power spectrum of the cross-polar power P^{ϕ} in the rotated basis is:

$$\begin{aligned} Y\left(P^{\phi}\right) &= \Im\left(P^{\phi}\right)\Im^*\left(P^{\phi}\right) = \left(\sum_i a_i \Im(P_i)\right)\left(\sum_i a_i \Im^*(P_i)\right) \\ &= \sum_i a_i^2 \Im(P_i)\Im^*(P_i) + \sum_{ij} a_i a_j \Im(P_i)\Im^*(P_j) \end{aligned}$$

The term in the double sum corresponds in the space domain to the cross-correlation between different P_i terms. Experimental measures point out that the energy of these cross-spectra is one order of magnitude less than the energy of the spectra related to the terms in the first sum. Therefore, the power spectrum in the rotated basis P^{ϕ} can be approximated by a linear combination of the power spectra of the diagonal elements of the covariance matrix in the HV basis (power terms) and the correlation terms of the co-polar elements $S_{hh}\ S_{vv}{}^*$and between the cross-polar and co-polar elements $S_{hv}\ S_{vv}{}^*$, $S_{hh}\ S_{hv}{}^*$:

$$\begin{aligned} Y\left(P^{\phi}\right) &= a_1^2 Y\left(|S_{hv}|^2\right) + a_2^2\left(\frac{1}{2}\left(Y\left(|S_{hh}|^2\right)+Y\left(|S_{vv}|^2\right)\right) - Y\left(\Re\left[S_{hh}S_{vv}^*\right]\right)\right) \\ &+ a_3^2\left(Y\left(\Re\left[S_{hv}S_{vv}^*\right]\right) - Y\left(\Re\left[S_{hv}S_{hh}^*\right]\right)\right) \end{aligned} \tag{6.49}$$

where: $a_1 = 2\cos 2\phi^2$ $a_2 = \sin 2\phi^2$ $a_3 = \sin 4\phi$

Since the wavelet transform is a linear operator, the power spectrum of the transformed signal is:

$$\hat{w}(\omega) = -i\frac{\omega}{4}\left(\frac{\sin\left(\frac{\omega}{4}\right)}{\frac{\omega}{4}}\right)^4 e^{-i\frac{\omega}{2}} \tag{6.50}$$

$$Y_{W_s}(\omega) = \left|s^{-\frac{1}{2}}\hat{w}_s(\omega)\right|^2 Y\left(\mathrm{P}^{\psi}\right)(\omega) \tag{6.51}$$

where $\hat{w}(\omega)$ is the Fourier transform of the wavelet.

The wavelet variance is computed from the autocorrelation function (ACF) of the output signal at lag $\tau = 0$, this being given for a wide sense stationary (WSS) process by the inverse Fourier transform of the power spectrum $\Upsilon_{Ws}(\omega)$:

$$W_{P\psi}{}^{2}(s,x) = \mathcal{R}(0) = \mathfrak{I}^{-1}\left(Y_{W_s}(\omega)\right)\Big|_{\tau=0} = \frac{1}{2\pi}\int Y_{W_s}(\omega)\,d\omega \tag{6.52}$$

The coefficients a_j^2 in Equation (6.49) are plotted in Figure 6.52 with respect to the linear polarization basis orientation angle ϕ. The function a_1^2 has maxima at $\phi = 0°, 90°$. This coefficient weighs the integral of the power spectrum of the cross-pol power element on the HV basis. The function a_2^2 has a maximum at $\phi = 45°$ and it weighs the integral of the power spectra of the co-pol power terms, and the power spectrum of their cross-correlation. The term a_3^2 presents maxima at $\phi = 20°, 70°$ and it weighs the (weaker) power spectra of the co-pol to cross-pol correlation terms.

Therefore, we would expect the wavelet signature at a given scale (frequency) to present a maximum near $\phi = 0°, 90°$ when the power spectrum of the cross-pol power channels dominates in the neighborhood of that frequency. Because of the duality between the autocorrelation function and the power spectrum, this means that a maximum at $\phi = 0°, 90°$ points to the existence of strong spatial correlations at that scale in the cross-pol power element. Analogously, a maximum near $\phi = 45°$ will point to stronger correlations in the co-pol channels.

A shift of the maximum from these positions is linked to contributions by the $\langle hh\ hv^*\rangle$ and $\langle vv\ hv^*\rangle$ elements power spectra. As discussed in more detail in Chapter 7, these elements are associated with the azimuthal reflection symmetry of the illuminated target. In the contest of the WASP signature, a shift of the maximum should therefore be interpreted as the presence of spatial correlations by a non-symmetric target.

The mathematics in Equations (6.49)–(6.52) allow us now to examine a few test cases where we compute the WASP starting from initial conditions set by the autocorrelation functions of the covariance matrix elements in the (H, H) basis.

For the (H, V) covariance matrix power terms we assume correlated clutter with exponential cosine correlation functions:

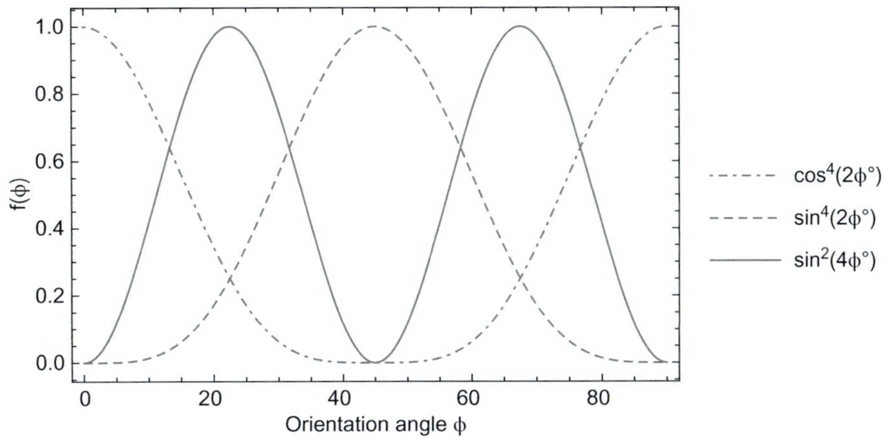

FIGURE 6.52 Graph of the coefficients a_j^2 in Equation (6.49). The dependence of the wavelet variance on the power spectra of the covariance matrix elements on the (H, V) basis is encapsulated in these coefficients.

$$\mathcal{R}(\tau) = \langle I \rangle^2 + \frac{\langle I \rangle^2}{\upsilon} e^{-\frac{1}{\Gamma}|x|} (1 + \cos \Omega x)$$

where $\frac{1}{\upsilon} = \frac{VAR(I)}{\langle I \rangle^2}$.

For the cross-correlation terms we assume Gaussian ACFs:

$$\mathcal{R}(\tau) = \langle I \rangle^2 + \frac{\langle I \rangle^2}{\upsilon} e^{-\frac{\tau^2}{\beta^2}}$$

Parameters of the ACFs are summarized in Table 6.2, and the ACFs for the power elements $\langle Shh^2 \rangle$ $\langle Svv^2 \rangle$ $\langle Shv^2 \rangle$ are plotted in Figure 6.53.

The cross-pol WASP computed by the numerical model with these initial conditions and as a function of scale and linear polarization orientation angle ϕ is shown in Figure 6.54.

A cross-section highlighting the dependence on scale at wave orientation angle $\phi = 0°, 22°, 45°$ is shown in Figure 6.55, while the dependence on wave orientation angle at scales $s = 2°, 2^3, 2^4$ is shown in Figure 6.56.

Regarding the wavelet variance in the scale domain as a function of the orientation angle ϕ the following can be noted. At $\phi = 0°$ (solid line) the wavelet statistic reflects the exponential ACF of the HV power with a sill around scale $s = 10$ (asymptote of the structure function corresponding to loss of correlation). As the orientation angle is increased toward $\phi = 45°$the periodicity of length 6 induced by the cosine term starts to appear, being revealed by a local maximum followed by a minimum of the structure function.

TABLE 6.2
Parameters of the Covariance Matrix power terms ACFs

	$\langle I \rangle^2$	υ	Γ	Ω
HH	5 10^4	0.2	0.1	$2\pi/6$
HV	4 10^3	0.2	0.8	$\pi/6.25$
VV	2 10^4	0.2	0.1	$2\pi/6$

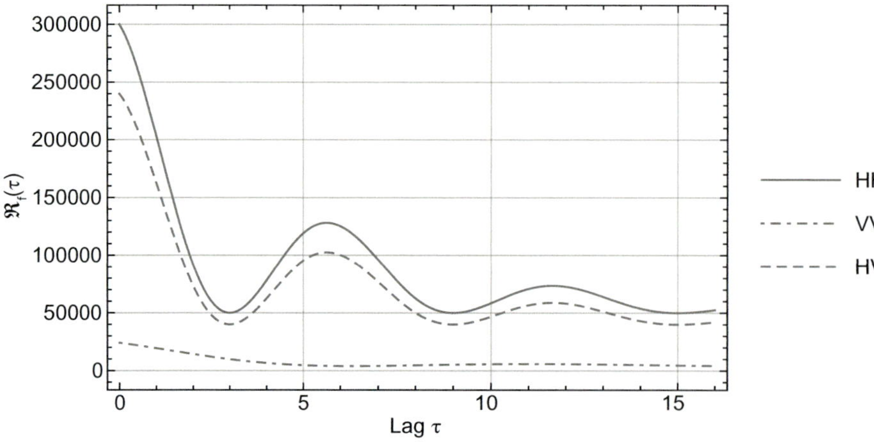

FIGURE 6.53 Autocorrelation functions of the power elements $\langle Shh \rangle$ $\langle Svv \rangle$ $\langle Shv \rangle$.

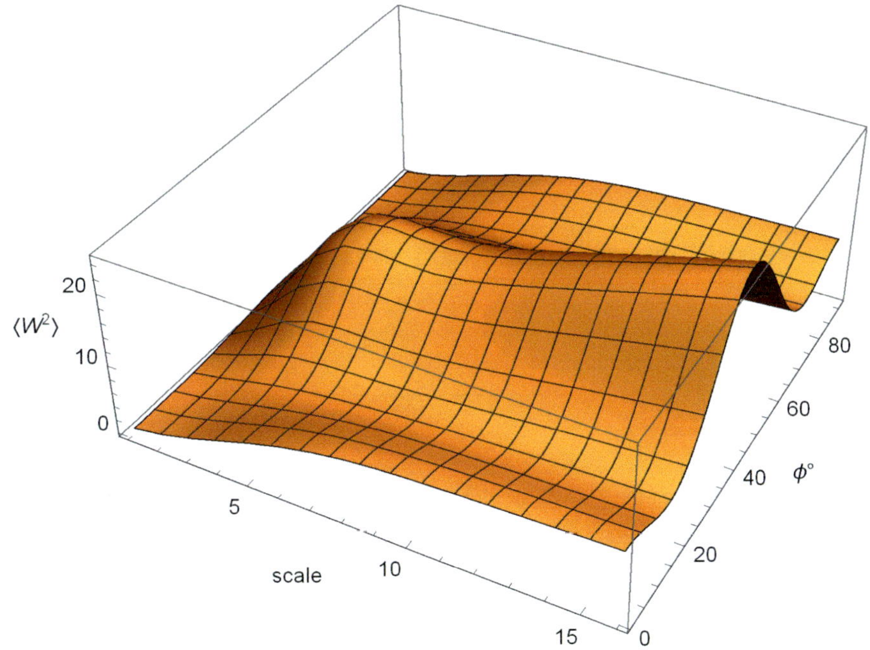

FIGURE 6.54 The cross-pol WASP computed by the numerical model with the initial conditions in Table 6.1.

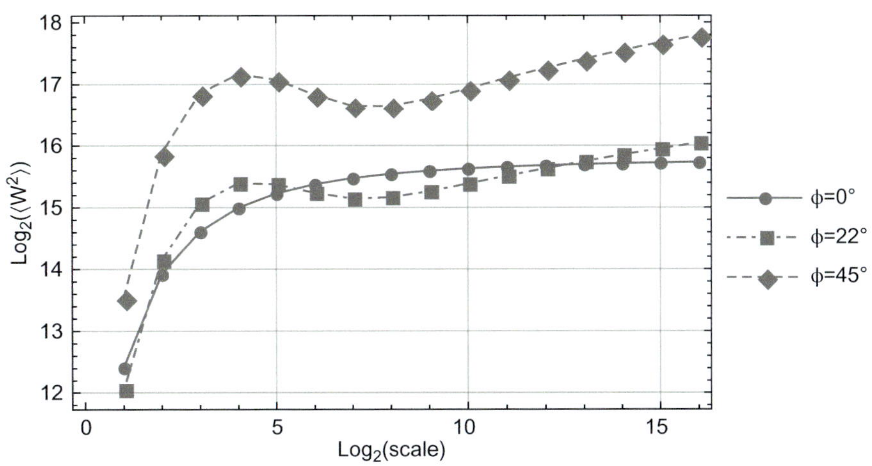

FIGURE 6.55 Cross sections of the 3D WASP highlighting the dependence on scale at orientation angle $\phi = 0°, 22°, 45°$.

Regarding the wavelet variance in the orientation angle domain, note that the functional dependence is similar at all scales, with an absolute maximum at $\phi = 45°$, corresponding to the longer correlation length in the co-polar elements (HH, VV).

The co-polar case is considered next.

The co-polar power term in the rotated basis is:

$$P^{\phi} = \langle \Omega_1' \Omega_1' \rangle^* = |\mathrm{Shh}|^2 \cos[\phi]^4 + 2\langle \mathrm{Shh\,Shv}^* \rangle \cos[\phi]^2 \sin[2\phi] + |\mathrm{Shv}|^2 \sin[2\phi]^2$$
$$+ \frac{1}{2}\langle \mathrm{Shh\,Svv}^* \rangle \sin[2\phi]^2 + 2\langle \mathrm{Shv\,Svv}^* \rangle \sin[\phi]^2 \sin[2\phi] + |\mathrm{Svv}|^2 \sin[\phi]^4$$

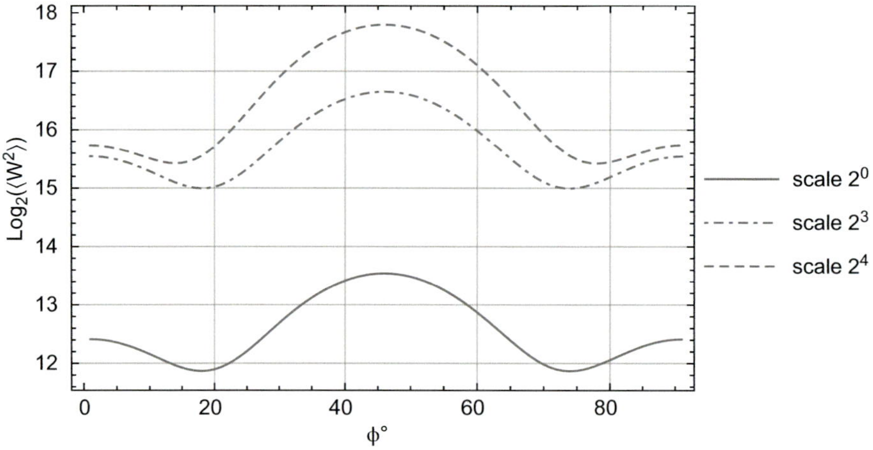

FIGURE 6.56 Cross sections of the 3D WASP highlighting the dependence on orientation angle ϕ at scales $s = 2^0, 2^3, 2^4$.

The power spectrum of the co-polar power in the rotated basis is:

$$\Upsilon\left(P^{\psi}\right) = 2a_3^4\Upsilon\left(\left|S_{hv}\right|^2\right) + a_2^4\Upsilon\left(\left|S_{vv}\right|^2\right) + a_1^4\Upsilon\left(\left|S_{hh}\right|^2\right) + \frac{1}{4}a_3^4\Upsilon\left(\Re\left[S_{hh}S_{vv}^*\right]\right)$$
$$+a_1^2a_3^2\Upsilon\left(\Re\left[S_{hv}S_{hh}^*\right]\right) + a_2^2a_3^2\Upsilon\left(\Re\left[S_{hv}S_{vv}^*\right]\right) \tag{6.53}$$

where $a_1 = \cos\phi^2$, $a_2 = \sin\phi^2$, $a_3 = \sin 2\phi$.

The coefficients $a_i$related to the power spectra $\Upsilon(|S_{hh}|^2)$, $\Upsilon(|S_{hv}|^2)$, $\Upsilon(|S_{vv}|^2)$ are plotted in Figure 6.57.

For the co-polar case we see that, concerning the influence of the power terms spectra, the HH term dominates at $\phi = 0°$, it is gradually mixed with the HV term for increasing orientation angle, until the HV terms dominates at $\phi = 45°$. For $\phi > 45°$ the VV term comes increasingly into play, at first with a superposition with the HV term, and finally dominating at $\phi = 90°$.

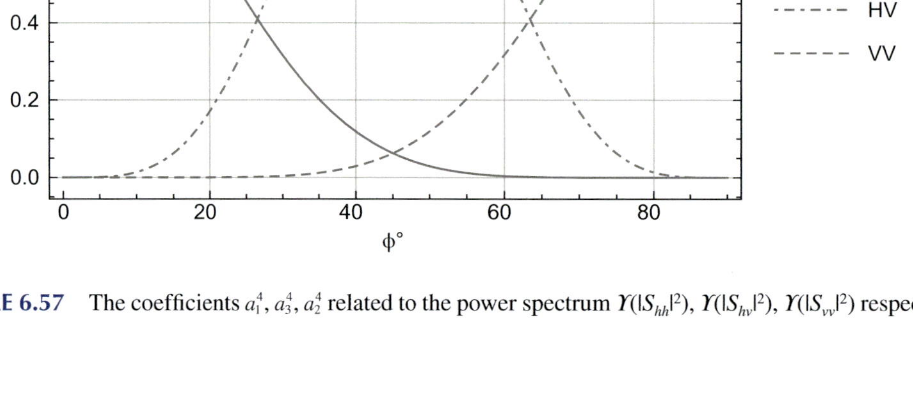

FIGURE 6.57 The coefficients a_1^4, a_3^4, a_2^4 related to the power spectrum $\Upsilon(|S_{hh}|^2)$, $\Upsilon(|S_{hv}|^2)$, $\Upsilon(|S_{vv}|^2)$ respectively.

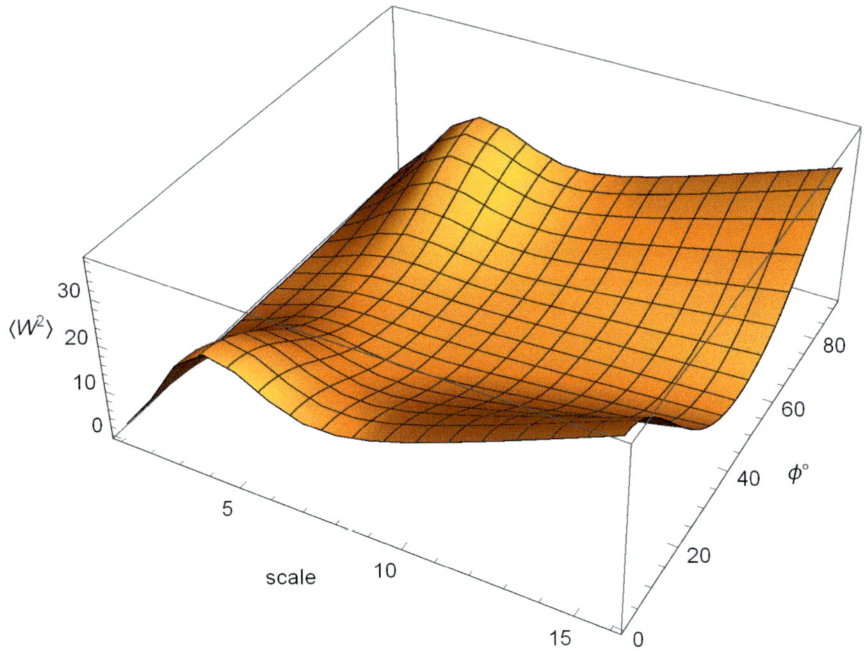

FIGURE 6.58 The co-pol WASP computed by the numerical model with the initial conditions in Table 6.1.

The co-polar WASP will therefore not be symmetric with respect to $\phi = 45°$, because the HH and VV power spectra can be different. Whereas, in the cross-pol case, because of the reciprocity of the target, the HV and VH spectra, corresponding to $\phi = 0°$, $\phi = 90°$ respectively, are equal.

The co-polar WASP, computed with the same initial conditions of the cross-polar case, is shown in Figure 6.58. A cross-section highlighting the dependence on scale at wave orientation angle $\phi = 0°$, 22°, 45°, 90° is shown in Figure 6.59, while the dependence on wave orientation angle at scales $s = 2°, 2^3, 2^4$ is shown in Figure 6.60.

As we have anticipated from the functional dependence of the coefficients in the linear combination of the power spectra (Equation 6.53), the co-polar WASP is not symmetric in the orientation angle domain with respect to $\phi = 45°$.

The cross-section in the scale domain (Figure 6.59) confirms the considerations we have proposed above, namely: at $\phi = 0°$ the WASP reflects the auto-correlation function (exponential cosine) we have set for the HH power term; at $\phi = 45°$ we find the signature of the shorter correlation length exponential ACF of the HV power term; and at $\phi = 90°$ the shape of an exponential cosine ACF appears, similar to the one at $\phi = 0°$, this being because of the VV power term.

Regarding the dependence of the WASP on orientation angle, we note an absolute maximum at $\phi = 0°$ at all scales, this corresponding to the influence of the HH power spectrum, and a local maximum at $\phi = 90°$, this corresponding to the VV power spectrum.

We shall now discuss the case of an elliptical basis with $\phi = 0°$ and the cross-pol case.

The cross-pol power term in the elliptical basis with orientation angle $\phi = 0°$ is:

$$P^{\phi} = \langle \Omega_2' \Omega_2' \rangle^* = 2|\mathrm{Shv}|^2 \cos[\tau]^2 + 2|\mathrm{Shh}|^2 \cos[\tau]^2 \sin[\tau]^2$$
$$+4\mathcal{R}e\langle \mathrm{Shh\,Svv}^* \rangle \cos[\tau]^2 + |\mathrm{Svv}|^2 \cos[\tau]^2 \sin[\tau]^2$$

Note that the only off-diagonal element of the covariance matrix is ⟨Shh Svv*⟩. In the elliptical basis cross-correlation elements ⟨Shh Shv*⟩ and ⟨Shv Svv*⟩ do not appear, and therefore the WASP on this basis will not be sensitive to effects related to azimuthal symmetry.

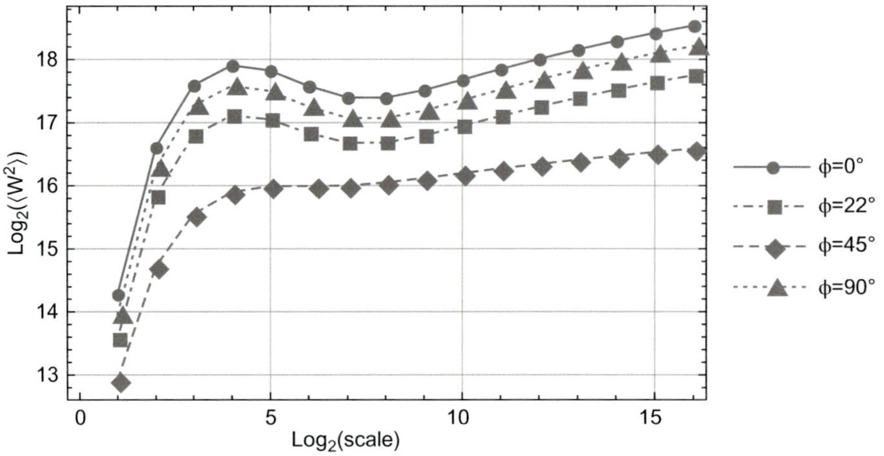

FIGURE 6.59 Cross-sections of the co-pol WASP highlighting the dependence on scale at orientation angle $\phi = 0°, 22°, 45°$.

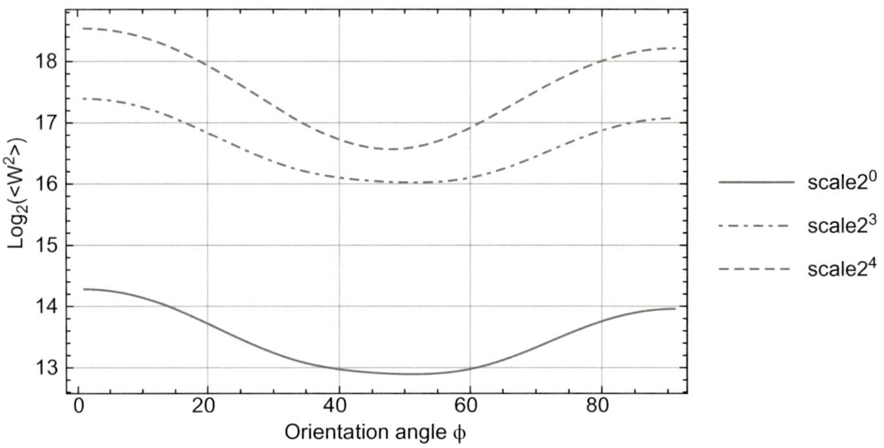

FIGURE 6.60 Cross-sections of the 3D WASP highlighting the dependence on orientation angle ϕ at scales $s = 2°, 2^3, 2^4$.

The power spectrum of the cross-pol power in the elliptical basis is:

$$Y\left(P^{\psi}\right) = 2a_1^4\, Y\left|\text{Shv}\right|^2 + \frac{1}{4}a_2^4 Y\left|\text{Shh}\right|^2 + \frac{1}{4}a_2^4 Y\left|\text{Svv}\right|^2 - a_2^4 Y\,\mathcal{R}e\left\langle \text{Shh}\,\text{Svv}^*\right\rangle$$

where $a_1 = \cos 2\tau$, $a_2 = \sin 2\tau$.

The coefficient a_1^4 weighs the power spectrum of the $|\text{Shv}|^2$ element, while the a_2^4 coefficient weighs the sum of the $\Upsilon|\text{Shh}|^2$ and $\Upsilon|\text{Svv}|^2$ power spectra and the spectrum of the real part of the off diagonal covariance matrix element ⟨Shh Svv*⟩. The dependence of these coefficients on the ellipticity angle τ is shown in the graph of Figure 6.61.

We note a local maximum at linear polarization $\tau = 0°$, where the power spectrum of the cross-pol element $\Upsilon|\text{Shv}|^2$ dominates. With increasing ellipticity, the power spectra of the co-pol terms weigh in, until at circular polarization $\tau = \mp 45°$ (left and right) the total power spectrum is affected only by the co-pol elements power spectra.

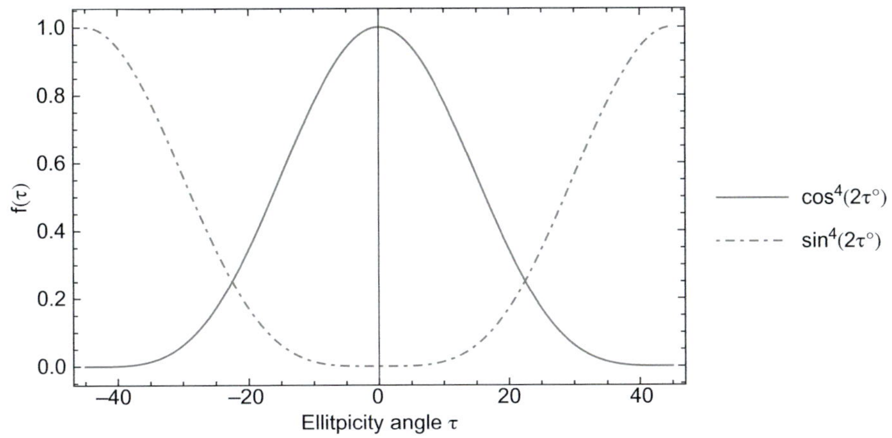

FIGURE 6.61 Dependence on the ellipticity angle τ of the coefficients a_1^4 and a_2^4.

The cross-pol WASP for an elliptical polarization basis is shown in Figure 6.62. Note the symmetry with respect to circular left and right polarization and the corresponding maxima of the wavelet variance given by the ACFs of the co-pol power terms.

The cross-section of the 3D WASP shown in Figure 6.63 highlights how the correlation properties that are set as initial conditions are reflected in the wavelet variance. We see that at cross-pol linear polarization, the signature points out the exponential ACF with short correlation length of the HV power term, while at circular polarization we observe the higher variance and the shape of the exponential cosine ACF with larger correlation length provided by the co-pol power elements.

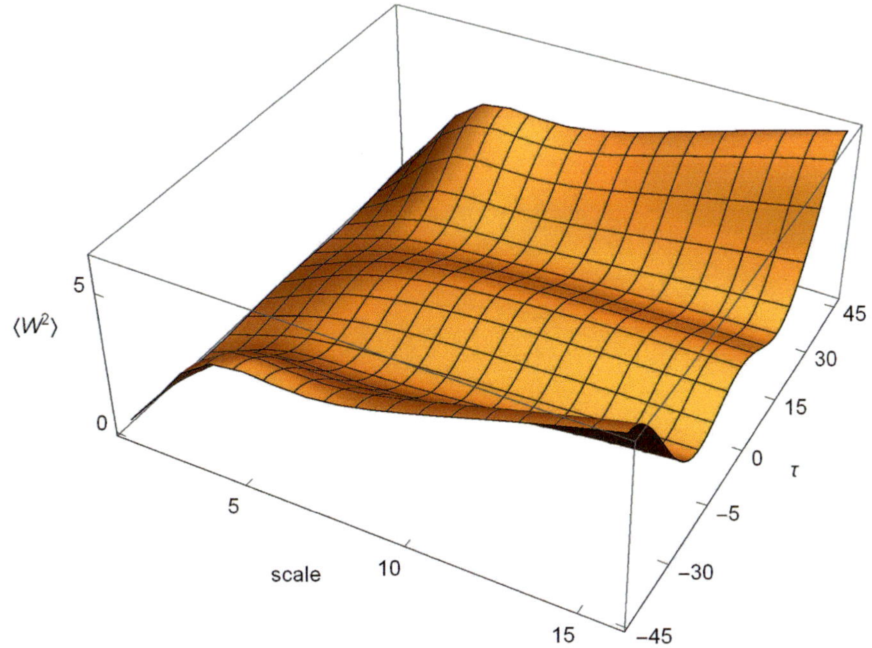

FIGURE 6.62 The cross-pol WASP for an elliptical polarization basis.

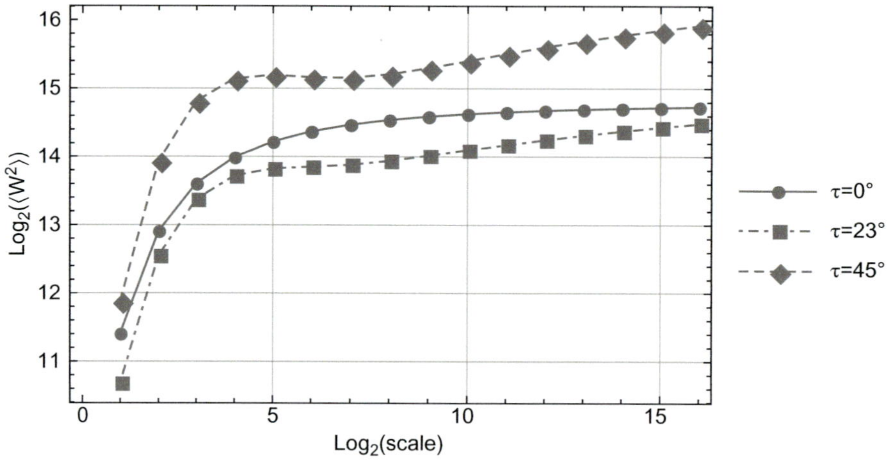

FIGURE 6.63 Cross-sections of the cross-pol WASP for an elliptical basis highlighting the dependence on scale at ellipticity angle $\tau = 0°, 22°, 45°$.

The cross-section in the scale domain (Figure 6.64) reveals the symmetry of the signature with respect to ellipticity angle, with wavelet variance maxima at linear and circular polarization.

The set of WASP analyses presented so far are indeed different pictures, pictures taken from slightly different viewing angles, of the same subject. The underlying mechanism boils down to the fact that the spatial statistic (textural properties) of the observed random field when measured on an orthogonal H, V basis are embedded in the elements of the covariance matrix on that basis. When the basis is changed, when the target is observed in a different way, the statistic (textural properties) of the covariance elements are mixed together and yield a different textural measure when the wavelet transform is applied. The resulting wavelet measure will act as a summary of the two-point statistic of all covariance matrix elements, and therefore afford the advantage of feature reduction. Moreover, the choice of the polarization basis will allow for enhancing effects that are more important for an application. For example, the choice of the linear cross-pol basis will provide a way to identify effects related to azimuthal symmetry.

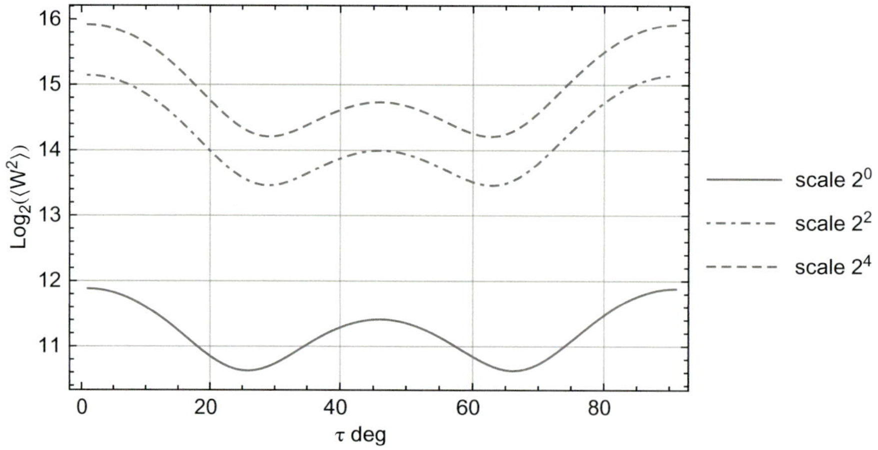

FIGURE 6.64 Cross-sections of the cross-pol WASP for an elliptical polarization basis highlighting the dependence on ellipticity angle τ at scales $s = 2°, 2^3, 2^4$.

WASP analysis of SAR full-pol data has been undertaken in different application contexts (see [48–51]). Next, we shall illustrate an example that is relevant for the main theme of the book.

6.11.2 WASP Analysis of a PALSAR Full-Pol Data Set

Whereas the WASP numerical model was based on the continuous wavelet transform, full-pol SAR data require an implementation in the discrete setting, i.e. the use of the discrete wavelet transform. We need first to explain how this is done.

In the case of full-pol PALSAR, data are delivered in complex scattering matrix form:

$$\underline{\Omega} = \begin{bmatrix} S_{hh} & \sqrt{2}\,S_{hv} & S_{vv} \end{bmatrix}$$

Data are first multi-Looked by averaging in range and azimuth (4 Looks) and transformed into a covariance matrix on the orthogonal (H, V) basis:

$$\underline{C} = \begin{bmatrix} \langle S_{hh}S_{hh}^* \rangle & \langle S_{hh}S_{hv}^* \rangle & \langle S_{hh}S_{vv}^* \rangle \\ \langle S_{hv}S_{hh}^* \rangle & \langle S_{hv}S_{hv}^* \rangle & \langle S_{hv}S_{vv}^* \rangle \\ \langle S_{vv}S_{hh}^* \rangle & \langle S_{vv}S_{hv}^* \rangle & \langle S_{vv}S_{vv}^* \rangle \end{bmatrix}$$

The covariance matrix can be mapped on a different basis by the linear operator:

$$T(\rho) = \frac{1}{(1+\rho\rho^*)} \begin{bmatrix} 1 & \sqrt{2}\rho & \rho^2 \\ -\sqrt{2}\rho & (1-\rho\rho^*) & \sqrt{2}\rho \\ \rho^{*2} & -\sqrt{2}\rho^* & 1 \end{bmatrix}$$

where $\rho = \dfrac{\cos 2\tau \sin 2\phi + j \sin\tau}{1+\cos 2\tau \cos 2\phi}$, ϕ and τ are the orientation and ellipticity angles of the polarization ellipse.

Since the wavelet frame operator $F(s,n)$ is also linear, the T, F operators commute. Thus:

$$c(s,n) = \mathrm{F}(s,n)T(\rho)\underline{C} = T(\rho)\mathrm{F}(s,n)\underline{C}$$

This point suggests a computationally efficient way of computing the WASP. First, apply the frame operator to the covariance matrix elements. This operation is tantamount to a series of nine wavelet coefficients, $c_{j=1..9}(s,n)$, since the covariance matrix can be represented by nine real numbers. Following this, the coefficients related to a specific power term (e.g. cross-pol) are mapped into the required (ϕ,τ) space on the new basis.

We shall now illustrate this technique with an experiment and using a fine beam fully polarimetric (FBFP) slant range ALOS-PALSAR dataset acquired over an area of the Congo River Floodplain (center coordinate 0° 30′ 75″ N, 18° 16′ 21.9″ E). The data set was provided by JAXA in the framework of the ALOS PI program. The original slant range scattering matrix data set (complex I-Q) was converted to covariance matrix representation and multi-Looked by four in cross-range. Therefore, spatial sampling in range and cross-range is approximately 9.36 m x 13.47 m. A Pauli decomposition of the dataset over the area of interest is shown in Figure 6.65 as a color composite, where the green, red and blue channels are assigned to volume scattering, double bounce and single bounce scattering mechanisms, respectively.

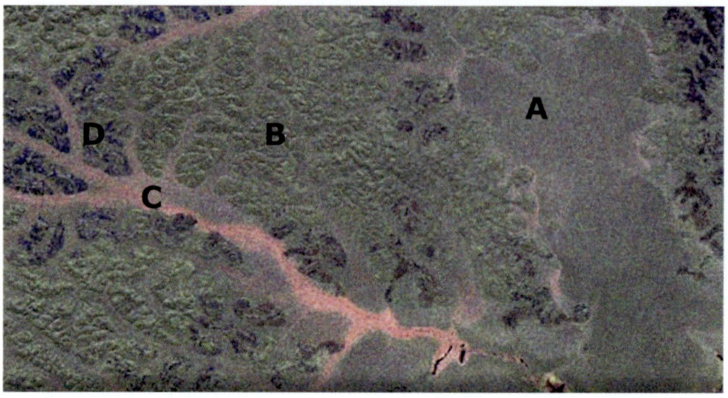

FIGURE 6.65 The PALSAR full-pol data set acquired over the Congo floodplain. The image is a Pauli decomposition of the covariance matrix data (averaged by four in cross-range), represented in an RGB composite with green, red and blue channels assigned to volume scattering, double bounce and single bounce scattering mechanisms, respectively. Areas corresponding to the thematic classes are marked as follows: (A) swamp forest, (B) lowland primary forest, (C) flooded forest, (D) gallery forest. The image is averaged by four in cross-range with respect to the WASP data for illustration purposes. Palsar PLR_SLC dataset courtesy of JAXA PI program.

The thematic context is similar to the one discussed in Section 6.8 and includes the swamp forest, the lowland primary forest, the flooded swamp forest and a formation of gallery forests. Gallery forests are stretches of forest flowing into regions sparsely populated with trees (such as savannas). Areas of interest (ROI) around the four classes are selected by visual inspection and expert knowledge, and are marked by letters in Figure 6.65: (A) swamp forest, (B) lowland primary forest, (C) flooded swamp forest, (D) gallery forest.

The WASP signatures of the four classes are shown in Figure 6.66. These 3D representations suggest immediately that the highest polarimetric diversity is provided by the swamp forest signature, while the most textured classes are the lowland and gallery forest classes.

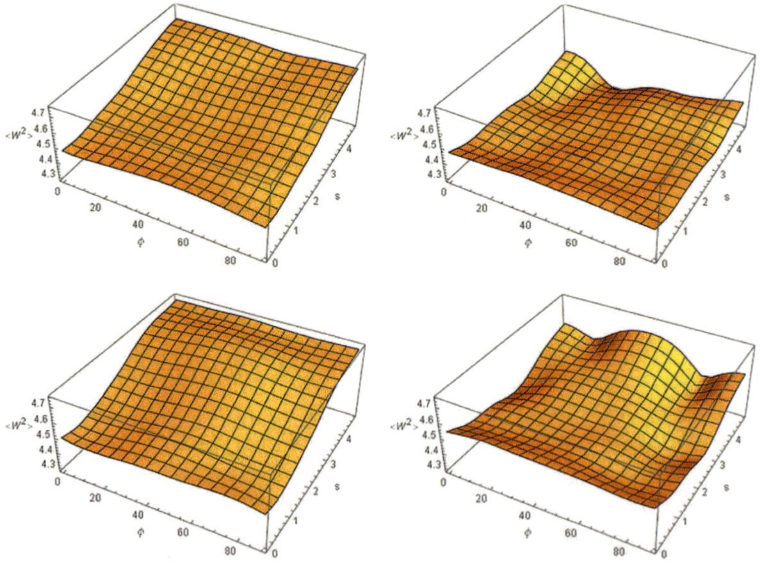

FIGURE 6.66 WASP signatures for the four classes (clockwise from top: lowland, swamp, gallery, flooded).

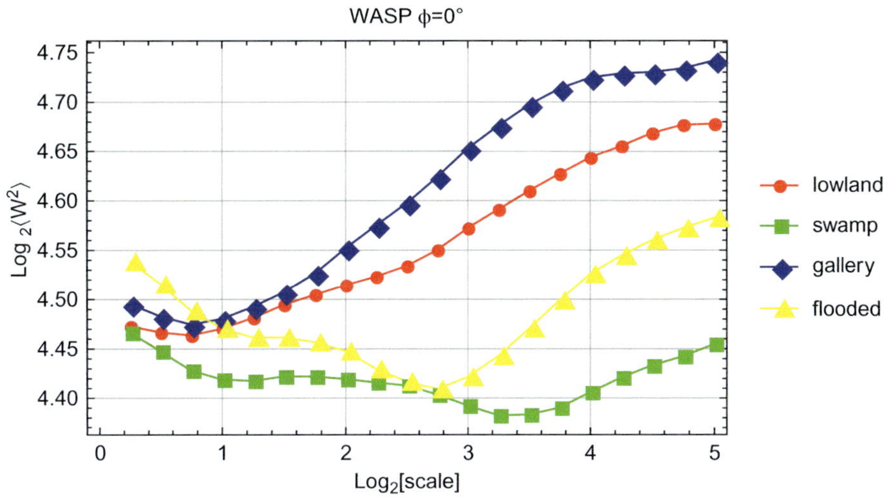

FIGURE 6.67 WASP scale dependence at orientation angle $\phi = 0°$ *(HV)*.

The scaling dependence of the wavelet variance within the ROIs is highlighted by the WASP cross-sections at selected polarization bases in Figure 6.67 ($\phi = 0°$) and Figure 6.68 ($\phi = 45°$).

At all orientation angles, the gallery forest signature yields the highest energy in the power spectrum of all classes in all scales and polarizations. This is therefore the most textured class. At =45°, it also presents strong correlations up to a sill around scale 2^4 with the onset of periodicity thereafter.

In contrast, the swamp forest signature is almost flat in all polarization states, suggesting the presence of a pure white noise random field. This is therefore the less textured class. As we pointed out in previous experiments (see Section 6.8) this statistic is evidence of the smoother upper canopy distribution with respect to the lowland forest.

The scaling behavior of the lowland rainforest signature (red line) indicates persistent correlations up to scale 2^5 (typical of a WSS process with exponential ACF), and no major dependence on polarization state (see Figure 6.68).

The scaling law of the flooded forest is remarkably different between the polarization states. At $\phi = 45°$ the trend is almost identical to the signature of the lowland forest, while at $\phi = 0°$ the trend becomes similar to the swamp forest signature for short scales up to $2^{2.5}$, but showing slightly higher correlations for longer scales. This fact confirms that the flooding conditions increase the double-bounce scattering component and hence affect the polarization configuration where the co-polar components come into play ($\phi = 45°$).

The polarimetric dependence of the signatures is depicted in Figure 6.69 with WASP cross-sections at scale $s = 2^1, 2^2, 2^5$. The lowland and the swamp forest signatures present the least polarimetric diversity on all scales. We surmise that at L-band the penetration depth is such as to drive the spatial statistic mainly from volume scattering, where the scattering center orientation and shape cannot contribute major diversity at different wave orientation angles.

While the gallery forest signature presents at short scale a marked maximum at $\phi = 45°$, where the co-pol power elements dominate the power spectrum. This effect is caused by the presence of strong single-bounce return from the bare soil or cultivated soil between the remnant forested ribbons. Indeed, the Pauli decomposition image confirms the presence of single bounce backscatter (blue color). The polarimetric diversity of the signature, however, decays fast with increasing scale, probably because at a larger scale the spatial differences between the patches of soil and vegetation are washed out.

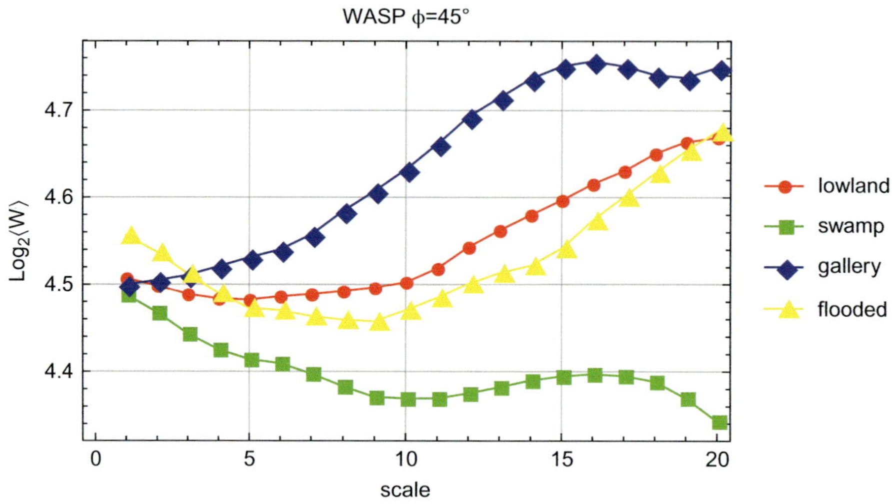

FIGURE 6.68 WASP scale dependence at orientation angle $\phi = 45°$.

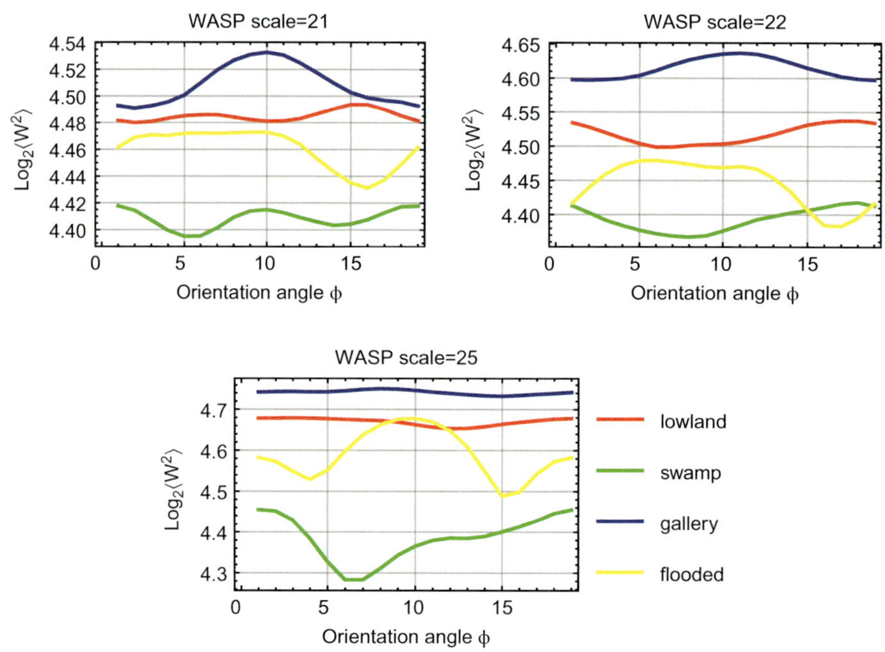

FIGURE 6.69 Polarimetric diversity afforded by the four classes WASP signatures at scales $s = 2^1, 2^2, 2^5$.

The flooded forest signature displays an opposing trend, with the greatest polarimetric diversity at large scale $s = 2^5 \approx 300\ m$. In flooding conditions there is a strong dihedral retro-reflection between the water surface and the forest trunks. This scattering component enhances the co-pol HH power with respect to the cross-pol HV power (volume scattering) and as a consequence also amplifies the difference between the wavelet statistic at orientation angle $\phi = 45°$ and $\phi = 0°$.

NOTE

1. Mathematica is a software tool by Wolfram Research, Inc. for doing mathematics.

REFERENCES

1. J.W. Goodman, "Statistical properties of Laser speckle patterns," in L. Speckle, J.C. Dainty (editors), *Related Phenomena*, Springer-Verlag, 1975.
2. E. Jakeman, P.N. Pusey, "A model for non-Rayleigh sea echo," *IEEE Trans. On Antennas and Propagation*, vol. AP-24, no. 6, Nov. 1976.
3. C.J. Oliver, "The representation of correlated clutter textures in coherent images," *Inverse Problems*, 4, 1988, 843–866.
4. C.J. Oliver, "A model for non-Rayleigh scattering statistics," *Optica Acta*, 1984, vol. 31, no. 6, 701–722.
5. C.J. Oliver, "The interpretation and simulation of clutter textures in coherent images," *Inverse Problems*, 2, (1986), 481–512.
6. M.L. Williams et al., Influence of canopy shape on SAR speckle distribution over woodland, *Proceedings of IEEE IGARSS* 1997, 755–757.
7. A.K. Fung, K.S. Chen, *Microwave Scattering and Emission Models for Users*, Artech House, 2010.
8. S.R. Cloude, *Polarisation, Applications in Remote Sensing*, Oxford University Press, 2010.
9. M. Petrou, P.G. Sevilla, *Dealing with Texture*, John Wiley & Sons, 2006.
10. R.M. Haralick, K. Shanmugan, I. Dinstein, "Textural features for image classification," *IEEE Trans. Syst. Man. Cybern.*, vol. SMC-8, no. 6, 610–621, Nov. 1973.
11. P.C. Chen, T. Pavlidis, "Segmentation by texture using correlation," *IEEE Trans. Patt. Anal. Machine Intell.*, vol. PAMI-5, 64–69, Jan. 1983.
12. R.L. Kashyap, R. Chelappa, A. Khotanzad, "Texture classification using features derived from random fields models," *Patt. Recogn. Lett.*, vol. 1, 43–50, 1982.
13. F.T. Ulaby, F. Kouyate, B. Brisco, T.H. Williams, "Textural information in SAR images," *IEEE Trans. Geosci. Remote Sens.*, vol. GRS-24, no. 2, 235–245, Mar. 1986.
14. E. Nezry, A. Lopes, R. Touzi, "Detection of structural and textural features for SAR images filtering," *Proceedings of IGARSS '91*, vol. 4, 2169–2170, 1991.
15. J.M. Coggins, A.K. Jain, "A spatial filtering approach to texture analysis," *Patt. Recogn. Lett.*, vol. 3, no. 3, 195–203, 1985.
16. A.C. Bovik, M. Clark, S. Geisler, "Multichannel texture analysis using localized spatial filters," *IEEE Trans. Patt. Anal. Machine Intell.*, vol. 12, no. 1, 55–73, Jan. 1990.
17. P. Kumar, E. Foufoula-Georgiou, "Wavelet analysis for geophysical applications," *Rev. Geophys.*, vol. 35, no. 4, 385–412, Nov. 1997.
18. E. Foufoula-Georgiou, P. Kumar (editors), *Wavelets in Geophysics*, Academic Press Inc, 1994.
19. E. Chandrasekhar, V.P. Dimri, V.M. Gadre, *Wavelets and Fractals in Earth System Sciences*, CRC Press, 2013.
20. S. Mallat, *A Wavelet Tour of Signal Processing*, 2nd ed., pp. 67–124, Academic Press, San Diego, 1999, ISBN 0-12-466606-X.
21. I. Daubechies, *Ten Lectures on Wavelets*, Society for Industrial and Applied Mathematics, Philadelphia, PA, 1992, ch. 2.
22. M. Vetterli, J. Kovacevic, V.K. Goyal, *Foundations of Signal Processing*, Cambridge University Press, 2014.
23. B.B. Mandelbrot, *The Fractal Geometry of Nature*, W.H. Freeman, New York, 1999.
24. K. Falconer, *Fractals: A Very Short Introduction*, Oxford University Press, 2013.
25. G.W. Wornell, "Wavelet-based representations for the 1/f family of fractal processes," *Proceedings of the IEEE*, vol. 81, no, 10, 1993.
26. C. Oliver, S. Quegan, *Understanding Synthetic Aperture Radar Images*, Artech House Inc., 1998.
27. A. Papoulis, *Probability, Random Variables and Stochastic Processes*, McGraw-Hill, Inc. 1991, ch. 10–2.
28. M. Simard, G. De Grandi, K.P.B. Thomson, G.B. Benie, "Analysis of speckle noise contribution on wavelet decomposition of SAR images," *IEEE Transactions on Geoscience and Remote Sensing*, vol. 36, no. 6, 1953–1962, Nov. 1998.
29. M. Simard, G.F. De Grandi, K. Thomson, "Adaptation of the Wavelet Transform for the Construction of Multi-scale Texture Maps of SAR images," *Canadian Journal of Remote Sensing*, vol. 24, no. 3, 264–285, Sept. 1998.
30. S.M. Kay, *Fundamentals of Statistical Signal Processing-Estimation Theory*, Prentice Hall, London, UK, 1993, ch.9, 294–299.
31. M.J. Shensa, "The discrete wavelet transform: Wedding the à trous and mallat algorithms," *IEEE Trans. On Signal Processing*, vol. 40, no. 10, pp. 2464–2482, Oct. 1992.

32. S. Mallat, S. Zhong, "Characterization of signals from multiscale edges," *IEEE Transactions on Pattern Analysis and Machine Intelligence*, vol. 14, no. 7, 710–732, Jul. 1992.
33. G. De Grandi, R.M. Lucas, J. Kropacek, "Analysis by Wavelet Frames of Spatial Statistics in SAR Data for Characterizing Structural Properties of Forests," *IEEE Trans. on Geoscience and Remote Sensing*, vol. 47, no. 2, 2009.
34. C. de Wasseige, J. Flynn, D. Louppe, F. Hiol, P. Mayaux, *Les forêts du bassin du Congo – État des Forêts*, Weyrich ed., 2013. ISBN: 978-2-87489-298-1.
35. E.C. De Grandi, E. Mitchard, I.H. Woodhouse, G.F. De Grandi, "Spatial wavelet statistics of SAR backscatter for characterizing degraded forest: A case study from cameroon," *IEEE Journal of Selected Topics in Applied Earth Observations and Remote Sensing*, vol. 8, no. 7, Jul. 2015.
36. G.F. De Grandi, M. Leysen, J.S. Lee, D. Schuler, "*Radar reflectivity estimation using multiple SAR scenes of the same target: Techniques and applications*," *Proceedings of IEEE IGARSS '97*, pp. 1044, Aug. 3–8 1997, Singapore.
37. Elsa C. De Grandi, E. Mitchard, D. Hoekman, "Wavelet based analysis of TanDEM-X and LiDAR DEMs across a tropical vegetation heterogeneity gradient driven by fire disturbance in indonesia," *Remote Sens.*, 8, 641, 2016.
38. E.C. De Grandi, "Spatial and temporal statistic of SAR and InSAR observations for providing indicators of tropical forest structural changes due to forest disturbance," PhD Thesis, University of Edinburgh, School of GeoSciesces, 2017.
39. SARMAP, 2012. SARscape. [Online]. Available at: http://www.sarmap.ch/pdf/SARscapeTechnical.pdf.
40. M. Unser, "Local linear transforms for texture measurements," *Signal Processing*, vol. 11, no. 1, pp. 61–79, Jul. 1986.
41. M. Unser, M. Eden, "Multiresolution feature extraction and selection for texture segmentation," *IEEE Trans. Patt. Anal. Machine Intell.*, vol. 11, no. 7, 717–728, Jul. 1989.
42. M. Unser, "Texture classification and segmentation using wavelet frames," *IEEE Trans. Image Processing*, vol. 4, no. 11, 1549–1560, Nov. 1995.
43. R.O. Duda, P.E. Hart, D.G. Stork, *Pattern Classification*, John Wiley & Sons, Inc., 2001.
44. G.C. Dargie, S.L. Lewis, I.T. Lawson, E.T. Mitchard, S.E. Page, Y.E. Bocko, S.A. Ifo, "Age, extent and carbon storage of the central Congo Basin peatland complex," *Nature*, 542(7639), 86–90, 2017.
45. Y.C. Tzeng, K.S. Chen, W. Kao, A.K. Fung, "A dynamic learning neural network for remote sensing applications," *IEEE Trans. Geosci. Remote Sens.*, vol. 32, no. 5, 1096–1102, Sept. 1994.
46. J.J. Van Zyl, H.A. Zebker, C. Elachi, "Imaging radar polarization signatures: Theory and observations," *Radio Science*, vol. 22, 529–543, 1987.
47. F.T. Ulaby, J.J. Van Zyl, "Wave properties and polarization," in F.T. Ulaby, C. Elachi (editors), *Radar Polarimetry for Geoscience Applications*, Artech House Inc, 1991.
48. G.F. De Grandi, J.S. Lee, D.L. Schuler, P. Siqueira, T. Ainsworth, M. Simard, "*Multiresolution analysis of polarimetric SAR data using wavelets*," *Proceedings of IEEE IGARSS '02*, CD-ROM paper TU08_260, Toronto, Canada, 2002.
49. G.F. De Grandi, D. Hoekman, J.-S. Lee, D. Schuler, T. Ainsworth, "A wavelet multiresolution technique for polarimetric texture analysis and segmentation of SAR images," *Proceedings of Geoscience and Remote Sensing Symposium*, vol. 1, 2004.
50. A. Gambardella, M. Migliaccio, G. De Grandi, "Wavelet polarimetric SAR signature analysis of sea oil spills and look-alike features," *Geoscience and Remote Sensing Symposium*, 983–986, 2007.
51. G.F. De Grandi, D. Hoekman, J-S. Lee, D. Schuler, T. Ainsworth, "*A Wavelet multiresolution technique for polarimetric texture analysis and segmentation of SAR images*," *Proceedings of IEEE IGARSS '04*, Anchorage, AK, USA, 20–24 Sept. 2004, CD-ROM paper 2TU_31_06.
52. C.J. Oliver, "On the simulation of coherent clutter textures with arbitrary spectra," *Inverse Problems*, 3, 1987, 463–475.
53. M. Vetterli, J. Kovacevic, *Wavelets and Subband Coding*, Prentice Hall PTR, Englewood Cliffs, NJ, 1995, ISBN 0-13-097080-8.
54. S.G. Mallat, "A theory for multi-resolution signal decomposition: The wavelet representation," *IEEE Trans. on Pattern and Machine Intell.*, vol. 11, 674, July 1989.
55. J.J. Van M. Unser, "Texture classification and segmentation using wavelet frames," *IEEE Trans. Image Processing*, vol. 4, no. 11, 1549–1560, Nov. 1995.
56. K. Tragl, "Polarimetric radar backscattering from reciprocal random targets," *IEEE Trans. Geosci. Remote Sens.*, vol. 28, no. 5, 856–864, Sept. 1990.

7 Hitting Corners

The Lipschitz Regularity, a Measure of Discontinuities in Radar Images Connected with Forest Spatial Distribution

7.1 INTRODUCTION

The celebrity pianist and conductor Daniel Barenboim wrote: "The power of music lies in its ability to speak to all aspects of the human being, the animal, the emotional, the intellectual, and the spiritual. Music teaches us, in short, that everything is connected."

We shall try here to establish yet another connection, a connection leading from the abstract mathematical theories of function regularity, singular distributions and wavelet frames, up to the characterization of discontinuities in a natural or man-made target, as seen by a remote-sensing radar. In our case, it is the ability of mathematics to speak to aspects of the human consciousness, to aspects of the reality that lies outside us, such as forests and grasslands, to aspects of an artifact created by the human mind, an orbiting radar. This connection prompts some deep questions: how did Mr Lipschitz imagine that a weird jumping mathematical function could be characterized by two numbers why we find beauty in the underpinning theory of Sobolev spaces; why we find shadows of mathematical objects in nature, and do those abstract mathematical conceptions have a Platonic reality?

The author seems to put forward excuses for having proposed a chapter that is a bit far-fetched, a bit too abstract to be in line with the rest of the book. Yet, as a reward for patiently following the journey through the connection, the reader will realize when studying Chapter 8 that this material has laid the ground for the development of a novel image processing algorithm and its application to a classification study of the tropical forest in South America.

7.2 THE LIPSCHITZ CONDITION

In mathematical analysis, a hierarchy of functions in terms of their increasing "degree of smoothness" can be defined in the following way: L_p (measurable and integrable) functions, bounded functions, continuous functions, functions satisfying a Lipschitz condition, differentiable functions, n-times differentiable functions, infinitely differentiable functions, analytic functions, functions expressible in power series (entire functions), polynomials, and constants.

Here we are interested in finding measures of discontinuities (edges, point targets) in SAR backscatter fields (or coherence), because these can provide valuable information for the detection of thematic features, such as the margin between forest and grassland, or discontinuities in the forest structure, which in turn could be tell-tale signs of disturbances, such as logging or fires.

For our purpose, we focus attention on the upper part of the hierarchy of functions, on the class of functions that satisfy a Lipschitz condition. The reason for this is, as will become apparent when considering the precise definition of the Lipschitz condition, that this class of functions include precise mathematical definitions of discontinuities and singularities, such as unit step and Dirac delta functions. These mathematical objects are the idealized shadows (in a Platonic sense) of the features we want to characterize in the backscatter signal domain, such as margins between clear-cut and forest, or man-made constructions within the forest. Yet these mathematical objects can be transposed, by a suitable wavelet transform, into the discrete setting of a SAR backscatter image, where measures are then provided of the discontinuities that provide a thematic interest.

To understand how this path develops, let us start by considering the basic definition of the Lipschitz condition.

A function $f(x)$ defined on an interval I is said to satisfy a Lipschitz condition of order α if two constants can be found, α and K, such that for all points x_1 and x_2 in the interval the following condition holds [1]:

$$\left|f\left(x_1\right)-f(x_{2)}\right| \leq K\left|x_1-x_2\right|^{\alpha} \tag{7.1}$$

This class of functions is designated as $Lip_K\ \alpha$.

For example, the Heaviside Theta function (or unit step), defined as 0 for $x < 0$ and 1 for $x > 0$, is discontinuous at $x = 0$ but bounded with $K = 1$ and $\alpha = 0$ in the interval [−1, 1], and it is Lip 0.

Therefore the Lipschitz condition also places an upper bound on functions that contain a discontinuity (e.g. are not differentiable everywhere in an interval), and the exponent characterizes the type of discontinuity, as we shall see in a moment.

To go forward and introduce the link between the Lipschitz condition and the wavelet transform, let us consider the following argument.

Consider the function Ax^{α} with $0 < \alpha < 1$. This function is $Lip_A\ \alpha$, as proved in the following way [1].

Take $h> 0$ and $x > 0$. Then: $\frac{d}{dx}\left[\left(x+h\right)^{\alpha}-x^{\alpha}\right]=\alpha\left[\left(x+h\right)^{\alpha-1}-x^{\alpha-1}\right]<0$ since the exponents are negative. Therefore, $(x + h)^{\alpha} - x^{\alpha}$ is decreasing for all $x> = 0$. Hence, $(x + h)^{\alpha} - x^{\alpha} \leq h^{\alpha}$.

Now consider a wavelet transform, where the wavelet acts as a differential operator. The convolution of the function $f(x)$ with the wavelet dilated at scale s results in a (locally averaged) difference of two points of the function that are separated by the scale s:

$$W\left(x,s\right) \cong \left(x+s\right)^{\alpha}-x^{\alpha} \leq s^{\alpha}$$

This relation suggests that the evolution with scale of the wavelet transform amplitude may be used to estimate the Lip exponent. Since the wavelet transform can be realized in a discrete setting (for example, by wavelet frames) a bridge can then be established between an ideal mathematical concept and a signal processing method, which will allow us to extract regularity information from SAR data.

Our argument sets the direction to arrive at the rigorous theory underpinning the connection between wavelets and Lipschitz regularity, and indeed, even more generally, between wavelets and regularity measures of a wider (smoother) class of functions, such as n-differentiable ones. The theory at the basis of our analysis was developed by Stephan Mallat and is available in the literature [2–4]. We refer the reader to these works for detailed mathematical formal constructions and proofs of theorems. Here, we merely summarize the main ideas and the results, to set the groundwork for our analysis.

The starting point is to consider, in the continuous domain, wavelets that possess n vanishing moments. Then a theorem in [4] (Theorem 6.2) proves that the wavelet can be written as the nth derivative of a function $\vartheta(t)$ with fast decay (such as a Gaussian) and whose integral is $\int\vartheta(t)dt \neq 0$. In this case, the resulting wavelet transform acts as a multi-scale differential operator.

The condition on wavelet moments is:

$$\int x^k\psi\left(x\right)dx=0\ for\ 0<k<n \tag{7.2}$$

Given a wavelet that is the nth derivative of a smoothing function $\vartheta(x)$:

$$\psi(x) = (-1)^n \frac{d^n \vartheta(x)}{dx^n} \tag{7.3}$$

Then the wavelet transform, from Equation (7.3), and inverting the convolution and differential operators, is:

$$Wf(u,s) = f \oplus \psi_s(u) = s^n \left(f \oplus \frac{d^n \vartheta(x)}{dx^n} \right)(u) = s^n \frac{d^n}{dx^n}(f \oplus \theta_s)(u) \tag{7.4}$$

where $\psi_s(x) = \frac{1}{\sqrt{s}} \psi\left(\frac{-x}{s}\right)$ is the wavelet dilated by s, and u is the variable of the convolution operator. This equation tells us that, under the condition Equations (7.2) and (7.3), the wavelet transform will yield at scale s the n-th derivative of the signal averaged in a neighborhood proportional to scale s.

For example, taking $n = 1$, the transform amplitude will be the first derivative of the signal averaged around a point u by the smoothing kernel in a neighborhood proportional to the scale s. The maxima of the transform amplitude will then correspond to sharp discontinuities (edges).

The decay of the wavelet modulus with scale can be related to the Lipschitz condition in a neighborhood $x\epsilon]ab[$ by a theorem [4] (Theorem 6.3) stating that, given a wavelet with n vanishing moments, for $\alpha \le n$ a function is uniformly Lipschitz α in a neighborhood $x\epsilon]ab[$ if there exists a constant K such that the wavelet modulus satisfies in $]ab[$:

$$|Wf(x)| \le Ks^{\alpha + \frac{1}{2}} \tag{7.5}$$

For dyadic scales, this equation is equivalent to:

$$\log_2 |W_{2^j} f(x)| \le \log_2 K + \left(\alpha + \frac{1}{2} \right) j \tag{7.6}$$

Notice that the term ½ in Equations (7.5) and (7.6) stems from the normalization $\frac{1}{\sqrt{s}}$ of the continuous wavelet transform, this being norm preserving. Alternatively, the term ½ can be cancelled in Equations (7.5) and (7.6) by assuming the $\frac{1}{s}$ normalization in the transform.

Therefore the Lipschitz parameters K and α for an isolated singularity can be estimated by a linear fitting of the wavelet modulus maxima $|W_{2^j}|_{max}$ in a base-2 logarithmic scale. The theorem leading to Equation (7.5) provides us with the basic mechanism to estimate the Lipschitz parameters associated with an isolated singularity. As will become clear through the examples reported in the next section, an increasing trend of the wavelet moduli with scale indicate a positive Lipschitz regularity (e.g. an edge), while a negative trend is a tell-tale of a more singular structure (e.g. a point target).

Yet the representation of a signal by a function of a real variable is a mathematical idealization. All physical devices generate signals of finite resolution, and SAR is no exception. Let us consider a finite approximation to a pure singularity by modeling it by means of an averaging Gaussian kernel g_σ with variance σ^2. Let us assume that the wavelet has one non-vanishing moment and be defined as the derivative of a smoothing function θ_s, such as a Gaussian with variance β^2. Thus:

$$Wf(x,s) = s\frac{d}{dx}(f \oplus g_\sigma \oplus \theta_s)(x) = s\frac{d}{dx}(f \oplus \theta_{\sigma s})(x) = s\left(\frac{df}{dx} \oplus \theta_{\sigma s} \right)(x) \tag{7.7}$$

with

$$\theta_{\sigma s} = \frac{1}{s}\frac{1}{\sqrt{2\pi\left(s^2\beta^2+\sigma^2\right)}}e^{-\frac{x^2}{2\left(s^2\beta^2+\sigma^2\right)}}$$

The trend with scale of the wavelet maxima corresponding to sharp variations of $f(x)$ therefore becomes non-linear with scale for a smoothed singularity (see the terms in $\theta_{\sigma s}$ in comparison with θ_s in Equation (7.4). Therefore the linear regression in the log domain Equation (7.5) cannot be used to estimate the Lipschitz exponent in this case. Theorem (7.5) does not hold true, and it has to be modified as follows.

We can interpret Equation (7.7) as the wavelet transform of the original pure singularity $f(x)$ obtained with a different scaling law, provided by the function $\theta_{\sigma s}$ instead of θ_s. In this case, Equation (7.5) can now be applied to give [4] (Equation 7.6):

$$\left|Wf\left(x,s\right)\right| \le Ks^{\alpha+\frac{1}{2}}\left(1+\frac{\sigma^2}{s^2\beta^2}\right)^{\frac{\alpha-1}{2}} \tag{7.8}$$

This result encapsulates the fact that, for large scales, $s \gg \frac{\sigma}{\beta}$ the Gaussian smoothing is not felt and the wavelet moduli decay as $f\left(x,s\right) \le Ks^{\alpha+\frac{1}{2}}$, while at short scales $s < \frac{\sigma}{\beta}$ the Gaussian averaging conceals the underlying singularity, the function $f(x)$ becomes differentiable, and the wavelet modulus decays as $\left|\mathrm{W}f\left(x,s\right)\right| \le Ks^{n+\frac{1}{2}}$, where n is the number of vanishing moments.

Nonlinear regression of the wavelet modulus maxima with the function in Equation (7.8) is used in this case to estimate the Lipschitz parameters K, α, σ.

7.3 SINGULAR FUNCTIONS AND LIP PARAMETERS ESTIMATED BY WAVELET MAXIMA TRAJECTORIES IN THE SCALE DOMAIN

To get to grips with the Lipschitz condition and the connection with wavelets, let us start by studying some discontinuities represented by mathematical functions defined in the one-dimensional Euclidean real space $\mathcal{R}_1$.

For mathematical convenience, we consider a wavelet defined as the first derivative of a Gaussian function with zero mean and unit variance (see Figure 7.1):

$$f\left(t\right) = \frac{1}{\sqrt{2\pi}}\mathrm{e}^{\left(\frac{-t^2}{2}\right)}$$

$$Wf\left(t\right) = f' = -\frac{\mathrm{e}^{-\frac{t^2}{2}}t}{\sqrt{2\pi}} \tag{7.9}$$

The wavelet has one consecutive vanishing moment.

$$m_1\left(Wf\right) = \int_{-\infty}^{\infty} t\left(-\frac{\mathrm{e}^{-\frac{t^2}{2}}t}{\sqrt{2\pi}}\right)\mathrm{d}t = -1$$

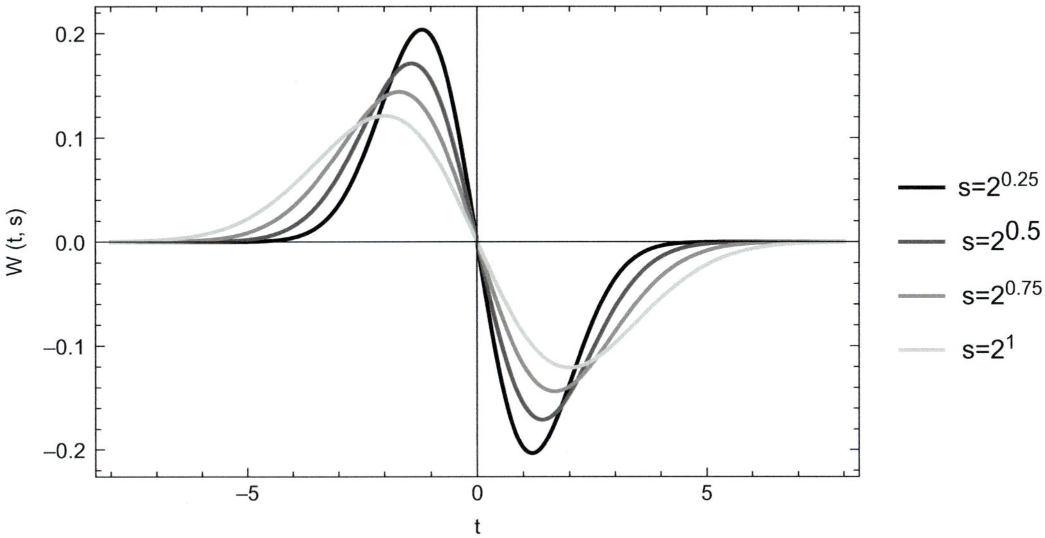

FIGURE 7.1 A wavelet which is the derivative of a Gaussian function, as a function of the scales corresponding to the first four voices.

$$m_2\left(Wf\right)=\int_{-\infty}^{\infty}t^2\left(-\frac{e^{-\frac{t^2}{2}}t}{\sqrt{2\pi}}\right)dt=0$$

$$m_n\left(Wf\right)=\frac{2^{-\frac{1}{2}+\frac{n}{2}}\left(-1+\left(-1\right)^n\right)\Gamma\left[1+\frac{n}{2}\right]}{\sqrt{\pi}}=\left\{-3,\ 0,\ -15,\ 0,\ -105,\ 0,\ldots.\right\}\ n>2$$

7.3.1 Step Function

As an appetizer, let us consider the Heaviside Theta function, also denoted in signal processing as the unit step function (see Figure 7.2).

A closed form solution for the wavelet transform can easily be derived in the frequency Fourier domain. The Fourier transform of the step function is:

$$\Im\left[\vartheta\left(x\right)\right]=\frac{1}{i\,\omega}+\pi\delta\left[\omega\right]$$

Where δ[ω] is the Dirac delta functional.

The Fourier transform of the wavelet dilated at scale s is:

$$\Im\left[\frac{-t/s}{\sqrt{2\pi}}e^{\left(\frac{-\left(\frac{t}{s}\right)^2}{2}\right)}\right]=ie^{-\frac{1}{2}s^2\omega^2}s^2\omega \tag{7.10}$$

Since convolution in the time domain is mapped on to multiplication in the Fourier domain, the wavelet transform of the step function using the 1/s normalization is:

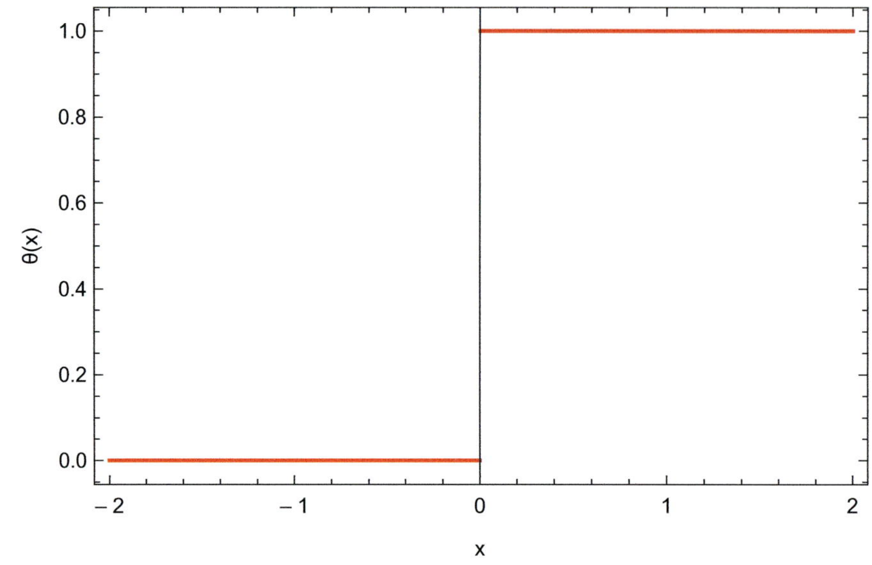

FIGURE 7.2 The Heaviside Theta function (denoted as the unit step function in signal processing).

$$W(t,s)=\frac{1}{s}\Im^{-1}\left[\frac{1}{i\omega}+\pi\delta[\omega]\right]\left[\mathrm{ie}^{-\frac{1}{2}s^2\omega^2}s^2\omega\right]=\frac{\mathrm{e}^{-\frac{t^2}{2s^2}}}{\sqrt{2\pi}}$$

Clearly, the scale s does not affect the amplitude of the wavelet coefficients, but only the decay rate from the maximum. The transform amplitude $W(t,s)$ as a function of t and parametrized by $s=\{2^{1/4},2^{1/2},2^{3/2},2\}$ is shown in Figure 7.3. The trajectory of the maxima with scale is a constant $Max[Wf(t,s)](s)=\mathrm{K}$, therefore from Equation (7.5) the step function is a *Lip* 0 function.

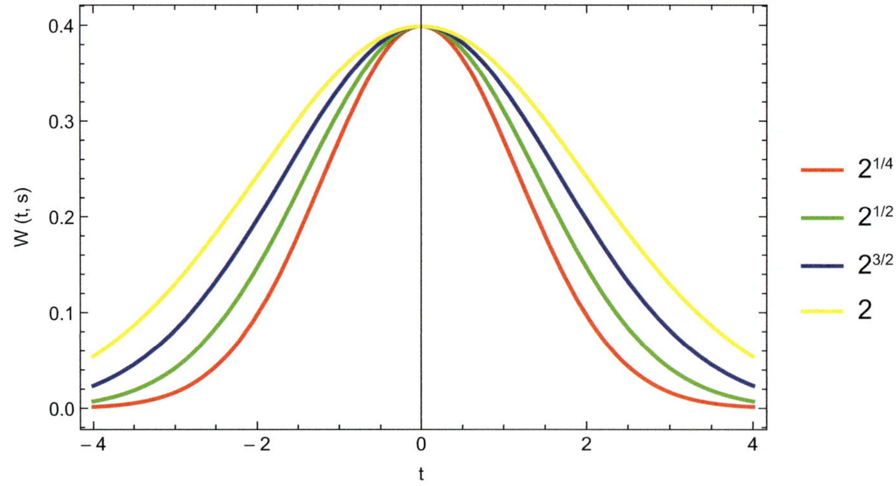

FIGURE 7.3 The wavelet transform amplitude of the Heaviside function, parametrized by scales $s=\{2^{1/4},2^{1/2},2^{3/2},2\}$.

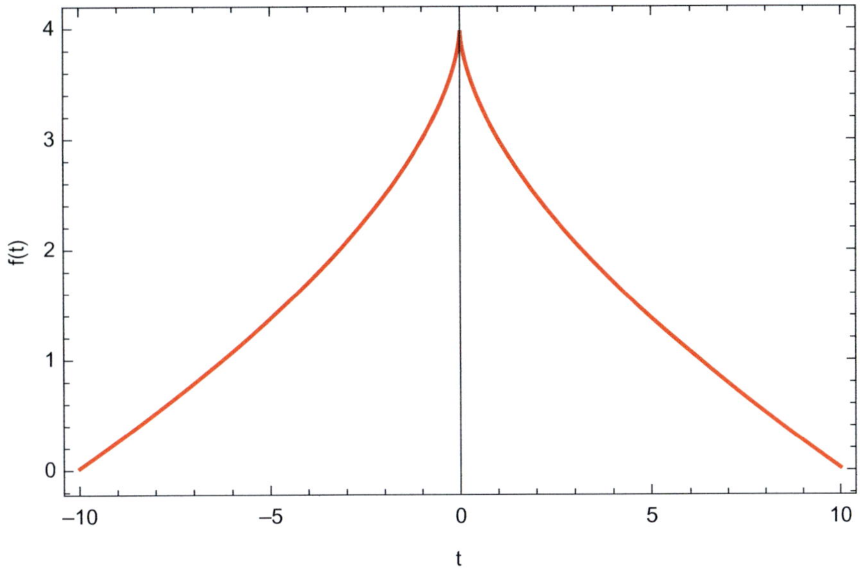

FIGURE 7.4 A cusp.

7.3.2 Cusp

Let us next consider a cusp (see Figure 7.4) defined by the following function:

$$f(t) = A - \text{Abs}\left[t^{0.6}\right]$$

The Fourier transform is:

$$\Im\left[f(t)\right] = \frac{1.44574}{\text{Abs}\left[\omega\right]^{1.6}} + 2A\pi\,\text{DiracDelta}\left[\omega\right]$$

From Equation (7.10) and by inverse Fourier transform the wavelet transform amplitude is:

$$W(t,s) = 0.48\sqrt{\frac{1}{s^2}}\left(s^2\right)^{0.3} t\,\text{Hypergeometric1F1}\left[0.7, \frac{3}{2}, -\frac{t^2}{2s^2}\right]$$

The amplitude $W(t, s)$ parametrized by $s = \{2^{1/4}, 2^{1/2}, 2^{3/2}, 2\}$ is shown in Figure 7.5. The amplitude maxima (computed at $t > 0$) in $\log_2$ scale and their linear fitting are plotted in Figure 7.6. From the linear fitting slope coefficient and Equation (7.6) we estimate *Lip* $\alpha = 0.6$.

The transform amplitude in the scale–space domain is also represented in Figure 7.7, and as a density plot in Figure 7.8. Notice how the wavelet maxima converge at fine scales towards the singularity. Yet as the scale increases, the maxima trajectories do not follow a straight path in the scale domain. Therefore the Lipschitz parameter estimator must include a suitable algorithm for following the maxima trajectory. This point will be expanded in the next chapter, where the Lipschitz regularity will be exploited for the generation of piece-wise smooth backscatter images.

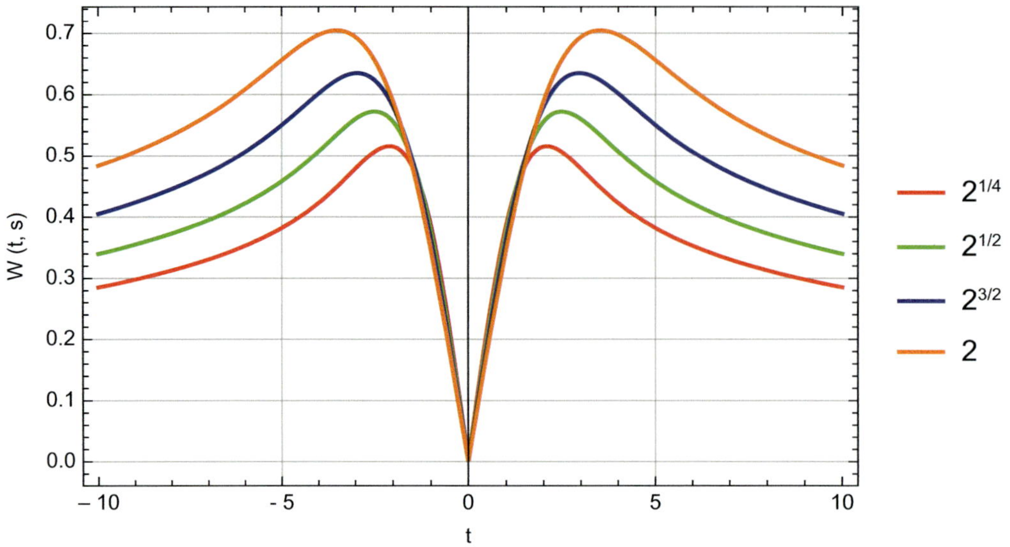

FIGURE 7.5 The wavelet transform amplitude of a cusp parametrized by scales $s = \{2^{1/4}, 2^{1/2}, 2^{3/2}, 2\}$.

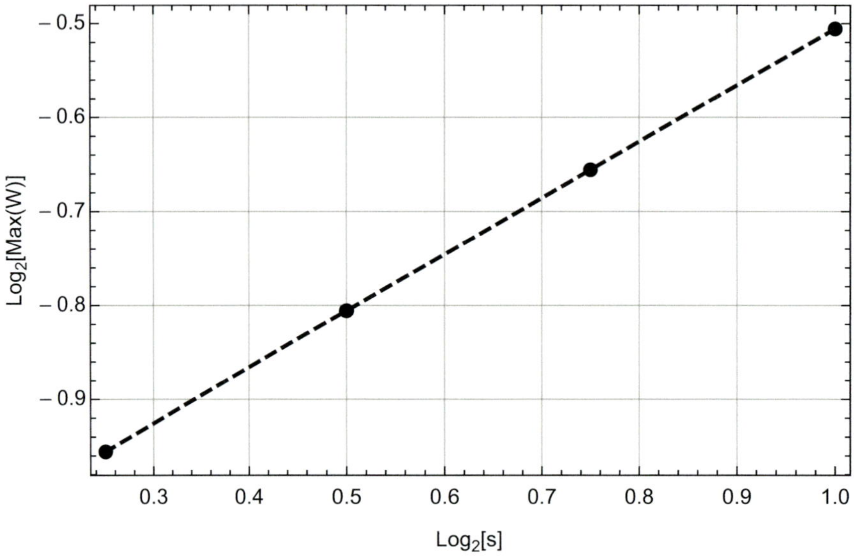

FIGURE 7.6 The maxima of the wavelet transform in Figure 7.5 and the linear fit in *a* log_2 domain.

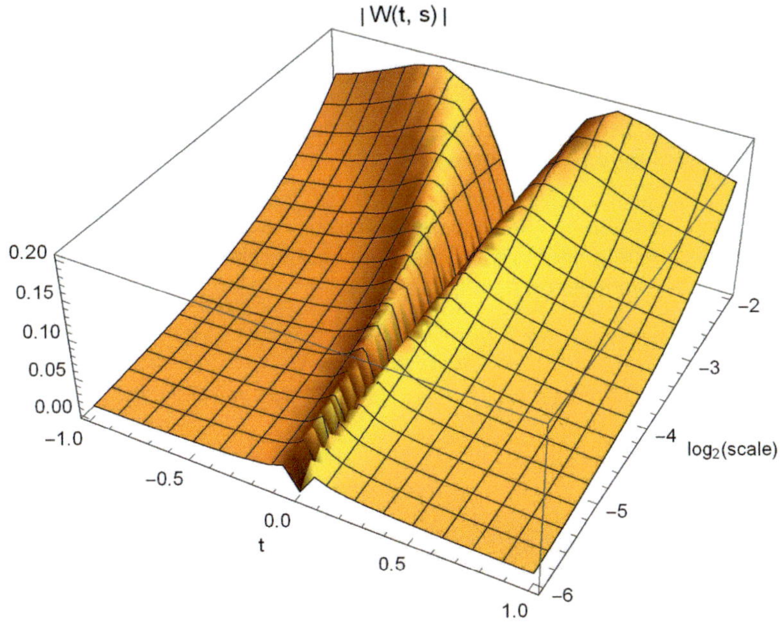

FIGURE 7.7 The cusp wavelet transform amplitude in the scale–space domain.

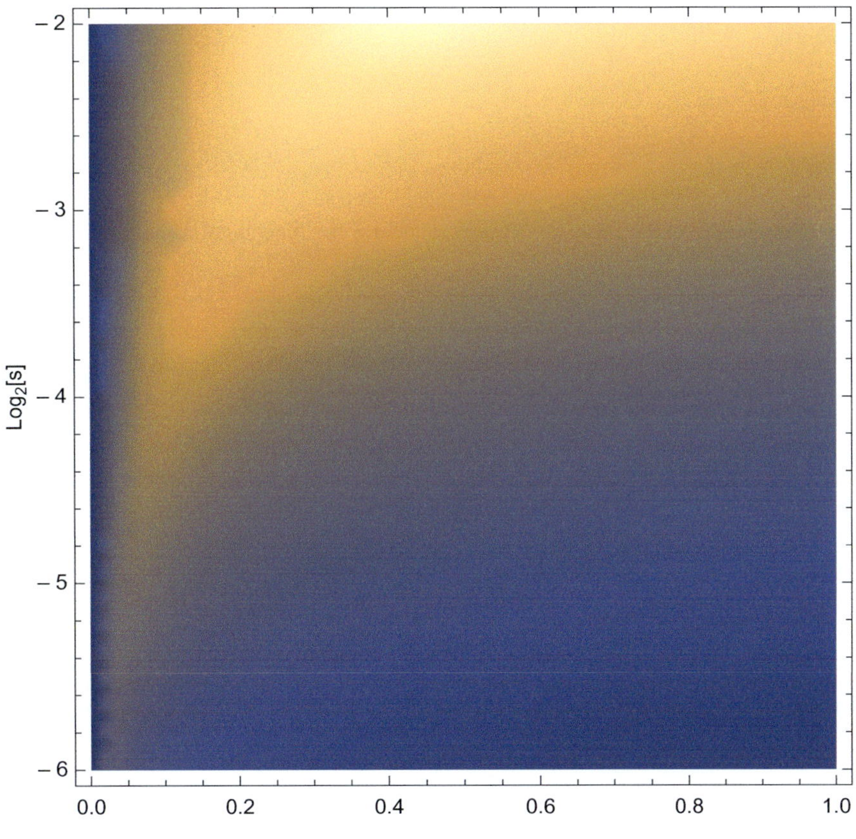

FIGURE 7.8 The cusp wavelet transform $W(t,s)$ represented as a density plot.

7.3.3 Impulse

We now turn our attention to signals of short duration and high energy, called impulses in signal processing. These signals appear in SAR imagery as a result of backscatter from hard targets (e.g. buildings, poles), but arise also in connection with the presence of speckle noise, or with intermittency caused by non-stationary targets. Interestingly, these discontinuities give rise to negative Lip α parameters.

From the mathematical point of view, the discontinuities are described by the Dirac delta functional $\delta(t)$. The t in this notation does not indicate a function of the variable t, but that the functional (function of functions) δ is defined on the t axis.

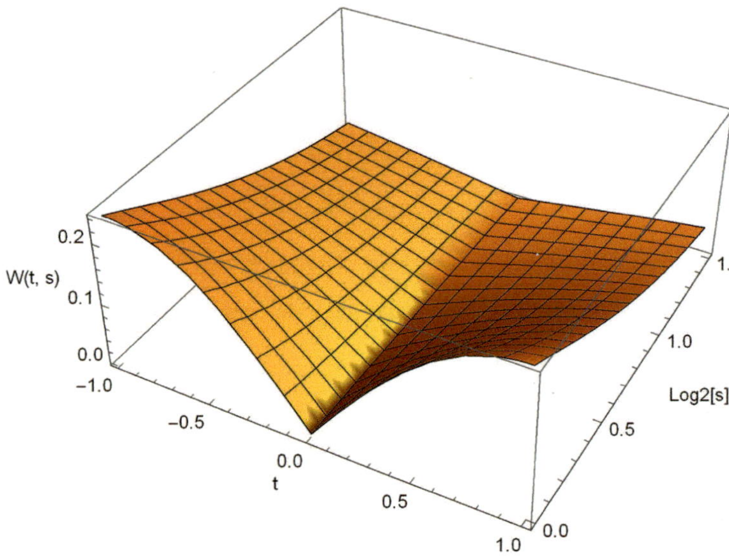

FIGURE 7.9 The modulus of the Dirac $\delta(t)$ wavelet transform as a function of the variable t and scale ($\log_2 s$).

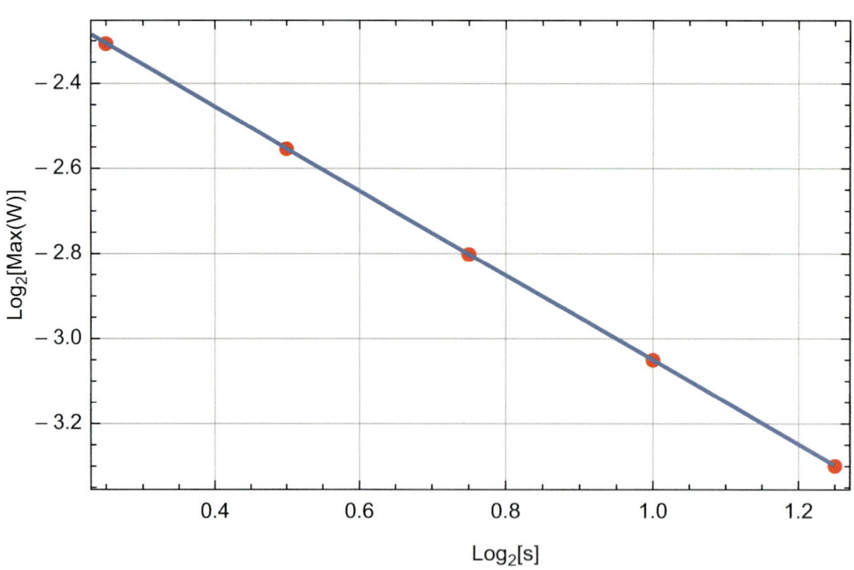

FIGURE 7.10 The Dirac $\delta(t)$ wavelet transform moduli maxima as a function of scale. The fitting line in the scale domain is overlaid.

The wavelet transform of the δ functional using the wavelet in Equation (7.10) can be derived in the frequency domain. Since $\Im(\delta(t)) = 1$, the wavelet transform is just the replica of the scaled wavelet:

$$W(t,s) = -\frac{e^{-\frac{t^2}{2s^2}}t}{\sqrt{2\pi}s^2}$$

The modulus $|W(t,s)|$ as a function of t and $\log_2 s$ is reported in Figure 7.9. It can be noticed how the amplitude maxima converge slowly for very short scales to the discontinuity at $t = 0$. Contrary to the case of the previously analyzed discontinuities, the convergence happens only in the limit to zero scale.

Moreover, the maxima are now decreasing with scale. The maxima related to samples at scales $2^{1/4}$, $2^{1/2}$, $2^{3/2}$, 2^1, $2^{5/4}$, $2^{3/2}$ are plotted in Figure 7.10 together with the fitting line. By the slope coefficient of the linear regression we estimate *Lip* $\alpha = -1$.

However, the question naturally arises as to whether we can talk of Lipschitz measures of a Dirac δ provided by the wavelet maxima in the way of Equation (7.1), and resulting in Lip -1, as suggested by the maxima linear fitting above. One problem arises considering that the Lipschitz condition, as defined in Equation (7.1), requires the definition of an interval around the discontinuity. In the δ case this interval would have measure zero (indeed, the wavelet maxima converge to the singularity only at scale zero). Therefore, while the wavelet transform provides a mapping in the (space–scale) domain through the sifting property of the delta functional, it is still not possible on mathematically rigorous grounds, to associate this mapping with a Lipschitz condition.

Despite this obstacle, it is still possible via a heuristic argument that we shall next present taking a leaf from distribution theory to talk about negative Lipschitz discontinuities that are approximations of the ideal δ functional.

Indeed, in the theory of distributions, δ is defined by an integral over a space of testing functions [5]. Since the testing functions constitute an approximation to δ, we shall use these testing functions to develop further our analysis of impulses.

The testing functions $\zeta(t)$ are defined as:

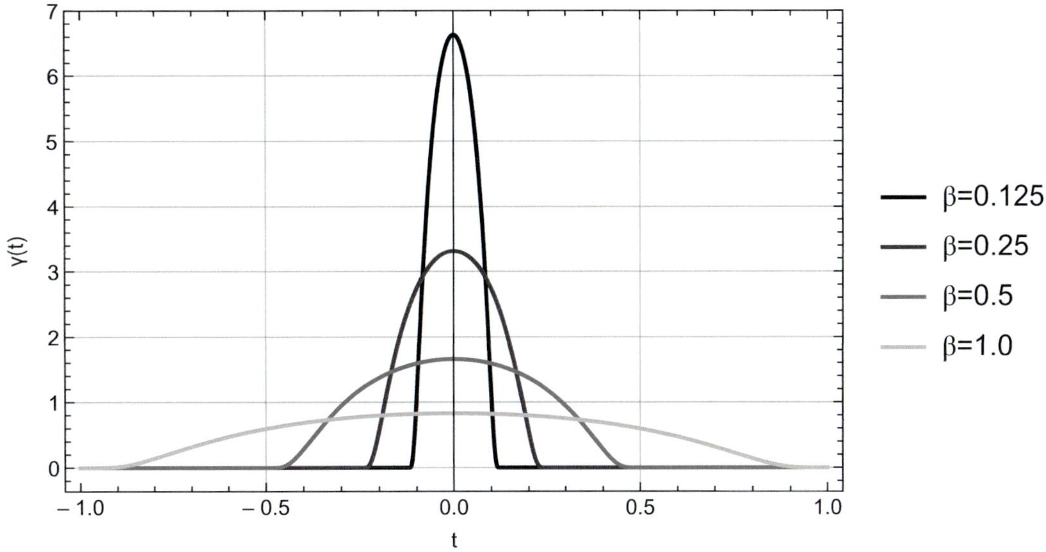

FIGURE 7.11 A set of testing functions with different β parameters.

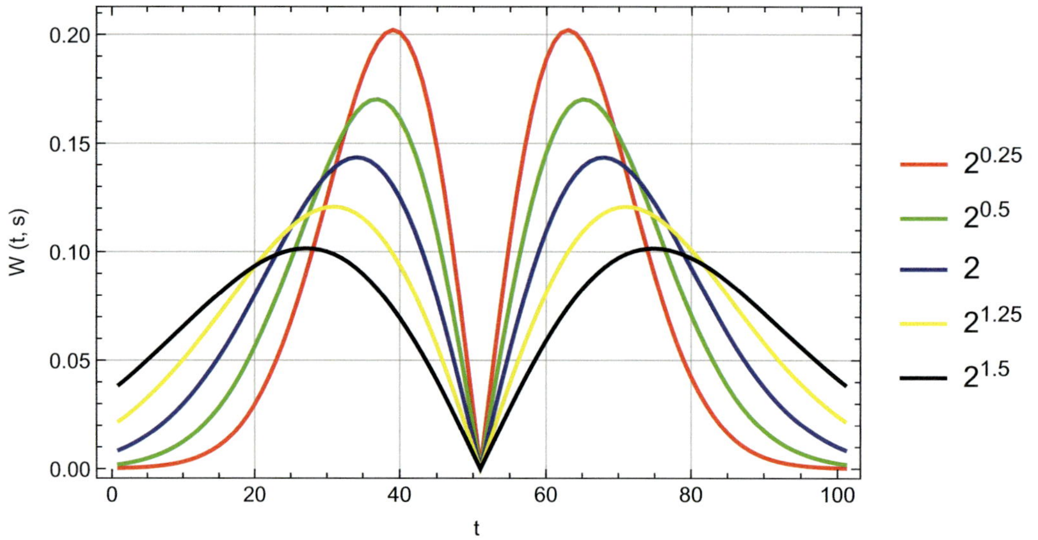

FIGURE 7.12 The wavelet transform moduli of the testing functions parametrized by scales $2^{1/4}$, $2^{1/2}$, $2^{3/4}$, 2^{1}, $2^{5/4}$ and $\beta = 0.25$.

$$\zeta(t) = \begin{cases} 0 & |t| \geq 1 \\ e^{\frac{1}{t^2-1}} & |t| < 1 \end{cases}$$

A dilated and normalized version is:

$$\gamma_\beta(t) = \frac{\zeta(t/\beta)}{\int \zeta(t/\beta)\,dt} \quad \beta > 0$$

These functions have short support, and are infinitely smooth and differentiable. A set of testing functions with different values of the β parameter is shown in Figure 7.11. The functions' sharpness depends on the parameter β. This example clearly suggests how the testing functions approximate in the limit of $\beta \to 0$ a δ functional. These functions are, in a sense, a less idealized model of the actual discontinuities that are present in a SAR signal. They are normal infinitely differentiable functions, and therefore afford the advantage of being amenable to Lipschitz measures by wavelet maxima trajectories in the way of Equation (7.1).

The wavelet transform for a discrete set of scales can be computed by integration:

$$W(t,s) = \int -\gamma_\beta(x) \frac{-(x-t)/s}{\sqrt{2\pi}} e^{\left(\frac{-\left(\frac{x-t}{s}\right)^2}{2}\right)} dx \tag{7.11}$$

The wavelet transform moduli parametrized by scales $2^{1/4}$, $2^{1/2}$, $2^{3/4}$, 2^{1}, $2^{5/4}$ and $\beta = 0.25$ are shown in Figure 7.12. By linear fit of the respective modulus maxima, we estimate *Lip* $\alpha = -0.98$. We notice that this singularity still results in the negative Lipschitz measure, albeit with a smaller absolute value.

To further study the dependence of the Lip measure on the testing function dilation parameter β, we compute the transform modulus in Equation (7.11) for a set of β values. The result is shown in the graph in Figure 7.13. We learn here that negative Lip measures always characterize the shape of the impulse discontinuity, namely smooth, infinitely differentiable functions with finite support,

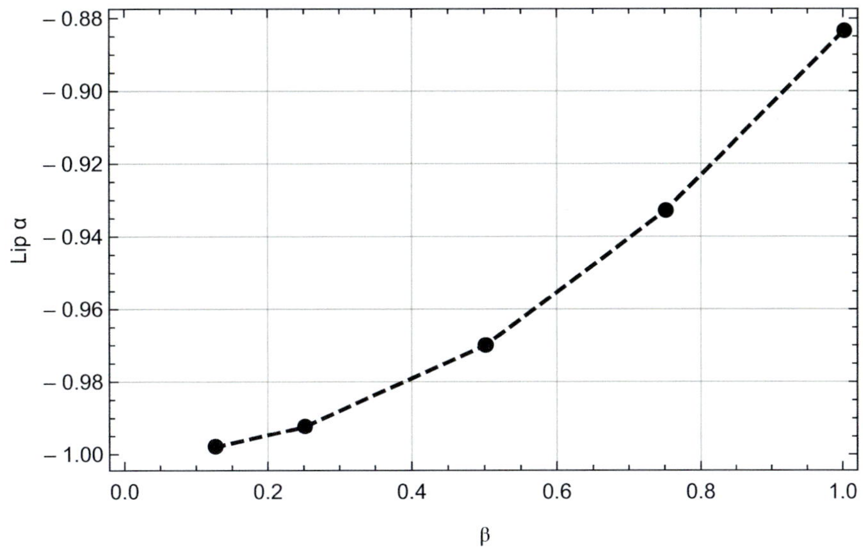

FIGURE 7.13 The measure Lip_α of the testing function as a function of the sharpness (parameter β in $\gamma_\beta(t)$).

independently of their sharpness (dilation). Yet, the magnitude of the Lip parameter decreases as the sharpness of the singularity increases, and indeed $\alpha \to -1$ for $\beta \to 0$ since the testing functions approximate the Dirac $\delta(t)$. This observation has a practical impact on the use of Lipschitz conditions in SAR filtering algorithms, since impulses are associated at short scale to speckle noise.

7.3.4 Smoothed Singularity

A smoothed singularity is modeled by the convolution of a step function with a smoothing Gaussian kernel:

$$g(t) = \frac{1}{\sigma\sqrt{2\pi}} e^{-\frac{t^2}{2\sigma^2}}$$

Since $\Im g(t) = \frac{e^{-\frac{\omega^2}{2\sigma^2}}}{\sigma^2}$, the smoothed step function, computed in the frequency domain, is:

$$f(t,\sigma) = \Im^{-1}\left[\left(\frac{1}{I\omega} + \pi \text{DiracDelta}[\omega]\right)\frac{e^{-\frac{\omega^2}{2\sigma^2}}}{\sigma^2}\right] = \frac{1+\text{Erf}[\frac{t}{\sqrt{2}}]}{2\sigma^3}$$

As an example, $f(t,5)$ is shown in Figure 7.14.

In the same way as it is done in Equation (7.10), the wavelet transform can be computed in the frequency domain:

$$W(t,s,\sigma) = \frac{1}{s}\Im^{-1}\left[ie^{-\frac{1}{2}s^2\omega^2}\sqrt{\frac{1}{s^2}}s^3\omega\left(\frac{1}{I\omega} + \pi\text{DiracDelta}[\omega]\right)e^{-\frac{\omega^2}{2\sigma^2}}\right] = s\frac{e^{-\frac{t^2}{2(s^2+\sigma^2)}}}{\sqrt{2\pi}\sqrt{s^2+\sigma^2}}$$

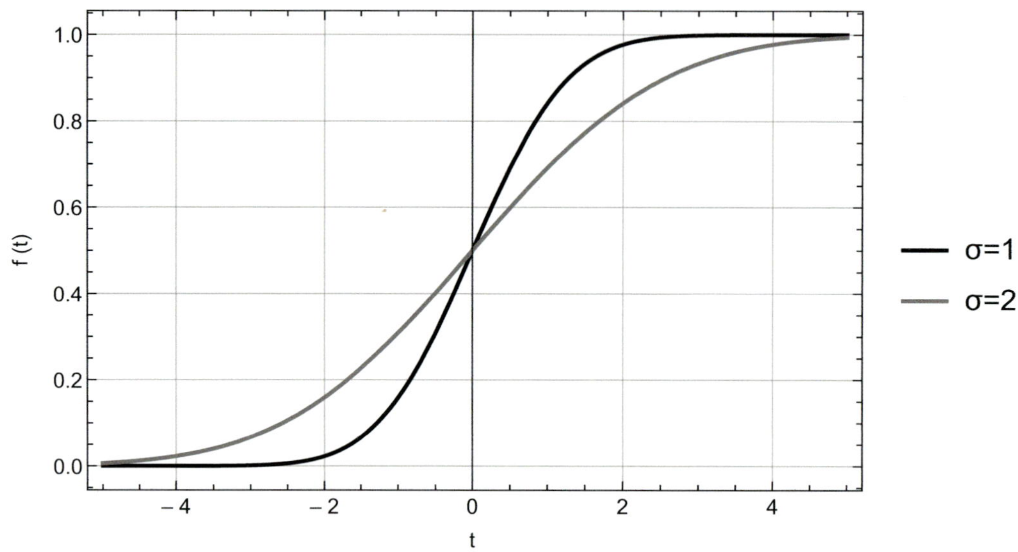

FIGURE 7.14 A smoothed step function, parametrized by the standard deviation σ of the smoothing Gaussian.

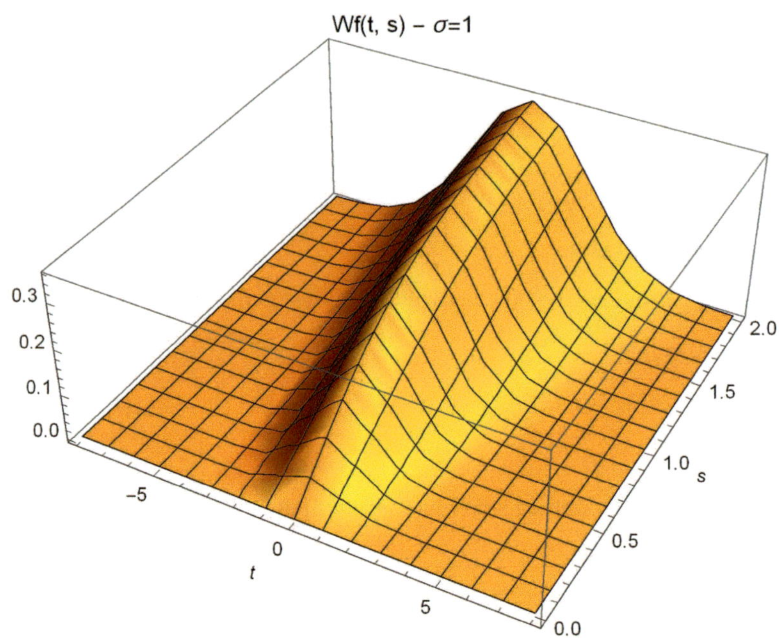

FIGURE 7.15 The wavelet transform of a smoothed step function with $\sigma = 1$ as a function of scale s and space t.

The wavelet transform $W(t, s, \sigma)$ for $\sigma = 1$ is shown as a function of space and scale in Figure 7.15 and for $\sigma = 2$ in Figure 7.16. Notice how the wavelet amplitude width in space increases with the smoothing factor, or, in other words, how the cone of influence grows as the discontinuity becomes less sharp. Notice also that, contrary to the case of a pure step function (see Figure 7.3), now the maxima (that occur at $t = 0$ for all scales) are not constant with scale, and importantly not linearly related to scale in the log domain.

The maxima as a function of scale s and smoothing factor σ are shown in Figure 7.17. At short scales ($s < 1$) the maxima decay non-linearly with scale. In this range of scales, the smoothing factor

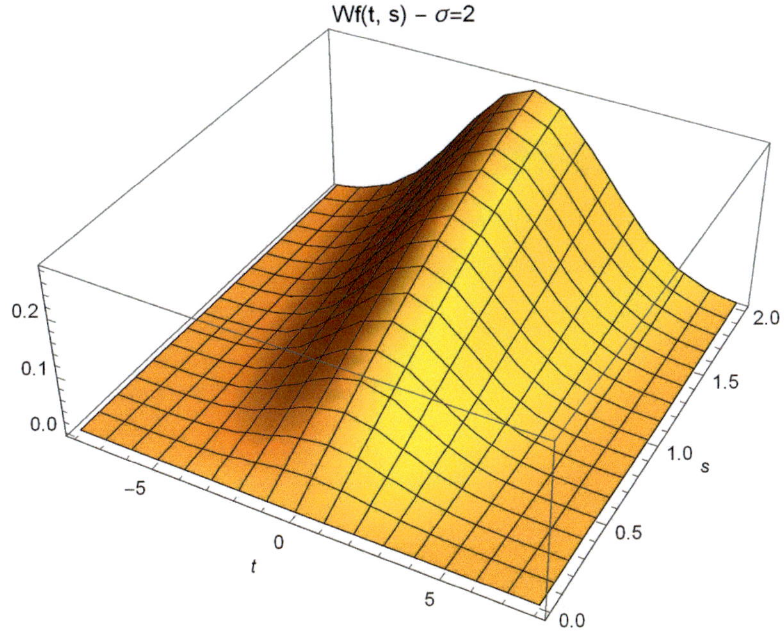

FIGURE 7.16 The wavelet transform of a smoothed step function with $\sigma = 2$ as a function of scale s and space t.

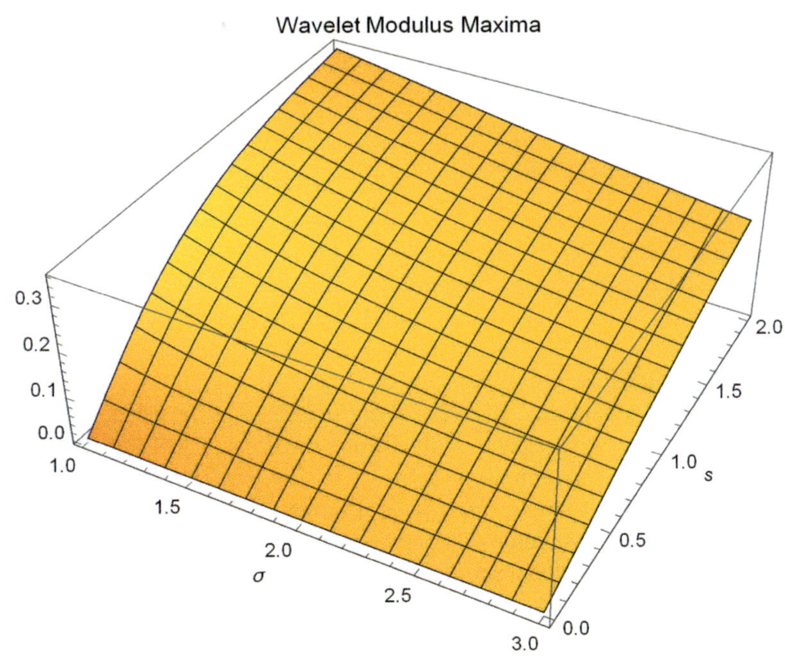

FIGURE 7.17 The wavelet maxima of the smoothed step function as a function of scale and smoothing factor σ.

of the discontinuity (σ) influences the wavelet modulus. Whereas, at larger scales the averaging performed by the dilated wavelet cancels the influence of the smoothing factor, and the wavelet maxima tend to reach a plateau, as in the case of the pure singularity.

The non-linear trend of the maxima can be better appreciated in the graph of Figure 7.18. Here, the set of wavelet modulus maxima (three dyadic scales and four voices) related to a discontinuity with smoothing $\sigma = 2$ (square marker) and to $\sigma = 1$ (round marker) are reported. The Lip estimator,

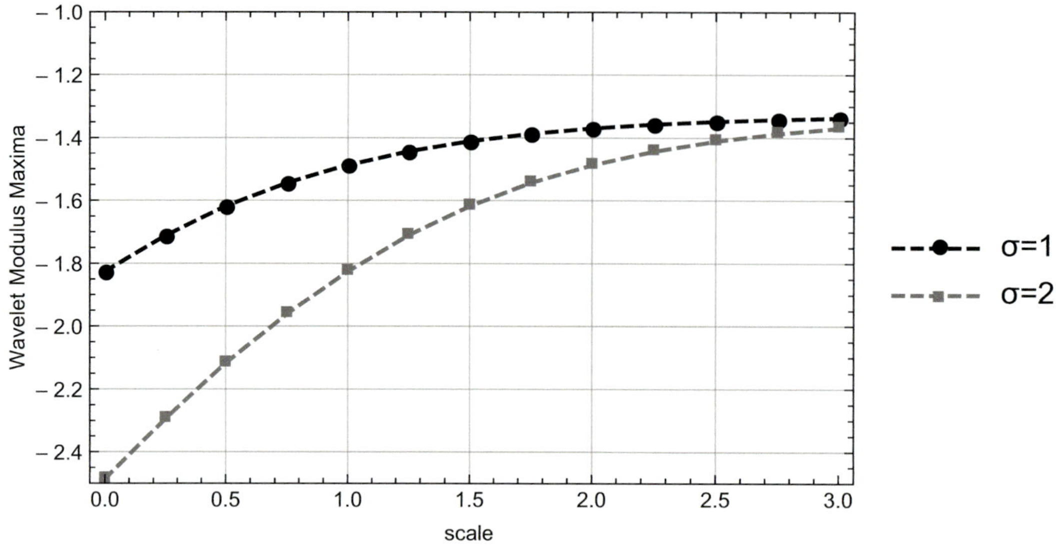

FIGURE 7.18 The smoothed singularity wavelet maxima as a function of scale, showing a non-linear trend with scale.

based on a non-linear minimization procedure of Equation (7.8), gives, respectively: *Lip* $\alpha = 0.0$, *Lip* $K = 0.39$, *Lip* $\sigma = 2$ and *Lip* $\alpha = 0.0$, *LipK* $= 0.39$, *Lip* $\sigma = 1.0$. The effect of smoothing on the ideal step function is therefore correctly reported by the *Lip* σ measure, while the *Lip* α and *Lip* K measures are not affected by the smoothing factor and remain sensitive to the underlying ideal singularity.

7.3.5 Non-Isolated Singularities

A wavelet with support $[-C, C]$, when dilated at scale s, affects all points in the space–scale domain (R, s) around a point $u \in R$ within a cone defined by $|x - u| \leq C\, s$. This region in (R, s) is called the cone of influence ([4]; see Chapter 6, Section 6.1.3). When a singularity is isolated, as in the cases we have studied so far, the cone converges at short scales to the singularity point $u \in R$, and all wavelet maxima values within the cone, being affected only by the singularity, are available for estimating the Lip parameters.

Yet, when two points of singularity, $u1$, $u2 \in R$ are separated by less than the width of the two cones of influence, the overlap in the space–scale distorts the trajectory of the maxima in a way that forbids the resolution of each discontinuity in terms of its Lip property. Rather, the merging of the two singularities is perceived as a single entity with different Lip properties.

This situation will be now illustrated by a simple test case. Let us consider the following signal, which presents two-step function type discontinuities (see Figure 7.19):

$$f(x,l) = \vartheta(-x + l) + \vartheta(x) \tag{7.12}$$

Where $\vartheta(x)$ is the Heaviside function, and l is the separation.

Using the mother wavelet in Equation (7.9) (derivative of a Gaussian), this having support $C = 4$, if we let $l = 8$ in $f(x, l)$ and wavelet transform, we obtain the results in the graph of Figure 7.20. The singularities being outside the respective cones of influence, the wavelet maxima are well resolved, they do not decay with scale and provide the correct *Lip* $\alpha = 0$ estimates for both singularities. The cones of influence are better highlighted in the $[x, \log_2 s]$ domain (see Figure 7.21). They converge at short scales to the points of singularity at $x = 0$, $x = 8$, and the maxima values do not decay with scale.

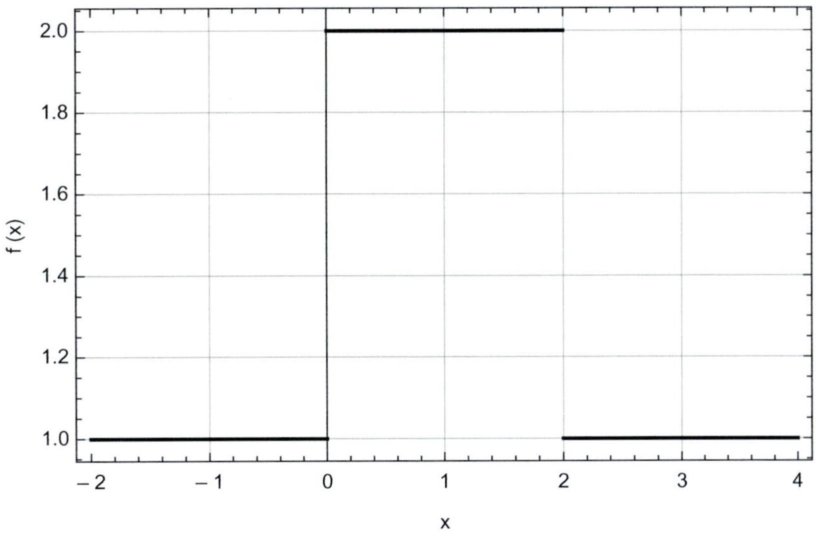

FIGURE 7.19 Two non-isolated singularities generated by shifted step functions.

Let us now bring the two singularities nearer and within the wavelet cone of influence, with $f(x,2)$. The wavelet transform moduli for scales $s = \{1,2,4,8\}$ are shown in Figure 7.22. The merging of the two cones of influence now distorts the trajectory of the maxima, these still converging to the two singularities at $[x = 0, x = 2]$, albeit now decreasing with scale.

The cones of influence in the $[x, \log_2 s]$ domain are shown in Figure 7.23, where we observe that not only the maxima magnitude decreases with scale, but also the trajectories are no longer linear with scale. This point has, of course, implications for the maxima follower algorithm that will be used for the Lip estimators.

In a nutshell, when the two singularities are not resolved, a single singularity is detected with different Lip properties. In this test case, from two unit step $Lip\alpha = 0$ edges, a single $Lip\,\alpha < 1$ impulse is sensed.

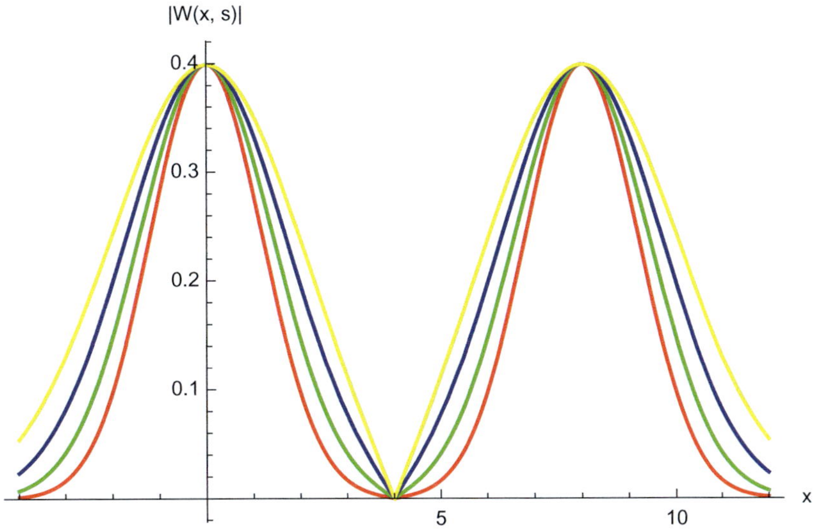

FIGURE 7.20 The wavelet transform of the shifted step function in Equation (7.12) with the separation $l = 8$.

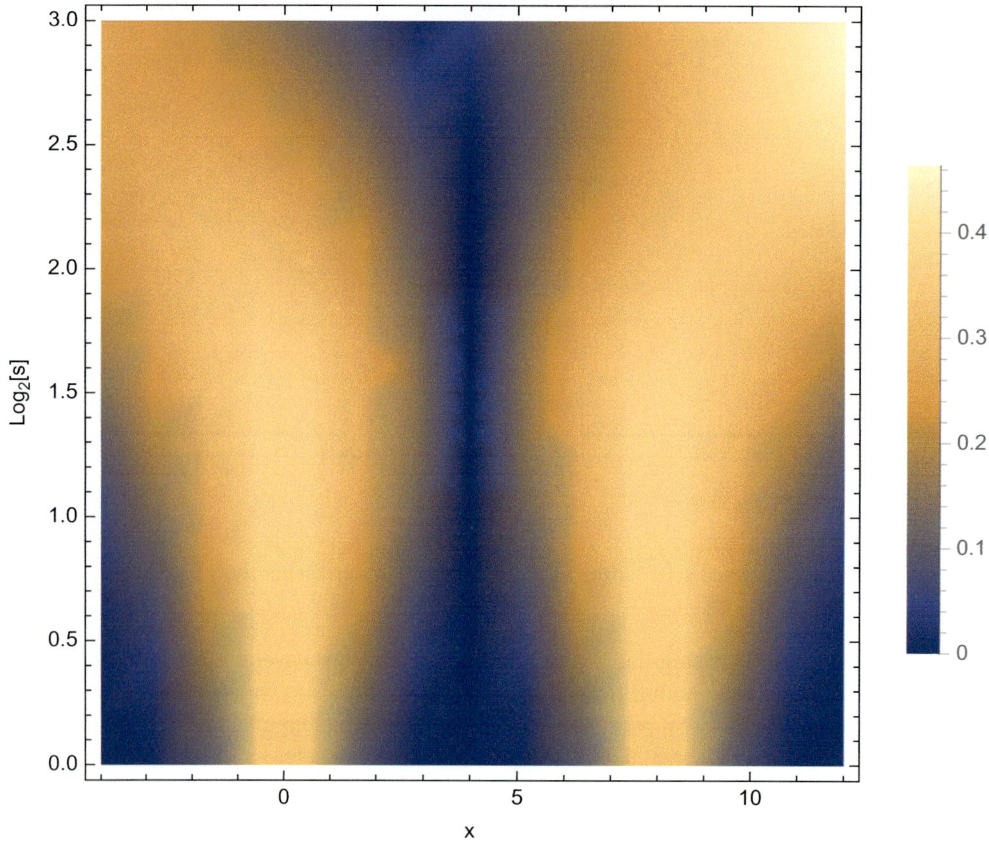

FIGURE 7.21 The cones of influence are highlighted in the scale–space graph of the wavelet transform amplitude of the shifted step functions.

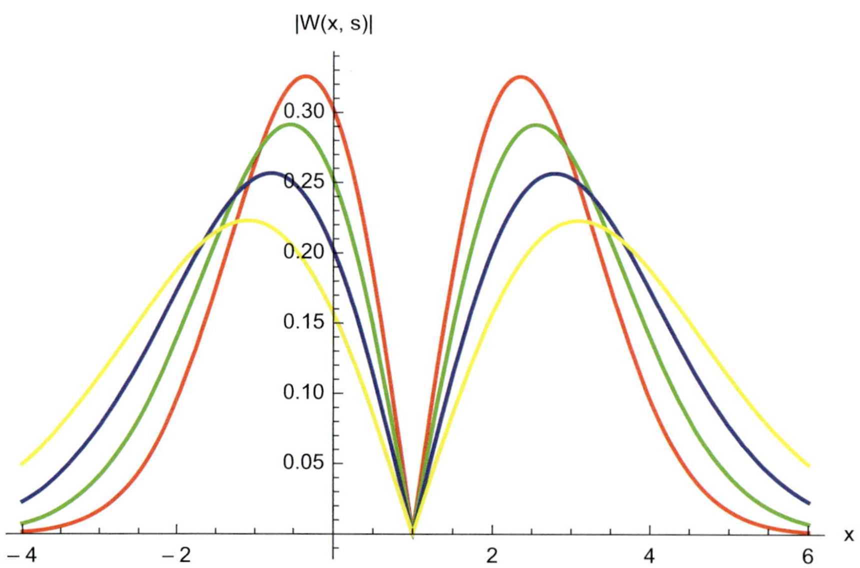

FIGURE 7.22 Wavelet transform moduli of the shifted step functions $f(x, 2)$ that are separated by less than the wavelet support.

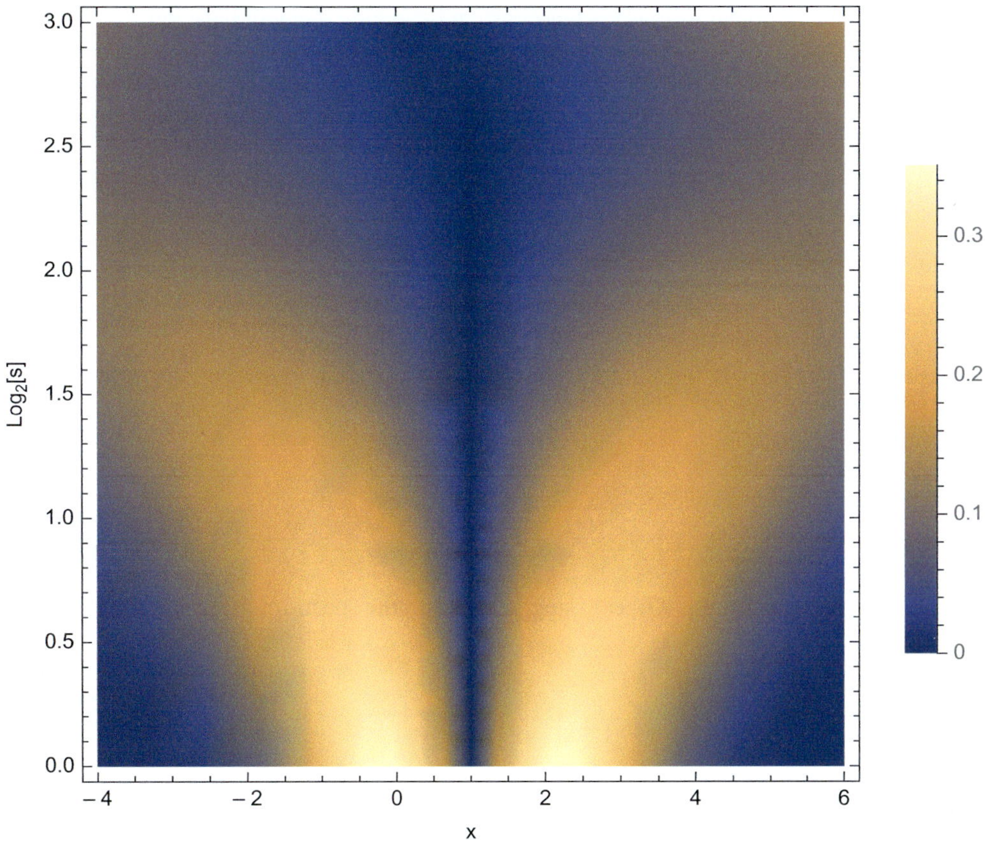

FIGURE 7.23 The cone of influence in the scale–space domain of the wavelet transform moduli in Figure 7.22.

7.3.6 Effect of Speckle

Speckle noise in homogeneous regions surrounding a discontinuity generates what we could call "micro discontinuities", with a spatial spacing depending on the correlation length of the fading variable. These discontinuities, when wavelet transformed, will result in number of wavelet modulus maxima, surrounding the more interesting signal, this being the maxima related to the discontinuity which we are interested in characterizing by a Lipschitz parameter. To gain an insight into the influence of speckle on the estimation of the Lip parameters, we propose here some simple back-of-the-envelope calculations that will enable us to establish the probability of detecting the wavelet maximum of interest as a function of speckle strength and the underlying discontinuity shape.

Let us consider a simple edge model composed of a step function surrounded by two homogeneous (stationary) linear regions. This could correspond in practice to the margin between a clear-cut and a forest, a situation that will be also considered next, both as a simulation and as a result from real SAR data. Given the scattering coefficients σ_1, σ_2 of the two bounding regions, the edge is modeled as:

$$f(t) = I1 + (I2 - I1)\vartheta(t)$$

$$I1 = 10^{\sigma 1_{dB}/10}, I2 = 10^{\sigma 2_{dB}/10}$$

where I_1, I_2 are intensities, with $I_2 > I_1$, and $\vartheta(t)$ is a step function.

Being speckle a multiplicative noise, let us consider the region that will have more of an effect on the detection capability, i.e. I_2. Let us suppose that the speckle random process in the region is stationary with N-Look intensity and an exponential autocorrelation function (ACF). Given the mean intensity value $\langle I_2 \rangle$, the number of Looks and the correlation length, we can calculate the variance of the process and the variance of the related wavelet coefficients in the following way.

For N-Looks intensity, the ratio of mean to variance is a constant and equal to the number of Looks:

$$Var(I) = \frac{\langle I \rangle^2}{nl}$$

Now, since our wavelet is a differential operator, we can scale down to zero mean, therefore $Var(I) = \langle I^2 \rangle$.

We assume an exponential ACF $I(t) = \langle I^2 \rangle\, e^{-\beta t}$ with a Lorentzian spectrum:

$$G(\omega) = \frac{4\langle I^2 \rangle\, \beta^2}{\beta^2 + \omega^2}$$

Since the wavelet transform is a linear integral operator, the spectrum of the transformed speckle process $W(t, s)$ at scale s is:

$$Swav(\omega, s) = \left| \mathrm{i} e^{-\frac{1}{2}s^2\omega^2} s^2\omega \right|^2$$

$$Snoise(\omega) = Swav(\omega) G(\omega)$$

where we have used the wavelet in Equation (7.9), which is the derivative of a Gaussian.

The variance of the zero mean wavelet coefficients is given by $ACF(\tau)$ for $\tau = 0$:

$$\left\langle W(t,s)^2 \right\rangle = \int Snoise(\omega)\, d\omega$$

The Lip estimators are based on the wavelet moduli. To account for the statistic of $|W(t, s)|$ we assume that the wavelet transform values are normally distributed (an approximation with validity increasing with scale). The wavelet moduli will then follow a folded normal distribution [6]. The resulting signal mean $\mu(s)$ and standard deviation $SD(s)$ are:

$$\mu(s) = \sqrt{\left\langle W(t,s)^2 \right\rangle} \sqrt{\frac{2}{\pi}}$$

$$SD(s) = \sqrt{\left\langle W(t,s)^2 \right\rangle - \mu(s)^2}$$

The wavelet transform of the step function (a convolution in the time domain) is computed in the frequency domain by multiplication of the Fourier transforms (see Equation (7.10)):

$$W\vartheta(t,s) = \Im^{-1}\left(\Im\left(w(t,s) \right) \Im(f(t)) \right)$$

Where $\Im\left(w(t,s)\right) = \mathrm{i} e^{-\frac{1}{2}s^2\omega^2} s^2\omega$ is the Fourier transform of the wavelet dilated at scale s.

To assess statistically the capability of discriminating the wavelet maxima trajectory associated with the singular function with respect to the maxima generated by the speckle random process, we resort to a distance measure called "deflection coefficient" or "effect size" (see also Chapter 10). This measure, in turn, can be interpreted in a hypothesis testing framework as a mean-shifted Gauss–Gauss problem, the latter enabling us to establish familiar quantities in detection problems, such as the probability of false alarm and probability of detection. The deflection coefficient in our case is:

$$d(s) = \frac{|W\vartheta(t,s)| - \mu(s)}{SD(s)}$$

From a given probability of false alarm P_{FA} and from the PDF of H_0 (null hypothesis) the threshold γ is determined by:

$$\gamma = CDF(\mathbb{N}(0,1)^{-1}(1-P_{FA})$$

From the PDF of H_1 the probability of detection P_D is:

$$P_D(s) = 1 - CDF(\mathbb{N}(0,1), \gamma - d(s))$$

Where $CDF(N(0,1))$ is the cumulative distribution function of a standard normal distribution, $CDF(N(0,1)^{-1}$ is the inverse cumulative distribution function.

Results of the detection problem are reported in Figure 7.24, in terms of probability of detection as a function of scale, given a $P_{FA} = 0.05$, and parametrized by number of looks. The edge is modeled with $I1 = -18.5\ dB$ and $I2 = -7.1\ dB$, these being typical scattering coefficient values also used in the simulation exercise described in the next section.

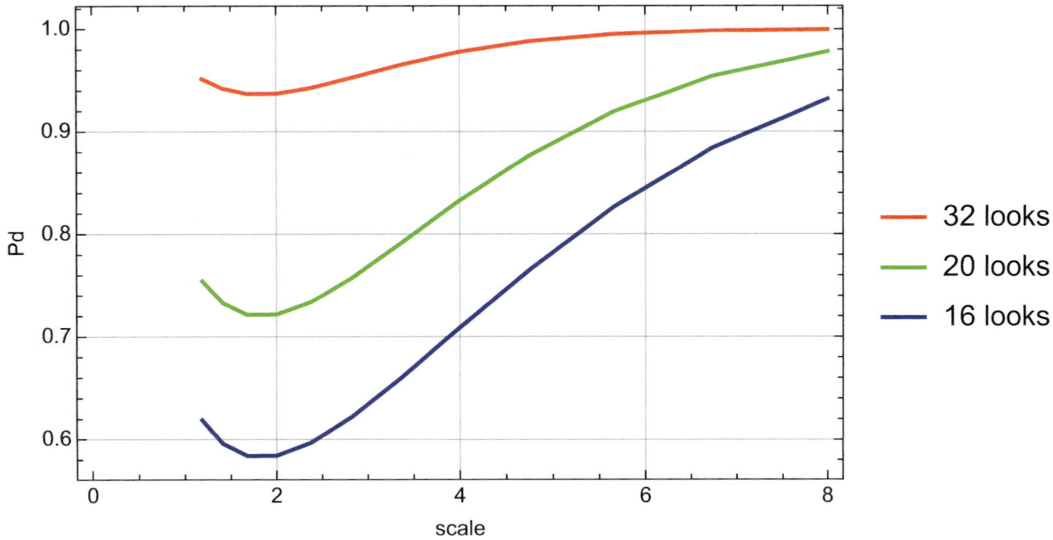

FIGURE 7.24 Probability of detection P_d as a function of scale by wavelet maxima trajectory of the step edge with speckle noise given $P_{FA} = 0.05$. P_d curves are parametrized by the number of looks of the intensity speckle pattern.

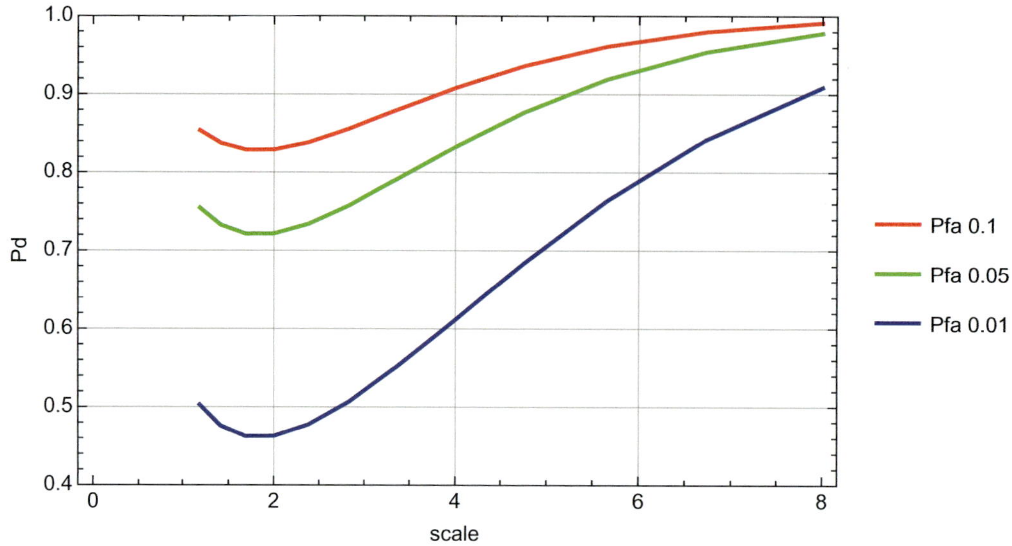

FIGURE 7.25 Probability of detection P_d as a function of scale by wavelet maxima trajectory of the step edge with speckle noise and number of Looks $n = 20$. P_d curves are parametrized by the probability of false alarm P_{FA}.

Note that the probability of detection at first decreases at short scales ($s < 2$). This is because the noise component (speckle) has high spatial frequency components that mainly affect the transform at short scales. This point suggests using scales $s > 2$ for the Lip parameters estimation. At longer scales the P_D increases significantly because of the smoothing effect of the dilated wavelet, which mainly affects the transitions of the signal associated with speckle.

Importantly, these transitions that occur with a spacing of the order of the speckle correlation length, are non-isolated singularities with *Lip* $\alpha \to -1$ for a wavelet with support greater than the correlation length. Therefore, wavelet modulus maxima associated with speckle decrease with scale, while wavelet maxima associated with the step function remain constant with scale, since it is a *Lip* $\alpha = 0$ discontinuity.

Moreover, the importance of speckle reduction for Lip estimation is clearly highlighted by this analysis. For example, in the experimental Lip analysis reported later in this chapter, the output of the low-pass filter at scale 2 of the SAR image wavelet frame discrete transform is used as input to the Lip estimator.

Finally, the detection performance as a function of P_{FA} is shown in Figure 7.25 for 20-Looks speckle strength.

7.4 A MONTE CARLO SIMULATOR OF POLARIMETRIC SAR BACKSCATTER DISCONTINUITIES

Enough of mathematical idealizations. Let us now move from continuous functional spaces to the discrete setting. This step will entail passage from the continuous wavelet transform to the discrete transform and a representation using wavelet frames (see Appendix: A wavelet tour). Moreover, we shall need to get nearer to SAR signals. In transition to the usage of real SAR signals, we shall introduce here a simulator of SAR backscatter discontinuities based on a combination of wave scattering modeling based on radiative transfer theory (see Chapter 5) and a numerical Monte Carlo generator to introduce a random process component. This step will allow us to conduct experiments in a controlled situation, to test the Lip parameters estimation in the discrete setting, and finally will give leeway to the introduction of the Lip polarimetric signature concept.

We are concerned with the backscatter discontinuity generated along a transect by the interface between two media, giving rise to two different scattering processes. This configuration may portray, for example, the margin between a clear-cut, or bare soil with low vegetation, and the onset of intact primary forest.

The description of the two media, in terms of geometric and dielectric parameters of the single scattering elements (see Chapter 5), is fed into the UTA wave scattering model [7,8]. The scattering model generates a set of backscattering polarimetric descriptors (in the form of Mueller matrices [9–11]) for one carrier frequency and a range of incidence angle values. The two sets of Mueller matrices (one for each media) are then chained together along a one-dimensional vector to simulate the discontinuity. The transition between the two scattering situations is modelled at the point of discontinuity by a weighted mixture of the two Mueller matrices. The weight factors are derived by a smoothed step function $\vartheta(x)$.Thus:

$$M(j) = M1(j)(\vartheta(-j)) + M2(j)\vartheta(j)$$

where $M1$ and $M2$ are the Mueller matrices of the two media, and j is the index of the vector M, defined on the interval $[-a, +a]$.

This model is a first-order approximation of the scattering region between the two media, and it is appropriate for low spatial resolution (order of 20 m), where averaging of intermittency within the resolution elements that straddle the discontinuity can be performed.

However, at higher spatial resolution (say order of 1–4 m), other effects should be taken into account. For example, a sort of "wall" effect could be surmised when the wave hits the margin of the forest, an effect that can be modeled by a rotation of the volume scattering with respect to the angle of incidence. In the opposite direction, when the wave is partially attenuated by the forest volume before illuminating the following targets, a shadowing effect will occur. These effects will be described in more detail in the next section.

For each resolution element in the vector, the Mueller matrices $M(j)$ are converted to covariance matrices $C(i,j)$:

$$C(i,j) = \begin{bmatrix} M(2,2) & M(2,3) + iM(2,4) & \overline{C(1,3)} \\ \overline{C(1,2)} & M(2,1) & \frac{1}{2}\left[\left(M(3,3) + M(4,4)\right) - i\left(M(4,3) - M(3,4)\right)\right] \\ \overline{C(2,3)} & M(1,3) - iM(1,4) & M(1,1) \end{bmatrix} \quad (7.13)$$

Backscatter intensity $I(j,\psi)$ is generated by polarization synthesis on a cross-pol linear basis using a set of polarimetric ellipse orientation angles $\psi \in [0, \pi]$. This configuration is the same as the one used in the definition of the wavelet polarimetric signature WASP (see Chapter 6), and allows us to include both the cross-polarized and co-polarized backscattering coefficients in the same functional dependence on only one parameter, this being the orientation ψ.

We recall that the covariance matrix C_{hv} on the H, V basis is mapped into the linear basis with orientation angle ψ by the linear transformation:

$$C_\phi = T(\rho) \cdot C_{hv} \cdot T^{\#}(\rho) \quad (7.14)$$

$$T(\rho) = \frac{1}{\left(1 + \rho\rho^*\right)} \begin{vmatrix} 1 & \sqrt{2}\rho & \rho^2 \\ -\sqrt{2}\rho^* & \left(1 + \rho\rho^*\right) & \sqrt{2}\rho \\ \rho^{*2} & -\sqrt{2}\rho^* & 1 \end{vmatrix} \quad (7.15)$$

For a linear basis:

$$\rho = \frac{\sin 2\psi}{1+\cos 2\psi} \tag{7.16}$$

$$T(\rho) = \begin{vmatrix} \cos\psi^2 & \frac{\sqrt{2}}{2}\sin 2\psi & \sin\psi^2 \\ -\frac{\sqrt{2}}{2}\sin 2\psi & \cos 2\psi & \frac{\sqrt{2}}{2}\sin 2\psi \\ \sin\psi^2 & -\frac{\sqrt{2}}{2}\sin 2\psi & \cos\psi^2 \end{vmatrix} \tag{7.17}$$

For each polarization state, described by the covariance matrix C_ϕ (Hermitian positive definite), a Monte Carlo method generates a polarimetric feature vector $\underline{\Omega} = \begin{bmatrix} S_{hh} & \sqrt{2}\,S_{hv} & S_{vv} \end{bmatrix}$ with correlated multivariate circular Gaussian (complex) random deviates. In this way, the second-order statistic of the complex deviates in $\underline{\Omega}$ are described by the covariance matrix C_ψ.

In more detail, this passage is accomplished as follows (see appendix 15.a in [12]):

A real covariance matrix is derived from the real and imaginary parts of the complex covariance matrix:

$$C_{real} = \frac{1}{2}\begin{bmatrix} A & -B \\ B & A \end{bmatrix}$$

$$C_{complex} = A + i\,B$$

By Cholesky factorization [13] the matrix C_{real} is represented by an upper triangular matrix:

$$C_{real} = C_{ut}^T C_{ut}$$

Finally, N realizations of the complex feature vector $\underline{\Omega}$ (real and imaginary parts) are computed from the Cholesky matrix C_{ut} and a matrix N $[N, 6]$ of independent normally distributed random vectors:

$$\underline{\Omega} = \mathbb{N}\cdot C_{ut}{}^T$$

A discrete multi-voice wavelet transform is applied to the set of vectors $I(j,\psi) = |S_{hv}|^2$. The wavelet (a frame) is the derivative of a box spline of order 3, and acts as a differential operator (see Section 6.7.1). Details of the wavelet's properties and the constructions of the dilated versions at fractional scales are given in the Appendix.

Wavelet modulus maxima are detected and their trajectories followed in the scale domain by an algorithm described in Chapter 8. Finally, the Lip parameters α, K, σ (regularity, swing and smoothing variance) are estimated from the maxima chain by the minimization of the root square differences between the maxima and the function in Equation (7.8):

$$f_{\psi\vartheta}(j,\alpha.K,\sigma) = \sqrt{\sum_j\left(\log_2\left|W_{2^j}f(\psi,\vartheta)\right|_{\max} - \log_2 K + \left(\alpha+\frac{1}{2}\right)j - \left(\frac{1-\alpha}{2}\right)\log_2\left(1-\frac{\sigma^2}{2^{2j}}\right)\right)^2} \tag{7.18}$$

Minimization of this objective function in the multi-dimensional space [α. K,σ] yields estimates of the Lip parameters for each polarization state ψ. The minimization is performed by a downhill simplex method (Nelder-Mead method [14]).

The fact that Lip parameters are estimated for each realization of the discontinuity, i.e. for each incidence angle ϑ and polarization orientation angle ψ, suggests the definition of Lip signatures, these being functional representations of each Lip parameter in the polarization state, incidence angle domain:

$$Lip(\alpha,K,\sigma)=f\left(I_{xpol},\psi,\theta\right)$$

In this way, these representations' graphs portray the dependence of the radiometric discontinuity on the incident wave polarization and the scattering geometry.

7.5 EXPERIMENTS USING SIMULATED SIGNALS

7.5.1 Toy Signals with Simple Discontinuities

We are now in a position to play with a set of simulated experiments. To set the groundwork for the interpretation of the results, and to check the sensitivity of the estimation method in the discrete setting, where both space and scale are sampled, let us first consider a toy signal composed of a smoothed step function with overlaid speckle noise. Simulated edge data (a vector of 128 intensity samples) for a 4-Looks speckle strength case are reported in Figure 7.26, where the dependence on incidence angle is highlighted for a single polarization (HV), and in Figure 7.27, where the dependence on the polarization orientation angle is considered for incidence angle $\theta = 36°$. The same test cases but for a 32-Looks speckle strength are shown in Figure 7.28 and Figure 7.29.

We first consider the naked discontinuity (no speckle) with a smoothing spline filter of half-width of 5 pixels, and then add multiplicative noise to check the influence of speckle. For the wavelet

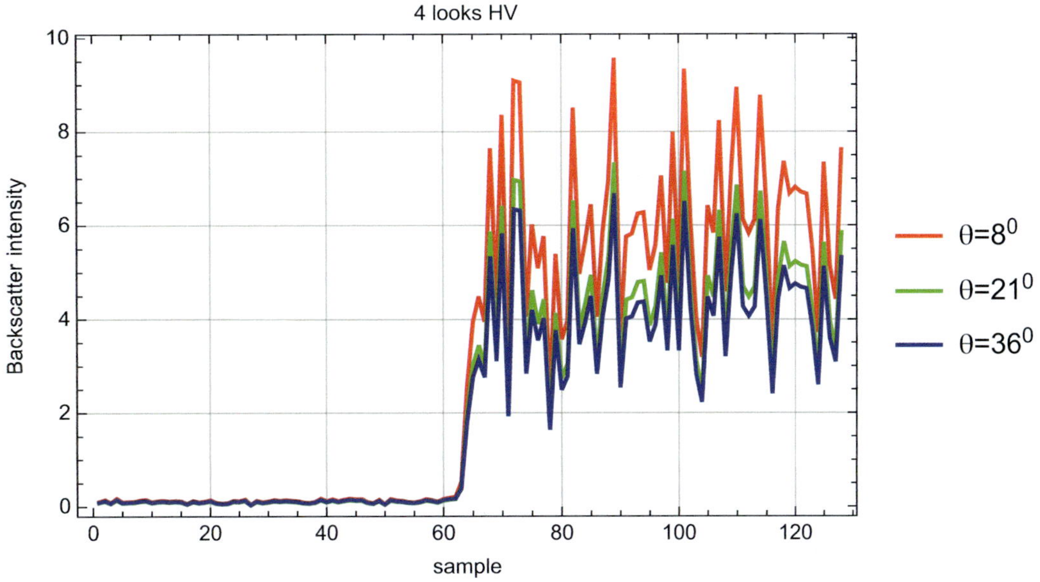

FIGURE 7.26 A discrete signal composed of a smoothed step function with overlaid 4-Look speckle noise. The dependence on incidence angle is highlighted for a single polarization (HV).

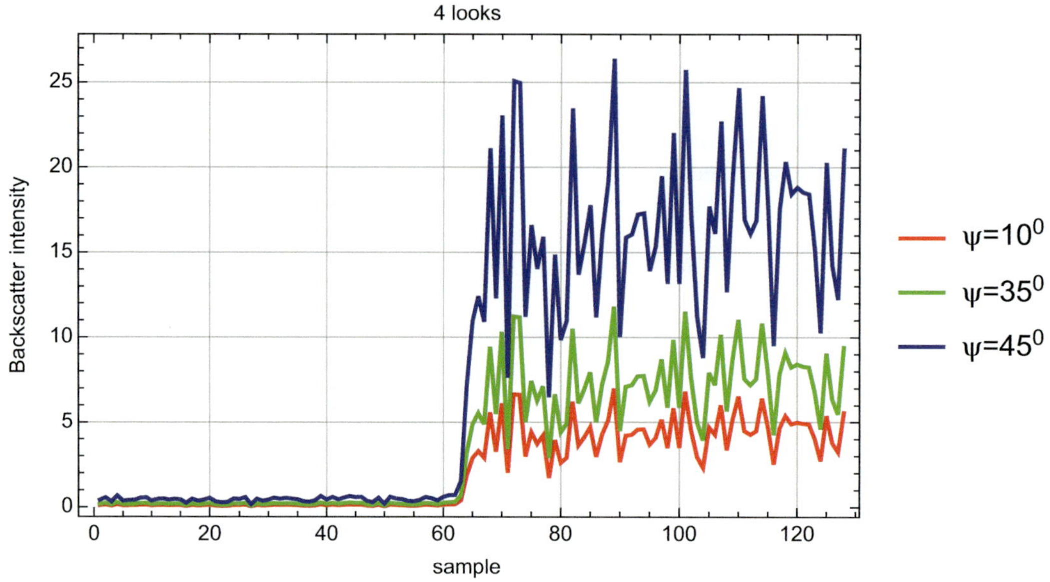

FIGURE 7.27 The same signal as in Figure 7.26 parametrized by different co-polar linear polarization basis with orientation angles $\psi = 10°, 35°, 45°$.

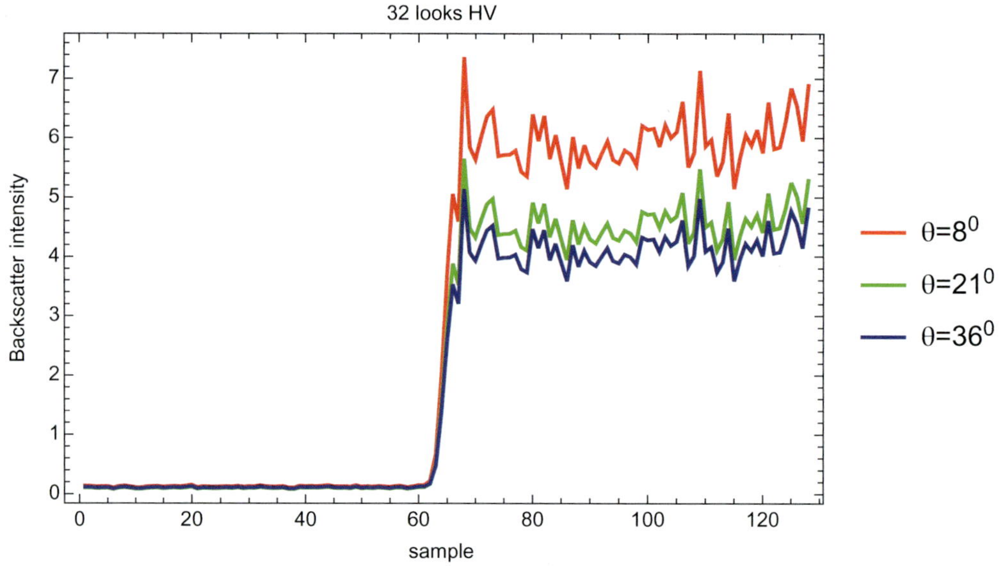

FIGURE 7.28 The step edge signal with overlaid 32-Look speckle and as a function of the incidence angle.

frame transform we use four dyadic scales and four voices, the configuration adopted for the simulated and real data experiments described in the following sections.

As to the naked discontinuity, a graph of the wavelet modulus maxima as a function of scale ($\log_2 maxima$, linear scale) is presented in Figure 7.30, together with the fitted function Equation (7.8) generated by the non-linear regression (see Equation (7.18)). The Lip estimates are: $Lip\ \alpha = 0.07$, $\log_2 K = 264$, $Lip\ \sigma = 1.31$. When the spline filter half-width is increased to 10, we obtain $Lip\ \alpha = 0.02$, $\log_2 K = 264$, $Lip\ \sigma = 1.8$. Therefore we conclude that for a pure, smooth edge the Lip estimation mechanism works fine with $Lip\ \alpha \cong 0$, near an ideal step function, and it also follows the departure from a pure step function with the correct $Lip\ \sigma$ value.

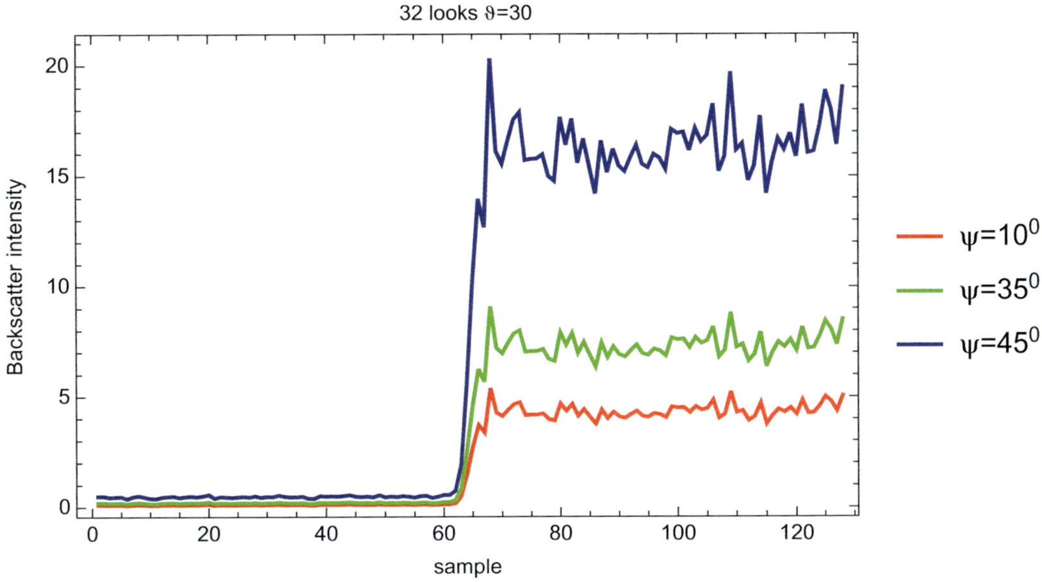

FIGURE 7.29 The 32-Look step edge signal as a function of the polarization basis.

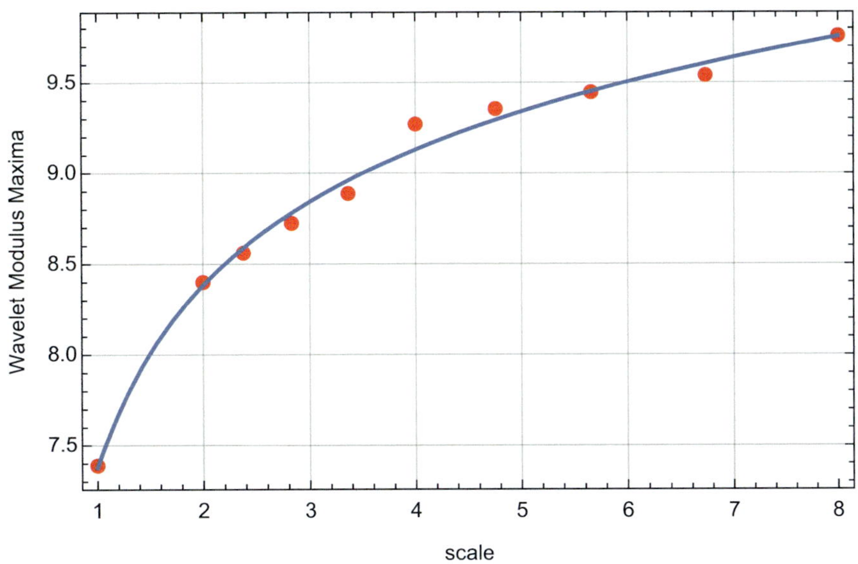

FIGURE 7.30 The wavelet modulus maxima (red markers) of the naked singularity in Figure 7.26 and the fitted non-linear function of the Lip parameters (K, α, σ).

Let us now check what happens when we add speckle to the signal with smoothing variance of 5 (the signal is multiplied by a 16-Looks fading variable Figure 7.26). Graphs of the wavelet moduli for a few selected scales are shown in Figure 7.31, and maxima as a function of scale is shown in Figure 7.32. Now, both the *Lip* α and *Lip* σ are underestimated, while *Lip* K is overestimated (the Lip estimates are: *Lip* $\alpha = -0.07$, $\log_2 K = 335$, *Lip* $\sigma = 1.0$).

A clue for an explanation of this effect can be found by examining the difference between the wavelet moduli at the first, shortest scales of the pure discontinuity and of the noisy one. The moduli of the noisy edge are clearly influenced by the ones on the right of the discontinuity, these being generated at short scales by the discontinuities intrinsic to the noise variation in the high backscatter part

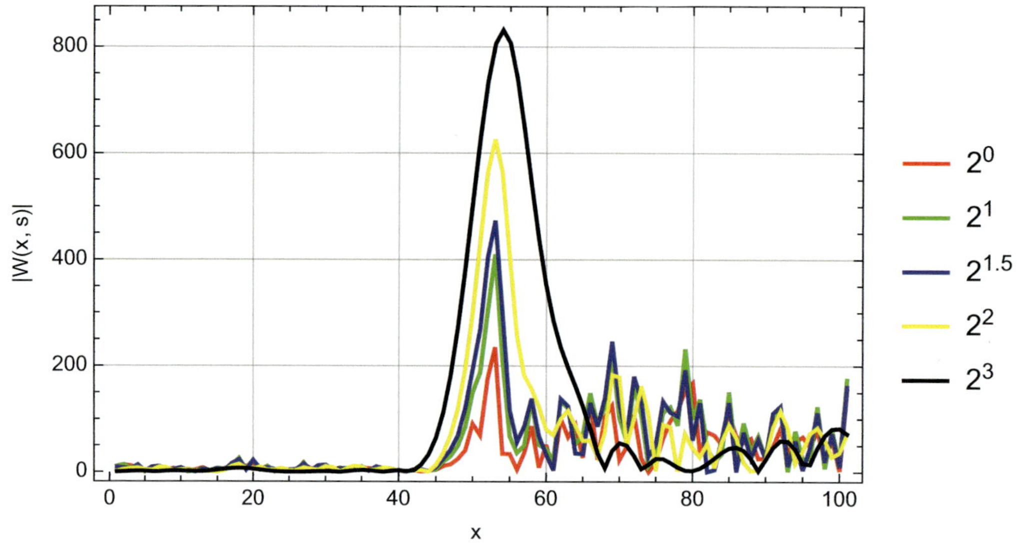

FIGURE 7.31 Wavelet moduli for a few selected scales of the speckled step function.

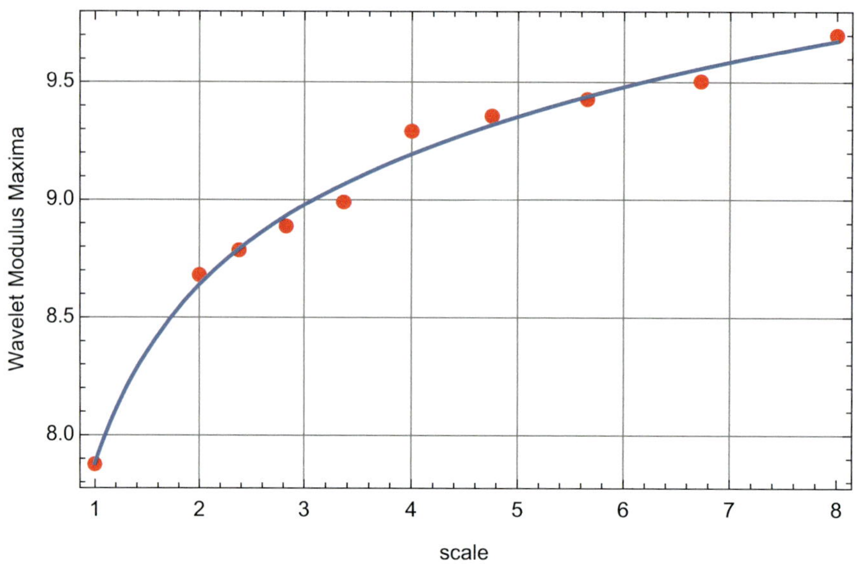

FIGURE 7.32 Wavelet moduli maxima of the speckled step function as a function of scale and the fitted nonlinear function used for the Lip parameters estimation.

of the signal (see Figure 7.31). In particular, the moduli at scale 2^0 of the pure and noisy discontinuity are compared in Figure 7.33, and the ones at scale 2^3 in Figure 7.34. The high frequency speckle noise clearly distorts the waveform at short scale, while as the scale is enlarged the waveforms tend to be equalized in the two cases. Speckle noise therefore impacts not only the capability of detecting the relevant wavelet maxima, as it was considered from the theoretical standpoint in Section 7.3.6, but also influences the Lip estimator's error.

Let us finally consider one more typical discontinuity that often appears in real SAR data: a smoothed impulse. The impulse is just one sample of the signal at a larger DN value: $f(n) = 0$, $[n \neq k]$; $f(n) = A$, $[n = k]$. Convolving this signal with a low pass filter, we replicate the impulse response function of the filter and obtain a smoothed version of the impulse.

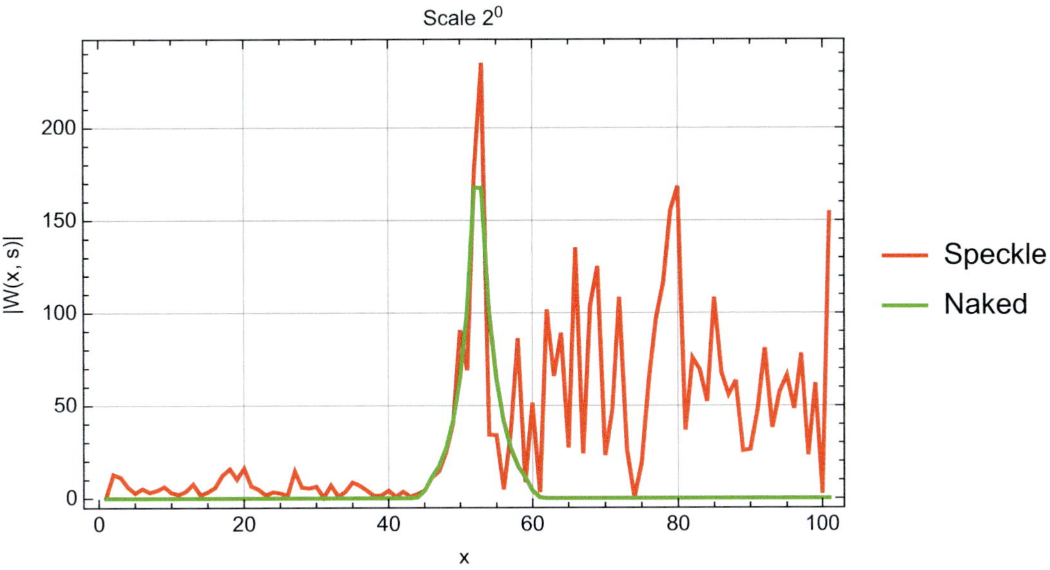

FIGURE 7.33 Comparison of the wavelet moduli at scale 2° of the naked and speckled singularity.

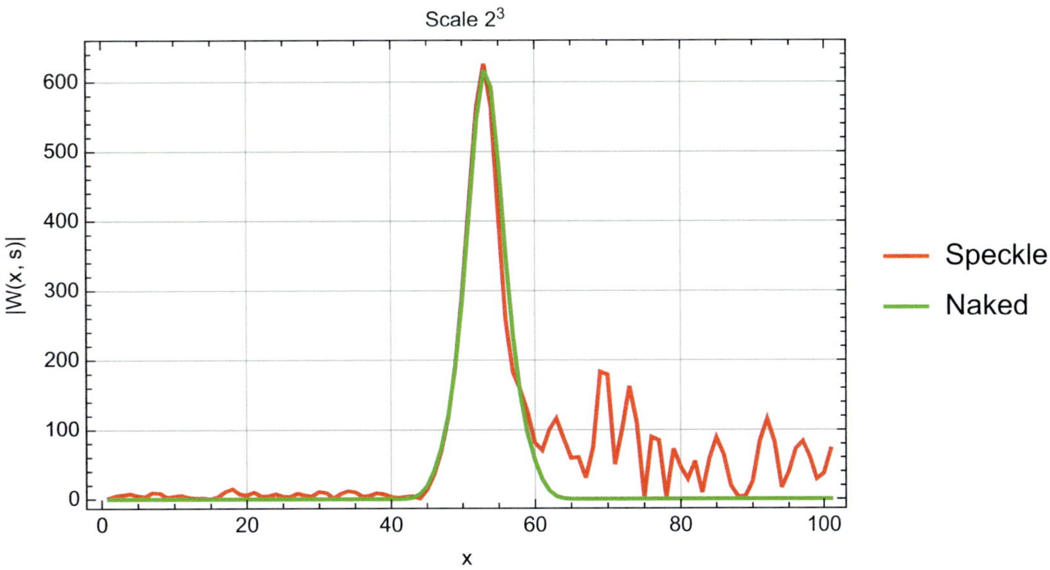

FIGURE 7.34 Comparison of the wavelet moduli at scale 2^3 of the naked and speckled singularity.

The maxima of the naked impulse $f(n)$ are plotted in Figure 7.35. The trajectories are decreasing rapidly and the Lip estimates are: $Lip\ \alpha = -1.0$, $\log_2 K = 4470$, $Lip\ \sigma = 1.0$. The estimator therefore gives results in line with what is expected theoretically for a Dirac distribution (see Section 7.3.3).

For the smoothed impulse we achieve the results plotted in Figure 7.36. Now the maxima trajectories are decreasing (and hence feeling the impulse type discontinuity) after scale 2^2. At shorter scales there is sensitivity to the smoothing filter, and the maxima are increasing. As a consequence, $Lip\ \alpha$ is still negative, but higher than for the naked impulse, $Lip\ K$ decreases, and $Lip\ \sigma$ increases: $Lip\ \alpha = -0.67$, $\log_2 K = 1013$, $Lip\ \sigma = 1.63$. These figures are in line with what is expected theoretically.

Finally, the effect of speckle is only minor, as it should be, because the impulse is a hard target. The maxima are reported in Figure 7.37, and the Lip estimates are only slightly different: $Lip\ \alpha = -0.73$, $\log_2 K = 1214$, $Lip\ \sigma = 1.4$. This point suggests the possibility of using the information

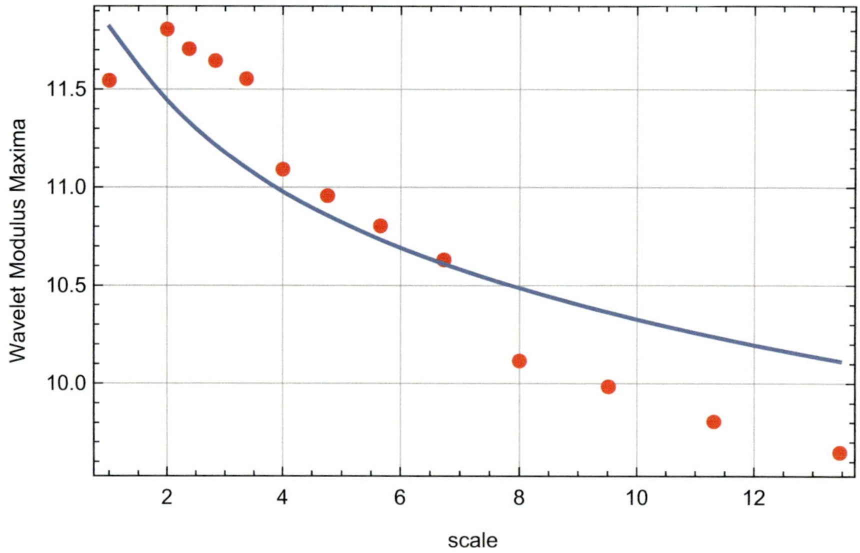

FIGURE 7.35 The wavelet modulus maxima trajectory in scale for the naked impulse. The fitted non-linear function of the Lip estimator is overlaid.

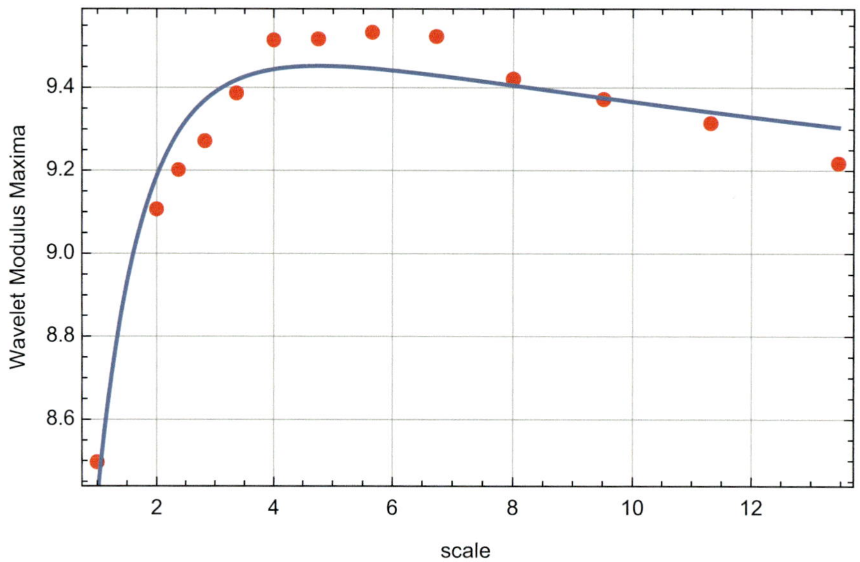

FIGURE 7.36 The wavelet modulus maxima trajectory in scale for the naked smoothed impulse. The fitted non-linear function of the Lip estimator is overlaid.

carried by wavelet coefficient trajectories in the scale domain for implementing hard (impulse type) target detectors in multi-temporal filters, and piece-wise smooth approximations of radar imagery (see Chapter 8).

Before continuing with the modeling experiments, let us first consider an alternative interpretation of the K and α parameters that enter into the signatures. This interpretation does not rest directly on the Lipschitz condition concept, but rather concerns the way these parameters are derived in a particular wavelet representation. Let us remember that Lip parameters are estimated by trajectories of wavelet modulus maxima in the scale domain, with a wavelet that is the derivative of a smoothing

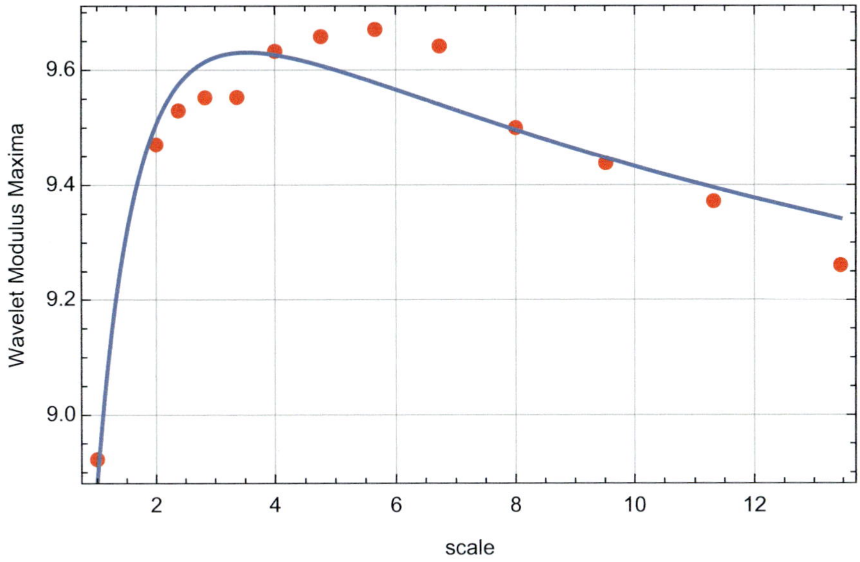

FIGURE 7.37 The wavelet modulus maxima trajectory in scale for the naked smoothed impulse when multiplicative speckle noise is added. The fitted non-linear function of the Lip estimator is overlaid.

function θ_s and, therefore, wavelet transform acts as a differential operator. Thus the wavelet coefficients at scale s correspond to the derivative of the signal smoothed at that scale:

$$W_f(s,b) = f(x) \otimes \left(s \frac{d\theta_s}{dx} \right)(x) = s \frac{d}{dx}(f \otimes \theta_s)(x)$$

Where θ_s is the smoothing function dilated at scale s.

First approximation estimates of the Lip parameters can be obtained by linear regression in the log domain of the wavelet maxima (see Equation (7.6)). Let us suppose that we are dealing with a $Lip\alpha = 0$ discontinuity (e.g. a step function). Then the K parameter, the intercept of the regression line, will correspond to the magnitude of the derivative, and therefore will be a measure of the sharpness (rate of change) of the discontinuity. Since the slope of the regression is zero, the sharpness will be the same at all scales, whereas a slope $\alpha \neq 0$ will indicate a variation of the signal's rate of change when the signal is smoothed at different scales.

7.5.2 Margin between a Clear-Cut and a Dense Forest

In this experiment, we want to simulate the radiometric discontinuity that would appear at the margin between a clear-cut (with low shrubby vegetation) and a dense tropical forest. This situation is what one would expect, for example, in areas of the tropical forest that were reclaimed for agriculture expansion. At C-band (or higher frequency) the main backscatter contribution in the forest is provided by leaves in the upper vegetation layer, which also results in strong attenuation and weak interaction with the woody components. As a consequence, data acquired in this frequency range are less interesting from the polarimetry standpoint. Therefore we study here the case of lower frequency SAR data (L-band, 1.2 GHz).

An example of simulated edge data (a vector of 128 backscattering coefficient samples), L-band, HH, 4-Looks speckle strength is shown in Figure 7.38. The trend with incidence angle of the backscattering coefficients computed by the UTA model (HH, HV, VV) related to the forest segment is presented in Figure 7.39. The backscattering values for the grassland segment are some 14 dB

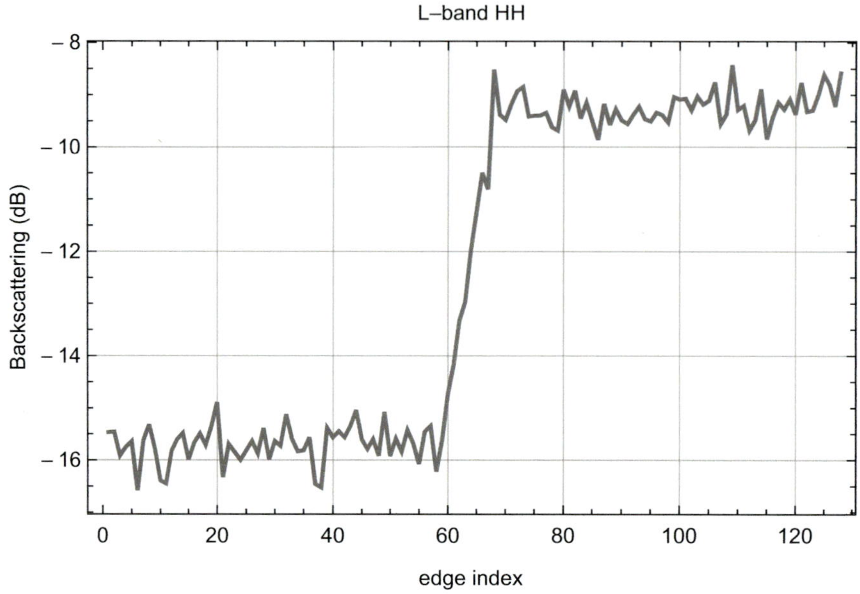

FIGURE 7.38 An example of the simulated edge data (L-band, HH, 4-Looks) depicting the margin between grassland and forest.

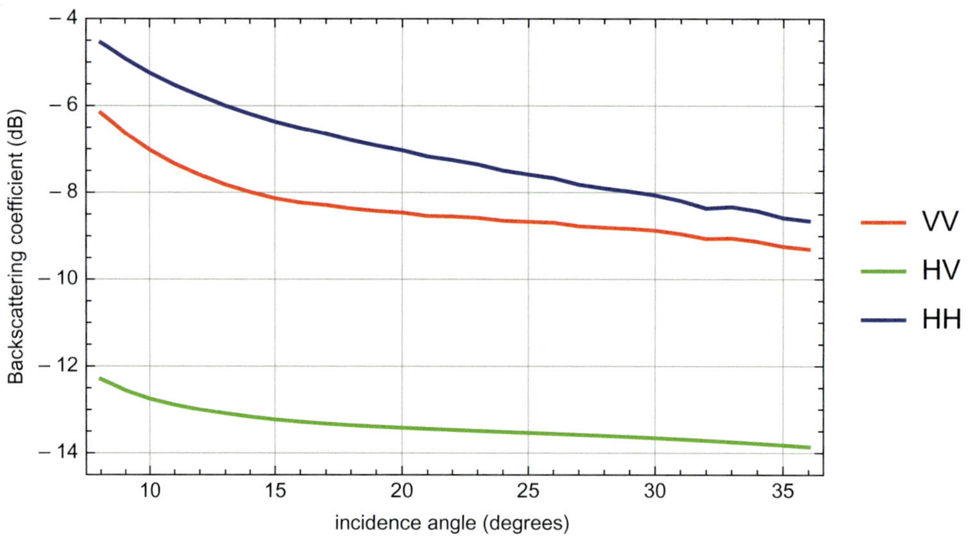

FIGURE 7.39 Backscattering coefficients as a function of incidence angle computed by the UTA model for the dense forest segment.

lower at HH, VV), (20 dB lower for HV) and are almost independent of the range of incidence angle considered (10 to 35).

Regarding the dependence on incidence angle in the forest segment, notice how the backscatter intensity decreases with incidence angle, which is a typical behavior in volume scattering, where the wave path through the volume is longer with shallower angles and the attenuation increases.

Indeed, the edge feature that bears on the Lip parameters is the relative dependence of the intensities in the two segments belonging to the grassland and forest classes. It is therefore interesting to analyze the dependence of the radiometric difference (edge amplitude) within the discontinuity on the observation parameters (i.e. polarization state and incidence angle). For the purpose, we define a differential polarimetric response as the dB difference (intensity ratio in dB) between the co-polar scattering coefficients at the beginning and ending sections of the discontinuity:

$$D\sigma_{copol}(\phi,\theta) = \log_{10} 4\pi \cos\theta\left[T(\phi,\chi=0)C_1(1,1)\right] - \log_{10} 4\pi \cos\theta\left[T(\phi,\chi=0)C_2(1,1)\right]$$

Where ϕ is the orientation angle, θ is the incidence angle, $C_1(1,1)$ is an element of the covariance matrix provided in the (H, V) basis by the UTA model for the forest simulation category (see Equation (7.13)), $C_2(1,1)$ is the matrix for the grass category, and $T(\phi,\chi = 0)$ is the polarimetric rotation matrix in Equation (7.15)

Similarly, the cross-pol differential response is defined as a dB difference (intensity ratio in dB):

$$D\sigma_{xpol}(\phi,\theta) = \log_{10} 4\pi \cos\theta\left[T(\phi,\chi=0)C_1(2,2)\right] - \log_{10} 4\pi \cos\theta\left[T(\phi,\chi=0)C_2(2,2)\right]$$

To include the dependence on the ellipticity angle χ, the co-polar ellipticity differential response is defined:

$$D\sigma_{ell}(\chi,\theta) = \log_{10} 4\pi \cos\theta\left[T(\phi=0,\chi)C_1(1,1)\right] - \log_{10} 4\pi \cos\theta\left[T(\phi=0,\chi)C_2(1,1)\right]$$

where $T(\phi = 0,\chi)$ is the matrix in (7.15) with $= \rho = \dfrac{i \sin 2\chi}{1+\cos 2\chi}$.

The cross-polar elliptical polarization differential response is defined as:

$$D\sigma_{ell}(\chi,\theta) = \log_{10} 4\pi \cos\theta\left[T(\phi=0,\chi)C_1(2,2)\right] - \log_{10} 4\pi \cos\theta\left[T(\phi=0,\chi)C_2(2,2)\right]$$

The co-polar differential response is shown in Figure 7.40. For all incidence angles, the maximum differential response occurs at $\phi = 0$ (HH). The HH difference depends strongly on incidence angle

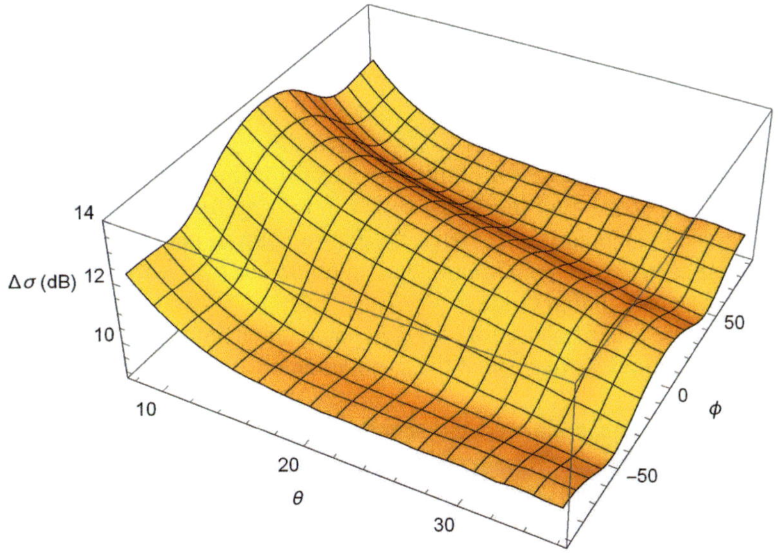

FIGURE 7.40 The co-polar differential response. The signature represents the difference of the co-polar backscattering coefficients in a linear basis (in dB) between the forest category and the grassland category as a function of the incidence angle θ and the polarization orientation angle ϕ.

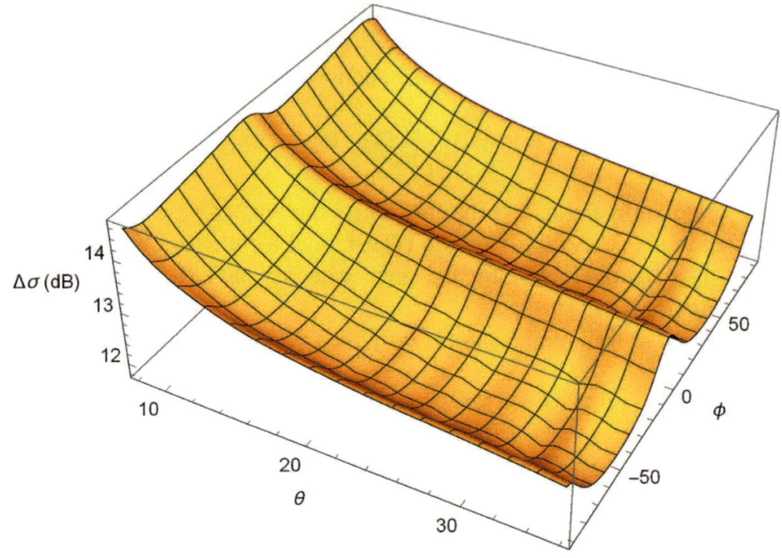

FIGURE 7.41 The cross-polar differential response. The signature represents the difference of the cross-polar backscattering coefficients (in dB) between the forest category and the grassland category as a function of the incidence angle θ and the polarization orientation angle ϕ.

[8° < θ < 36°], and decays non-linearly with increasing angle. As the rotation angle ϕ increases, the dependence on incidence angle becomes weaker for angles $\vartheta > 20°$. Regarding the polarization dependence, the signature presents three local maxima at $\phi = 0°$, and $\phi = \mp 90°$ (HH and VV) with the differential response at HH always greater than the one at VV.

The cross-pol response is shown in Figure 7.41. Here, the dependence on incidence angle is very weak at orientations $\phi = 0°, \mp 90^{\circ}$ (as expected, since depolarization mainly stems from volume scattering), and increases for $\phi \to \mp 45°$ The difference maxima now occur at orientation $\phi = \mp 45°$, this corresponding to a combination of co-polar power terms and their phase difference:

$$T\left(\phi = 45,\chi\right)C_1\left(2,2\right) = \frac{1}{2}hh^2 + \frac{1}{2}vv^2 + \mathrm{Re}\left(hh \cdot vv^*\right)$$

The co-polar differential response for elliptical polarization and $\phi = 0$ is shown in Figure 7.42. The edge amplitude increases for all incidence angles in moving from linear, through elliptical, up to circular polarization. In all polarization states, the edge amplitude decreases with incidence angle. Circular polarization achieves the maximum edge amplitude at $\vartheta = 8$, this being greater than all values of the linear co-polar configuration.

The cross-polar differential response for elliptical polarization and $\phi = 0$ is shown in Figure 7.43. The maximum edge amplitude is reached at linear polarization. The response decreases from linear to circular polarization.

It is also interesting to consider the edge discontinuity in terms of a signal over noise problem; that is, to consider the realization of the first segment (grassland) as the noise and the second segment (the forest) as the signal. We would then be interested in finding the optimal polarization state to maximize the ratio between the two classes. This type of problem was addressed in a slightly different context (target detection) in [15]. Here, given the covariance matrices of signal and clutter C_s and C_c, a matched filter (PMF) $\bar{h}$ (linear processor) was sought that maximizes the signal-to-noise ratio:

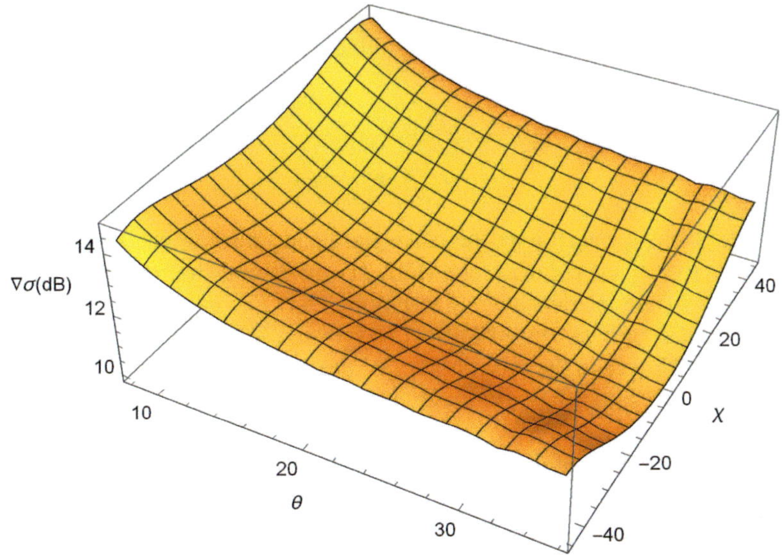

FIGURE 7.42 Differential response for the co-polar elliptical polarization and orientation $\phi = 0$.

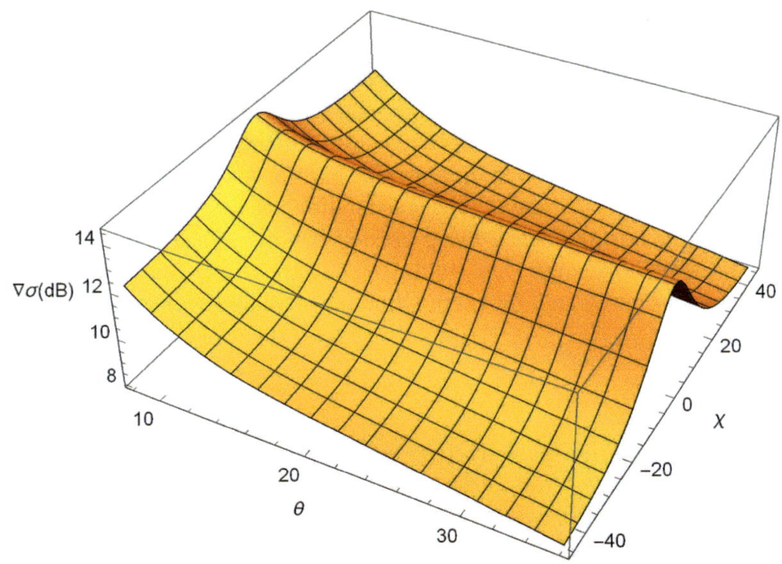

FIGURE 7.43 Differential response for the cross-polar elliptical polarization and orientation $\phi = 0$.

$$\frac{S}{N} = \frac{h^{*T} C_s h}{h^{*T} C_c h}$$

In line with this approach, we seek the scattering vector w that maximizes the contrast within the discontinuity by the solution of an eigenvalue–eigenvector problem:

$$C_2^{-1} C_1 w = \lambda w$$

where C_2 is the covariance matrix of class grassland, and C_1 is the covariance matrix of class forest.

The scattering vector w_m associated with the maximum eigenvalue λ_m provides, through the covariance matrices of the discontinuity, the intensities related to this optimal scattering mechanism:

$$I_1 = w^{T*} C_1 w$$

$$I_2 = w^{T*} C_2 w$$

Ratios of the class forest return over the class grassland return (signal-to-clutter ratio) for the linear co-polar and cross-polar polarizations (HH, HV, VV) and the matched filter outcome are reported as a function of the incidence angle in Figure 7.44. The parametric matched filter (PMF) achieves an improvement of between 90% and 110% over VV for all incidence angles, and from 20% to 80% over HH. However, the cross-polar HV performs similarly well for incidence angles greater than 25, and only 2% to 8% worse for angles of less than 25.

Having characterized the polarimetric response of the simulated edge data, we can now proceed to examine the related estimators of the Lip parameters, these being obtained by the minimization of the function in Equation (7.18). Simulated data with 4-Looks speckle patterns are used. We consider here, for one wave frequency (e.g. L-band), the dependencies of the Lip parameters estimate (α, K, σ) on the incidence angle and on the polarimetric orientation angle ϕ, when synthesis is performed in a linear ($\chi = 0$) cross-polar configuration:

$$Lip\left(\alpha, K, \sigma\right) = f\left(I_{xpol} \phi, \theta\right)$$

The *Lip* α signature, which shows the dependence of the α parameter on incidence angle and orientation angle ϕ, is shown in Figure 7.45. A box-and-whisker chart of all data in the (θ, ϕ) space gives for the quartiles: Q2 (median) = −0.116124; Q1 = −0.116652; and Q3 = −0.115857. The quartile deviation (a measure of spread in the signature) is: $\triangle = 4.0\ 10^{-4}$. Since the tolerance of the Powell method was set at 10^{-6} and the number of iterations never exceeded 14, we conclude that variations in the signature are significant, being greater than numerical errors of the estimator. The main information provided is that the ideal (in the mathematical sense) underpinning singularity is Lip $\alpha \cong 0.$, therefore near a unit step function, while the signature variation, being of the order of

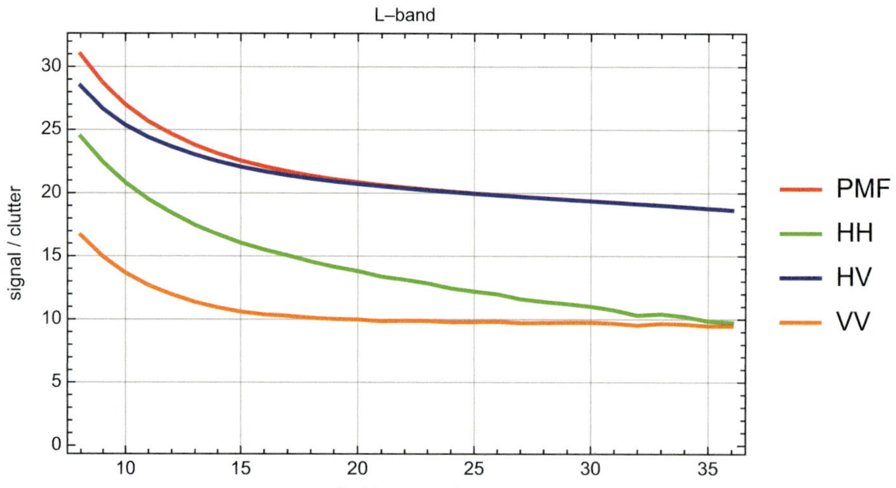

FIGURE 7.44 Signal-to-noise ratio achieved by the HH, HV, VV backscattering coefficients and the parametric matched filter (PMF).

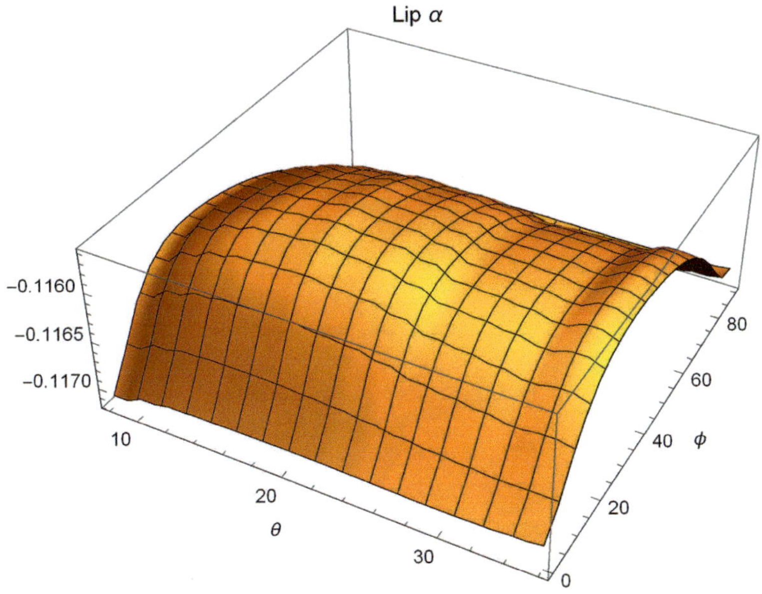

FIGURE 7.45 The *Lip α* signature, providing the dependence of the estimator on the incidence angle θ and the polarization orientation ϕ.

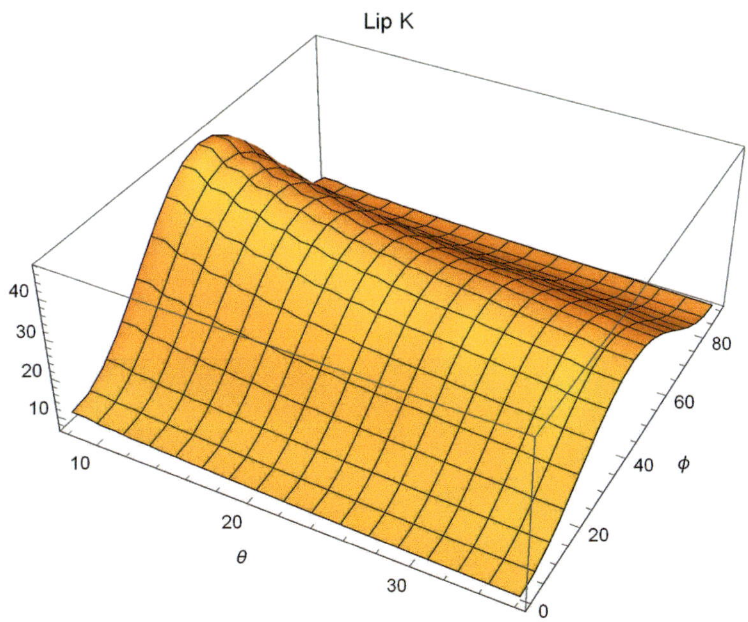

FIGURE 7.46 The *Lip K* signature.

0.8% relative to the median, does not carry additional useful information on the Lipchitz measure of the discontinuity.

More interesting is the *Lip K* signature (see Figure 7.46), that we now turn to analyze. A box plot of all samples in the signature gives Q1 = 10.84, Q2 = 19.61, and Q3 =28.34. The relative spread of the distribution (Q3–Q1)/Q2 is now 44%. The standard error of the estimator (standard deviation over mean) is shown in the graph of Figure 7.47.

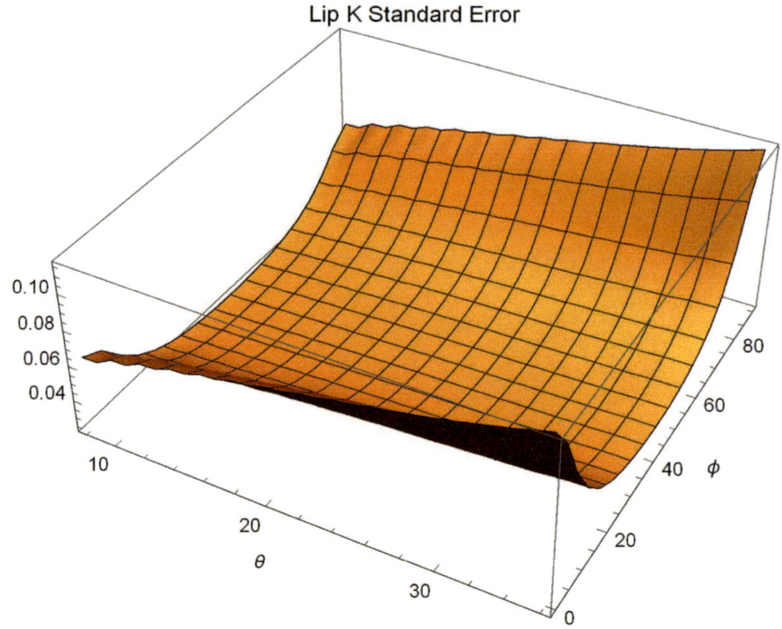

FIGURE 7.47 Standard error of the *Lip K* estimator.

The dependence of *Lip K* on incidence angle for three polarization orientation angles ($\phi = 0^{\circ}$, $\phi = 45^{\circ}$, and $\phi = 90^{\circ}$) is highlighted by cross-sections of the signature (see Figure 7.48). First, we note that the absolute magnitude (independent of angle) of *Lip K* increases with orientation angle from 0° to 90° (i.e. in the transition from purely cross-polar channel to a combination of the co-polar channels).

For all polarizations, *Lip K* decreases with incidence angle, an indication that the rate of change (sharpness) of the edge decreases with incidence angle at all observation scales (being the edge nearly *Lip* $\alpha = 0$). However, the relative rate of decrease is sensitive to polarization, this being 33% at

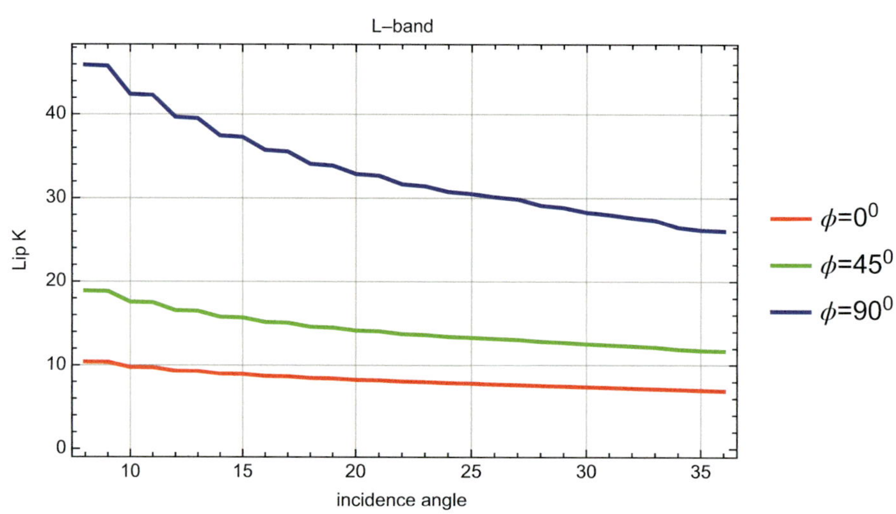

FIGURE 7.48 Cross-sections of the *Lip K* highlighting the dependence of the signature on incidence angle for three selected polarization orientation angles.

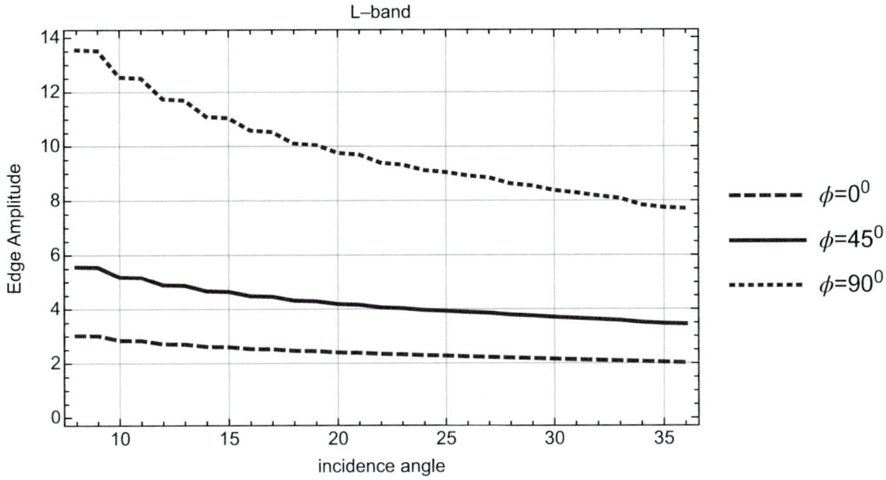

FIGURE 7.49 Dependence of the edge amplitude (difference between class forest and class grassland mean backscatter) on incidence angle at three selected polarizations.

$\phi = 0^\circ$ (HV) and 43% at $\phi = 90^\circ$ (combination of the co-polar channels). This feature reflects that fact that volume scattering is always less dependent on incidence angle, while mechanisms affecting the co-polar channels, such as surface backscatter and multiple scattering between trunks and ground, are more sensitive to incidence angle.

By comparison, the graph in Figure 7.49 shows the dependence of the edge amplitude (difference between class forest and class grassland mean backscatter). The edge amplitude follows the same dependence as *Lip K*, yet the coupling between edge width and amplitude, as characterized by the ratio *Lip K/amplitude*, presents a different trend (see Figure 7.50).

At all polarizations there is an inversion of the dependence of the relative *Lip K* on incidence angle around $\theta = 15^\circ$, with decreasing values before this angle, and increasing values for larger angles. This trend can be summarized, stating that at shallow incidence angles, even if the amplitude of the edge decreases, the relative sharpness of the edge increases. Moreover, the magnitude of coupling between edge width and amplitude depends on polarization in an opposite way with respect to

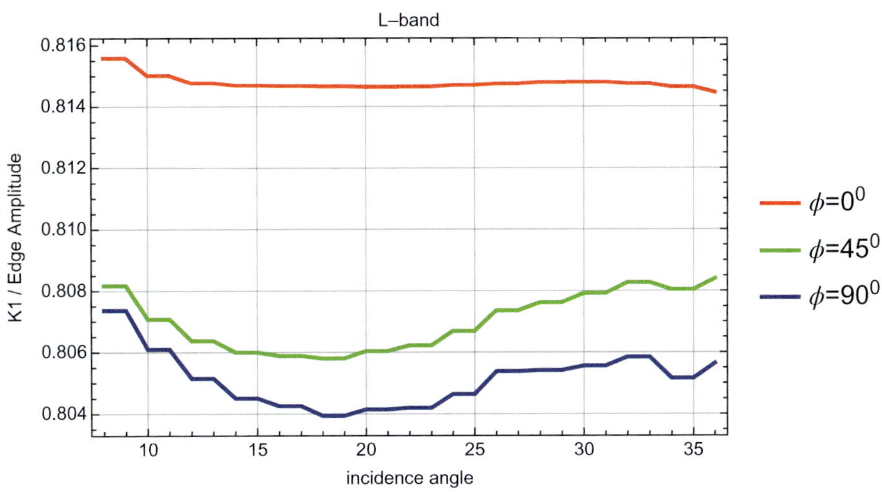

FIGURE 7.50 Ratio of *Lip K* over edge amplitude.

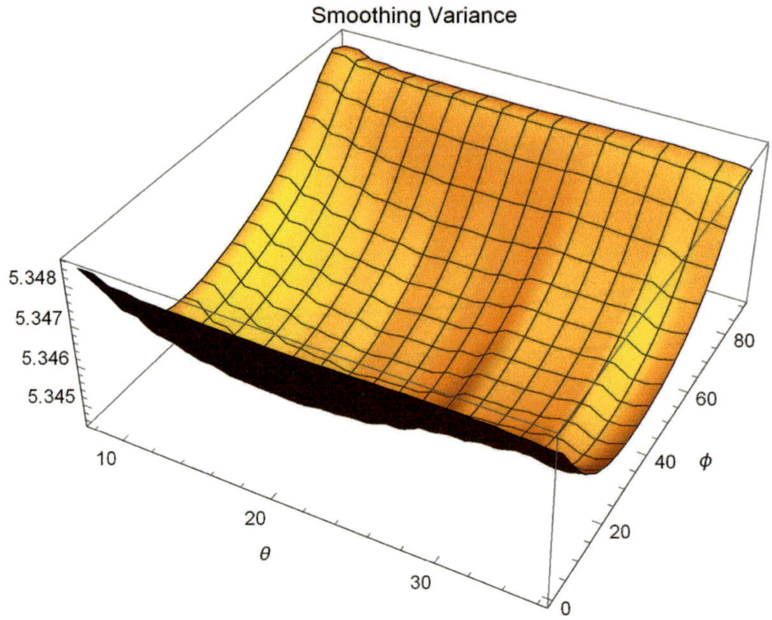

FIGURE 7.51 The smoothing variance *Lip* σ signature.

the magnitude of the dependence of *Lip K* and edge amplitude, whereas the absolute magnitude of *Lip K* and amplitude increases from orientation 0° to 90°, the absolute magnitude of relative *Lip K* decreases. Therefore, in the present case, where the edge is composed of two segments with dominating volume scattering, the cross-polar channel (HV) provides the least edge amplitude, albeit the best compromise between edge width and amplitude.

The smoothing variance signature (*Lip* σ) is reported in Figure 7.51. Descriptive statistic gives the following quartiles: 5.3452, 5.34588, 5.34722. The spread relative to the median is 0.04%. The most interesting variation is the one related to orientation angle. We note that, at all incidence angle, the variance increases (the edge becomes more smoothed) from $\phi = 45^{\circ}$ (co-polar terms) to the maximum at $\phi = 0^{\circ}$ (HV).

7.5.3 Edge on Tilted Terrain

Symmetry is an important concept that pervades physical laws [16]. Symmetry of an object in 3-dimensional space, such as an atom, a molecule or ions in a crystal, is a fingerprint of the object. The symmetry of the object has impact on the properties of scattering radiation; for example, it determines which of the elements of the scattering tensor will be different from zero. By this principle, optical spectroscopy, such as that based on Raman scattering, is used to unravel the structural aspects of atoms, inorganic molecules and ions, while electron microscopy is applied to study complex organic molecules such as the ribosome.

One of the authors fell in love with symmetry during his Master's studies, being inspired by a small but wonderful book on Raman scattering [17]. He encountered symmetry again in microwave remote sensing, while studying together with colleagues at the US Naval Research Laboratory, a method for estimating terrain slope by radar polarimetry [18,19].

In this study, a dependency was found between the terrain slope in the azimuth (cross-range) direction and a shift of the synthesized co-polar power maxima in the orientation angle direction.

We take a leaf here from that work, and revisit those findings in a different context, i.e. to study the effect of terrain slope on the Lip estimators.

A physical interpretation of the maxima shift phenomenon calls into play the azimuthal reflection symmetry properties of the target. The polarimetric response from reflection symmetric targets has been studied, for example in [20,21], where it is demonstrated theoretically that for azimuthally symmetric targets the correlations between co- and cross- polarized backscattering components is zero.

In simple, intuitive terms, we can summarize this effect and relate it to our problem in the following way. Consider a deterministic target which is symmetric with respect to a plane (the reflection plane – RP) containing the incident radar beam. A complementary problem is defined in wave scattering when the sources (charges and currents) and the fields are the images of the original problem with respect to some operator. In our case, the operator is the reflection with respect to RP. If the charge and current distribution is the same for the complementary problem, i.e. the target is symmetric, the H polarized scattered fields will be an odd function of the azimuthal angle, and the *V* polarized field will be an even function.

This stems from the invariance of the Maxwell equations with respect to reflection operators. Since the excitation sources are the same for the original and complementary problem, this also implies that the co-pol scattering matrix elements will be even functions, and the cross-pol scattering elements will be odd functions. Now consider a random target and compute the ensemble average $\langle S_{hh}\, S_{hv}\rangle$ over all azimuthal angles; there will always be a contribution coming from a resolution element in the complementary problem that will cancel out the respective contribution from the original problem. In other words, the cross-correlations terms in the covariance matrix will be zero.

On the other hand, if the target is tilted with respect to RP, and loses reflection symmetry, the sources in the complementary problem will be different from the original problem, the terms in the ensemble average will not be balanced out, and the cross-correlation terms will tend to increase. We shall now look at the influence of those terms on the cross-pol power synthesized on a rotated linear polarization basis.

Given a target feature vector on the orthogonal (H, V) basis:

$$\underline{\Omega} = \left[S_{hh}\ \sqrt{2} S_{hv}\ S_{vv} \right]$$

The rotated vector at polarization orientation angle ψ is:

$$\underline{\Omega}' = \underline{T}(\rho) \cdot \underline{\Omega}$$

where ρ is the polarization ratio:

$$\rho = \frac{E_v}{E_h} = \frac{\cos 2\tau \sin 2\psi + j \sin 2\tau}{1 + \cos 2\tau \cos 2\psi}$$

Where ψ, τ are orientation and ellipticity angles of the polarization ellipse.
For a linear polarization basis:

$$\underline{T}(\rho) = \begin{bmatrix} \cos\psi^2 & \sqrt{2}/2 \sin 2\psi & \sin\psi^2 \\ -\sqrt{2}/2 \sin 2\psi & \cos 2\psi & \sqrt{2}/2 \sin 2\psi \\ \sin\psi^2 & -\sqrt{2}/2 \sin 2\psi & \cos\psi^2 \end{bmatrix} \tag{7.19}$$

The cross-pol component $\underline{\Omega}'_2$ of the rotated feature vector in a linear polarization basis is:

$$\underline{\Omega}'_2 = \frac{1}{\sqrt{2}} \left(2 S_{hv} \cos 2\psi + \left(S_{vv} - S_{hh} \right) \sin 2\psi \right)$$

The cross-polarized power element in the rotated vector is:

$$\Omega_2'\Omega_2'^* = 2|S_{hv}|^2 \cos 2\psi^2 + \frac{1}{2}\left(|S_{hh}|^2 + |S_{vv}|^2\right)\sin 2\psi^2 + \Re e\left[S_{hv}S_{vv}^*\right]\sin 4\psi$$
$$-\Re e\left[S_{hv}S_{hh}^*\right]\sin 4\psi - \Re e\left[S_{hh}S_{vv}^*\right]\sin 2\psi^2$$

Therefore, the cross-pol power is a linear combination of trigonometric functions. For a reflection symmetric target ($S_{hv}S_{vv}^* \to 0\, S_{hv}S_{hh}^* \to 0$) the function $\cos 2\psi^2$ has maxima at $\psi = 0 + n\frac{\pi}{2}$, while $\sin 2\psi^2$ has maxima at $\frac{\pi}{4} + n\frac{\pi}{2}$. Since $|S_{hv}|^2 < |S_{hh}|^2 + |S_{vv}|^2$, the signature for this target will always have a maximum at $\frac{\pi}{4} + n\frac{\pi}{2}$.

For a non-symmetric target, the terms $\sin 4\psi$ weigh in, these having maxima at $\frac{\pi}{8} + n\frac{\pi}{2}$ and thus enabling a shift of the signature maxima away from $\frac{\pi}{4} + n\frac{\pi}{2}$.

To get a feeling for the shift magnitude as a function of the relative weight of the power terms and the cross-correlation terms, let us consider an approximation of cross-polar power equation:

$$\mathrm{P}_{xpol}(\psi) = t_{HH}\sin 2\psi^2 + t_{HV}\cos\psi^2 + t_{corr} 2\sin 2\psi\cos 2\psi$$

where t_{HH} includes all co-polar terms, t_{HV} the cross-polar term, and t_{corr} the cross-correlation terms.

The derivative with respect to ψ is:

$$d\mathrm{P}_{xpol}(\psi)/d\psi = 4t_{corr}\cos 2\psi^2 + 2t_{HH} 2\sin 2\psi\cos 2\psi - 2t_{HV} 2\sin 2\psi\cos 2\psi - 4t_{corr}\sin 2\psi^2$$

The first zero of the derivative by solution of $d\mathrm{P}_{xpol}(\psi)/d\psi = 0$ is:

$$\psi = \frac{1}{4}\tan^{-1}\left[\frac{t_{corr}}{t_{HH} - t_{HV}}\right] + n\pi$$

This zero corresponds to the first minimum of $\mathrm{P}_{xpol}(\psi)$, and will allow us to obtain a figure of the signature shift with respect to the first minimum with $P_{corr} = 0$; that is, $\psi = 0$.

Let us consider the covariance matrix used in the simulation exercise concerning the margin between grassland and the primary forest. The estimated ψ shifts (degree) are reported in Table 7.1 for two incidence angles $\vartheta = 8°, 36°$ and two cases: first the original cross-correlation term,

TABLE 7.1
Estimated Shifts of the ψ Orientation Angle for given Ratios of the Co-pol/Cross-pol Correlation terms Over the Power terms of a Sample Covariance matrix

Incidence angle	t_{corr}/t_{HH}	$\Delta\psi$
8	0.03	0.5
36	0.001	0.02
8	0.03 × 10	4.9
36	0.001 × 10	0.26

corresponding to zero azimuth slope, and second when the cross-correlation term is boosted by a factor of 10.

Some observations. The weight of the cross-correlation terms with respect to the co-pol terms decreases with incidence angle. As a consequence, the shift $\triangle\psi$ of the cross-polar power also decreases with incidence angle.

Even for no azimuthal terrain slope, the forest being intrinsically an azimuthally non-symmetric target (because of different scattering elements in the illuminated volume with respect to the reflection plane) there is a shift to be expected in the signature maximum, albeit a very small one.

Since the rotation operator used for synthesizing the cross-polar power Equation (7.19) is the same as the one used for the Lip signature Equation (7.17), we expect to see similar consequences of the terrain slope with respect to the Lip estimators. To prove this point, we have run the simulator using the same covariance matrices as in the case presented in Section 7.5.2, but boosting the cross-correlation terms by a factor of 10.

The Lip *K* signatures for tilted terrain (green line) and flat terrain (red line) at incidence angle $\vartheta = 8°$ and at $\vartheta = 36°$ are shown in Figures 7.52 and 7.53, respectively. In line with the results shown in Table 7.1, a maximum shift $\triangle\psi = 5°$ is observed in the steep incidence angle case, while there is only a fraction of degree shift in the shallow incidence angle case.

The *Lipα* signatures (see Figures 7.54 and 7.55) indicate a much stronger effect of the terrain slope. The maxima for flat terrain occur at $\psi = 45$ for a shallow angle, while the maximum is already shifted at $\psi = 30$ for a steep incidence angle, this being attributed to the intrinsic asymmetry of the forest target. The distortion of the maxima toward a lower incidence angle is amplified in the case of the terrain azimuthal slope.

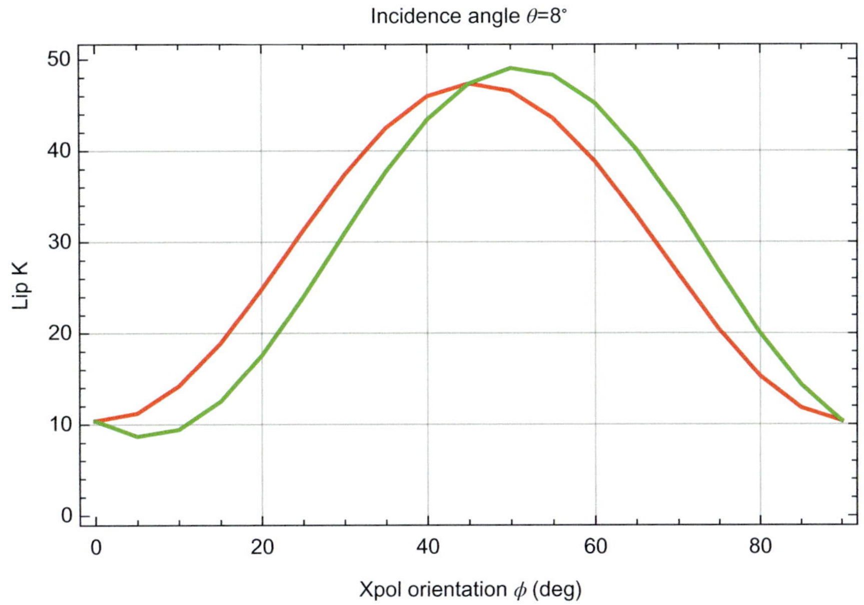

FIGURE 7.52 *Lip K* signature at steep incidence angle $\theta = 8°$ for tilted terrain in the azimuth direction (green line) and for flat terrain (red line).

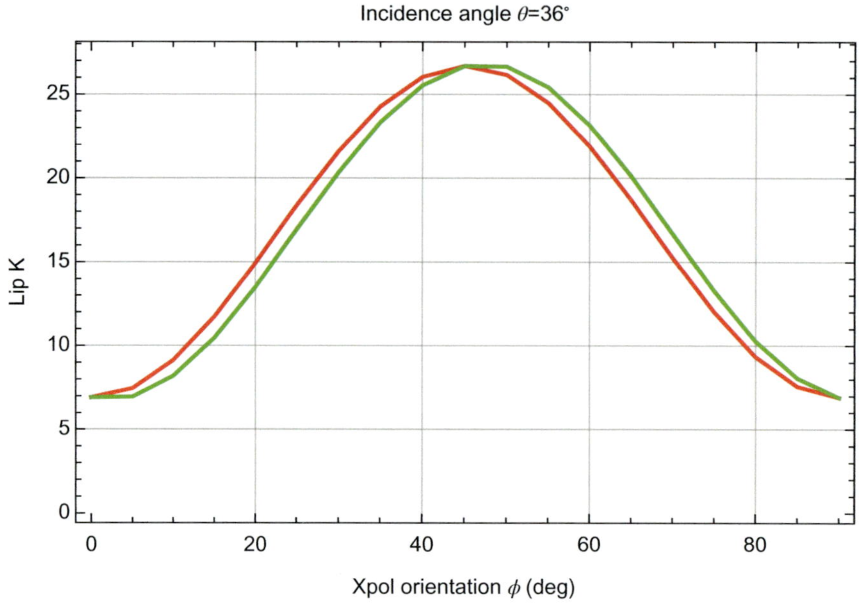

FIGURE 7.53 *Lip K* signature at shallow incidence angle $\theta = 36°$ for tilted terrain in the azimuth direction (green line) and for flat terrain (red line).

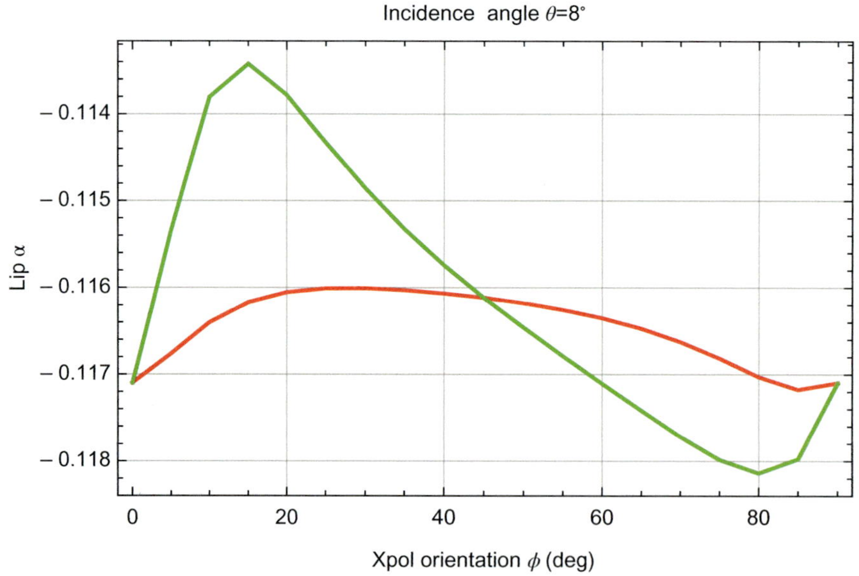

FIGURE 7.54 *Lip α* signature at steep incidence angle $\theta = 8°$ for tilted terrain in the azimuth direction (green line) and for flat terrain (red line).

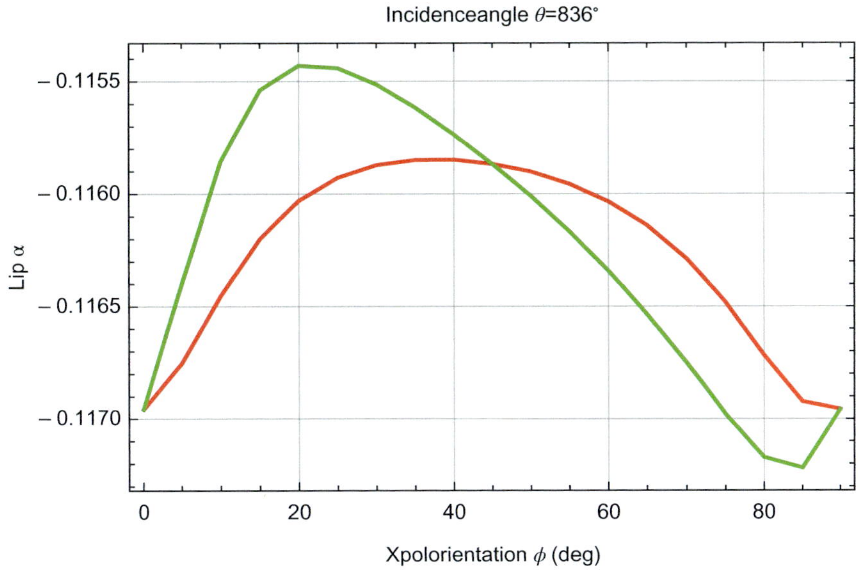

FIGURE 7.55 *Lip α* signature at shallow incidence angle $\theta = 36°$ for tilted terrain in the azimuth direction (green line) and for flat terrain (red line).

7.6 LIPSCHITZ REGULARITY IN REAL SAR DATA

After having established a background of mathematical theory and signal processing simulation tools for Lipschitz regularity estimators, it is now high time to look at edges in a real tropical forest, at least as far as the radar eyes may capture them. For the purpose, we will set up a number of test cases, using backscatter intensity and InSAR coherence amplitude data sets.

7.6.1 TanDEM-X Backscatter Data

For the first test case, we use TanDEM-X data sets acquired over a site in the Democratic Republic of Congo, centered around the Lulanga River, and the town of Basankusu ($1°13'20''N$, $19°48'10''E$). Data were provided by DLR in the framework of the announcement of opportunity AO2010, project VEGE0330. The X-band data were acquired with incidence angle $\vartheta = 23.3°$, baseline 150.9 m, and were processed to achieve a a ground- range pixel size of 4.6 m. It is the same dataset used for the experiment described in Section 6.8.2.

A cut of the whole dataset shows the area of interest (Figure 7.56), with an agricultural mosaic at the margin of the primary lowland forest. The cut corresponds to the agricultural mosaic marked NU in Figure 6.28 of Chapter 6. The original data are 4-Looks intensity, and they were filtered by the multi-temporal technique [22] that takes advantage of multiple acquisitions of the same area, thus achieving a 24-Looks dataset.

Transects spanning various types of discontinuities were extracted by visual inspection, and are presented in Figure 7.56 as oriented lines with a label for reference.

Transect A depicts the case of the passage from a clear-cut with some low-vegetation area to the dense forest. It is therefore similar to the situation studied in the simulation exercise. However, here we take special note that the edge is facing the range direction of the radar beam. As we shall see, this setting will influence the shape of the backscatter profile, and the shape of the wavelet transform coefficients. Remember that here we are using a discrete multi-voice wavelet transform, and therefore both the space and scale domain are discretized. That is why we talk of wavelet coefficients, and not of wavelet values.

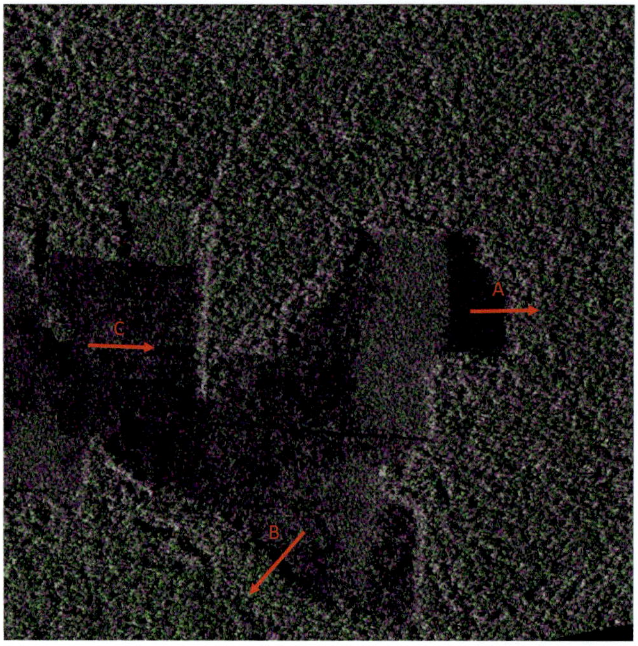

FIGURE 7.56 The area of interest for the Lip experiment in the TanDEM-X data set acquired around the Lulanga River, in the Democratic Republic of Congo. The data set was provided by DLR in the framework of the AO2010, project VEGE0330.

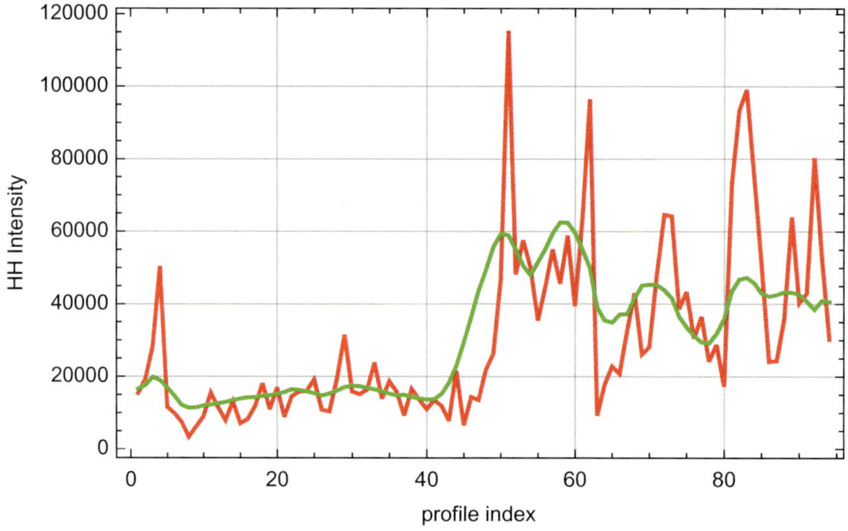

FIGURE 7.57 Intensity backscatter profile (HH) measured along the line transect *A* in Figure 7.56. The red line is the 4-Looks signal, and the green line is the smooth signal at the output of the scale 2 wavelet filter bank.

The HH intensity profile is reported in Figure 7.57. Note how at the onset of the transition from the two scattering situations (low volume over surface scattering and high dense volume scattering), around pixel number 50, a very sharp and strong peak occurs (intermittency). We postulate that this effect arises within the resolution elements that hit the discontinuity because there is a forward scattering component from the surface that illuminates the vegetation volume from the side, without being attenuated from the upper canopy layer, as in the case of direct incidence over the vegetation volume. The double bounce scattering from this radiation component and the forest elements (trunks and leaves) boosts the backscatter component because of direct illumination of the forest volume. Similar strong backscatter variation can be noted further down the transect. This highly non-stationary effect is of a different nature, and it is a result of the presence of strong heterogeneity of the forest upper canopy vertical structure. This kind of micro-topography effect is also discussed in Chapter 6, Section 6.2.

A discrete multi-voice wavelet transform with four dyadic scales and four voices is applied to the transect data. The first three voices of the first dyadic scale are shorter than the resolution of the transect data; they do not provide information and are discarded. Therefore 13 scales are used, these being: $\log_2 s$ = [0, 1.0, 1.25, 1.5, 1.75, 2.0, 2.25, 2.5, 2.75, 3.0, 3.25, 3.5, 3.75]. This wavelet

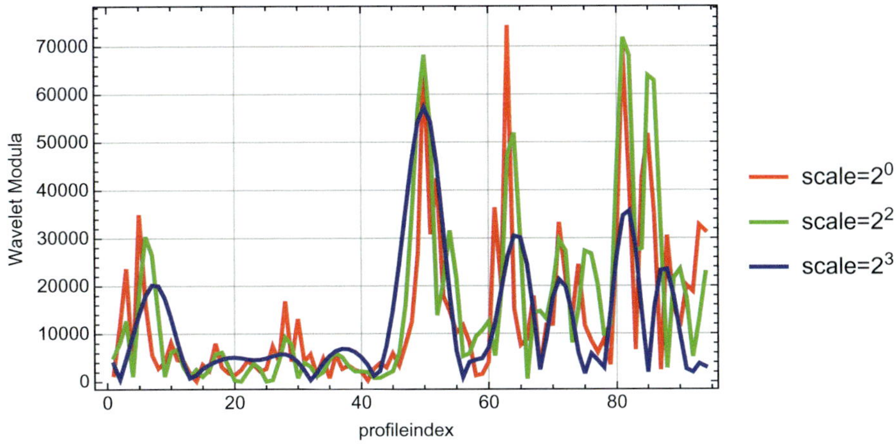

FIGURE 7.58 Wavelet moduli at scale 2^1, 2^2, 2^3 related to the clear-cut/forest profile in Figure 7.57.

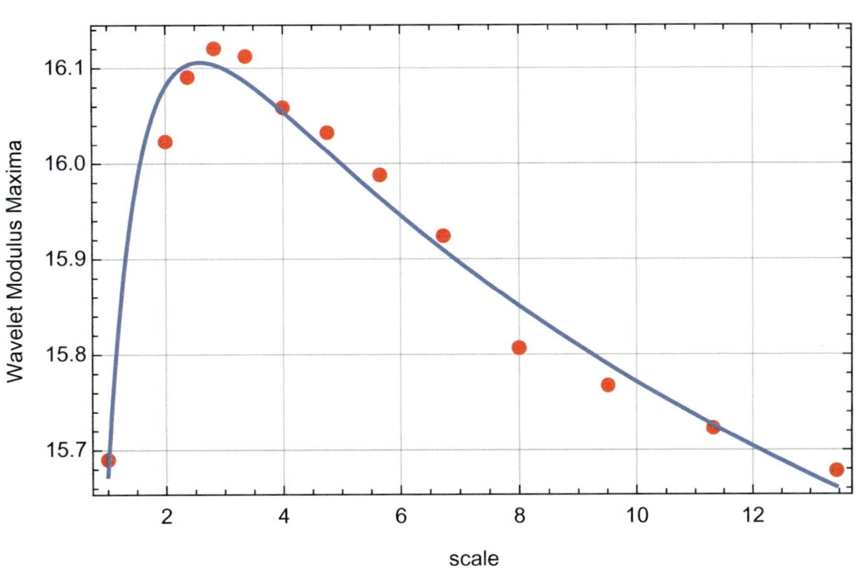

FIGURE 7.59 Wavelet modulus maxima as a function of scale related to the profile in Figure 7.57. The fitted function Equation (7.8) of the Lip estimator is overlaid.

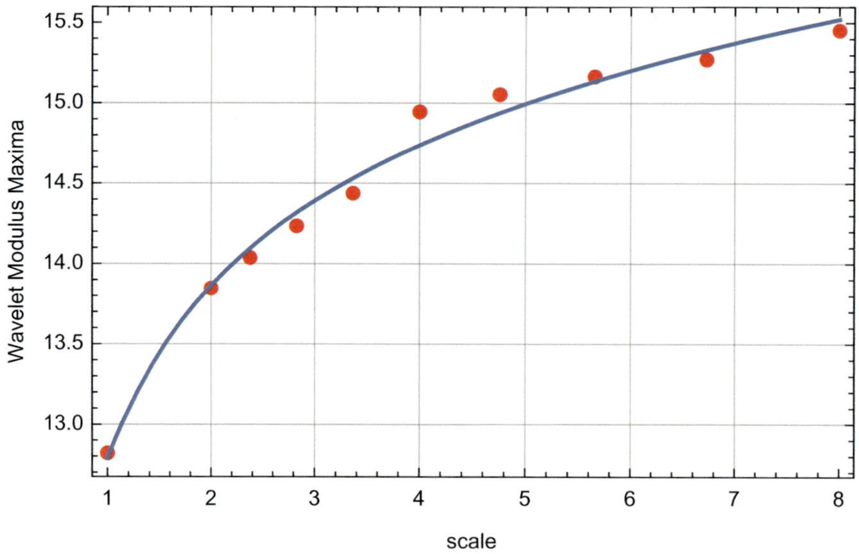

FIGURE 7.60 Wavelet modulus maxima of the smoothed HH signal at scale 2^2 as a function of scale. The fitted function Equation (7.8) of the Lip estimator is overlaid.

frame transform, based on the à trous algorithm and a differential operator, is the same as one used in the polarimetric simulator experiments (see Section 7.4 and Appendix).

The moduli of wavelet coefficients for scales 2^1, 2^2, 2^3 are reported in Figure 7.58. The wavelet coefficients are well resolved around the edge (margin clear-cut forest) at transect pixel 40, as well as around the next discontinuities generated by the forest canopy heterogeneity.

The wavelet modulus maxima detected around the edge, together with the fitted values by the regression performed by the Lip estimator, are reported in Figure 7.59. The maxima are decreasing after scale $2^{1.5}$, and the estimator gives: $Lip\ \alpha = -0.77$, $\log_2 Lip\ K = 16.7$, $Lip\ \sigma = 1.1$. Therefore, because of the strong intensity peak, the discontinuity is characterized as a $Lip\alpha \rightarrow -1$, with small smoothing variance, hence almost as an impulse type discontinuity.

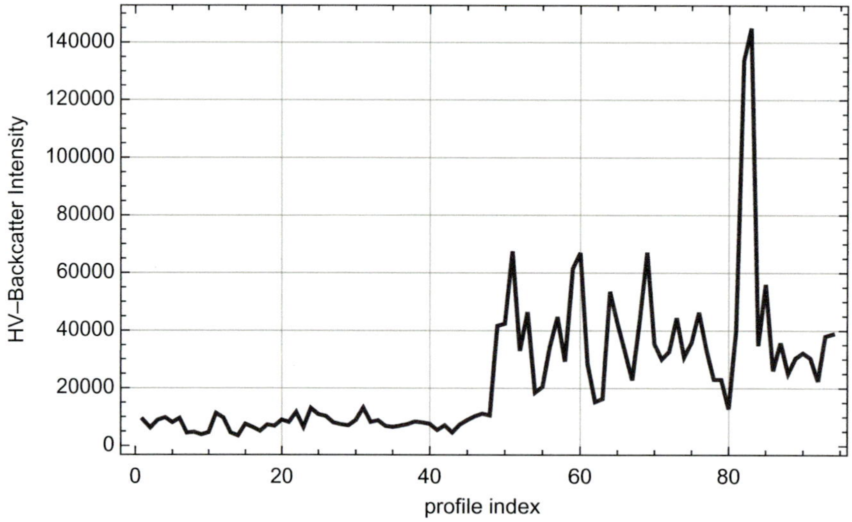

FIGURE 7.61 HV backscatter intensity profile for transect *A*.

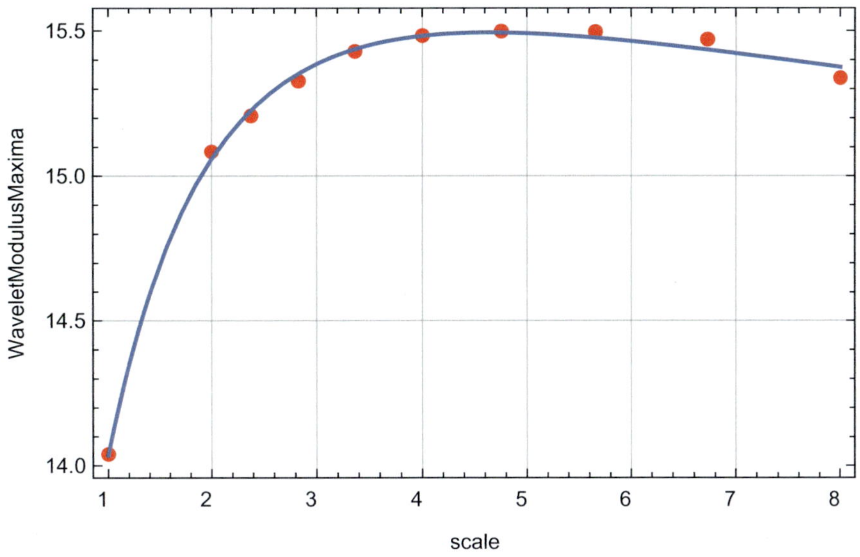

FIGURE 7.62 Wavelet modulus maxima trajectory in scale for the HV profile in Figure 7.61.

Sensitivity to the underlying step function type discontinuity can be gained by low-pass filtering the signal. This can be achieved, for example, by taking the output of the à trous filter at scale 2^2. The corresponding maxima trajectory are reported in Figure 7.60, and the estimator gives: $=Lip\ \alpha = 0.2$, $\log_2 Lip\ K = 13.3$, $Lip\ \sigma = 1.24$, which is typical of a smoothed step function.

Next we consider the same transect A but with HV intensity data (see Figure 7.61). Similar to the HH case, the point at the margin of the forest presents a sharp discontinuity, this corresponding to the passage from mainly surface to volume scattering. The segment within the forest area presents comparable mean intensity for HH and HV. However, the quartile deviation is much greater for the HH case, again reflecting the difference between two mechanisms at X-band. In the HH case, it is the upper canopy layer, limited by short penetration depth, that affects the scattering through the local incidence angle variation; for HV it is the variation of volume scattering in space, i.e. of scattering elements density, shape and orientation within the illuminated volume, that generates the intensity discontinuities.

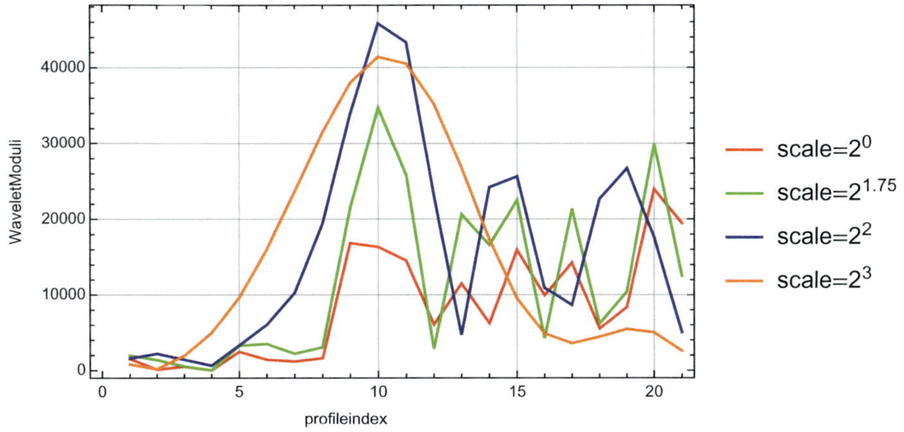

FIGURE 7.63 The wavelet moduli as a function of the profile index and parametrized by $log_2 s = [0.0, 1.75,\ \ 2.0, 3.0]$

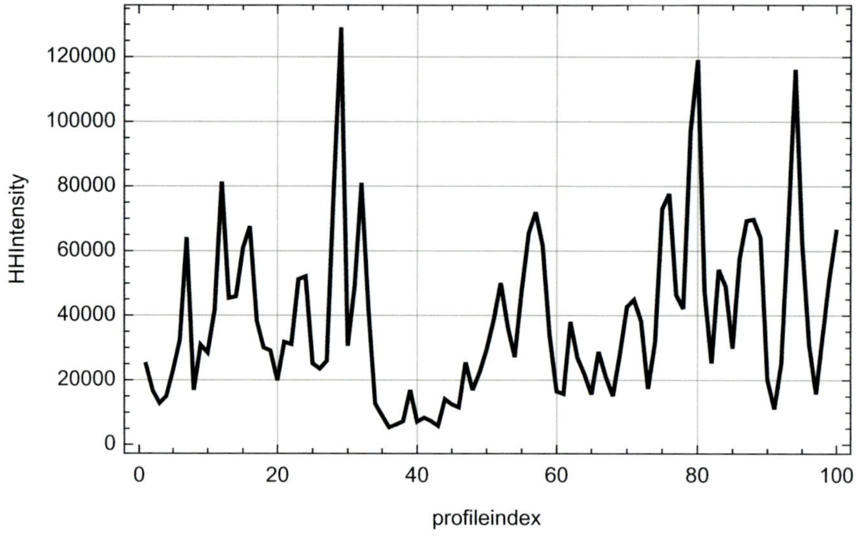

FIGURE 7.64 HH intensity profile measured along a transect extending from an agricultural field into the forest. The transect is laid in the opposite direction with respect to the SAR range direction, therefore it crosses an area of shadowing near the margin with the forest.

The wavelet modulus maxima detected around the edge, together with the fitted values by the regression performed by the Lip estimator, are reported in Figure 7.62. The Lip estimator gives: $\alpha = -0.94$, $\log_2 Lip\ K = 16.6$, $Lip\ \sigma = 2.37$. As can be noted from the maxima graph, the maxima trajectory tends to decrease after scale 2^2. As in the HH case, the discontinuity is interpreted as a negative $Lip\ \alpha \rightarrow -1$, hence almost as an impulse type discontinuity.

This effect is caused by the strong variations at the right of the edge, these inducing a merge of the cones of influence at scales $>2^2$, as highlighted in the graph of Figure 7.63, where the wavelet moduli for scales $\log_2 s = [0.0, 1.75,\ 2.0,, 3.0]$ are reported. The wavelet coefficients for the first four shorter scales are well resolved and the maxima tend only to slightly increase, but starting from $s = 2^3$ an influence from the variations at the right weighs in, and the maxima start to decrease, resulting in the negative $Lip\ \alpha$ estimate.

We shall now consider an upended situation with respect to the former one, namely when the transect crossing the border between non-forest and forest is directed opposite to the radar beam. In this case, the forest volume will act at the discontinuity as a two-way absorbing layer before the radiation reaches the ground, thus generating shadowing.

Transect B depicts this situation (see Figure 7.64). It starts from an area at the right of the margin, which is cultivated with a cassava (manioc plant) nursery, and continues into the remnant of the forest. The transect direction is opposite to the radar range direction.

Note how, in the section covering the agricultural field (pixels 1 to 40), the HH backscatter mean and variance are comparable to those of the section within the forest. Indeed, the plants in the nursey are laid out in a regular pattern, which is well reflected in the HH image, because of strong scattering from the stalks and the terrain.

Near the margin with forest (pixel 35), the backscatter drops into the forest shadowing, generating the first discontinuity. At the end of the shadowed zone, the backscatter increases and a second discontinuity appears. The wavelet moduli at scale 1 and at scale 2^3 are overlaid on the intensity profile, are shown in Figure 7.65).

Let us concentrate on the first discontinuity, that presents a stronger and less noisy response in the wavelet domain. The relayed wavelet modulus maxima detected around the edge, together with the fitted values, are reported in Figure 7.66. The Lip estimator gives: $Lip = -0.78$, $\log_2 Lip\ K = 16.98$, $Lip\ \sigma = 2.53$. The discontinuity at the onset of shadowing is interpreted as an impulse type

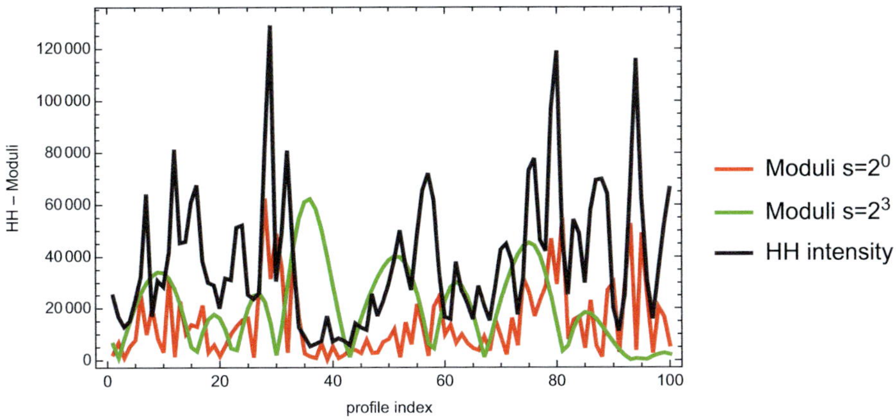

FIGURE 7.65 Wavelet moduli of the shadowed edge at scale 2^0 and 2^3 overlaid to the HH intensity profile.

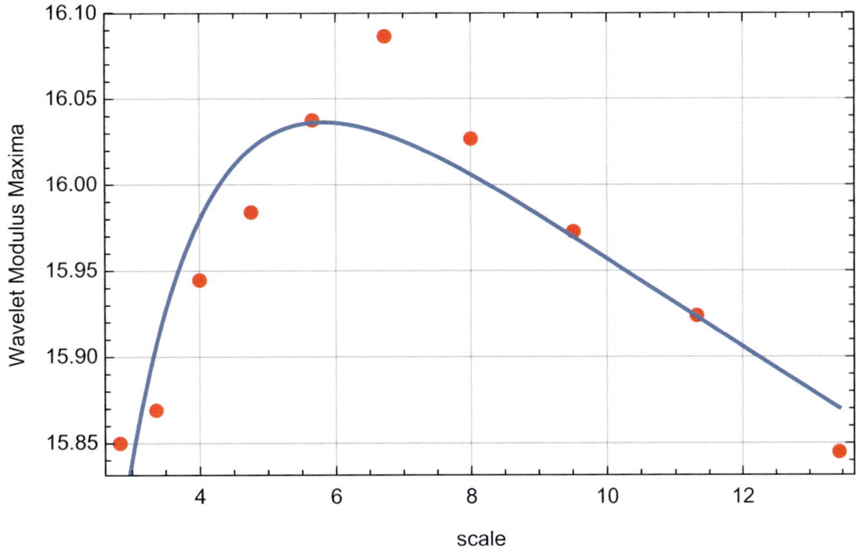

FIGURE 7.66 Wavelet modulus maxima corresponding to the first discontinuity at the onset of shadowing.

discontinuity *Lip* $\alpha \to -1$. Similar results are obtained for the second edge at the end of the shadowing region. It is one more example of how strong oscillations conceal the underlying step function discontinuity.

The same transect provides different results at HV polarization. The profile is shown in Figure 7.67. Observe how the first discontinuity at the onset of shadowing is now comparable in strength to the one at the end of the shadowing region. Moduli at scales 2^1 and 2^4, overlaid on the HV intensity profile are shown in Figure 7.68. Now both discontinuities are well resolved. Related maxima are reported in Figures 7.69 and 7.70. The Lip estimators for the two discontinuities give, respectively: *Lip* $\alpha = -0.42$, $\log_2$*Lip* $K = 14.2$, *Lip* $\sigma = 1.02$ and *Lip* $\alpha = -0.10$, $\log_2$*Lip* $K = 14.08$, *Lip* $\sigma = 1.18$. The discontinuities are therefore comparable, and are typical of a smooth step function-type discontinuity.

The last experiment concerns the capability of resolving and measuring in the Lipschitz sense the discontinuity provided by a single plant canopy within the nursery. Transect C is used for the scope (see Figure 7.71 for HH and HV polarization). There is a strong intensity response at HH provided by direct return from the stalk and double-bounce from the soil/stalk.

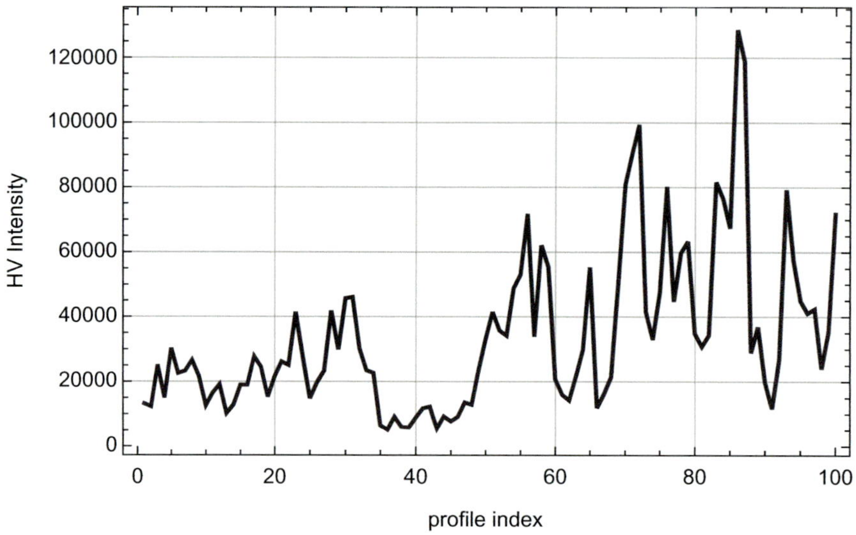

FIGURE 7.67 HV intensity profile measured along the same transect as in Figure 7.64.

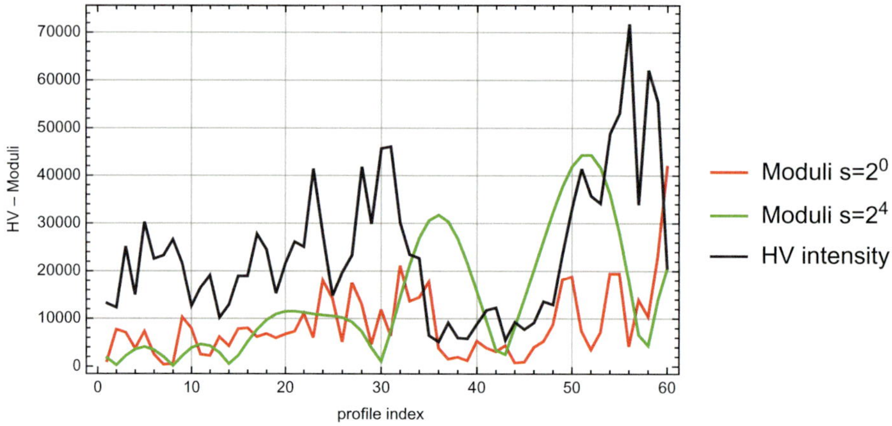

FIGURE 7.68 Moduli at scales 2^1 and 2^4, overlaid on the HV intensity profile within a subset of the profile straddling the shadowing region.

The related modulus maxima for HH polarization are shown in Figure 7.72. The trajectories present a negative slope with scale. The Lip estimator gives: *Lip* $\alpha = -1.0$, $\log_2$*Lip* $K = 17.28$, *Lip* $\sigma = 0$. Therefore a pure Dirac type discontinuity is detected.

The HV intensity profile indicates clearly that the strong discontinuity associated with one plant is not detected as a result of the small variation in volume scattering. There are discontinuities in the rightmost part of the transect, these being probably related to lower sparse vegetation. These variations can be Lip measured, and result in the maxima reported in Figure 7.73, while the estimator gives: *Lip* $\alpha = -1$, $\log_2$*Lip* $K = 15$, *Lip* $\sigma = 1.5$.

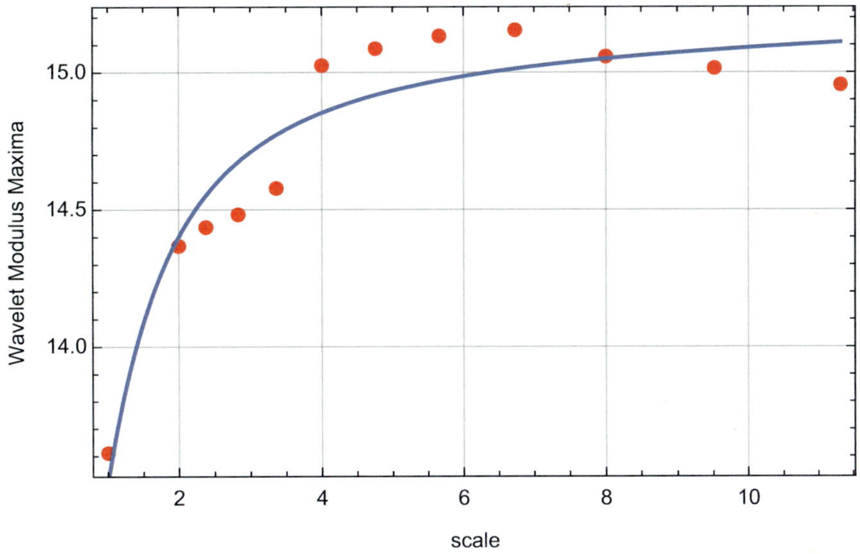

FIGURE 7.69 Wavelet modulus maxima trajectories and Lip estimator fitted function related to the first edge in the HV profile of Figure 7.67.

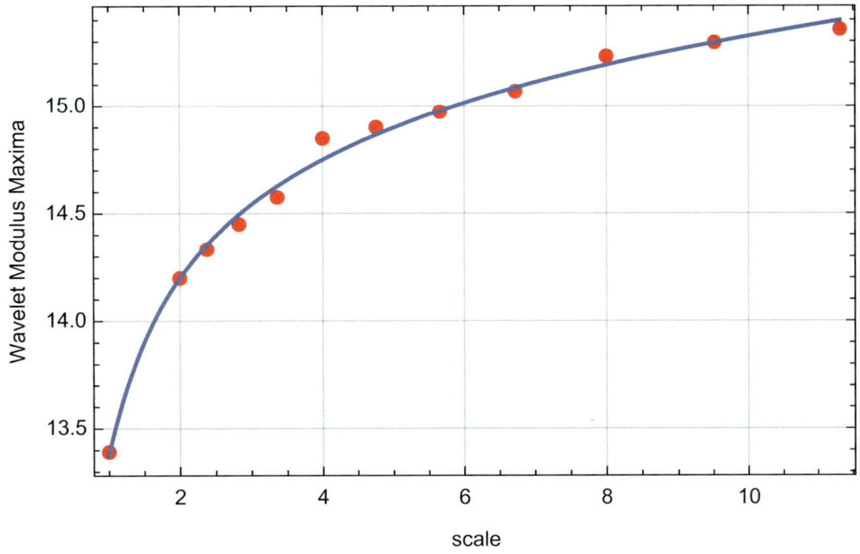

FIGURE 7.70 Wavelet modulus maxima trajectories and Lip estimator fitted function related to the second edge in the HV profile of Figure 7.67.

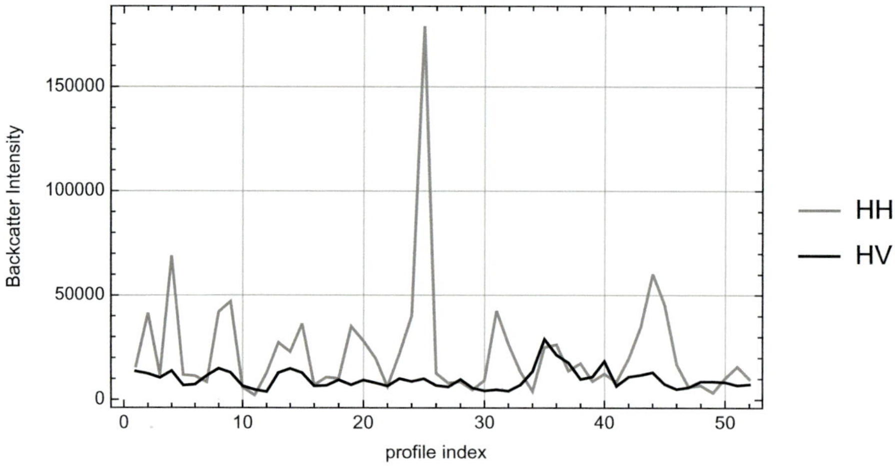

FIGURE 7.71 HH and HV backscatter intensity profiles for a transect intersecting a single tree in the nursery.

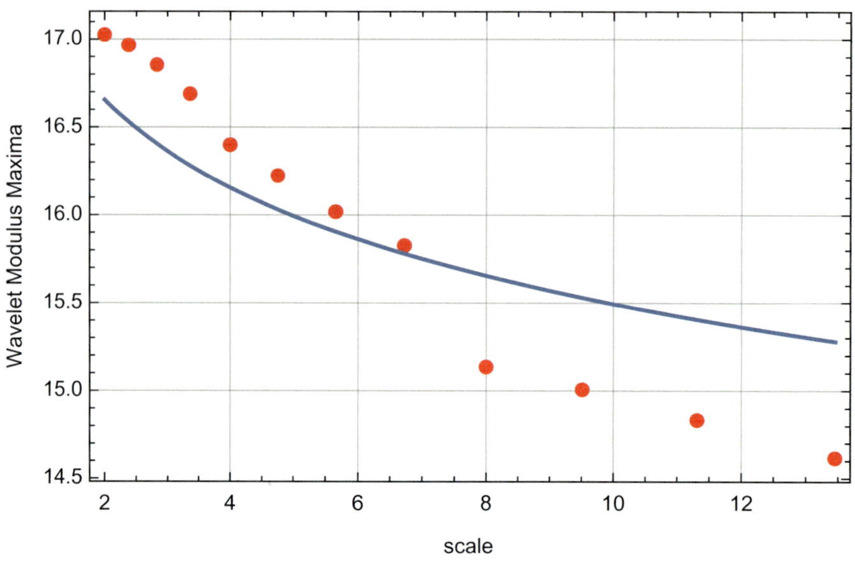

FIGURE 7.72 Modulus maxima related to the profile in Figure 7.71 for HH polarization.

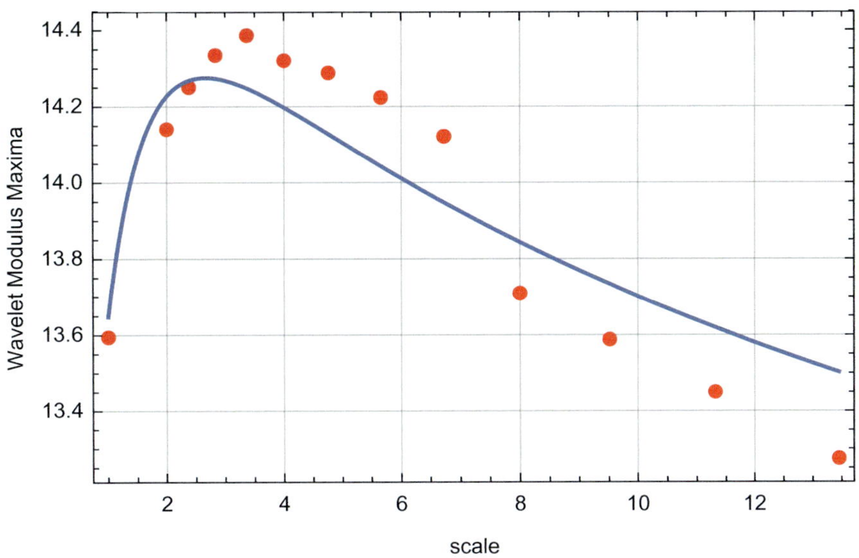

FIGURE 7.73 Wavelet modulus maxima related to the discontinuities in the rightmost part of the HV transect in Figure 7.71.

7.6.2 TanDEM-X Coherence Data

Interferometric coherence amplitude (hereafter coherence) images also present discontinuities at the boundaries between clear-cut areas (high coherence) and forest (volume decorrelation, low coherence), but also at the margin in a degraded forest between areas where different canopy development occurs, and the forest vertical structure function varies spatially. Lip measures are therefore also very useful to quantify these situations. We will give here two examples, once more based on TanDEM-X data.

The first focuses on clear-cut forest discontinuity. The dataset is the same as that used in the former backscatter experiments. The coherence amplitude image is shown in Figure 7.74, where the high coherence values correspond to the cassava nursery and other fields reclaimed for agriculture, and the low coherence values to the forest.

FIGURE 7.74 Coherence amplitude image (HH) over the same area as shown in Figure 7.56.

The HH coherence amplitude signal along the transect depicting the clear-cut/forest discontinuity is reported in the graph of Figure 7.75. Note the presence of strong noise in the low coherence segment, which spells difficulties in coherence related estimation problems (e.g. vegetation height). The issue of the characterization of the statistical properties of the coherence random process will be addressed in Chapter 8, where a method for filtering the coherence field will be proposed.

The wavelet moduli in the neighborhood of the edge for scales $s = 2^0, 2^1, 2^{1.75}, 2^{2.5}, 2^3$ are plotted in Figure 7.76. Observe how the maxima amplitude in the segment corresponding to the low coherence value (noisy part, transect indexes 25 to 45 in the graph) is comparable to or even greater than the amplitude corresponding to the discontinuity (transect index 20 in the graph). This means that an algorithm for the automatic detection of the edge based on Lip estimates would not stand much of a chance of succeeding.

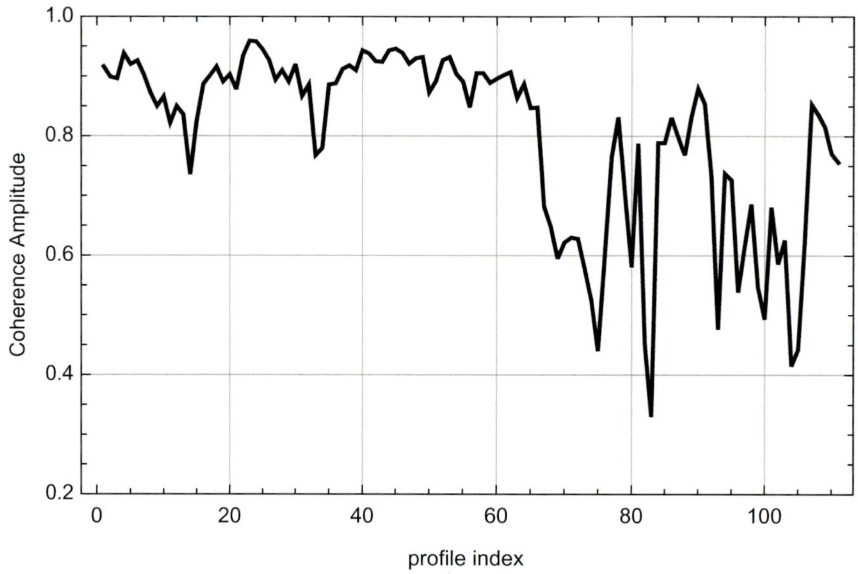

FIGURE 7.75 The HH coherence amplitude profile for the clear-cut/forest discontinuity.

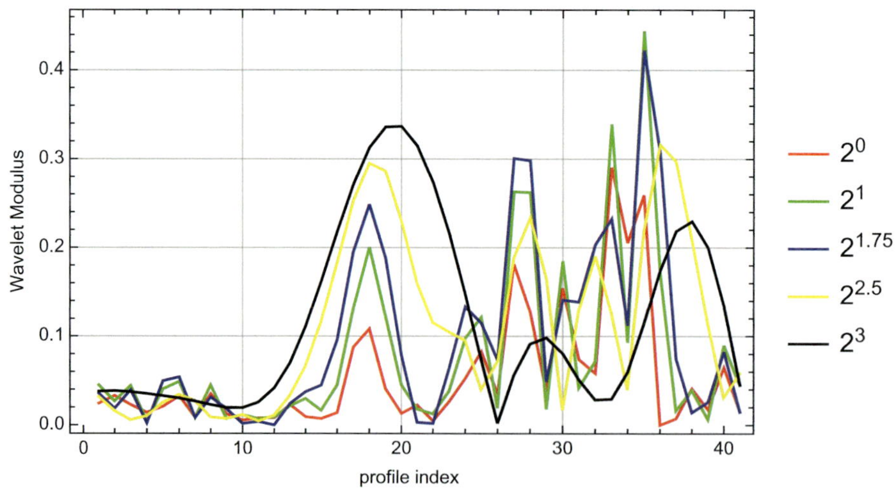

FIGURE 7.76 Wavelet modulus at scales $\{2^0, 2^1, 2^{1.75}, 2^{2.5}, 2^3\}$.

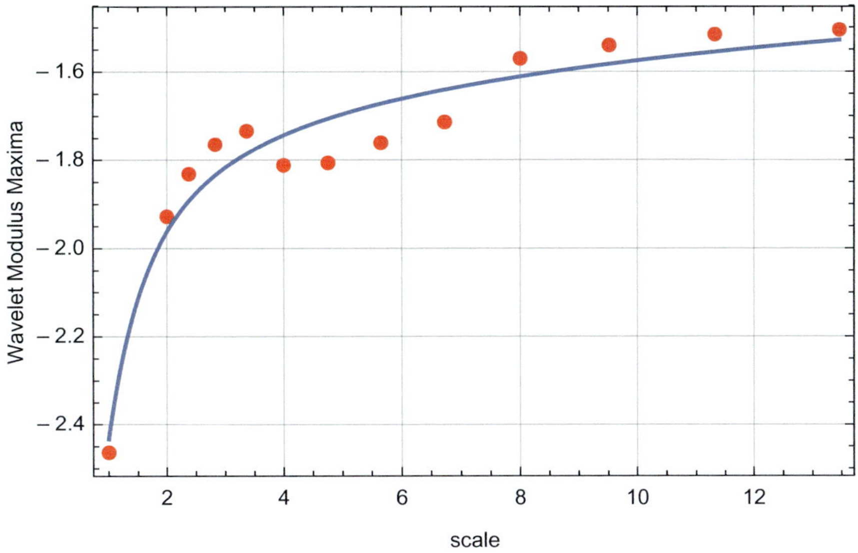

FIGURE 7.77 Wavelet modulus maxima trajectory in scale and fitted Lip estimator function for the coherence clear-cut/forest discontinuity.

FIGURE 7.78 A cut of the filtered coherence amplitude image (HH) depicting the transition between the intact forest (low coherence area at left) and the degraded forest (higher coherence area at right). TanDEM-X AO VEGE6702 dataset over the Sungai Wain Protection Forest (see also Chapter 9).

However, selecting by visual inspection the maxima related to the edge, we obtain the following results (see also the maxima and the fitted maxima values in Figure 7.77): $\alpha = -0.39$, $\log_2 Lip\,K = -1.9$, $Lip\,\sigma = 0.83$. These results are in line with the typical figure of a noisy smooth-step function.

The second experiment leads us to consider a very important thematic case, namely the interface between the primary intact forest and the degraded forest. The degraded forest is the end-point of a vegetation dynamic process, whereby from a disturbance event that caused the decrease (or full destruction) of some forest biophysical properties (e.g. biomass by clearing or by a fire event), a new, albeit different, state is created where some of the properties are reconstructed. Mapping the extent and the properties (e.g. biomass, tree height) of this new state, and its evolution in time, is a very important issue in environmental studies. Indeed, bio-physical parameters of the degraded forest state are still much less accurately quantified than the corresponding intact forest's parameters.

The whole issue was considered in Chapter 6, where an experiment is described concerning a fire-induced disturbance in the Sungai Wain Protection Forest in East Kalimantan, Indonesian Borneo. Here, we take a leaf from that study, showing how TanDEM-X coherence data, when properly noise filtered, can provide ways of detecting and estimating Lip measures of the margin between the degraded and the intact forest.

The TanDEM-X data acquired over the Sungai Wain Protection Forest, (provided by DLR in the context of the announcement of opportunity AO VEGE6702) consisted of eight dual pol (HH, HV) sets, with a ground resolution of approximately 3 m, spanning a period from August 2013 and December 2014. The coherence data sets were processed using the multi-temporal filter described in Chapter 9. A cut of the filtered HH coherence amplitude image (see Figure 7.78) depicts part of the interface between the degraded (at the right in the image) and the intact forest (at the left). From a pattern recognition point of view, the two classes (intact and degraded forest) are clearly dis-homogeneous and spatially intertwined. Still, the degraded forest area presents locally and in mean value higher coherence amplitude. This is because of a sparser and lower canopy cover. In some segments of the line following the margin between the two classes it is possible to identify sharp discontinuities. We analyze here one of these discontinuities, by tracing a transect across the edge.

The transects (from degraded to intact forest) holding the original and the filtered coherence values are plotted in Figure 7.79. Low-pass filtering of the amplitude signal reveals that a transition zone between the high and low coherence values can be defined between transect indexes 35 and 45 (see vertical arrows in the graph).

Let us examine the original coherence signal. Clearly, the high noise level of the low coherence segment (indexes 45 to 70) generates a series of discontinuities whose strength would mask the edge belonging to the transition zone. This situation is reflected in the wavelet moduli (see Figure 7.80). The maxima search algorithm detects the main discontinuity around transect index 43, this corresponding to the first strong discontinuity related to high coherence variance. Moreover, in the neighborhood of this position there is strong influence of similar discontinuities, with overlap of the cones of influence even at short scales. In this condition, with highly non-isolated singularity, the Lip parameters' estimator fails to find a solution to the minimization problem (Equation 7.8).

The situation looks improved for the filtered data. In this case, the search algorithm finds the candidate maximum at transect index 35; that is, in correspondence at the beginning of the transition zone (see arrow in Figure 7.81). In other words, it detects a discontinuity related to the transition

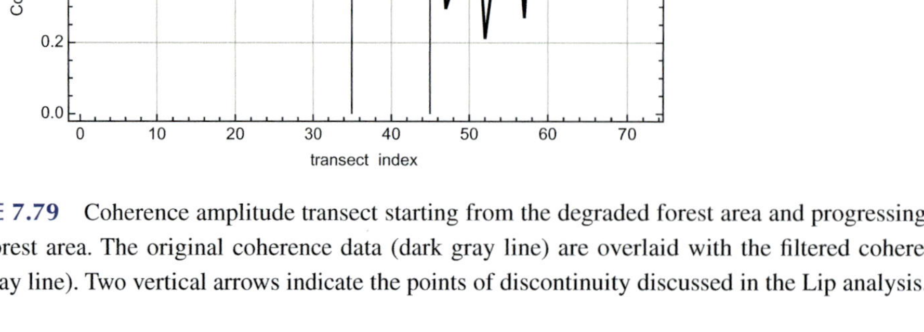

FIGURE 7.79 Coherence amplitude transect starting from the degraded forest area and progressing into the intact forest area. The original coherence data (dark gray line) are overlaid with the filtered coherence data (light gray line). Two vertical arrows indicate the points of discontinuity discussed in the Lip analysis.

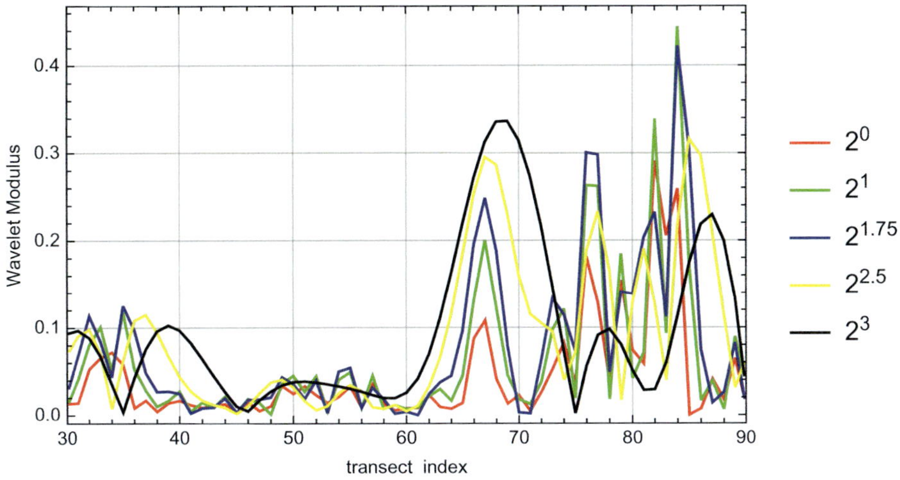

FIGURE 7.80 Coherence amplitude wavelet moduli related to the segments straddling the margin between the degraded forest and the intact forest. Clearly, strong discontinuities in the intact forest segment would mask the edge between the two classes.

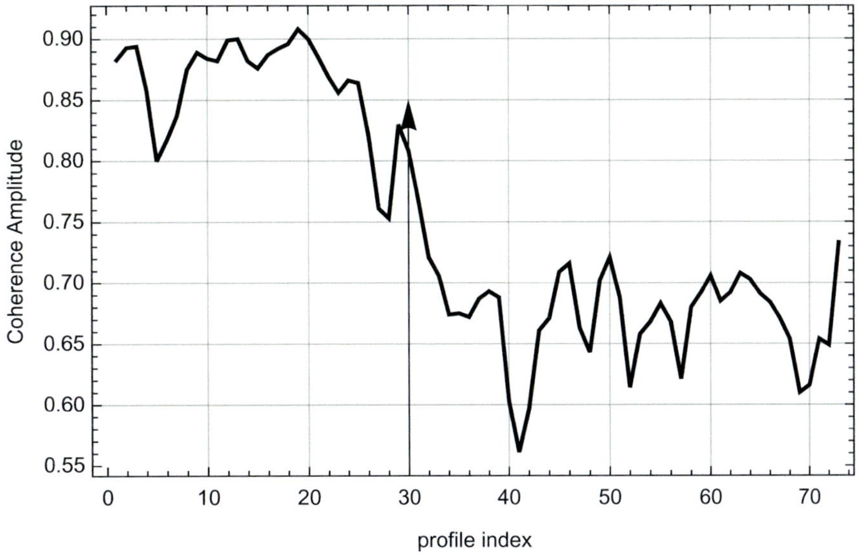

FIGURE 7.81 Filtered coherence amplitude profile. The vertical arrow marks the point where a maximum search algorithm spots the edge between the degraded and the intact forest.

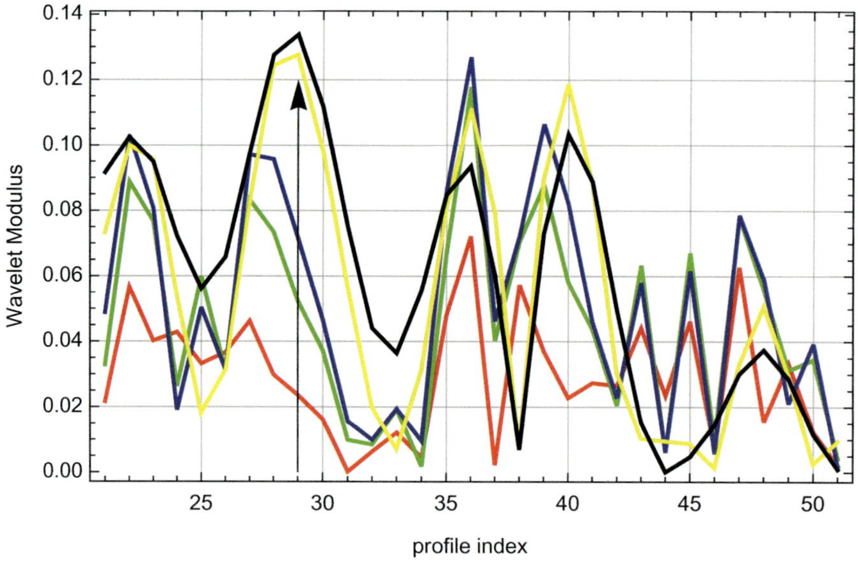

FIGURE 7.82 Filtered coherence wavelet moduli within a subset of the transect straddling the margin between the degraded and intact forest.

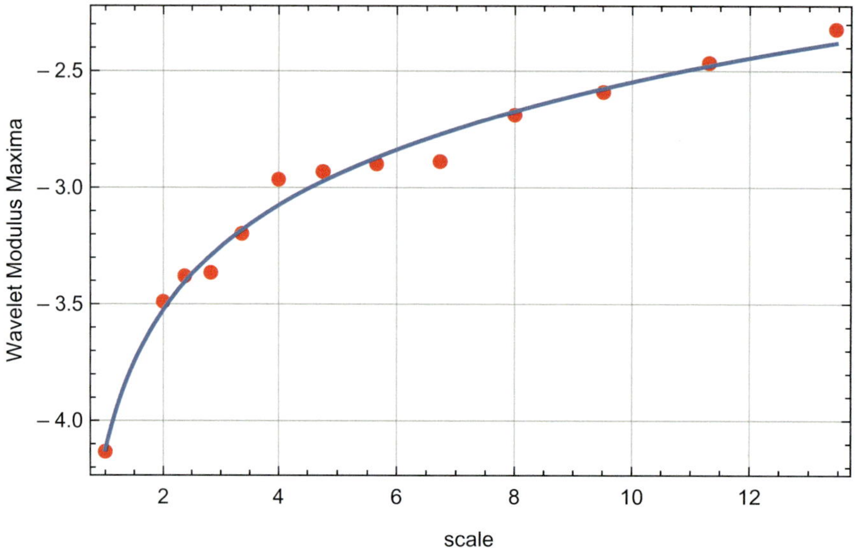

FIGURE 7.83 Filtered coherence wavelet maxima trajectory with overlaid Lip estimator fitting function.

of interest, namely the margin between high and low coherence areas. The corresponding wavelet moduli are reported in Figure 7.82. Notice how the wavelet moduli related to the edge (around position 30) are well separated from the neighboring ones (greater index), despite being of comparable strength. In this setting, the minimization algorithm converges nicely (see plot of maxima and fitted values in Figure 7.83) and the Lip estimator gives $\alpha = -0.11$, $\log_2 Lip\ K = -3.8$, $Lip\ \sigma = 0.67$, pointing toward a smoothed step function.

7.7 IMAGE-WIDE REPRESENTATIONS OF LIPSCHITZ PARAMETERS

So far, we have considered radiometric discontinuities along a one-dimensional transect, this being defined by visual inspection of the SAR image. We propose here a method that will allow us to build an image-wide unsupervised approximate representation of Lipschitz measures. The idea is as follows.

Each resolution element in a SAR dataset (e.g. backscatter or coherence), together with a set of neighboring elements, can be considered as a local discontinuity. If the element belongs to a constant reflectivity homogeneous area, then the discontinuity with respect to the neighborhood will be caused by intrinsic random reflectivity and noise (speckle) variations, and will be randomly distributed with radial direction around the point. If the element belongs to a non-stationary area – for example, a point on a ridge – then the discontinuity will be caused by a deterministic change of reflectivity with high spatial correlation in a specific direction along a chain of resolution elements.

This image model suggests that by applying a two-dimensional multi-scale differential operator we could obtain approximations of local (pixel-wise) Lipschitz measures. Suppose we use a 2D dyadic discrete wavelet transform. For each point P(x, y) in space the transform will generate the gradient of the reflectivity when smoothed at a given scale s. Therefore the gradient's modulus will explore, as the scale is increased, the differential properties of larger neighborhoods of the reflectivity element. The gradient's angle will provide information on the spatial orientation of the reflectivity change in the neighborhood. Linear fitting of the trajectories along the scale of the gradient's moduli will provide approximations of the Lip parameters in the same way as was done in the mono-dimensional analysis using Equation (7.6).

Thus, image wide Lip parameters are computed according to the following steps:

$$c(i,j,s) = WT\left(I(i,j),s\right) = \vec{\nabla}(I_s)$$

$$Lip\alpha, LipK = Fit\left[\log_2\left|\vec{\nabla}(I_s)\right|, \alpha \log_2 s + K\right]$$

$$\vartheta = \tan^{-1}\left(\nabla_x I, \nabla_y I\right)$$

where $WT(I(i,j),s)$ is the discrete wavelet transform of image I, (i,j) are spatial indexes, and $s = \{2^{\circ}, 2^1, 2^2, 2^3\}$ are dyadic scales.

The results are then combined in multi-band datasets, dubbed Lip maps, that include:

$$LipMap\ (i,j,b1) = -Lip\alpha,\ LipMap\ (i,j,b2) = LipK,\ LipMap\ (i,j,b3) = \vartheta_{s=2}$$

The $Lip\ \alpha$ is inverted, to enhance the magnitude of the band where strong impulse type $Lip\ \alpha = -1$ discontinuity are detected.

For dual-pol data, an alternative representation can be used to highlight the difference between polarizations:

$$LipMap\ (i,j,b1) = -Lip\alpha\ HH,\ LipMap\ (i,j,b2) = LipK\ HH,\ LipMap\ (i,j,b3) = LipK\ HV$$

For visual interpretation, the Lip map bands can be combined in an RGB color composite. Examples are reported next.

The maps are computed using the same TanDEM-X data sets that were used for the Lip estimators in Section 7.1. A cut of the RGB map using HH backscatter data is shown in Figure 7.84. Here $LipMap(-\alpha)$ is assigned to the red channel, $LipMap(K)$ to the green, and $LipMap(\vartheta)$ to the blue.

Note the reddish circular spots that signal the presence of strong $Lip\alpha \rightarrow -1$ targets. These interestingly corresponding to a strong HH return from single plant canopy. Note also the bluish

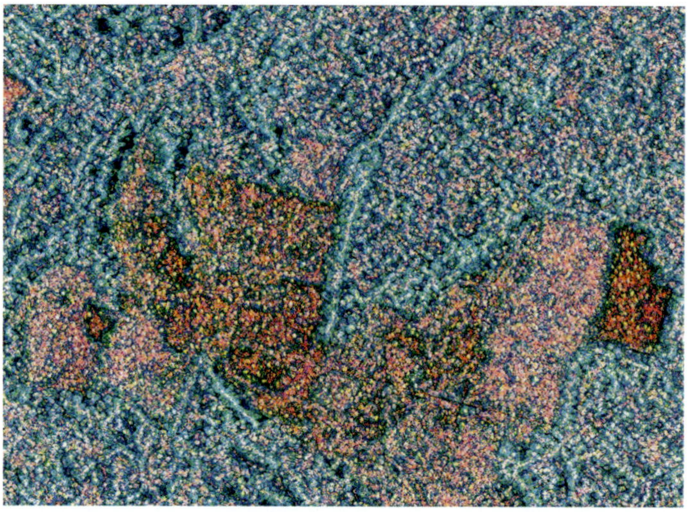

FIGURE 7.84 Lip map of the TanDEM-X HH backscatter image (same area as in Figure 7.56).

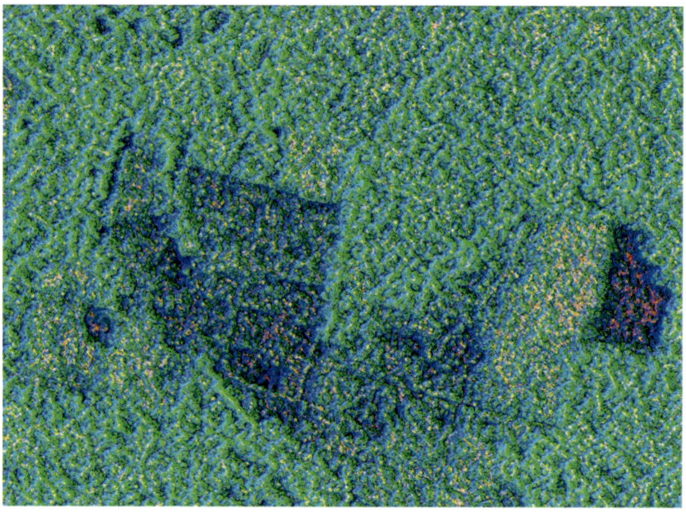

FIGURE 7.85 Lip map of the TanDEM-X HV backscatter image (same area as in Figure 7.56 and Figure 7.84).

patterns; these correspond to spatially correlated features of the gradient angle, and therefore depict discontinuities such as edges at the margin of forest with a coherent spatial development. A remarkable example is the thin linear feature toward the bottom of the image, this being a road.

In the second example (Figure 7.85), the Lip representation for the HV backscatter data is shown. In this case the edges between the bare soil areas (surface scattering and low HV return) and the forest (volume scattering, depolarization and large HV return) provide strong correlated structures, which are well captured by the gradient angle (blue band). In the large rectangular area at the right (secondary forest) the red spots appear that are signs of the *Lip* $\alpha \rightarrow -1$ features because of sparse canopy cover. Note the thin linear feature toward the bottom of the image, corresponding to a road, which is even better delineated than in the HH case, because of the greater contrast between surface and volume scattering.

Beyond thematic visual interpretation, the Lip map can be used in support to image processing tasks, such as edge detectors. A case in point will be developed in the next chapter, where Lip measures will be at the core of an algorithm for the generation of piece-wise smooth approximations of SAR reflectivity images.

REFERENCES

1. P.J. Davis, "*Interpolation and Approximation*", New York, NY: Dover Publications Inc., 1975ch. 1.
2. S. Mallat, and W. Hwang, "Singularity detection and processing with wavelets *IEEE Transaction on Information Theory*, vol. 38, no. 2, pp. 617–642, 1992.
3. S. Mallat, and S. Zhong, "Characterization of signals from multiscale edges", *IEEE Transaction on Pattern Analysis and Machine Intelligence*, vol. 14, no. 7, pp. 710–732, 1992.
4. Mallat, *A Wavelet Tour of Signal Processing*, 2nd ed., San Diego: Academic Press, pp. 163–219, 1999.
5. A.H. Zemanian, *Distribution Theory and Transform Analysis – An Introduction to Generalized Functions with Applications*, New York, NY: Dover Publications Inc. 1965.
6. M. Tsagris, C. Beneki, and H. Hassani, "On the folded normal distribution", *Mathematics*, 2014,2, 12–28; doi:10.3390/math2010012.
7. M.A. Karam, A.K. Fung, R.H. Lang, and N.S. Chuahan, "A microwave scattering model for layered vegetation", *IEEE Transaction on Geoscience and Remote Sensing.*, vol. 30, no. 4, pp. 767–784, 1992.
8. M.A. Karam, F. Amar, and A.K. Fung, "Electromagnetic wave scattering from a forest or vegetation canopy: Ongoing research at the University of Texas at Arlington", *IEEE Antenna and Propagation Magazine*, vol. 35, no. 2, pp. 18–26, 1993.
9. F.T. Ulaby, and J.J. Van Zyl, "Wave properties and polarization", in *Radar Polarimetry for Geoscience Applications*, F.T. Ulaby, C. Elachi (editors), Norwood, MA: Artech House Inc, 1991.
10. J-S. Lee, and E. Pottier, *Polarimetric Radar Imaging*, CRC Press, 2009.
11. H.C. van de Hulst, *Light Scattering by Small Particles*, New York, NY: Dover Publications Inc 1981.
12. S.M. Kay, *Fundamentals of Statistical Signal Processing-Estimation Theory*, London: Prentice Hall, 1993, ch.9, pp. 294–299.
13. A. Papoulis, *Probability, Random Variables, and Stochastic Processes*, New York, NY: McGraw-Hill, Inc., 1991.
14. Nelder, John A., and R. Mead (1965). "A simplex method for function minimization". *Computer Journal*, vol. 7, no. (4): 308–313. doi:10.1093/comjnl/7.4.308.
15. L.M. Novak, M.E. Setchin, and M.J. Cardullo, "Studies of target detection algorithms that use polarimetric radar data", *IEEE Transactions On Aerospace And Electronic Systems*, vol. AES-25, no. 2 March 1989.
16. H. Weyl, *Symmetry*, Princeton, NJ: Princeton University Press, 2016.
17. J.A. Koningstein, *Introduction to the Theory of Raman Effect*, Dordrecht, Netherlands: D. Reidel Publishing Company, 1972.
18. D. Schuler, J.S. Lee, and G.F. De Grandi, "The measurement of topography using polarimetric SAR images", *IEEE Transactions on Geoscience and Remote Sensing*, vol. 34, no. 5, September 1996.
19. G.F. De Grandi, D.L. Schuler, T.L. Ainsworth, and J.S. Lee, "*Topography sensing by polarimetric SAR: theoretical basis and application using orthogonal-pass Airsar data*", *IEEE IGARSS'97*, Singapore, 1987.
20. S.V. Nghiem, S.H. Yueh, R. Kwok, and F.K. Li, "Symmetry properties in polarimetric remote sensing", *Radio Science*, vol. 27, no. 5, 693–711, 1992.
21. S.H. Yueh, R. Kwok, and S.V. Nghiem, "Polarimetric scattering and emission properties of targets with reflection symmetry", *Radio Science*, vol. 29, no. 6, 1409–1420, 1994.
22. G.F. De Grandi, M. Leysen, J.S. Lee, and D. Schuler, "*Radar reflectivity estimation using multiple SAR scenes of the same target: techniques and applications*", *Proceedings IEEE IGARSS'97*, pp. 1044, August 3–8, 1997, Singapore.

8 The Beauty Farm

A Wavelet Method for Edge Preserving Piece-wise Smooth Approximations of Radar Images

In Chapter 7 we discussed the connection between a mathematical measure of the regularity of a class of functions, the Lipschitz exponent, and a signal representation provided by the wavelet transform. We discovered that following the decay of the wavelet transform modulus maxima, we could estimate the parameters that characterize Lipschitz functions.

Moreover, we extended our findings from the realm of functions defined in the domain of real numbers to the discrete case, by the introduction of a wavelet frame (a discretization of the wavelet transform) that acts as a differential operator. This step allowed us to move from the abstraction of mathematics to the characterization of discontinuities in physical phenomena, such as a SAR signal.

We now turn our attention to ways in which this technique could be exploited to support the passage from SAR observations to thematic products. In particular, we present a wavelet method for reconstructing piece-wise smooth approximations of the original SAR backscatter image while at the same time preserving discontinuities, such as edges and point targets. It is maintained that this is an important pre-processing step in connection with the segmentation and classification of radar imagery.

Basically, there are two approaches to segmentation and classification of SAR data. In one world, the algorithms specifically take into account the statistical characteristics of SAR data. They are therefore special purpose statistical decision engines. One example is a Bayesian maximum likelihood classifier proposed in [1]. Here, for example, it is assumed to deal with single look complex data that obey a complex multivariate Gaussian distribution. Another example is the Wishart classifier proposed by Lee in [2].

All these classifiers assume a scalar speckle multiplicative model with constant underlying radar reflectivity. They are non-contextual, or, in other words, they label an individual pixel based on some distance criterion with respect to prototypes provided by the user. Neighboring pixels are not taken into account. They are therefore most effective in cases where homogeneous areas with fully developed speckle dominate the image, but perform less well in situations of high fragmentation or where small features are present.

In the second world, there are algorithms that assume a simpler data model (in general piece-wise constant reflectivity with additive white Gaussian noise) but a more sophisticated way of taking into account local relationships among the picture elements. A particularly successful example of this class is the Pappas algorithm [3], a clustering technique that generalizes the classical K-means segmentation by introducing spatial constraints (based on a Gibbs field model of the label distribution) and taking into account local intensity variations of the image. The algorithm is therefore contextual and multi-scale.

Refinements of the classical Pappas algorithm in the direction of small but genuine feature preservation have led to the development of a family of clustering algorithms, such as MPAC [4], which constitute consolidated and well characterized engines, resting on a solid statistical foundation.

However, these algorithms are ill suited to deal with radar images because of the presence of speckle and, more generally, to deal with situations where classes of interest present a high within-class variability or texture. In other words, these methods were conceived with a specific class of imagery in mind, namely optical images (with additive noise) including objects with smooth surfaces.

To overcome this problem, the idea is to generate approximations of the radar images that fit the requirements of the clustering algorithms. In this way, one can take advantage of the more sophisticated features of the MPAC class algorithms – context and multi-scale handling – without modifying the underlying models (posterior probability density function of the labels given the data, and the a priori density of the region labels).

Let us now explain how this goal can be achieved by the combination of a proper image model, the discrete wavelet transform, and the connection with the Lipschitz regularity [5,6].

8.1 THE IMAGE MODEL AND A CONCEPTUAL VIEW OF THE METHOD

Let us consider multi-scale approximations of the toy signal portrayed in Figure 8.1. They are generated, for example, by successive application of the low pass filter in the filter bank processing scheme of the wavelet transform. The toy signal consists of a plateau rising from a flat background area.

From a visual perception point of view, sharp transitions in an image correspond to contours or boundaries of regions with which we associate a meaning. Of course, this intuitive definition cannot be cast in universal terms as an operational definition. What is a strong edge? What is an area of interest in terms of measures? Still, it is possible to associate measures of these quantities of interest

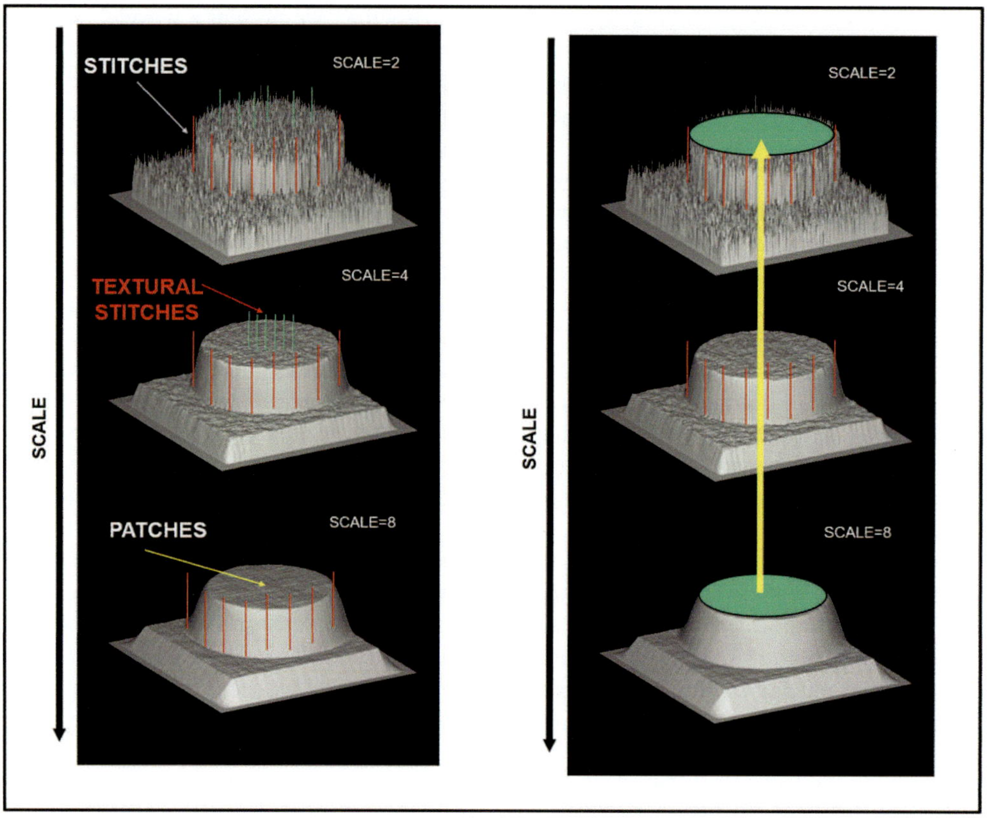

FIGURE 8.1 A conceptual view of the method.

for each particular situation where we can define a theoretical or empirical model of our signal. In our toy signal, the plateau is our area of interest.

At scale one in our decomposition, assuming that the signal is strongly contaminated by noise, we shall have sharp transitions both at the borders of the area of interest and within the area. The transitions within the area can be caused both by noise and intrinsic variations of the signal (texture). We call these points "stitches".

Suppose now that by some device one is able to generate derivatives of the signal, or the gradient field in the case of an image. Then the stitches will correspond to maxima of the gradient field moduli.

When we move to larger scales, and as the signal becomes smoother and smoother, stitches belonging to variations within the area (caused by noise and texture) will fade away, while the ones belonging to the contour of the area will persist. This is because the "type" of discontinuity caused by noise is different from the type related to the edge – in the first case it is like a spike, but in the second it is a step (see also Chapter 7). Moreover, the amplitude of the signal variations within the area is supposed to be less than the transition at the contour.

The signal within the area will tend to a constant, gradually losing all the textural information. We call the smooth part of the signal at a certain scale a "patch". Clearly, if we had a device to measure the gradient field on several scales, and we were armed with a model of the smoothed signal plus noise at different scales, we could exploit the position and decay of those points of fast transitions (the stitches) to isolate areas of interest.

It turns out that we can perform exactly that using a 2D (two-dimensional) wavelet frame representation, with the mother wavelet acting as a differential operator (see the Appendix for a formal definition).

The next interesting question is: Are we able to move the smooth approximation of our area of interest from a coarser scale to a finer one, but at the same time to preserving the sharpness of the boundaries that define the area? The answer is yes, because the wavelet basis provides a complete and stable representation of the signal. Given this representation, we can reconstruct by the inverse wavelet transform a modified version of the original image. In the reconstruction process, wavelet coefficients belonging to the unwanted transitions are trimmed, while the ones belonging to the contours of interest are preserved (see Figure 8.1).

We shall discuss next how this conceptual representation is cast into a computational engine.

8.2 THE COMPUTATIONAL ENGINE

We will first give an informal description of the main stepping stones that compose the engine. At the core of the engine, we find a wavelet modulus maxima follower algorithm. The maxima follower operation can be summarized as follows.

First, persistent maxima at the coarse scale of choice are searched. The maxima location problem is reduced to a one-dimensional problem by intersecting the gradient with a plane in the direction of the gradient angle. Wavelet moduli are first normalized by the smooth signal at the same scale in order to equalize the speckle multiplicative noise.

Once the modulus maxima are detected at the coarse scale, they need to be tracked in position and value in the finer scales. Maxima positions can change with scale according to the types of discontinuities (e.g. relative to the size of the support of the scaling function).

One more problem is that the wavelet coefficients become increasingly noisy as we move up toward the finer scales. At scale 2^1 the coefficients are too noisy to make any good use of them. The wavelet coefficients at this critical scale are therefore reconstructed by extrapolation from the selected maxima at scale 2^2, using the Lipschitz exponent estimated by the maxima trajectories with scale.

Summarizing, the wavelet maxima follower algorithm produces information on the position and amplitude at each scale of the wavelet modulus maxima that define contours of areas of interest.

This information is used for de-noising the fine scale wavelet coefficients, to refine the maxima thresholding of small features using a comparison between two scales and for the reconstruction algorithm.

The second step concerns the image reconstruction from coarse to fine scale. The reconstruction is achieved by means of an inverse wavelet transform, and exploiting the evolution with scale of wavelet modulus local maxima detected in the first step by the tracking algorithm.

Algorithms for the reconstruction of a signal from the extrema of its wavelet transform have been proposed in [7,8]. In these works, starting from values and positions of modulus local maxima, the wavelet transform is estimated at all points to be used as input to the inverse wavelet transform for signal reconstruction. In our setting, the wavelet transform is already known at all points; thus a simpler and computationally efficient signal reconstruction algorithm can be adopted.

In one dimension, the reconstruction algorithm proposed here searches right and left neighborhoods of the retained gradient modulus maxima characterized by a monotonic derivative. A regularized version of the wavelet modulus is used for the estimation of the two neighborhoods to avoid non-monotonicity caused by noise. Wavelet coefficients are retained in these neighborhoods and set to zero everywhere else (trimmed coefficients).

In the two-dimensional case (image), the left and right neighborhoods are searched along the cross-section of the regularized gradient modulus (2D) in the direction defined by the gradient angle. The values of the gradient modulus and angle within the neighborhoods are then used to compute the wavelet coefficients.

Finally, missing coefficients among the trimmed ones are reconstructed by interpolation from the nearest neighbors.

Image reconstruction at scale 2^0 is obtained by the inverse wavelet transform using the smoothed image at scale 2^s and the trimmed and interpolated wavelet coefficients at scales 2^s to 2^1.

Next, we shall give a more formal description of the flow of operations. The flow is broken down into successive phases. The intended functions are explained in words, while the relevant data structures are described using a pseudo-programming language notation.

1. Determination of the scale that is a suitable starting point for the smooth signal reconstruction. This choice is application dependent. It can be carried out by measuring the statistical properties of the smooth image at that scale (e.g. making sure that the variance in homogeneous areas is below a certain threshold). Or appealing to a theoretical model of the image. Let us indicate this scale as 2^{ns}.
2. Direct discrete wavelet transform from scale 2^1 to scale 2^{ns}. This step generates 2D sets of wavelet coefficients and smooth images **Wx[1..ns], Wy[1..ns], S[1..ns]** for each scale in the decomposition.
3. Computation of the gradient moduli and angle from the wavelet coefficients for each scale: **M[1..ns], A[1..ns]**.
4. Determination of the noise threshold **Th1** for the gradient modulus at scale 2^{ns} and the noise threshold **Th2** for the gradient modulus at scale 2^{ns-1}. For multiplicative noise, the gradient modulus normalized by the smooth signal is used. This step can be carried out interactively by visual inspection of the gradient in areas of interest. Alternatively, the threshold can be determined automatically by estimating the normalized standard deviation of the original SAR image (speckle strength) and using a theoretical model to estimate the amplitude of the wavelet coefficients relative to speckle at scale 2^{ns}.
5. Regularization of the wavelet coefficients and the gradient moduli at all scales using a smoothing spline filter. The strength of the filtering is modulated as a function of scale (the first two scales are smoothed less than the remaining ones). This step generates: **Wsmoothx[1..ns], Wsmoothy[1..ns], Msmooth[1..ns]**.
6. Thresholding of the smoothed gradient moduli at scale 2^{ns} and 2^{ns-1} using the threshold **Th1 and Th2** determined at point 4: **MsmoothTh[ns], MsmoothTh[ns–1]**.

7. Find positions of local maxima of **MsmoothTh[ns]**. These extrema correspond to strong transitions that propagate to scale 2^{ns}. Let us call the set of these extrema **MaxPersitentNS**.
8. Find positions of local maxima of **MsmoothTh[ns–1]**. These extrema correspond to strong transitions that propagate to scale 2^{ns-1}. Let us call the set of these extrema **MaxPersitentNSM1**.
9. Follow trajectories and values of gradient modulus maxima in set **MaxPersitentNS** from scale 2^{ns} to scale 2^2. Generates sets: **MaxFollowed[2..ns–1]**. Modula values are computed from the original gradient modula **M[2..ns]** to preserve their true dependency on scale, while maxima positions are computed using the regularized modula to improve the estimation accuracy.
10. Estimate Lipschitz exponent for each maxima trajectory.
11. Check if trajectories are monotonic. This step allows us to classify the discontinuities associated with the maxima trajectory into:
    a. Noise: monotonic decreasing
    b. Step edge: monotonic increasing
    c. Pulse: non monotonic.

    This information will be used later on in the construction of neighborhoods of the wavelet extrema.
12. Generate the set**: MaxShadowed** = **MaxPersitentNSM1 – MaxFollowed[ns–1]**. This is the set of maxima that are persistent down to scale 2^{ns-1} but do not propagate to scale 2^{ns}. They belong either to fine features with respect to scale s or to features that are closer than the resolution at scale ns.
13. Follow trajectories and values of gradient modulus maxima in set **MaxShadowed** from scale 2^{ns-1} to scale 2^2. Generates sets: **MaxShadowedFollowed[2..ns–2]**.
14. Estimate Lipchitz exponent for each maxima trajectory.
15. Check if trajectories are monotonic. Classify the discontinuities associated with the maxima trajectory as in Point 11.
16. Reconstruct wavelet coefficients in neighborhoods of the selected gradient modulus maxima positions from scale 2^{ns} to 2^2. This is a critical step that sets the groundwork for the signal reconstruction at high resolution. For each position of modula maxima, belonging to the union of sets **MaxFollowed** and **MaxShadowedFollowed**, left and right neighborhoods are searched for in the plane defined by the direction of the gradient. The neighborhoods are defined as a descent path along the regularized gradient modulus where the first derivative is negative. The type of discontinuity associated with the maximum, as derived at points 11 and 15, is checked. If the point of maximum belongs to type "noise" then it is discarded. If the point belongs to type "step", then trimmed **WxTr** and **WyTr** wavelet coefficients are computed along the neighborhood using the regularized gradient modulus and angle. If the point belongs to type "impulse", then the trimmed wavelet coefficients are computed using the original gradient modulus and angle. This distinction is a trade-off between the necessity to preserve point targets and to de-noise edges.
17. Reconstruct wavelet coefficients at scale 2^1 from those at scale 2^2 using the estimated Lipschitz exponent. As mentioned earlier, the coefficients at scale 2^1 are always too noisy to be used in the reconstruction; on the other hand, they hold important high frequency information that cannot be discarded. They are therefore extrapolated in the neighborhoods defined at point 16 for scale 2^2 using the regularized gradient modulus: **Msmooth[1] = Msmooth[2] $2^{-\text{Lip}}$**.
18. Inverse discrete wavelet transform from scale 2^{ns} to 2^1 using the smooth signal at scale 2^{ns} and the trimmed wavelet coefficients **WxTr** and **WyTr**.

8.3 PROBLEMS RELATED TO MULTIPLICATIVE SPECKLE NOISE

Multiplicative speckle noise poses a number of problems at different stages of the reconstruction method.

Speckle generates in homogeneous areas a strong gradient field at a fine scale, which is proportional to the mean value of the signal intensity and propagates through scales. If the speckle is white, it tends to behave like a series of Dirac δ *distributions*, and therefore the wavelet modulus maxima amplitude decays with scale (see Chapter 7). If the speckle is correlated, it tends to behave like a series of pulse functions, and the wavelet maxima increase with scale up to the point where the wavelet support becomes comparable to the correlation length, and then decrease, as we now demonstrate numerically.

Let us consider a single-Look intensity speckle pattern, where correlation is induced by the SAR imaging process, and the resulting autocorrelation function (ACF) is:

$$\langle I\rangle^2\left(1+\mathrm{Sinc}\left[\frac{t}{d}\right]\right)$$

where, $\langle I\rangle$is the mean speckle intensity, and d is the position of the first null of the sinc function, which is an indication of the average width of a speckle grain.

Since the wavelet transform is a linear operator, the wavelet transform variance can be computed from the zero-lag value of the ACF, this in turn being given in the Fourier domain by the product of the speckle pattern power spectrum and the amplitude squared of the wavelet high-pass filter spectrum.

$$\mathcal{F}\left(\mathcal{R}(x)\right)=\langle I\rangle^2\left(2\delta(\omega)+d\pi\frac{\Pi(\omega)}{d}\right)$$

$$\left|H(\omega)\right|^2=\mathrm{e}^{-s^2\omega^2}s^2 Abs\left[\omega\right]^2 \tag{8.1}$$

$$\left\langle W^2(x,s)\right\rangle=\mathcal{F}^{-1}\left[\left(\mathcal{F}\left(\mathcal{R}(x)\right)\left|H(\omega)\right|^2,(\omega,t=0)\right)\right]$$

$$\left\langle W^2(x,s)\right\rangle=\langle I\rangle^2\frac{1}{\pi}\int_0^{\frac{2\pi}{d}}\left(2\delta(\omega)+d\pi\frac{\Pi(\omega)}{d}\right)\mathrm{e}^{-s^2\omega^2}s^2 Abs\left[\omega\right]^2 d\omega \tag{8.2}$$

For $d = 1$:

$$\left\langle W^2(x,s)\right\rangle=\frac{1}{2}\langle I\rangle^2\left(-e^{-s^2}+\frac{\sqrt{\pi}\,Erf\left[s\right]}{2s}\right) \tag{8.3}$$

For $d = 2$:

$$\left\langle W^2(x,s)\right\rangle=\frac{1}{2}\langle I\rangle^2\left(-e^{-\frac{s^2}{4}}+\frac{\sqrt{\pi}\,Erf\left[\frac{s}{2}\right]}{s}\right) \tag{8.4}$$

From the sequence of expressions Equations (8.3) and (8.4) for the wavelet variance as a function of the speckle grain *d*, it can be seen that the decay of the gradient field with scale is not monotonic. The terms, including the error functions (second term in the summation), are decreasing functions with scale *s*. The terms with the exponential functions e^{-s} (first term in the summation) are increasing functions with s, and they tend asymptotically to zero with increasing scale. Yet the point of closest approach to the asymptote is shifted toward higher scales proportionally to the parameter d. This can be appreciated by the zeros of the second derivative:

$$\frac{d^2 e^{-s^2}}{ds^2} = 0 \overset{yields}{\rightarrow} s = 1/\sqrt{2}$$

$$\frac{d^2 e^{-s^2}}{ds^2} = 0 \overset{yields}{\rightarrow} s = \sqrt{2}$$

The wavelet variance for a correlated speckle pattern as a function of scale and parametrized by the speckle grain width Equation (8.2) is shown in Figure 8.2.

This analysis shows that speckle noise impacts the reconstruction methods in two ways. On the one hand, it affects the evolution of wavelet maxima with scale. To allow for distinguishing between maxima belonging to a strong impulsive target and noise (see phase 11 of the flow of operations), care must be taken to assure that, within the range of scales used for the maxima tracking, speckle is uncorrelated. This means that, if single- look slant range data are used, then data resampling by a non-integer interval must be applied as a pre-processing step.

On the other hand, Equation (8.2) indicates the wavelet variance associated with a speckle pattern to be proportional to the mean intensity $\langle I \rangle$. This point affects the capability of thresholding the wavelet coefficients in a consistent way in homogeneous areas with different mean intensity.

To overcome this problem and achieve a spatially adaptive threshold, the gradient modulus at scale s is equalized by the output of the low-pass filter (the à trous filter in the filter bank) at the same scale, following a technique that was proposed in [9,10](see also Chapter 6).

$$\left|\nabla wav(x,y,s)\right|^2 = \frac{w_x(x,y,s)^2 + w_y(x,y,s)^2}{S(x,y,s)^2} \tag{8.5}$$

For an area that is homogeneous at scale s, $S(x, y, s) \cong \langle I(x \pm s, y \pm s)\rangle$ is a local estimator of the mean intensity in the neighborhood of the point (x, y). Therefore, by Equation (8.2) we see that in this neighborhood the wavelet variance will be independent of intensity.

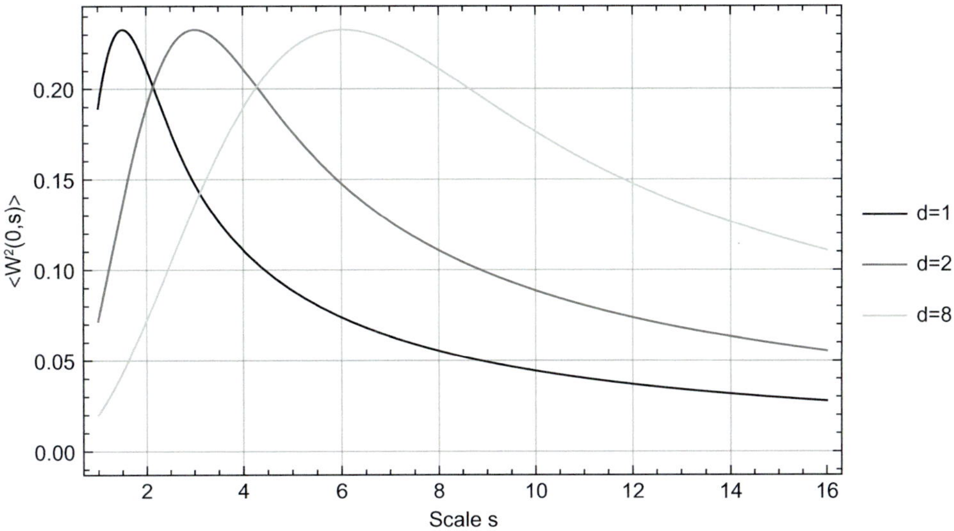

FIGURE 8.2 Wavelet variance of correlated speckle as a function of scale and speckle grain width.

This position thus solves the problem in stationary areas for trimming wavelet coefficients belonging to unwanted features, but raises one more question related to the identification of discontinuities by means of wavelet maxima evolution with scale. In other words, the question now is, since both numerator and denominator in Equation (8.5) are functions of scale, what happens to trajectories of equalized wavelet maxima when isolated singularities are present?

Let us check this issue by a limiting case, i.e. going to the continuous case and to functions defined on the domain of real numbers. Following the approach in the analysis of Lipschitz measures in Chapter 7, let us consider a step function and the wavelet being the derivative of a Gaussian function, whose Fourier representations are:

$$\mathcal{F}\left[\vartheta(x)\right] = -\frac{1}{i\omega} + (2A+1)\pi\delta\left[\omega\right]$$

$$Wf(\omega,s) = ie^{-\frac{1}{2}s^2\omega^2}s^2\omega \tag{8.6}$$

where A is the Heaviside function offset.

Let us also take as a low pass filter at scale s $S(\omega, s)$ a dilated Gaussian function with variance σ^2:

$$g(t,s) = \frac{1}{\sigma s\sqrt{2\pi}}e^{\left(\frac{-\left(\frac{t}{s}\right)^2}{2\sigma^2}\right)}$$

$$S(\omega,s) = \mathrm{e}^{-2s^2\omega^2}$$

The wavelet transform with normalization s^{-1} of the step function is:

$$\mathcal{F}^{-1}\left[\mathcal{F}\left[\vartheta(x)\right]Wf(\omega,s)\right] = \frac{\mathrm{e}^{-\frac{t^2}{2s^2}}}{\sqrt{2\pi}}$$

Here we see that the wavelet maxima ($t = 0$) do not decay with scale.

Let us now take the equalized wavelet transform. The smoothed Heaviside function at scale s is:

$$\mathcal{F}^{-1}\left[\mathcal{F}\left[\vartheta(x)\right]S(\omega,\ s)\right] = \frac{1}{2} + A + \frac{t\mathrm{Erf}[\frac{|t|}{2\sqrt{2}s}]}{2|t|}$$

The equalized wavelet transform is:

$$W_{equal}(t,s) = \frac{\mathrm{e}^{-\frac{t^2}{2s^2}}\sqrt{\frac{2}{\pi}}|tD|}{(1+2A)|t| + t\mathrm{Erf}[\frac{|t|}{2\sqrt{2}s}]}$$

Taking, for convenience, A=1:

$$\lim_{t\to 0}\frac{dW_{equal}(t,s)}{dt} = \frac{1}{9\pi s}$$

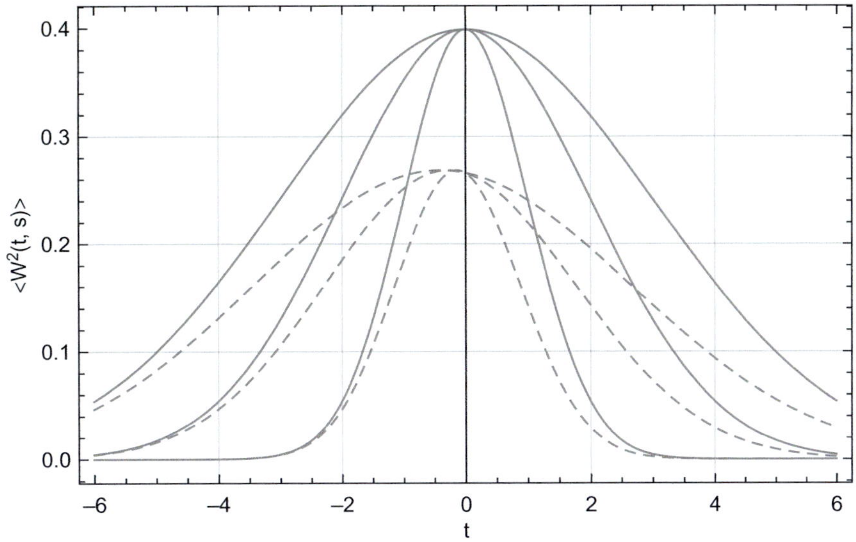

FIGURE 8.3 Unit step function wavelet transform as a function of time and parametrized by scale (solid line) compared to the equalized transform (dashed line). Note that the equalized maxima are shifted with respect to $t = 0$, but they do not depend on scale.

Hence we realize that the equalized wavelet modulus maxima do not occur at $t = 0$, as in the case of the original wavelet amplitude.

Still, numerical evaluation of the equalized maxima reveals they do not depend on scale, albeit they are shifted with respect to $t = 0$ (see Figure 8.3). Moreover, note that the equalized transform amplitude decreases with respect to the original one.

Let us now investigate what happens with a Dirac δ distribution. By the sifting property of δ distribution, the wavelet transform, using the wavelet in Equation (8.6), is:

$$W(t,s) = \frac{e^{-\frac{t^2}{2s^2}} t}{\sqrt{2\pi} s^2}$$

The smoothed δ *distribution* from the low-pass filter at scale s is:

$$\delta_{smooth}(t,s) = \frac{e^{-\frac{t^2}{8s^2}}}{2\sqrt{2\pi} s}$$

The equalized wavelet transform is:

$$W_{equal}(t,s) = \frac{2e^{-\frac{3t^2}{8s^2}} t}{s}$$

The derivative is:

$$\frac{dW_{equal}}{dt} = -\frac{2e^{-\frac{3t^2}{8s^2}}}{s} + \frac{3e^{-\frac{3t^2}{8s^2}} t^2}{2s^3}$$

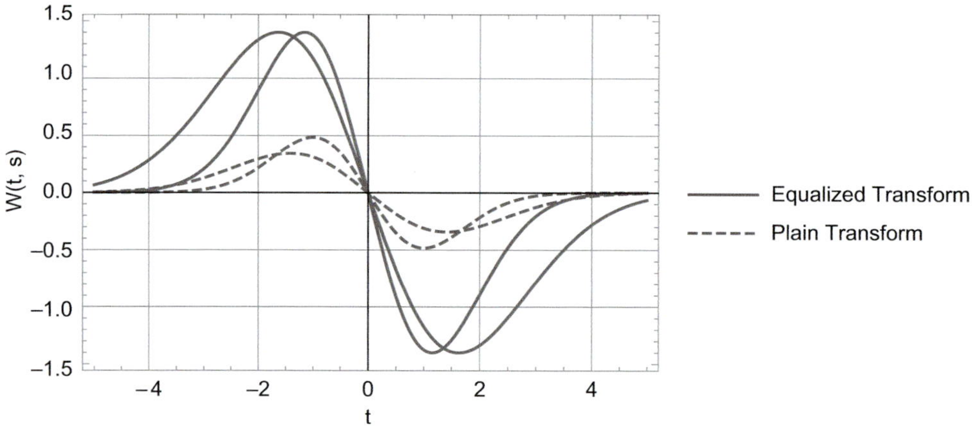

FIGURE 8.4 Wavelet transform of a Dirac δ distribution as a function of time and for scales $s = 2^0$, $s = 2^{0.5}$. Solid line: equalized transform; dashed line: normal transform.

Zeros of the derivative are: $t \to \pm\frac{2s}{\sqrt{3}}$, yielding the wavelet modulus maxima:

$$W_{equal}\left(\pm\frac{2s}{\sqrt{3}}, s\right) = \frac{4}{\sqrt{3e}}$$

The maxima are therefore independent of scale. For example, the wavelet transform at scales $s = 1$, $s = 2^{0.5}$ is shown in Figure 8.4 (solid line is the equalized transform; dashed line the plain transform). In conclusion, in the case of a δ singularity, the equalization boosts the transform amplitude, but cancels the characteristic dependence of the maxima on scale, which enables the discrimination between different types of singularities.

For these reasons, the equalization procedure is applied only for establishing an adaptive coefficient threshold (Step 4 of the flow of operations), while for operations involving estimates of the maxima trajectories (e.g. Step 9) the plain wavelet transform is used.

8.4 ISSUES RELATED TO TEXTURAL EDGES

In some applications, strong gradient fields that propagate through scales appear within a thematic class of interest, if the SAR backscatter field presents correlated textures, or if it is not statistically stationary. For example, in an application dealing with tropical forest, we are interested in the thematic class "degraded forest" (see Chapter 6). In the SAR backscatter domain, this class is not a pure class, because of the heterogeneity of the scene with diffuse cultivated fields patched within the forest canopy, there are local backscatter variations, well above the speckle noise level. These textural patterns generate large wavelet coefficients that propagate through scales, comparable to the ones pertaining to the margin toward the non-forest area. In other words, within the thematic class of interest there are textural edges comparable in strength to class boundary edges.

To achieve a smooth reconstruction of the textured area, wavelet maxima related to within-class variance should be removed, while maxima belonging to between-classes discontinuities must be preserved. In the reconstructed signal these mixed classes will be perceived as a piece-wise region with a radiometric intermediate behavior from forest class to deforestation.

A bespoke rule-based mechanism was devised to solve the problem in a specific thematic context, namely forest/on-forest mapping.

The approach is supervised, because it requires estimation of the backscatter spatial statistic within the class of interest.

In a nutshell, the mechanism searches within a neighborhood of each strong wavelet maxima at scale 2^{ns} and retrieves a local one-point statistics of the $S_{ns}(i,j)$ along right–left transects oriented along the gradient angle. $S_{ns}(i,j)$ is the image generated by the low-pass filter at scale ns in the discrete dyadic wavelet transform. The transects are either N pixels long, or equal to the distance between the current and the neighboring wavelet maxima.

Thus, two one-dimensional smoothed backscatter neighborhoods are defined for each maxima: $\Delta 1$, $\Delta 2$. Let us call $m1$, $m2$, $SD1$, $SD2$ the $\Delta 1$, $\Delta 2$ backscatter mean value and standard deviation.

The statistic of a reference set is estimated in a region classified by expert knowledge as belonging to the class that features textural edges. Let us call the reference set $\{m_i^{ref}, SD_i^{ref}, i = 1, 2\}$. Gradient maxima belonging to between-class edges are then detected by checking the balance between the left and right segments of the signal along the steepest descent. The ground being that around an inter-class edge, the backscatter statistic of the two segments will be significantly different with respect to the one estimated for a within-class edge. This criterion is encapsulated in the rule:

$$|m1 - m2| \neq |m_1^{ref} - m_2^{ref}|, |SD1 - SD2| \neq |SD_1^{ref} - SD_2^{ref}|$$

8.5 MAXIMA LINKING

As an added benefit, the wavelet modulus maxima algorithm provides, on top of smooth regions, also information on their boundaries (edges) that can be exploited by segmentation algorithms (e.g. the region growing segmentation used in the application to be discussed in the next heading).

The stitches derived by the maxima follower algorithm at scale 2 are distributed along contours of the regions of interest in a sparse arrangement. It is useful to create connections between neighboring maxima along a path where the gradient has similar properties (modulus and angle) with respect to the maxima. This closure operation allows for a better definition of the contours; moreover, a vector representation of the contours can be generated.

The stitch-connection algorithm uses a growing window to explore the neighborhood of each maximum. The window is grown until either at least one more maximum is included or a predetermined size is reached. If no more maxima are included, the maximum under consideration is considered to be either an isolated singularity or the terminal point of the current chain.

Otherwise, the nearest maximum is analyzed, and is included in the chain if two conditions are satisfied:

1. The gradient angles (extrapolated form scale 2^4) are aligned within a tolerance; and
2. The relative magnitudes of the moduli are comparable within prefixed bounds.

Each chain is represented by a linked list. When all maxima have been examined, a second pass is performed to check maxima labelled as terminal nodes of the chains, and to connect them if their gradient values satisfy the criteria mentioned above.

8.6 FROM THEORY TO PRACTICE: A TROPICAL FOREST COVER MAPPING EXERCISE USING SMOOTH APPROXIMATIONS OF GRFM SAR DATA

As an example of how the approach we have discussed so far can lend support to a thematic application, we shall outline next a research project aiming at land cover mapping over large areas of the South America tropical regions using the SAR mosaics developed within the GRFM initiative (see Chapter 2).

The project was carried out at the JRC in the 1990s. It relies, from the methodological point of view, on the generation of smooth SAR image approximations in a pre-processing step, followed by segmentation algorithms that are well adapted to the information content provided by the first step.

We shall only summarize the main results and the aspects related to the wavelet based method, while more details on the project can be found in [11,12].

The main objective of the project was to assess the capability of supporting by means of the GRFM mosaics a regional scale tropical forest mapping exercise in the Amazon. Quantitative measure of the capability was obtained by statistical error analysis, with estimates provided by LANDSAT imagery.

8.6.1 Processing Methods

It was realized at an early stage that conventional image classification methods (such as ISODATA) were ill suited for the purpose, because of the GRFM data statistical properties, including speckle noise and textures. One more difficulty was caused by the large data volume, this being necessary to extend the classification on a regional scale. Therefore, a bespoke processing chain had to be designed for thematic information extraction, including:

1. Piece-wise smooth image generation (wavelet method described in this chapter);
2. Segmentation by regional growing and supervised class gathering by expert knowledge; and
3. Two-stage nearest multiple prototype (NMP) classifier.

The techniques in points (2) and (3) are parallel paths to the classification problem's solution, and are pursued for evaluating the respective performances and trade-offs. They are briefly described next.

8.6.1.1 Region Growing

The region-growing algorithm exploits information from local statistics (mean and variance) of the smooth image (i.e. estimated in a neighborhood of a point in the field) and from the chained maxima (edges) supplied by the smooth image generator (see Section 8.5, Maxima Linking). We call an element of the linked maxima list a stitch.

From a random position, a seed region is grown, including recursively all-neighboring pixels that do not include a stitch. N pixels are counted in, with N large enough for obtaining a sufficient statistic (typically a 7×7 window).

The local statistic of the seed point is computed; the sample variance is used as the point estimate of the variance; and an interval estimate of the variance based on the assumption of normality of the data is derived. Also, given the sample mean, the theoretical variance for a homogeneous area is computed assuming multiplicative noise in the data. The signal model in our case foresees Raleigh distributed correlated speckle amplitude data, the statistical properties of which are modified by a series of low pass filters in the wavelet decomposition.

If the lower bound of the interval estimate of the local variance is greater than the theoretical variance of a homogeneous area, then the seed point is discarded since its local statistic is not representative of a homogeneous area. The seed point is also flagged as a non-classified point, while the points that were included in the region are kept, as candidates for future assignments.

If the local variance is compatible with the theoretical variance, pixels in the seed region are assigned to a new segment and the region is grown further, always recursively, assigning neighbors that do not belong to stitches.

Each new candidate pixel is compared with the local statistic and is accepted in the segment if its value falls between two standard deviations from the segment mean.

While the segment is grown also the local statistic (mean and variance) are updated until they reach asymptotic values.

Each assigned pixel is flagged as processed and assigned a progressive segment number in the output segment map. The growth process is stopped when no more neighbors can be accepted; then a new seed position is chosen as the nearest not-yet-processed pixel on the bisected distance of the original seed to image origin. The overall segmentation process is stopped when all the pixels have been flagged as processed.

Each segment is represented by a two-dimensional feature vector consisting of the per-segment mean and standard deviation of backscattered power. An unsupervised learning first stage iteratively gathers segments into segment categories based on a Jeffries-Matusita (JM) pairwise minimum distance criterion until a user-defined number of segment categories is reached.

In the supervised learning second stage, unsupervised segment categories are gathered into supervised classes of interest according to expert photo-interpreters.

8.6.1.2 NMP Classifier

The NMP first stage employs ELBG [13–15], a near-optimal vector quantization algorithm. ELBG is a vector quantization algorithm proposed in the pattern recognition literature as an improvement over the well-known LBG (i.e. hard c-means) vector quantization algorithm. It is a batch non-constructive vector quantizer, i.e. it tries to minimize distortion error, or mean square error (MSE), with a fixed number of code words that are user-defined.

At the second stage of the NMP classifier, template vectors are labeled, i.e. many-to-one relationships between unsupervised template vectors (code words) and land cover classes of interest are defined with the help of expert photo-interpreters.

8.6.2 Test Sites and Thematic Class Definition

Three test sites were selected covering different forest and savannah ecosystems along with different land uses. They are located in the state of Mato Grosso in west-central Brazil, in the state of Rondônia in northern Brazil, and in Colombia. The location of the three test sites was chosen taking as guidelines the Tropical Forest Ecosystem Environments monitoring by Satellites (TREES) [16,17] stratification of the tropical forests into "hot spots" and "cool spots" of deforestation.

The centers' geographic coordinates are:

1. Mato Grosso (−13.014, −55.035)
2. Rondônia (−13.014, −61.215)
3. Colombia (1.447, −75.109).

Each site covers approximately 31 000 square kilometers.

Moreover, the generalization ability of the per-pixel two-stage hybrid classifier was assessed over a fourth independent site in North Rondônia State (−11.569, −62.443).

The included land use patterns reflect the major forms of anthropogenic activity within the Amazon Basin (large-scale ranching, selective logging, shifting cultivation, organized colonization projects and mining). These practices result in different deforestation patterns. For example, in Rondônia, ranching and selective logging produce characteristic linear shapes, while in Colombia shifting cultivation and small plant nurseries produce diffuse patches of bare or cultivated fields against the background of the forest remnants.

Out of the complex ecological content of the Amazon landscape, the thematic target is condensed in this work in to four land-cover classes: forest, degraded forest, non-forest and water bodies.

The class "forest" includes the lowland primary forest (large phenological diversity and heterogeneous closed canopy), the upland forest with drier and open canopy, the swamp forest (dense and homogenous canopy), and the mangrove forest. The non-forest class includes bare and cultivated soil, vegetated areas not covered by forest, such as savanna, grassland, plantations and man-made structures.

The action term "degraded forest" is used to describe any intermediate state from forest class to deforestation. Degraded forest can also be a mosaic of isolated forest patches within non-forest regions or vice versa, isolated bare-soil regions surrounded by a forest area.

All the anthropogenic disturbances in the forest domain (owning to ranching, shifting cultivation, selective logging) fall into the classes of non-forest and degraded forest.

Let us have look at how some of the spatial arrangements resulting from deforestation appear in the radar imagery.

Characteristic deforestation patterns, their shape suggesting a herringbone, appear in the Mato Grosso site and are depicted in Figure 8.5A. The origin of these patterns is to be ascribed to the creation of access roads into the forest area by wood sawmill machinery. The creation of the first cleared terrain, appearing in the radar scene as a thin, straight line, will be followed in time by parallel clearings on each side of the road. Over the years, the clearings will expand from the road outward into the forest.

Massive clearings (low backscatter) with regular geometric shapes also appear in the Mato Grosso site (Figure 8.5B). They are a consequence of the conversion of primary forest to pasture for cattle ranching practices. Environmental damage by this practice includes carbon dioxide emissions, loss of biodiversity, soil degradation and water pollution.

Sparse and small clearings appear in the Colombia test site, these resulting from small-scale farming and ranching (Figure 8.5C).

8.6.3 Selected Results

Forest thematic maps derived from LANDSAT imagery by the NASA Tropical Rain Forest Information Centre (TRFIC) [18] were used as reference data for the validation of the GRFM SAR mosaics classifications.

TRFIC validated the accuracy of their classifications through a field-based accuracy assessment program. The map pixel spacing is 30 m and the geographic localization error is rated as 500 m. The classes in the TRFIC thematic product are: Forest, Deforested, Re-growing Forest, Water, Cloud,

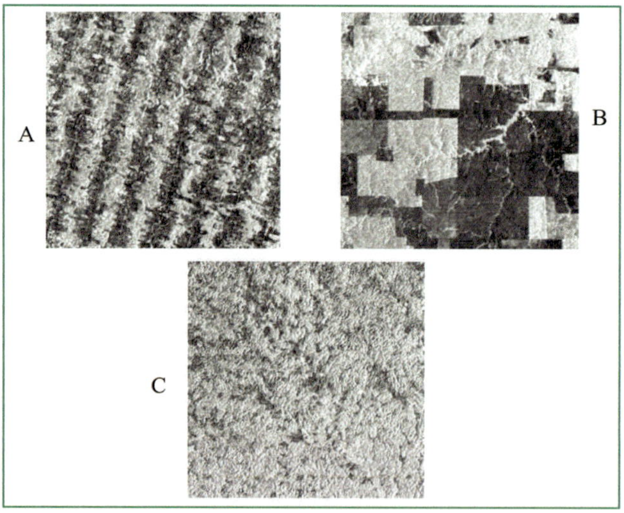

FIGURE 8.5 (A) The herringbone deforestation patterns in the Mato Grosso site. (B) Large clearings related to cattle ranching in the Mato Grosso site. (C) Sparse and small clearings in the Colombia test site. Copyright NASDA/MITI.

TABLE 8.1
Mapping of the TRFIC Thematic Classes on to the GRFM Classes

TRFIC thematic classes	SAR thematic classes
5. Cloud 6. Cloud Shadow	0. No correspondence
4. Water	1. Water
1. Forest	2. Forest
3. Re-growing Forest	3. Degraded forest
2. Deforested 7. Cerrado	4. Non-forest

Cloud Shadow, Cerrado. These classes were mapped into the GRFM thematic classes as indicated in Table 8.1.

To retain the same spatial location and scale in the comparison, maps derived from LANDSAT data were co-registered with the radar maps and scaled from 30 m to 100 m. Manually selected tie points were used for co-registration to avoid relative geographic spatial errors in the comparison.

To examine the mapping accuracy, error matrices $CM[5,5]$ between SAR and LANDSAT thematic data (TM) maps were computed, with entries corresponding to the GRFM thematic classes.

For the three test sites (Mato Grosso, Rondônia, Colombia) the overall accuracy between reference and radar maps was assessed. Moreover, the relative performance of the region growing and the hybrid classifiers was measured. Results for the thematic subset forest/non-forest are reported in Table 8.2

Of particular relevance for the assessment of the wavelet method (here dubbed wavelet segmentation), which is the subject of this chapter, are per-site analyses comparing the performance of two classification methods: an ISODATA unsupervised classifier with the original SAR data, and the hybrid classifier using the smooth SAR backscatter approximations.

The performances of the two classification techniques were assessed through Kappa analysis. Kappa analysis is a discrete multivariate technique that provides (through the estimator $\hat{K}$) from an error matrix a measure of agreement, or accuracy, between the remote sensing classification and a reference map [19,20]. The $\hat{K}$ statistic gives a stronger verdict than the overall accuracy (sum of the diagonal elements of the error matrix), since it also takes into account off-diagonal elements.

The $\hat{K}$ estimator is computed from the elements of the error matrix in the following way:

TABLE 8.2
Overall Accuracy for the Thematic Subset Forest/non-forest and the Two SAR Classification Methods

Site	SAR Region Growing	SAR Hybrid Classifier
Mato Grosso	91%	92%
Rondônia	73%	84%
Colombia	57%	82%

$$\hat{K} = \frac{n\sum_{i=1}^{k} n_{ii} - \sum_{i=1}^{k} n_{ia}n_{ai}}{n^2 - \sum_{i=1}^{k} n_{ia}n_{ai}}$$

Where: n is the total number of sample units, k is the number of classes, n_{ia} is the total number of samples in column i over all rows, and n_{ai} is the total number of samples in row i over all columns.

Confidence levels around $\hat{K}$ can be computed form the variance $V\hat{A}R(\hat{K})$ and the fact that the $\hat{K}$ statistic is asymptotically normal. In this way, the significance of the $\hat{K}$ statistic for a single error matrix can be assessed, and hence determine whether the agreement between the remote sensing and reference data is significantly better than a random classification.

The test statistic for assessing the significance of a single error matrix is:

$$Z = \frac{\hat{K}}{\sqrt{v\hat{a}r(\hat{K})}}$$

where Z is a standard normal deviate.

In the hypothesis testing setting: H_0: $K = 0$, H_1: $K \neq 0$. H_0 is rejected if $Z \geq Z_{\alpha/2}$ where $\alpha/2$ is the confidence level of a two tailed Z test with infinite degrees of freedom. For $\alpha = 0.05$ (0.025 in each tail) tor 95% confidence level, the critical value is $Z = 1.96$.

Moreover, the K analysis can be used to determine whether two error matrices are significantly different. In this case the test statistic is:

$$Z_{12} = \frac{\left|\hat{K}_1 - \hat{K}_2\right|}{\sqrt{v\hat{a}r(\hat{K}_1) + v\hat{a}r(\hat{K}_2)}}$$

Given the hypothesis H_0: $(K_1 - K_2) = 0$, H_1: $(K_1 - K_2) \neq 0$. H_0 is rejected if $Z \geq Z_{\alpha/2}$.

Table 8.3 presents, for the three sites, the results of the Kappa analysis on the error matrices derived from the two classification approaches: wavelet smooth approximations and the ISODATA algorithm.

Landis and Koch (1997) [21] characterized the $\hat{K}$ into three groupings: strong agreement $\hat{K} > 0.8$, moderate agreement $0.4 < \langle \hat{K} \rangle < 0.8$, and poor agreement $\hat{K} < 0.4$.

Maps from the wavelet method perform consistently better than ISODATA. In particular, the wavelet method reaches moderate agreement for sboth the Mato Grosso and Rondônia sites. ISODATA shows poor agreement for all sites apart from the Mato Grosso site, where the performance is in line with the wavelet method.

We remark, incidentally, that these results were derived for the GRFM mosaic SAR imagery, and are already smoothed in the pixel aggregation process from the original 4-Looks JERS-1 ground range images to 64-Looks. A much stronger difference between the accuracy of the two methods would results in the case of full resolution JERS-1 data.

From the Z statistic, since for all sites $Z>1.96$, we conclude that all classifications are in good agreement with the reference data.

From the Z_{12} statistic we infer that the error matrices related to the two methods are significantly different, the more so for the Rondônia and Colombia sites.

TABLE 8.3
Kappa Analysis for Assessing the Agreement of the GRFM Maps with Reference Data and the Significance of the Difference between Maps Generated using the Wavelet Smooth Approximation and the ISODATA Algorithm

Site	Method	$\hat{K}$	$\sqrt{\mathrm{var}(\hat{K})}$	Z	Z_{12}
Mato Grosso	Wavelets	0.605	0.021	28.30	27.43
	ISODATA	0.603	0.022	27.47	
Rondônia	Wavelets	0.407	0.024	17.24	200.50
	ISODATA	0.265	0.026	10.18	
Colombia	Wavelets	0.303	0.027	10.86	101.67
	ISODATA	0.205	0.31	6.61	

REFERENCES

1. J.A. Kong, A.A. Swartz, and H.A. Yueh, "Identification of terrain cover using the optimum polarimetric classifier", *Journal of Electromagnetic Waves and Applications*, vol. 2, no. 2, pp. 171–194, 1987.
2. J-S. Lee, M.R. Grunes, T.L. Ainsworth, L-J. Du, D.L. Schuler, and S.R. Cloude, "Unsupervised classification using polarimetric decomposition and the complex Wishart classifier", *IEEE Transactions on Geoscience and Remote Sensing*, vol. 37, no. 5, July 1998, pp. 2249–2258.
3. T.N. Pappas, "An adaptive clustering algorithm for image segmentation", *IEEE Transactions on Signal Processing*, vol. 40, no. 4, April 1992, pp. 901–914.
4. A. Baraldi, P. Blonda, F. Parmiggiani, and G. Satalino, "Contextual clustering for image segmentation", *Optical Engineering*, vol. 39, no. 4, pp. 1–17, April 2000.
5. F. De Grandi, J.S. Lee, M. Simard, and H. Wakabayashi, "*Speckle filtering, segmentation and classification of polarimetric SAR data: a unified approach based on the wavelet transform*", *Proceedings, IGARSS 2000*, Honolulu, HI, USA, paper TU08_05.
6. F. De Grandi, J.S. Lee, P. Siqueira, A. Baraldi, and M. Simard, "*Segmentation and labeling of polarimetric SAR data: Can wavelets help?*", *Proceedings IGARSS 2001*, Sydney, Australia, paper TU10_01.
7. S. Mallat, and S. Zhong, "*Characterization of signals from multiscale edges*", *IEEE Transactions on Pattern Analysis and Machine Intelligence*, vol. 14, no. 7, pp. 710–732, 1992.
8. R. A. Carmona, "Extrema reconstruction and spline smoothing: variations on an algorithm of Mallat & Zhong", in *Wavelets and Statistics, Lecture Notes in Statistics*, A. Antoniadis, and G. Oppenheim, (editors), New York: Springer Verlag, 1995, pp. 83–94.
9. M. Simard, G. De Grandi, K.P.B. Thomson, and G.B. Benie, "Analysis of speckle noise contribution on wavelet decomposition of SAR images", *IEEE Transactions on Geoscience and Remote Sensing*, vol. 36, no. 6, pp. 1953–1962, November 1998.
10. M. Simard, G.F. De Grandi, and K. Thomson, "*Adaptation of the wavelet transform for the construction of multi-scale texture maps of SAR images*", *Canadian Journal of Remote Sensing*, vol. 24, no. 3, pp. 264–285, September 1998.
11. M. Sgrenzaroli, G.F. De Grandi, H. Eva, and F. Achard, "Tropical forest cover monitoring: estimates and validation from the GRFM JERS-1 radar mosaics using wavelet zooming techniques", *International. Journal of Remote Sensing* (GRFM special issue), vol. 23, no. 7, April 2002.
12. M. Sgrenzaroli, A. Baraldi, G. De Grandi, H. Eva, and F. Achard, "A novel approach to the classification of regional scale radar mosaics for tropical vegetation mapping", *IEEE Transactions on Geoscience and Remote Sensing*, vol. 42, no. 11, 2654–2669, November 2004.
13. G. Patanè, and M. Russo, "The Enhanced-LBG algorithm", *Neural Networks*, vol. 14, no. 9, pp. 1219–1237, November 2001.

14. G. Patanè, and M. Russo, "ELBG implementation", *International Journal of Knowledge Based Intelligent Engineering Systems*, vol. 4, pp. 94–109, 2000.
15. G. Patanè, and M. Russo, "Fully automatic clustering system", *IEEE Transactions on Neural Networks*, vol. 13, no. 6, pp. 1285–1298, 2002.
16. J. P. Malingreau, F. Achard, G. D'Souza, H.J. Stibig, J. D'Souza, C. Estreguil, and H. Eva, "AVHRR for global tropical forest monitoring: The lessons of the TREES project", *Remote Sensing Reviews*, vol. 12, pp. 29–40, 1995.
17. Achard F., Eva H.D., Glinni A., Mayaux P., Stibig H.J., and Richards T., *Identification of Deforestation Hot Spot Areas in the Humid Tropics*, TREES Publications Series B4, European Commission, Luxembourg, EUR 18079 EN, 100, 1998.
18. TRFIC Tropical Rain Forest Information Center, NASA's Earth Science Information Partnership. Available at: http://bsrsi.msu.edu/trfic/index.html.
19. C. Bishop, *Neural Networks for Pattern Recognition*, Oxford: Oxford University Press, 1995.
20. R. Congalton, and K. Green, *Assessing the Accuracy of Remotely Sensed Data: Principles and Practices*, Boca Raton, FL: Lewis Publishers, 1999.
21. K. Landis, "The measurement of observer agreement for categorical data", *Biometrics*, vol. 33, pp. 159–174, 1977.
22. S. Mallat, and W. Hwang, "Singularity detection and processing with wavelets", *IEEE Transaction on Information Theory*, vol. 38, no. 2, pp. 617–642, 1992.
23. S. Mallat, *A Wavelet Tour of Signal Processing*, 2nd ed., San Diego, CA: Academic Press, pp. 163–219, 1999.

9 The Cleaning Service

A Multi-temporal InSAR Coherence Magnitude Filter

9.1 RATIONALE

To introduce the matter, let us consider the one-point statistic of the sample InSAR coherence amplitude (modulus) for a stationary process. The sample coherence amplitude (sometimes referred to in the literature as the degree of coherence) is derived by averaging a number of complex circular Gaussian signals received by the two antennas of the interferometer:

$$\gamma = \frac{\sum_{i=1}^{L} z_{1i} z_{2i}}{\sqrt{\sum_{i=1}^{L} |z_{1i}|^2} \sqrt{\sum_{i=1}^{L} |z_{2i}|^2}} \tag{9.1}$$

The sample coherence magnitude $|\gamma|$ is the maximum likelihood estimate of the true coherence magnitude $|\hat{\gamma}|$. The probability density function (PDF) of $|\gamma|$ given the true coherence $|\hat{\gamma}|$ and the number L of averaged samples as derived in [1] and reported in [2] (Appendix A3) is:

$$f(|\gamma|) = 2(L-1)\left(1-|\hat{\gamma}|^2\right)^L |\gamma|\left(1-|\gamma|^2\right)^{L-2} F_2^1\left(L,L;1,|\hat{\gamma}|^2|\gamma|\right) \tag{9.2}$$

where F_2^1 is the hypergeometric function.

The PDF for two values of the true coherence ($|\hat{\gamma}| = 0.8, |\hat{\gamma}| = 0.4$) is shown in Figure 9.1. Note how the variance increases for low coherence values.

From Equation (9.2) the expected value of the ($|\gamma|$) estimate can be derived [1,3]:

$$E((|\gamma|)) = \frac{\Gamma(L)\Gamma\left(\frac{3}{2}\right)}{\Gamma\left(L+\frac{1}{2}\right)} {}_2^3F\left(\frac{3}{2},L;L,L+{}^1\!/_2,1;|\hat{\gamma}|^2\right)$$

The expected value of the estimator $E((|\gamma|))$ as a function of the true coherence $|\hat{\gamma}|$ and parametrized by a few values of the number of samples L in the average is shown in Figure 9.2. For reference, the unbiased values of the estimates are also shown. Clearly, a significant bias is present with a small number of samples L, and with increasing importance for low coherence values.

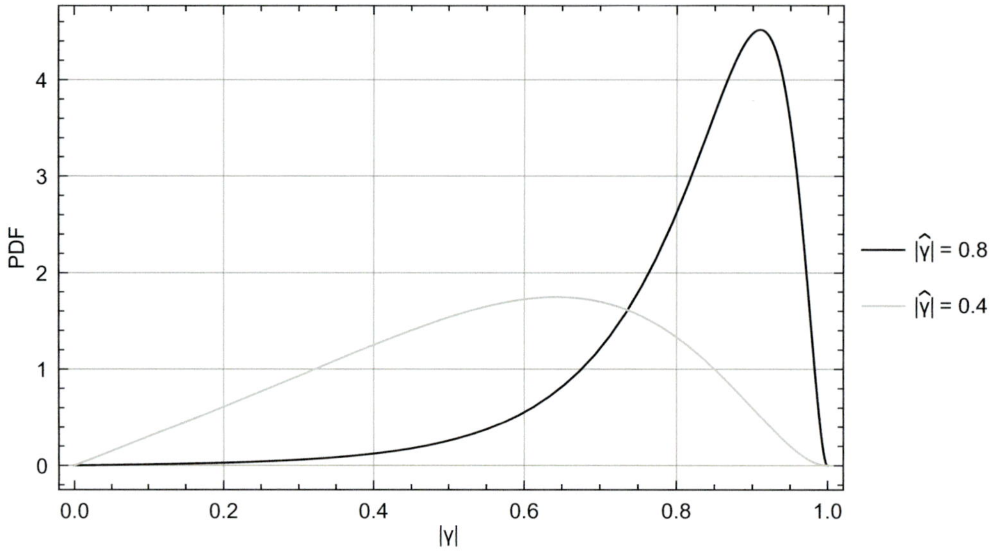

FIGURE 9.1 Probability density function of coherence magnitude.

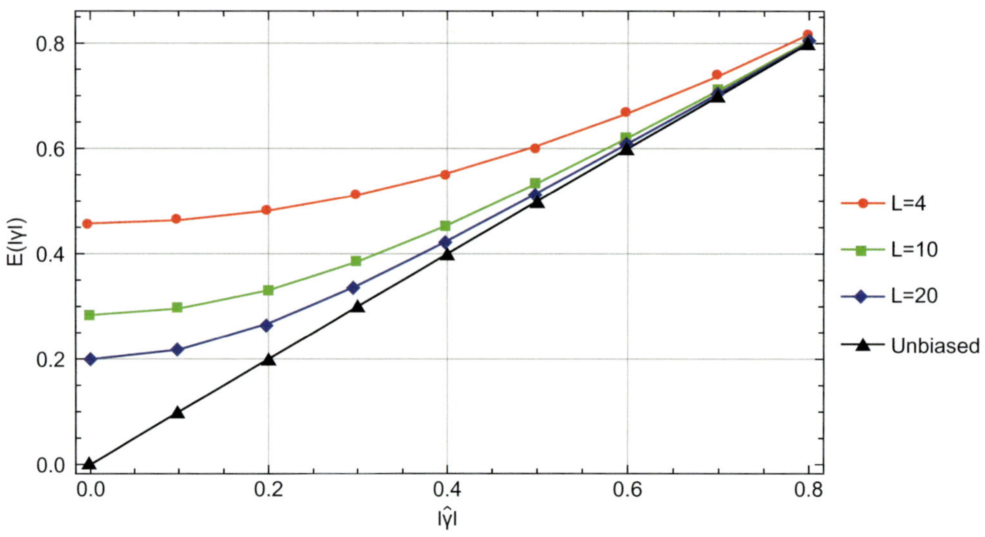

FIGURE 9.2 Coherence amplitude expected value as a function of the true coherence.

Reducing the bias by averaging more samples in the neighborhood of each data set element is a possible solution (multi-looking), yet the cost is to reduce the spatial resolution. We propose here a method that overcomes this difficulty, and achieves a better estimator expected value and variance by adding a time dimension, i.e. by adding to the average samples taken by corresponding realizations of the same random process (eventually with different moments) provided by data acquisitions of the same target at different dates.

For example, starting with a 2-Look coherence dataset and including eight multi-temporal sets in the filter, given the independence of the processes, we expect to achieve a bias of the expected coherence values of 8.4×10^{-2} (from 4.7×10^{-1}) at $|\gamma| = 0.2$ and 2.8×10^{-3} (from 5.7×10^{-2})at $|\gamma| = 0.8$.

Let us consider now what happens to the estimator variance. From Equation (9.2) the second moment of the estimate is:

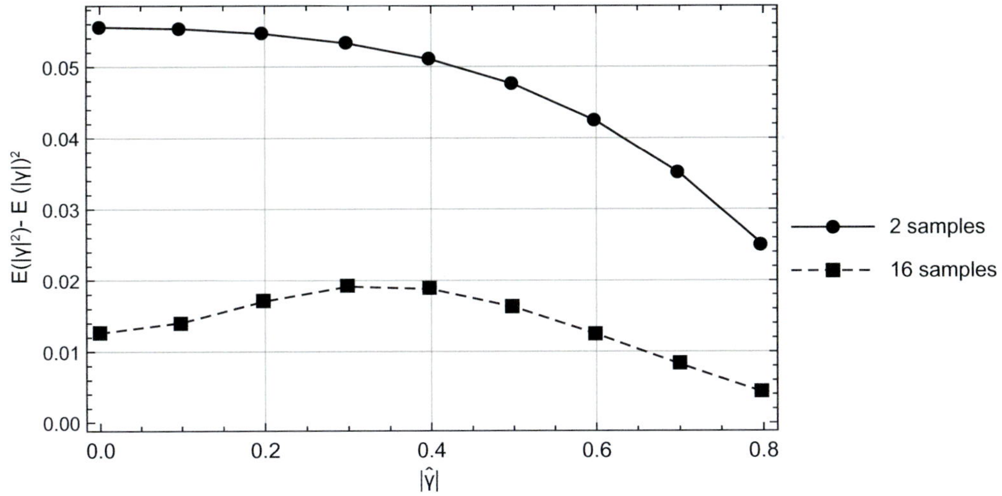

FIGURE 9.3 Coherence estimator variance as a function of the true coherence.

$$E\left(\left(|\gamma|^2\right)\right) = \frac{\Gamma(L)\Gamma(2)}{\Gamma(L+1)}\, {}_2^3F\left(2,L;L,L+\tfrac{1}{2},1;|\hat{\gamma}|^2\right)$$

A plot of the estimator variance $(E((|\gamma|^2)) - E((|\gamma|))^2$ for the case of two and 16 samples is shown in Figure 9.3. The graph indicates that the filter with 16 sets of one-Look data will achieve a variance reduction of the order of 20% to 35%, as well as providing a more consistent estimator throughout the range of true coherence values.

The dependence of the estimator variance on the number of samples L in the average is further illustrated in Figure 9.4, where the variance as a function of L is parametrized by two values of the true coherence $|\hat{\gamma}| = 0.4, 0.8$. The graph indicates that the variance reduction is especially important for low coherence values.

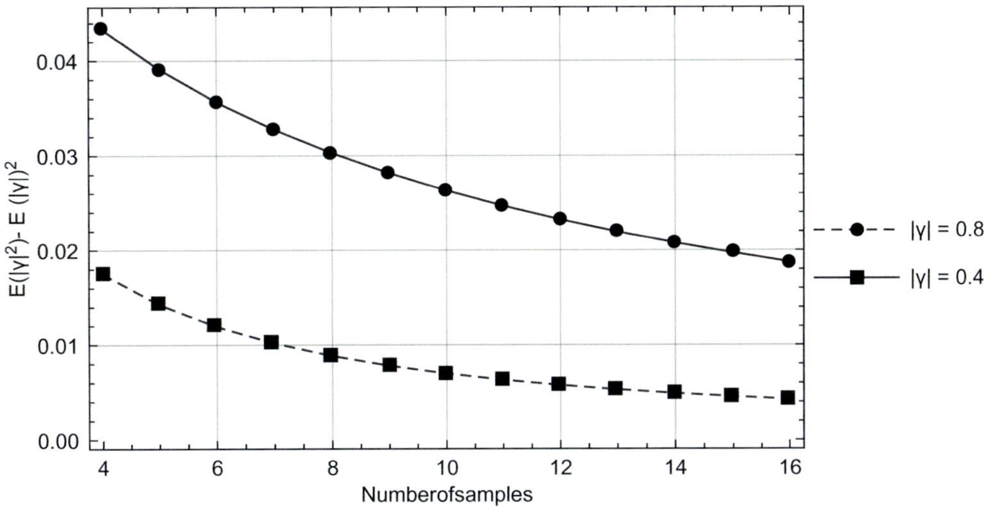

FIGURE 9.4 Estimator variance as a function of number of Looks.

Before delving into the description of the filter machinery, let us hint at the connection between coherence magnitude and forest properties. Let us consider the case of a single-pass interferometer, assume that baseline decorrelation was corrected for, and therefore the main source of decorrelation will be volume decorrelation. For illustration purposes, let us look at a simple random volume over-ground (RVOG) model with an exponential vertical structure function portraying the variation of radar reflectivity with height within the vegetation volume [2] (Chapter 5).

The complex coherence can be obtained in closed form by integrating overall contributions along the vertical profile $f(z) = e^{\frac{2\sigma_E z}{\cos\theta}}$:

$$\gamma = f(h, \sigma_E, \beta_z) = e^{i\beta_z z_0} \frac{p\left(e^{(p+i\beta_z)h} - 1\right)}{(p + i\beta_z)\left(e^{ph} - 1\right)} \; p = 2\sigma_E \sec(\theta) \tag{9.3}$$

Where σ_E the one-way power loss extinction coefficient in unit m^{-1}, θ is the incidence angle, $\beta_z = \dfrac{4\pi\, bs}{\lambda R \sin\theta}$ is a parameter that encapsulates the interferometer sensitivity to baseline bs, wavelength λ, and incidence angle θ.

From Equation (9.3) the coherence amplitude $|\gamma|^2$ can be derived:

$$|\gamma|^2 = \frac{-p^2 \cos(\beta_z h) + \cosh(ph)}{(\beta_z^2 + p^2)(\cosh(hp) - 1)} \; p = 2\sigma_E \sec(\theta) \tag{9.4}$$

This simple model shows that, having fixed the shape of the scattering elements distribution with vertical coordinate z, the coherence amplitude is dependent on the height of the vegetation volume and the power loss (extinction) of the wave through the volume. In turn, extinction and volume height are measures of the forest woody and leafy density and of the vertical developments. In this way, the coherence magnitude can provide measures related to forest biomass. See for example, [4–7]. These are the applications that the filter would facilitate.

To illustrate the dependence of coherence amplitude on vegetation volume height, we consider the case of an L-band single-pass space-borne interferometer (such as the forthcoming TanDEM-L mission). Setting a large baseline (800 m) and $\theta = 23^0$ the dependence of coherence amplitude on forest volume height from Equation (9.4) and for two extinction values $o_E = 0.17$ dB/m, 0.22 dB/m (these being typical of a dense forest) is represented in Figure 9.5.

Two remarks. As the extinction increases, sensitivity to volume height (derivative of the function $|\gamma| = f(h)$) decreases, since the wave penetration is shorter, and no matter how deep the volume, only an upper limited layer is sensed and generates volume decorrelation. Sensitivity also decreases and tends to zero for increasing volume height.

9.2 THE FILTER MACHINERY

The filter achieves coherence estimates with reduced variance using a co-registered time-series of InSAR coherence amplitude images, acquired over the same target. Each set in the series is pre-processed to obtain a co-registered stack of $|\gamma|$ data.

The filter works in a combined space–scale–time domain. In the scale–space domain, a wavelet soft thresholding technique provides, for each dataset, local estimates of $|\hat{\gamma}|$. Wavelet thresholding is equivalent to local signal averaging with a kernel that adapts to the signal regularity in the neighborhood of each sample [8]. It is therefore a powerful tool for noise reduction because it can cope with stationary areas as well as singularities, such as edges and point targets.

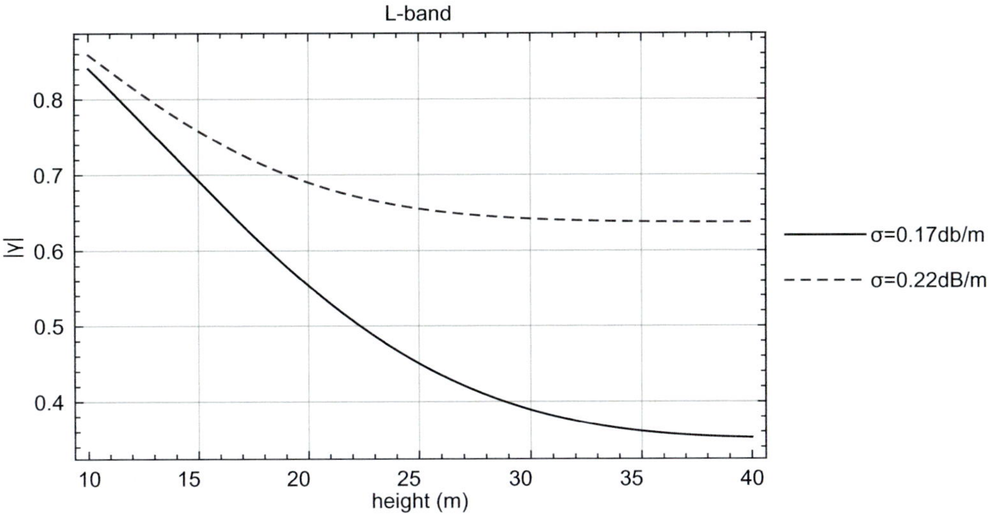

FIGURE 9.5 Dependence of coherence amplitude on forest volume height.

In more detail, we use an adaptive soft thresholding technique, in line with the one proposed in [9], where the wavelet coefficients are scaled by a value T proportional to the wavelet coefficients noise variance $T_j \propto \sigma_j$ at decomposition level j, where $\sigma^2 = f\,(E[|\gamma|)$ is a function of the local expected values of the coherence.

The relationship $\sigma^2 = f\,(E[|\gamma|)$ is estimated by sampling on a regular (line i, column j) grid the coherence dataset $\gamma(i,j)$ and the related wavelet coefficients $W_x^k(i,j)$and $W_y^k(i,j)$, these being the coefficients of a dyadic frame transform at decomposition level k. For each point in the grid (at coordinates (i,j)) the $E[\gamma]$ and the $E\left[W_{xy}^k(i,j)^2\right]$ are estimated by averaging within a 64×64 window. These estimates are then ordered by increasing $E[\gamma]$ and the points are fitted by a second order polynomial, $E[\gamma]_n$ being the independent variable and the noise variance $E\left[W_{xy}^k(i,j)^2\right]_n$ the dependent variable.

$$E\left[W_{xy}^k(i,j)^2\right]_n = a_k + b_k\,\mathrm{E}[\gamma] + c_k\,\mathrm{E}[\gamma]^2$$

In the inverse wavelet transform, the amplitude of the wavelet coefficients at each level k, and for each dataset in the series, is attenuated by a threshold $T_k(i,j) = \sqrt{a_k + b_k\,\mathrm{E}[\gamma] + c_k\,\mathrm{E}[\gamma]^2}$, where now local estimates $E[\gamma]$ are provided by the output of the low pass filter (smooth image) at the third decomposition level.

The local (in space) estimates $E[[\gamma](i,j)$ are then passed over to averages in time domain:

$$\tilde{\gamma}(i,j)_n = \sum_{t \neq n} \frac{E[\gamma]_n}{E[\gamma]_t}\,\gamma(i,j)_t$$

where n and t are indexes in the time domain, i.e. indexes of sets in the series.

This position is the same as the one adopted in [10], where a multi-temporal backscatter filter is discussed, and the one in [11], where a speckle filter for polarimetric SAR data is proposed. The estimates $\tilde{\gamma}(i,j)_n$ are a linear combination with unit coefficients of independent variables taken from the same distribution but with different mean values. The factors $\frac{E[\gamma]_n}{E[\gamma]_t}$ compensate for these differences.

9.3 GENERATION OF A TESTING DATASET

For assessing the filter performance in controlled conditions, a testing dataset was generated by a Monte Carlo technique. Vectors of pseudo-random uncorrelated variates $|\gamma|_n$ obeying the probability density function $PDF = f(|\gamma|)$ in Equation (9.2) were generated with $L = 4$ and for a set of true coherence values: $|\hat{\gamma}| = \{0.8, 0.6, 0.5, 0.4\}$.

The $|\gamma|_n$ variates are generated from uniformly distributed variates $u(n) \in [0, 1]$ by numerical integration and inversion of the distribution function $F(X)$ corresponding to the PDF $f(|\gamma|)$ in Equation (9.2):

$$|\gamma|_n = F^{-1}(u(n)) \text{ where } F(X) = \int_0^X f(|\gamma|)\, dx$$

A set of eight arrays (512 × 512 coherence amplitude images) are then filled with the $|\gamma|_n$ random deviates in different ways to simulate a time series of InSAR acquisitions over a natural target such as a forest, where patterns corresponding to disturbances are appearing at different dates. In more detail, each image is filled with a background of $|\hat{\gamma}| = 0.4$. This field simulates a stable area with volume decorrelation, such as a dense forest.

Two regions of interest (ROIs) are then created by shape files in each image in the series. These ROIs are intended to simulate areas whose coherence changes over time. The first region (ROI 1) is filled with variates corresponding to $[|\hat{\gamma}| = \{0.8, 08, 06, 06, 05, 05, 04, 04\}$. This ROI therefore simulates an area of clear-cut, where a process of forest regrowth is taking place, thus gradually decreasing correlation over time. The second region (ROI 2) is filled with variates corresponding to $[|\hat{\gamma}| = \{05, 05, 06, 06, 08, 08, 08, 08\}$. This ROI therefore mimics the inverse process, i.e. deforestation, as this increases coherence over time.

A selection of the simulated images (4-Looks) is shown in Figure 9.6, and the corresponding images after filtering in Figure 9.7. The degree of noise reduction and the effective preservations of

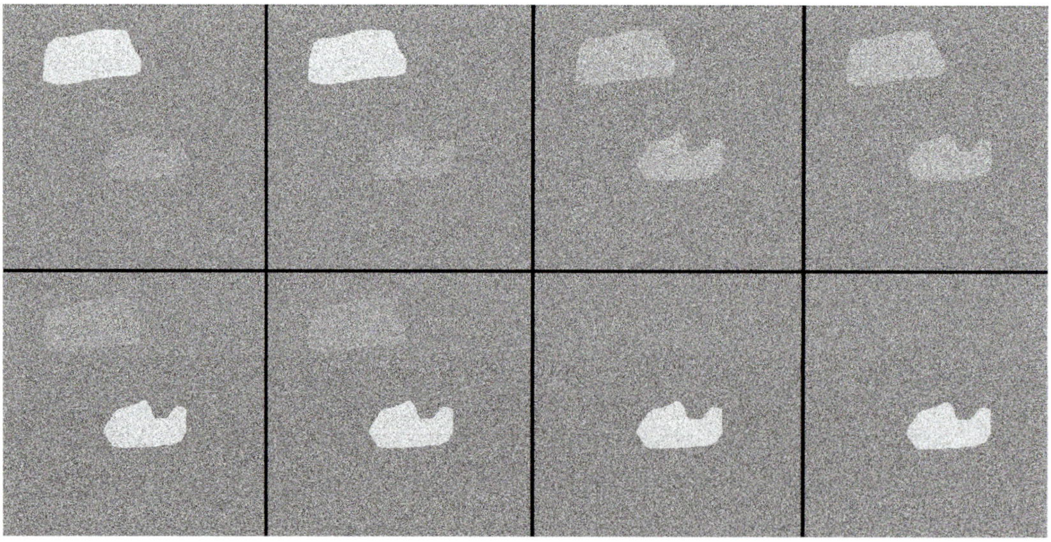

FIGURE 9.6 Set of simulated images with 4-Looks coherence amplitude random deviates.

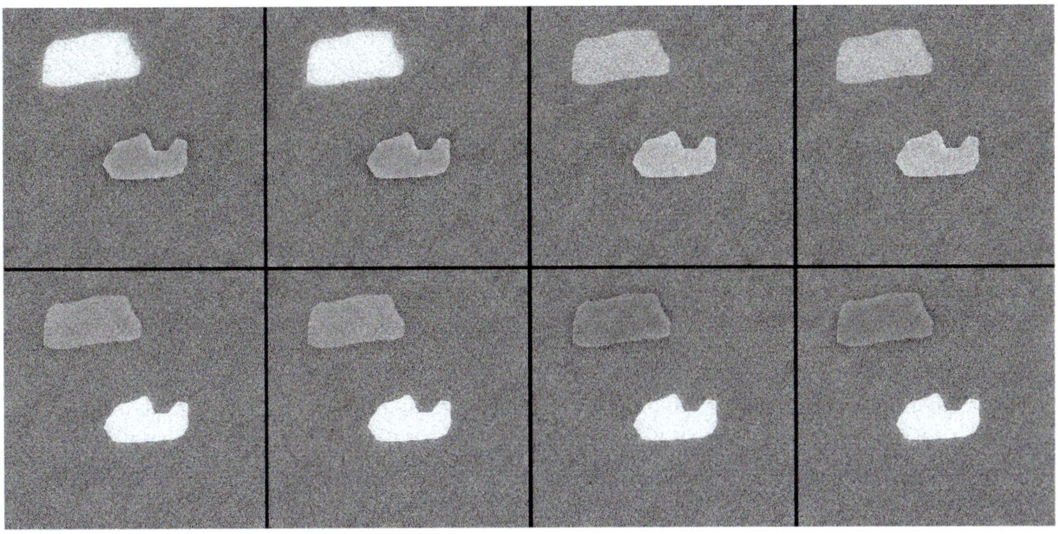

FIGURE 9.7 Set of simulated images after filtering.

the ROI's boundaries and coherence amplitude can already be appreciated qualitatively by visual inspection.

Quantitative measures of the filter performance are provided by sample statistic (mean value and variance) within the ROIs and for each dataset in the series. The results are summarized in Table 9.1.

For each ROI, the first column holds the true coherence $|\hat{\gamma}|$, the second column is the sample mean value of the filtered data, the third column holds the variance of the filtered data, and the fourth column is the variance of the original data. Each line in the table correspond to a dataset in the series. It is seen that the variance of coherence within the ROI is importantly reduced in the filtered data, and that the mean value approximates well the true value, the better the higher the true coherence.

The gain in variance reduction can be quantified by the relative difference of the filtered data variance and original data variance i.e. $(VAR(\gamma_{orig}) - VAR(\gamma_{filt}))/VAR(\gamma_{orig})$. The maximum relative difference goes from ≅90% at low coherence values (⟨γ⟩ ≅ 0.4 down to ≅60% at high coherence values ⟨γ⟩ ≅ 0.8.

TABLE 9.1
Sample Statistic (Mean Value and Variance) for ROIs in the Filtered and Original Simulated Images

ROI 1				ROI 2			
$\lvert\hat{\gamma}\rvert$	$\langle\gamma\rangle$	$VAR(\gamma_{fil})$	$VAR(\gamma_{oril})$	$\lvert\hat{\gamma}\rvert$	$\langle\gamma\rangle$	$VAR(\gamma_{fil})$	$VAR(\gamma_{ori})$
0.8	0.8065	0.0077	0.0178	0.5	0.6307	0.0032	0.0413
0.8	0.8050	0.0077	0.0181	0.5	0.6314	0.0032	0.0418
0.6	0.6737	0.0050	0.0367	0.6	0.6849	0.0035	0.0367
0.6	0.6741	0.0050	0.0362	0.6	0.6852	0.0035	0.0361
0.5	0.6211	0.0043	0.0419	0.8	0.8120	0.0051	0.0177
0.5	0.6193	0.0043	0.0410	0.8	0.8132	0.0050	0.0174
0.4	0.5753	0.0039	0.0440	0.8	0.8118	0.0051	0.0171
0.4	0.5763	0.0038	0.0433	0.8	0.8119	0.0051	0.0171

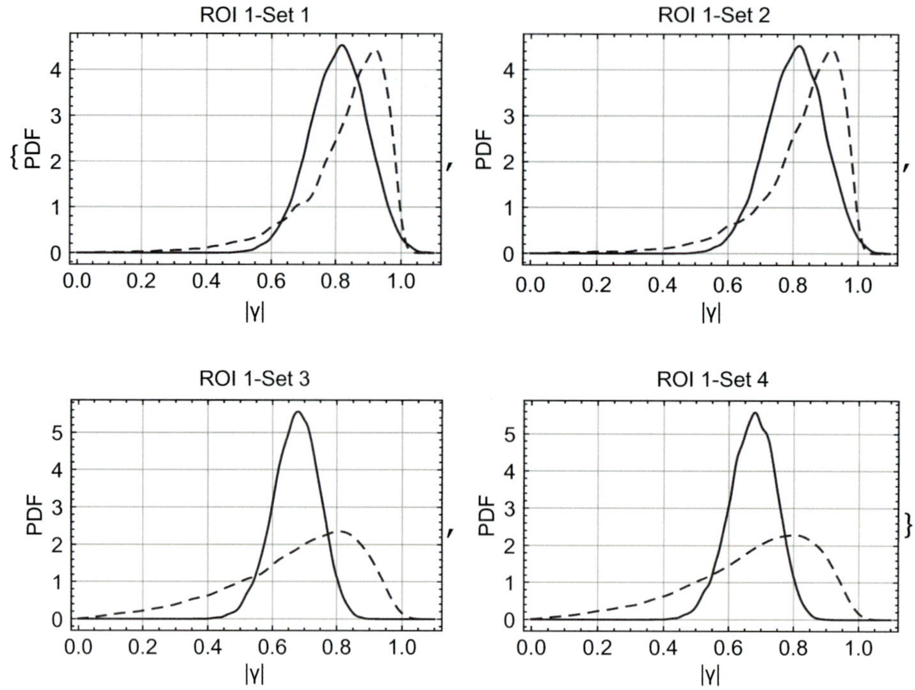

FIGURE 9.8 Density functions of ⟨γ⟩ samples within ROI 1 (original and filtered data).

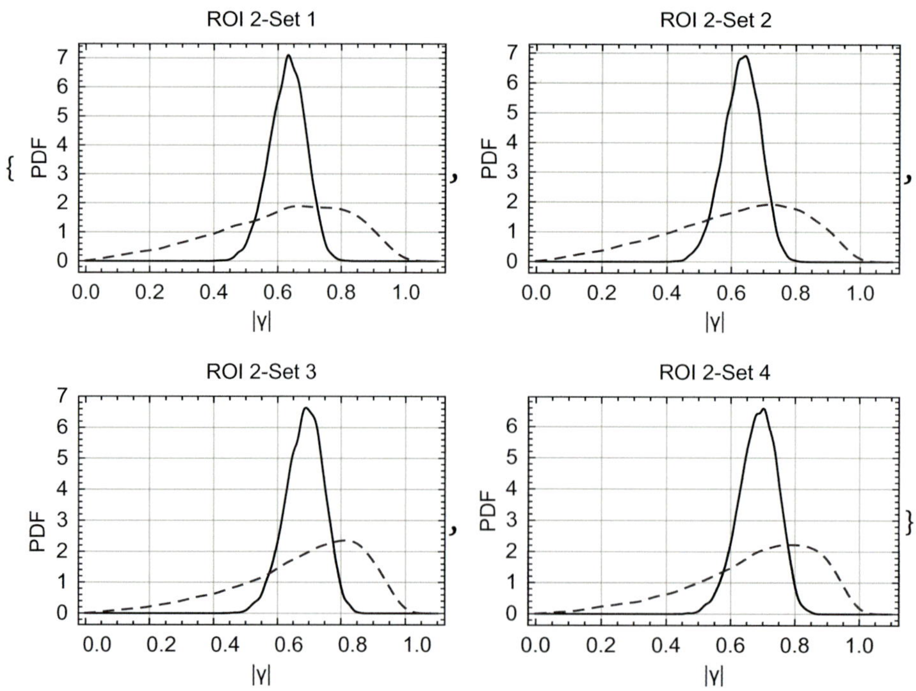

FIGURE 9.9 Density functions of ⟨γ⟩ samples within ROI 2 (original and filtered data).

It is also interesting to consider how the probability density function of coherence samples within ROIs changes after filtering. For the purpose, the PDFs of samples in each region and for each dataset in the series are shown in Figures 9.8 and 9.9. The solid line refers to the filtered data and the dashed line to the original data. The effects to be noted are that the PDF of the filtered data becomes more normal, as expected by averaging more samples; that the spread is reduced; and the mean value is shifted towards the true $\hat{\gamma}$.

9.4 TEST CASES USING TANDEM-X DATA

To illustrate the capabilities of the coherence filter in the context of forestry applications, we discuss next a test case developed using a time series of TanDEM-X acquisitions. The time series was supplied by DLR through a research project (AO VEGE6702) aimed at characterizing forest heterogeneity in a disturbance gradient of the Sungai Wain Protection Forest (SWPF). The forest is located to the north of the port town of Balikpapan (116° 46′ 14.08″ *E*, 0° 59′ 57.9″ *S*) in East Kalimantan (Indonesia). Matters of this research concerning forest structure measures by spatial statistics of InSAR and LiDAR digital elevation models are reported in [12], and are also discussed in Chapters 6 and 7.

The Sungai Wain forest comprises a core of intact lowland dipterocarp forest, surrounded by disturbed forest dominated by genus *Macaranga species* that developed after the 1997–1998 fires. It is one of the last areas of unburned primary rainforest in the Balikpapan–Samarinda area with an extremely rich biodiversity and a large number of rare and endangered wildlife species. The primary lowland dipterocarp forest is composed of trees ranging 10–50 m in height and having an above-ground biomass of 100–400 tons/ha. Cultivated areas are located to the north and to the east of the SWPF, with mainly abandoned mixed rice fields, grassland and scrub composed of a mix of vegetation at various stages of development.

The time series consists of seven sets acquired in StripMap bistatic, dual-pol (HH, VV) mode from March 2013 to August 2014.

The slant range datasets were processed using SARscape v5 [13] to obtain coherence amplitude images from the interferograms. Each coherence image is generated by averaging two samples in range and two samples in azimuth (4-Looks), resulting in slant range resolution of 3.79 m in range and 6.59 m in azimuth. Since the HH and VV sets hold independent samples of the same target, they were combined in a single series of 14 sets as input to the multi-temporal filter.

Roaming through the filtered imagery, we have selected a number of cases that depict well the capabilities of the approach in a thematic context.

The first example portrays the forest/non-forest boundary on the east side of the Sungai Wain Protection Forest. The HH set acquired on December 2014 (original and filtered coherence data) is shown in Figure 9.10. Local communities rely on small-scale agriculture, which has encroached on the Sungai Wain Protection Forest. Approximately 20 km^2 at the eastern border of the reserve has experienced illegal clearing since the 1970s for conversion to gardens. The clearings (higher coherence) are much better delimited in the TanDEM-X filtered set.

A more quantitative assessment is achieved by *K*-means clustering of the original and filtered sets. A run for five classes and four iterations using the first set of the series resulted in the maps (Figure 9.11) for the original coherence data and for the filtered data. It is already clear by visual inspection that the classes defined by clusters of the filtered data are much more spatially homogeneous, and delineate better the areas of clearing.

Sample mean and variance of all samples (original and filtered data) within each cluster are reported in Table 9.2. Obviously, the mean values for samples in the same areas of filtered and original data are very close, since they are the result of a large average. However, notice how the variance is reduced by an order of magnitude that was also foreseen by the simulation exercise.

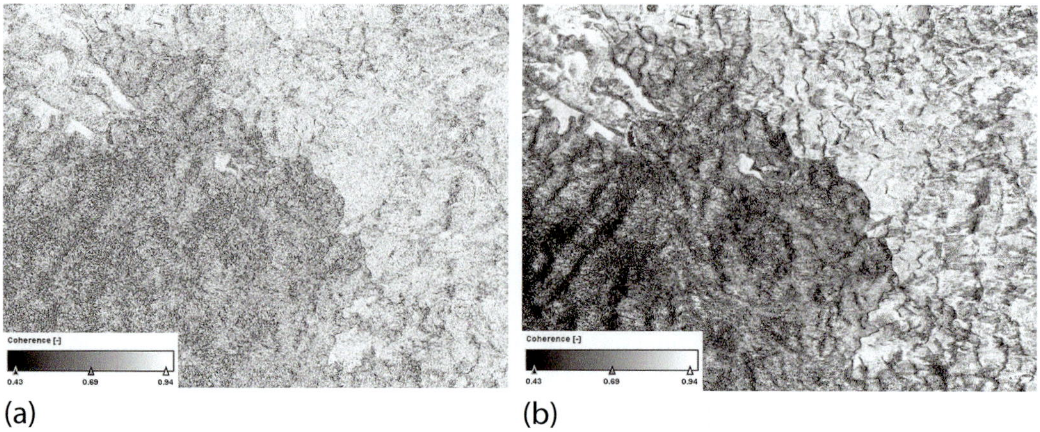

FIGURE 9.10 Coherence image depicting the forest/non-forest boundary on the east side of the Sungai Wain Reserve (original 4-Looks coherence data (a) and filtered data (b))

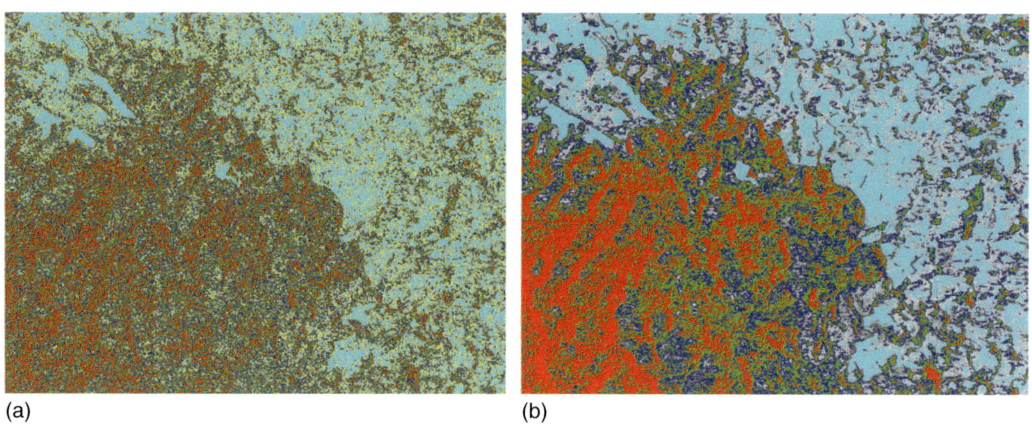

FIGURE 9.11 *K*-means clustering (five classes) of the original (a) and filtered (b) coherence data.

TABLE 9.2
KM Cluster Statistics

	Cluster 1	Cluster 2	Cluster 3	Cluster 4	Cluster 5
$\langle\gamma_{filt}\rangle$	0.5113	0.6307	0.7277	0.8141	0.8907
$\langle\gamma_{ori}\rangle$	0.5244	0.6073	0.7261	0.7933	0.8424
$VAR(\gamma_{filt})$	0.0026	0.0014	0.0014	0.0013	0.0010
$VAR(\gamma_{ori})$	0.0303	0.0238	0.0136	0.0077	0.0054

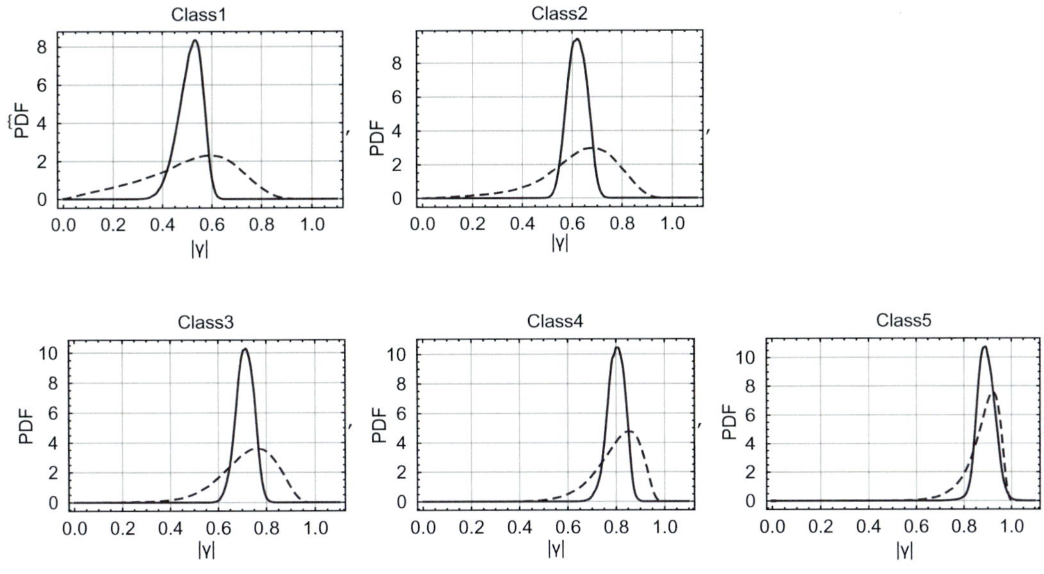

FIGURE 9.12 Probability density functions (PDFs) of samples within each cluster (original data represented by a dashed line overlaid on filtered data represented by a continuous line).

The statistics of the samples with the same label is further documented by the related density functions, these being reported in Figure 9.12, where for each class the distribution of the filtered and original data are overlaid for comparison (dashed line for the unfiltered data and continuous line for the filtered data). It is seen, in line with considerations put forward for the simulation exercise, that the effect of filtering is to normalize the distribution function, and reduce the variance, in greater proportion for low coherency values.

When all distribution functions for all classes are overlaid in a graph, the capability of the filtering approach to improve the class separability becomes very clear (see Figures 9.13 and 9.14).

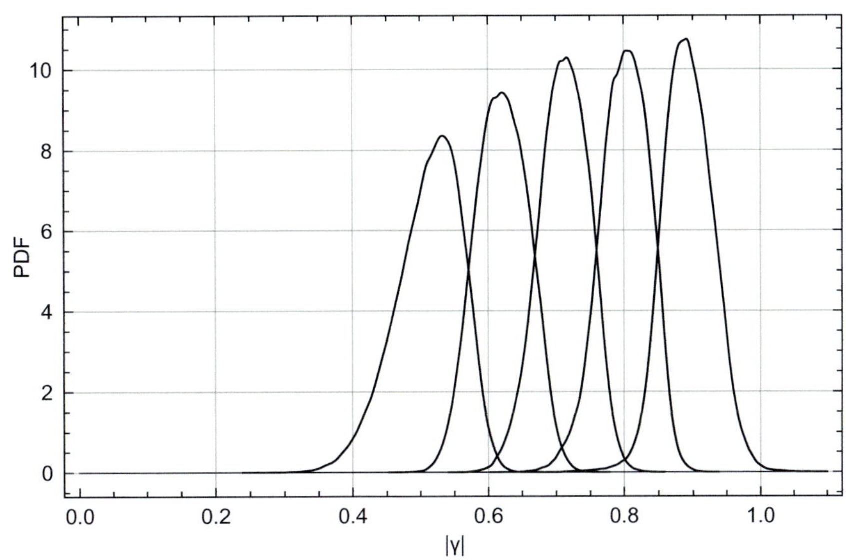

FIGURE 9.13 Overlay of PDFs of filtered |γ| samples within clusters (*K*-means segmentation using filtered data).

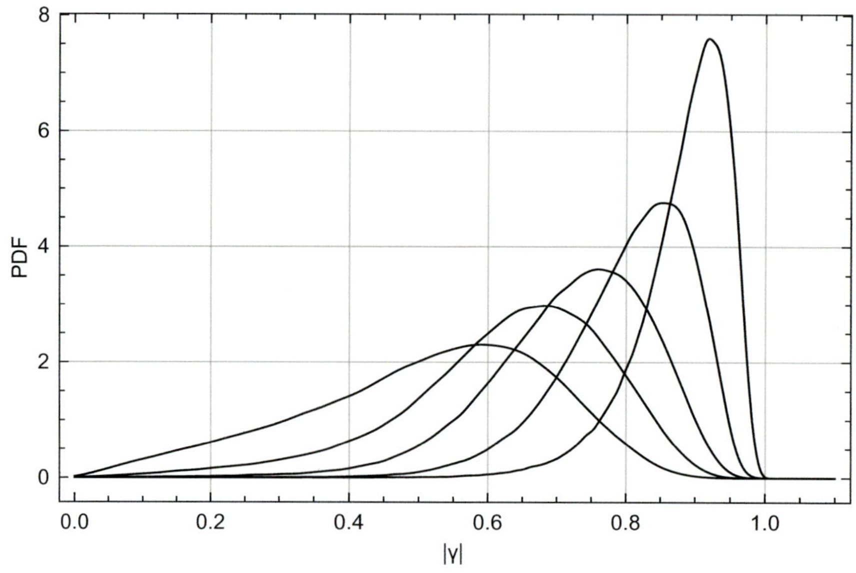

FIGURE 9.14 Overlay of PDFs of original |γ|samples within clusters (*K*-means segmentation using filtered data).

A completely different situation is observed in the case of clustering using the unfiltered data. In this case, partitioning of the space fails, and labels are not associated with clusters, but simply by slices of an overall distribution of random coherence values (see Figure 9.15)

Next, we quantify the cluster separability by the pairwise distance between the density functions (PDFs) of clusters (Jeffries–Matusita distance):

$$J_{ij} = \int \left(\sqrt{f(x|\omega_i)} - \sqrt{f(x|\omega_j)} \right)^2 dx \tag{9.5}$$

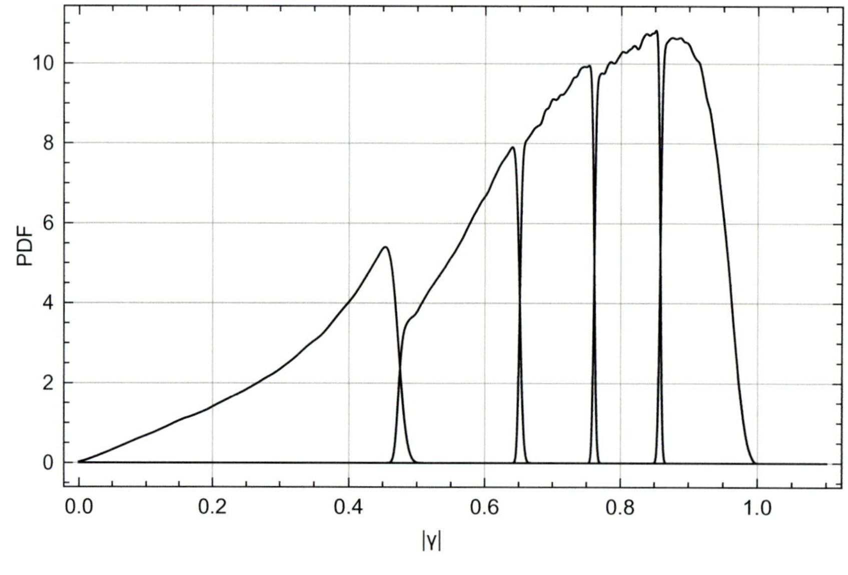

FIGURE 9.15 Statistic of *KM* segmentation classes using the original |γ| data.

where $f(x|\omega_i)$ $f(x|\omega_j)$ are the and j, respectively. Since the PDFs of the unfiltered data are non-Gaussian, the distance was computed by numerical evaluation of the PDFs and integration. The distance matrix for the five classes of the filtered data is reported in Table 9.3, while the one for unfiltered data is shown in Table 9.4.

The distance J_{ij} as defined in Equation (9.5) is asymptotic to $J_{ij} = 2$ for increasing class separation. Accordingly, it is seen in Table 9.3 (filtered data) that the classes with higher coherence values (mean values 0.8, 0.9) are very well separated from those with lower coherence values (mean values: 0.51, 0.61). The average pairwise *JM* distance is 1.62.

JM values for the un-filtered data indicate a much lower pairwise class separation, with average *JM* of 0.56.

To get some intuitive understanding of what these *JM* measures tell us in connection with class separation, we take a leaf from a classical problem in statistical detection theory, namely a Gauss–Gauss mean shift problem [14]. In this exercise, we pretend that our two clusters are described by zero mean unit variance Gaussian distributions. To mimic cluster separation, we shift progressively the mean $\mu_2$of the second PDF with respect to the mean μ_1 of the first PDF. In the statistical decision context, given a test statistic $T \in N$, we want to devise a detector based on T that will allow us to decide between two hypotheses, being the null hypothesis $H_0 \rightarrow T = \aleph(\mu_1, 1)$ and $H_1 \rightarrow T = \aleph(\mu_2, 1)$. In other words, to decide whether there is a change in the expected values of the two distributions.

The detector is designed using a measure referred to as the deflection coefficient, this being basically the difference of expected values in units of variance:

$$d = \frac{\left(E\left(T;H_1\right) - E\left(T;H_0\right)\right)^2}{(VAR\left(T;H0\right) + VAR\left(T;H_1\right)/2} = \frac{\left(\mu_2 - \mu_1\right)^2}{\sigma}$$

TABLE 9.3

JM Distance Matrix: Filtered Data

0.	1.212	1.947	1.999	1.997
1.21234	0.	1.061	1.888	1.988
1.94662	1.061	0.	1.065	1.881
1.99885	1.888	1.065	0.	1.187
1.997	1.988	1.880	1.187	0.

TABLE 9.4

JM Distance Matrix: Original Data

0.	0.123981	0.436861	0.911658	1.42238
0.123981	0.	0.114286	0.493275	1.07986
0.436861	0.114286	0.	0.166497	0.696869
0.911658	0.493275	0.166497	0.	0.253174
1.42238	1.07986	0.696869	0.253174	0.

The probability of detection can be interpreted as the probability that an expectation operator will provide the value of the shifted mean μ_2. It depends on the probability of false alarm P_{FA} and is a monotonic function of d [14]:

$$P_D = Q\left(Q^{-1}\left(P_{FA}\right) - \sqrt{d}\right)$$

where $Q = \int_x^{\infty} \aleph\left(0,1\right) dx$.

For each shift $d\mu$ of the Gaussian PDFs $\aleph(\mu_1, 1)$, $\aleph(\mu_1 + d\mu, 1)$ we can then compare measures of the cluster separation provided by two criteria: by the *JM* distance and the mean shift detector.

Results of this exercise are reported in Figure 9.16. The graph shows the relationship between the *JM* distance and the mean shift probability of detection, parametrized by the probability of false alarm. It can be seen that for lower P_{FA}values ($P_{FA} < 1\%$), the sensitivity $\frac{dP_d}{dJM} = f\left(JM\right)$ is weak for $JM < 1.5$, while for increasing P_{FA} the derivative tends to be more and more linear over the whole range of *JM* values.

The second example (Figure 9.17) depicts the forest/no- forest boundary to the north of the Sungai Wain Reserve (1° 3.148' S, 116° 51.939' E) (March 8, 2013, HH polarization). The area with high coherence is made up of mainly abandoned mixed rice fields, grassland dominated by the invasive *Imperata cylindrica* and mixed scrub. This is in contrast with secondary forest presenting low coherence values. The agriculture and mixed scrub areas are more clearly delineated in the filtered coherence data. Regarding the cluster statistic, similar considerations to the ones reported in the previous example apply.

The Sungai Wain Protection Forest was affected by severe fires in 1998 which damaged parts of the forest, burning several hectares of forest. Moreover, during the 1997–1998 droughts, forest fires entered the primary forest of the Sungai Wain Reserve from the neighboring logging concession and small agricultural fields around the reserve, damaging approximately 5000 hectares but the core area of the reserve was saved from the fires.

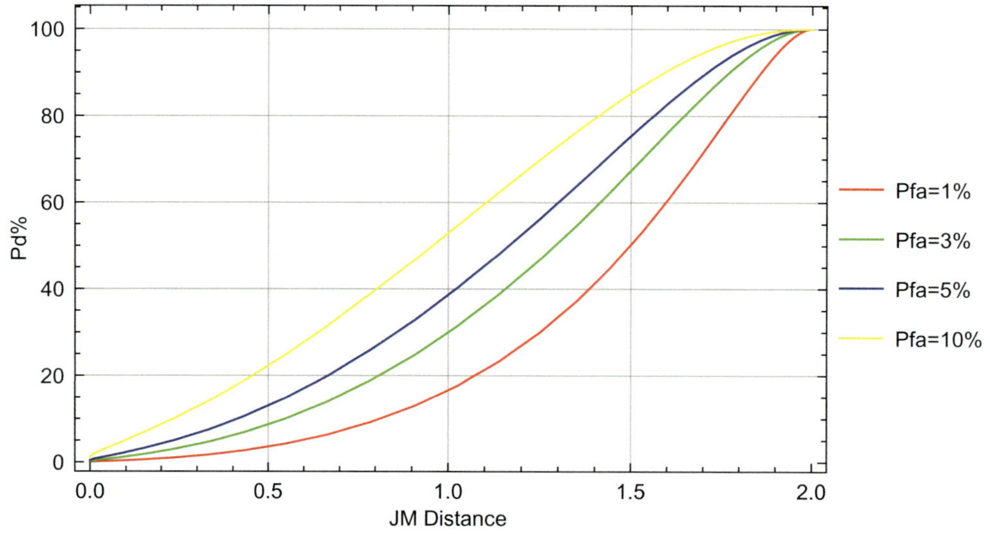

FIGURE 9.16 Relationship between JM distance and probability of detection in a Gauss–Gauss mean shift problem.

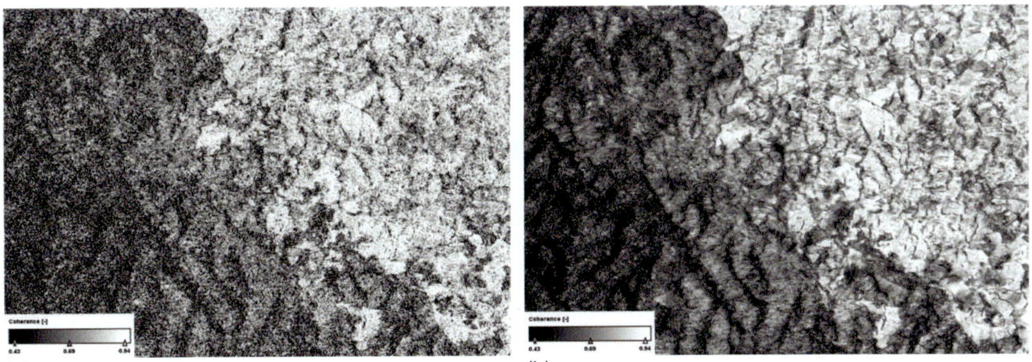

(a) (b)

FIGURE 9.17 Coherence image depicting the forest/non-forest boundary to the north of the Sungai Wain Reserve (original (a) and filtered data (b)).

The next example (Figure 9.18) shows how coherence amplitude can portray in the damaged areas the remnants of the primary dipterocarp forest (red outline) and secondary forest (blue outline) which developed after the fires. The filtered coherence shows much more clearly the boundary between dense primary forest and secondary forest.

It is also interesting to analyze transects of coherence data across the boundaries (discontinuities) between homogeneous targets (such forest/non-forest) or across non-homogeneous targets, such as buildings. The analysis is similar to the one performed in Chapter 7. In that case, the scope was to characterize the Lipschitz parameters of a coherence discontinuity, and check the difference between results with unfiltered and filtered data. Here, the purpose is to check whether the filter not only cleans the noise, but, importantly, at the same time preserves edges and other discontinuities.

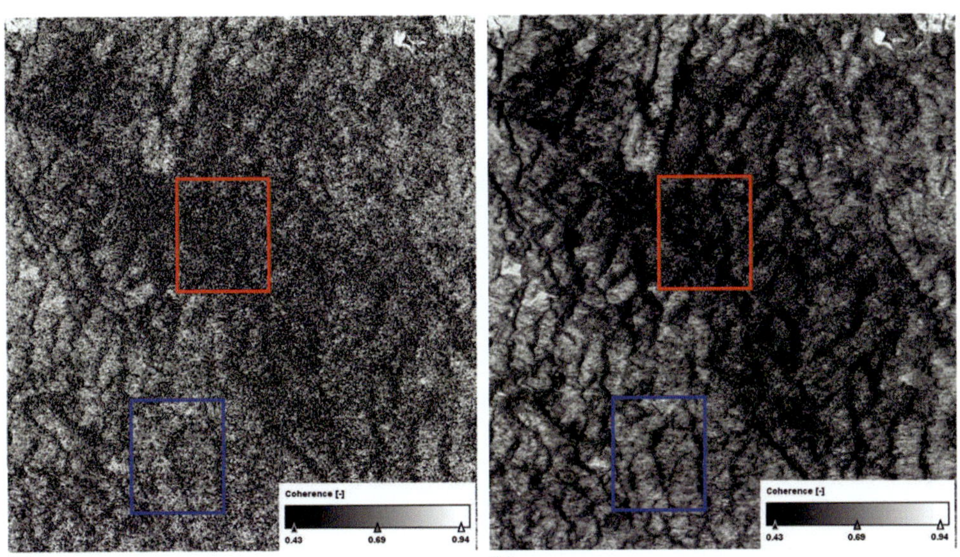

(a) (b)

FIGURE 9.18 Coherence images of an area in the Sungai Wain Protection Forest that was damaged by severe fires (original coherence data (a) and filtered data (b)). (The remnants of the primary forest are marked with a red outline, while the secondary forest is marked with a blue outline.)

In this analysis, we shall apply some test statistic either in the context of edge detection or of spatial frequency analysis and again check the improvements obtained by filtering the data.

The first exercise concerns the transition between forest (low coherence) and non-forest (high coherence). The transects (filtered and unfiltered, first band in the stack), defined by supervised inspection of the image cut of Figure 9.17, are plotted in Figure 9.19. A vertical line in the graph marks the edge between the two targets.

One-point statistic of the segments leading and trailing the edge are summarized in the box-whisker plots in Figure 9.20, where the median, 25% quantile, 75% quantile and lower and upper fence are shown. It is seen that the medians of the forest (0.52/0.53) and of non-forest (0.90/0.87) are well matched. However, the spread of the distributions, and importantly the maximum and minimum values difference, are significantly reduced in the filtered data.

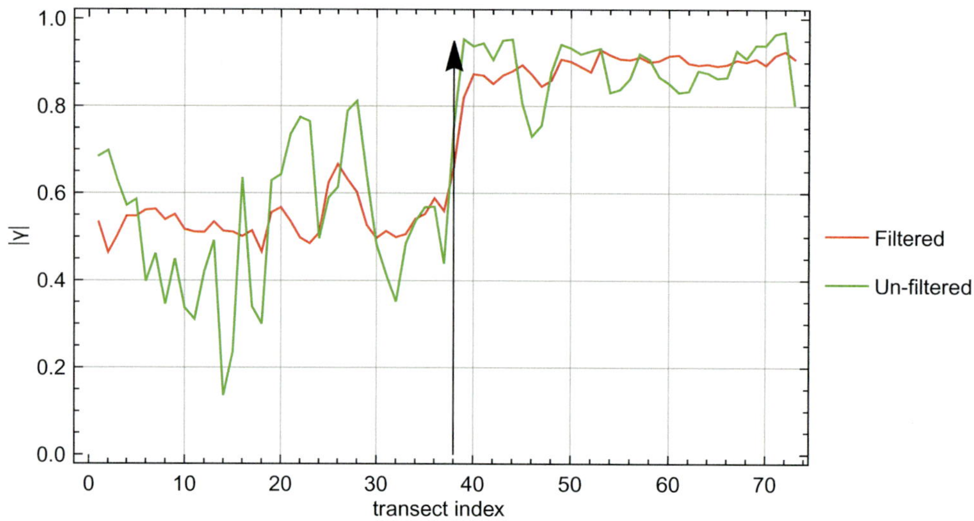

FIGURE 9.19 Coherence transects crossing the margin between forest and non-forest to the north of the Sungai Wain Protection Forest.

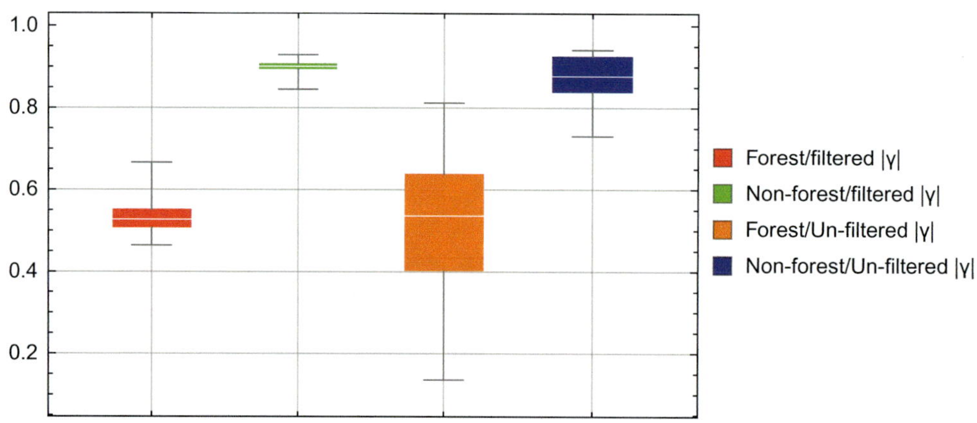

FIGURE 9.20 Box plots of |γ| samples in the leading and trailing parts within the transect around the forest/non-forest margin.

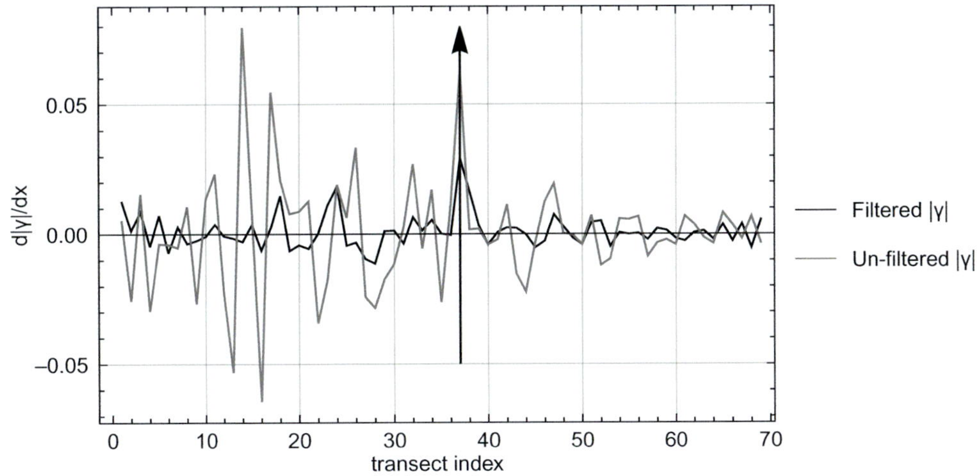

FIGURE 9.21 Signals resulting from the convolution of the |γ| values (non-filtered and filtered) with a differentiator kernel.

The gain in filtering can also be appreciated in an edge detection context. Convolution of the signals with a differentiator – a finite impulse response (FIR) filter with equiripple amplitude response – results in the signals shown in Figure 9.21.

A classical constant false alarm (CFAR) detector would rely on the interference power statistic estimated in the regions leading and trailing the cell under test to decide the detector threshold. Here, we are not interested in the detector, but use the same approach to develop a criterion of comparison between the detection capabilities before and after filtering the data. Thus we define the signal-to-noise ratio as:

$$SNR = 10\log_{10}\frac{A^2}{\sum_{j=a}^{b} x_j^2}$$

Where the summation of samples x_j is taken in the regions bordering the test cell, and A is the value of the test cell.

For the unfiltered data SNR = 6.3 dB while for the filtered data SNR = 12.0 dB. In this case, there is a gain of 6 dB.

We next examine a more delicate case, this concerning the interface between primary and secondary forest. In this case the coherence amplitude values are lower, and a much lower separation, since in both cases there is volume decorrelation. A transect crossing the border between regions of primary and secondary forest was defined visually in the data cut of Figure 9.18. The related signals of filtered (red) and un-filtered data (green) are shown in Figure 9.22, where the vertical arrow marks the edge between the two regions.

The one-point statistic of the two regions (before and after the edge) are reported by the box-plots in Figure 9.23. The median of the filtered and unfiltered data for the two classes are again well aligned (0.50 and 0.66 for filtered, 0.54, 0.67 for un-filtered); however the spread of the un-filtered data (quantiles and fences) is much larger, resulting in a JM distance between the two filtered signals of 1.67, and 1.37 for the un-filtered data.

Finally, we shall have a look at the capability of the coherence filter in enhancing the sharpness of contours of man-made structures. We pick as an example a cut of the first band in the coherence

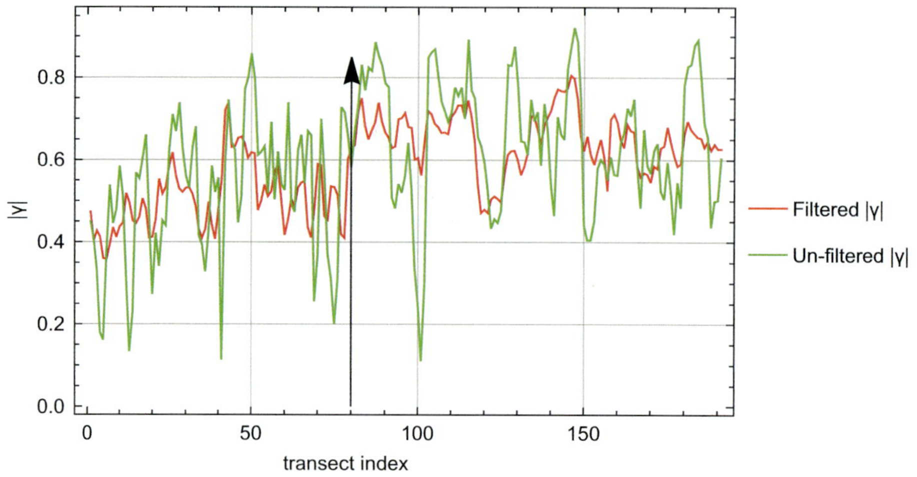

FIGURE 9.22 Transect crossing the margin between primary forest and secondary forest.

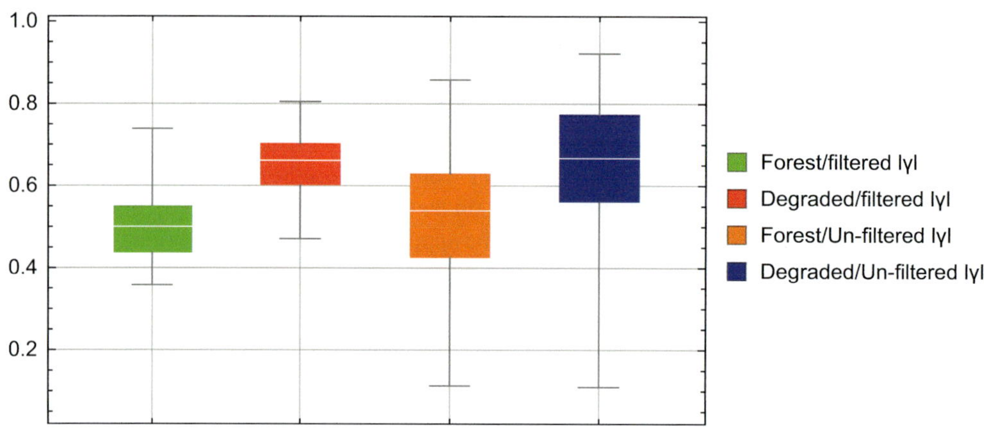

FIGURE 9.23 Statistic of the coherence samples in the segments straddling the margin between forest and secondary forest.

image stack portraying an oil refinery, located near the city of Balikpapan. Figure 9.24 shows the details of the refinery as captured by the filtered coherence (b) and the unfiltered coherence (a). A series of large, round storage tanks are clearly delineated in the filtered image, albeit hardly visible in the unfiltered image. Moreover, in the filtered image, single buildings with high coherence, as well as linear features such as roads, can be much better identified.

This visual impression is reinforced by applying a gradient operator to the two data sets. Results are shown in Figure 9.25, where edges delineating man-made features are much better resolved in the filtered image.

A statistical measure of the relative strength of the gradient images is supplied by the entropy, which serves as an indicator of the image randomness:

$$Z = -\sum_i p_i \log p_i$$

where p_i is the probability of a pixel having gray level i.

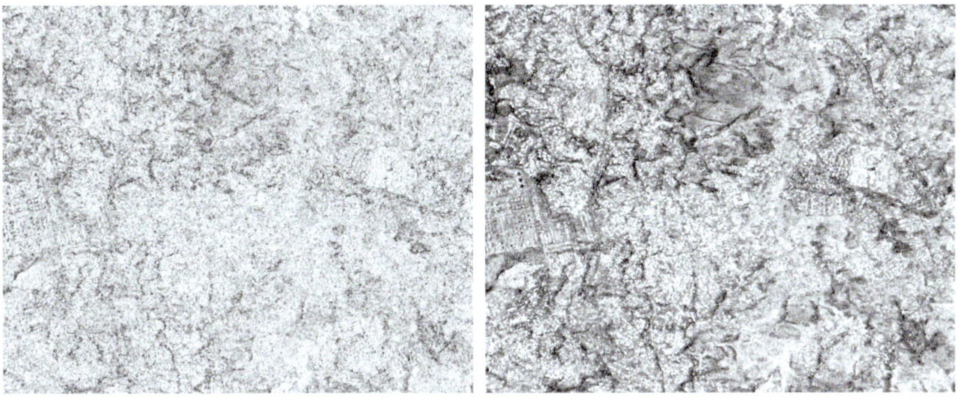

(a) (b)

FIGURE 9.24 Coherence image of an oil refinery near the town of Balikpapan (Indonesia) (1° 5′23″*S*, 116° 44′59″). Original data (a) and filtered data (b).

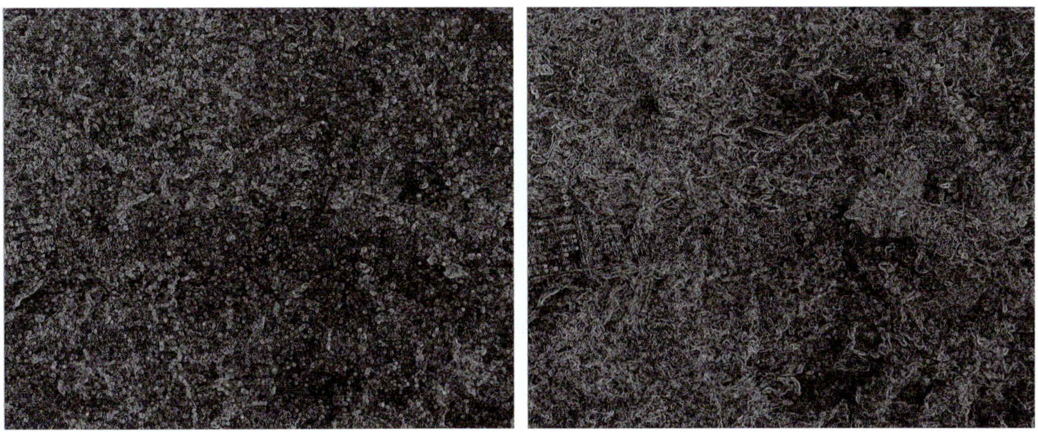

(a) (b)

FIGURE 9.25 Gradient operator applied to images in Figure 9.24. Original data (a) and filtered data (b).

Entropies of the filtered and non-filtered gradient are $Z_{filt} = 11.043$ and $Z_{unfilt} = 11.045$, indicating a lower level of randomness in the filtered dataset.

Analysis of spatial frequencies by means of a periodogram also provides quantitative measures of the enhancement obtained by the coherence filter. We recall that the periodogram gives for discrete data an estimate of the power spectrum, and can be computed by a Fourier series as:

$$S(\omega_k) = \left| \sum_n x_n e^{-i\omega_k n} \right|^2$$

Cuts of the periodograms of the filtered and un-filtered images in the range direction are reported in Figure 9.26. The tails at high frequencies ($\omega = \pi$, $\omega = 2\pi$) are also shown for clarity in Figure 9.27. While the spatial low frequencies (belonging to the man-made structures) are well preserved in the filtered image spectrum, power in the high frequency range generated by coherence noise is larger in the original image by roughly a factor of 10 dB.

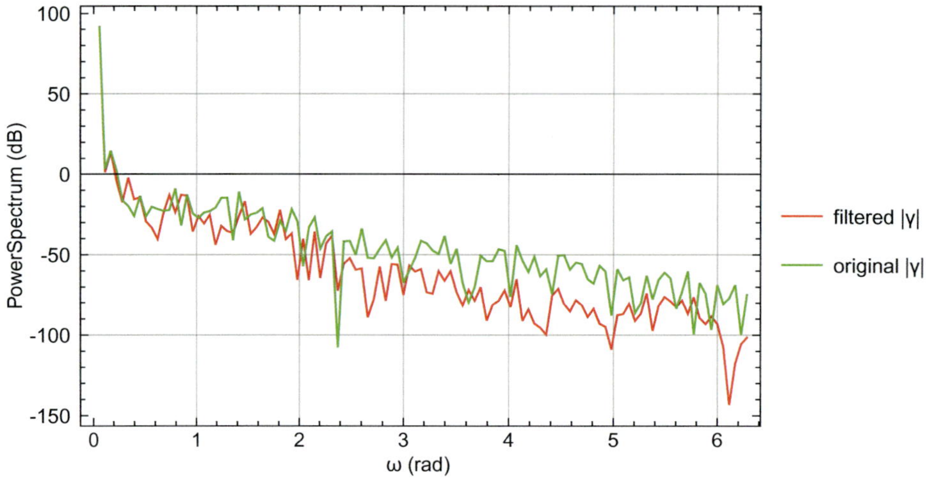

FIGURE 9.26 Cut along the range direction of the original (unfiltered) coherence image periodogram (green) and the filtered image periodogram (red).

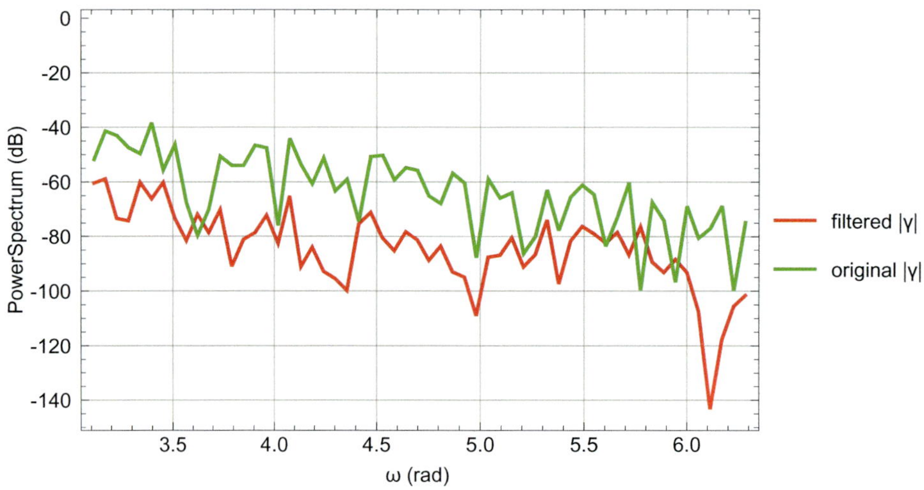

FIGURE 9.27 Blow up of the filtered and original (unfiltered) $|\gamma|$ periodogram in the range of frequencies $3\ rad < \omega < 6\ rad$.

9.5 TEMPORAL FEATURES

Last but not least, before leaving this chapter, we mention that the multi-temporal filter can also pave the way to improved time-series analysis, where clean trajectories of each resolution element (temporal features) can be exploited to track changes of the imaged ecosystems in time. As a proof of concept, we consider the case of temporal features derived in the stack of seven HH filtered coherence images. For each resolution element, a trajectory is thus composed of seven points in time. The following operators are then applied:

1. Linear trend: $\gamma t(t_i) = a + b\ t_i$
2. Residuals: $\sum_{t_i} \left(\gamma(t_i) - \gamma t(t_i)\right)^2$

3. Swing: $\left(max\left(\gamma\left(t_i\right)\right)-min\left(\gamma\left(t_i\right)\right)\right)/\overline{\gamma\left(t_i\right)}$

4. Variance differentiator: $\left\langle \frac{d\gamma\left(t_i\right)^2}{dt_i} \right\rangle$

where $\gamma t(t_i)$ are obtained by a linear regression of the coherence values at time t_i.

Features 1 and 2 are designed to check if there is a smooth increase or decrease of coherence over time. Feature 3 gives a normalized measure of the overall change of the trajectory. Feature 4 will check if, independently of the overall trend, which could also be flat in time, neighboring dates have variance or if there is intermittency in the series, e.g. some abrupt event.

An example of the feature-based analysis is reported next. The slope coefficient b of the linear trend (Feature 1) is shown in Figure 9.28, while Swing (Feature 3) is shown in Figure 9.29.

Note the patches of dark pixels, these belong to the negative values of the slope coefficient, and the corresponding high Swing values. These feature values therefore indicate a significant coherence decrease over an extended area. Coherence statistics computed within one of these areas are reported by histograms of the first coherence band (08 March 2013) and the last coherence band (29 August 2014) in Figure 9.30. Note how the first date histogram is roughly normal, with mean around 0.7, while the last date histogram becomes bi-modal, with the second mode around 0.4.

Finding indication of change then prompts the need for an interpretation, or to identify the physical phenomenon that triggered changes of the observations. This step is crucial to arrive at thematic conclusions, and requires either supporting evidence from a model of the observed phenomenon (e.g. water flow model), from field data or from additional remote sensing data. In our demonstration

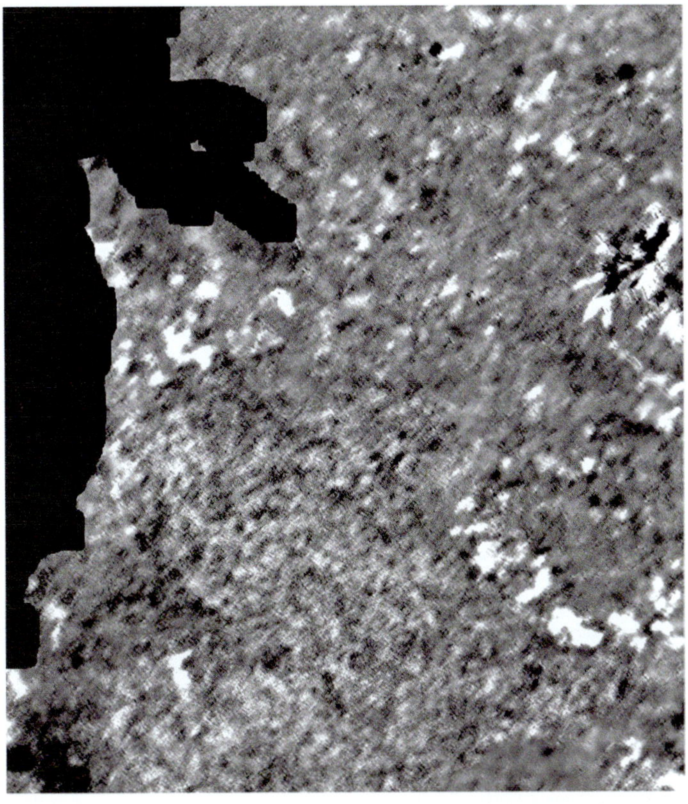

FIGURE 9.28 Example of temporal features analysis: linear trend slope coefficient (Feature 1).

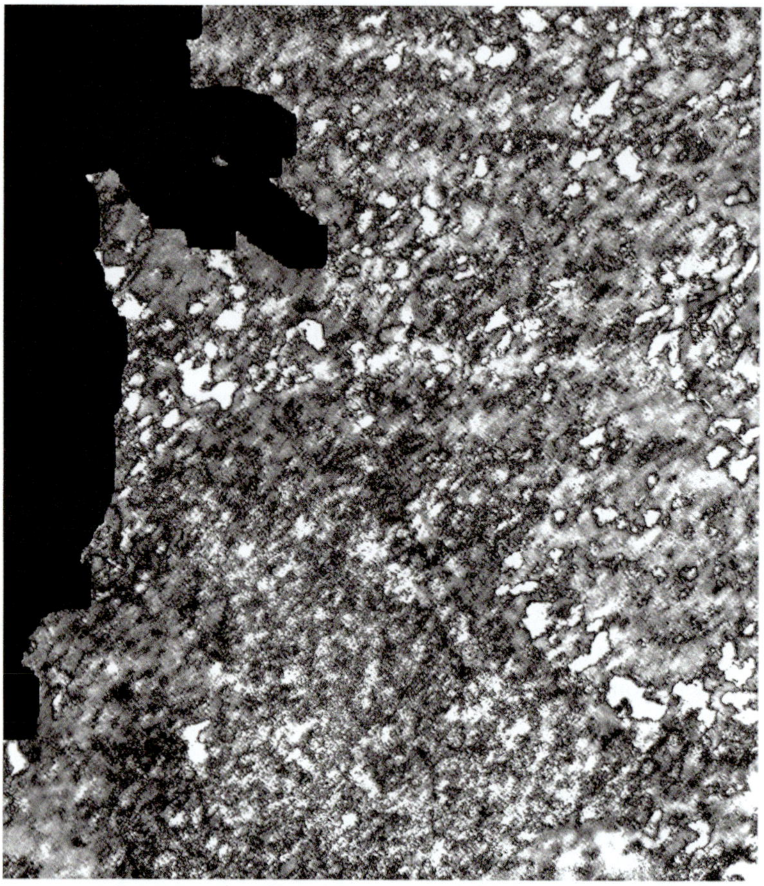

FIGURE 9.29 Normalized maximum change (Swing, Feature 3).

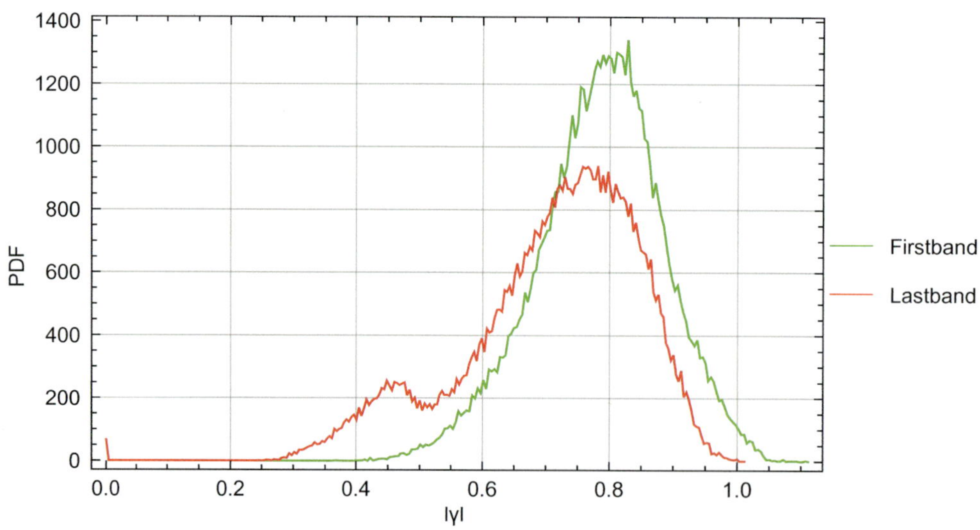

FIGURE 9.30 Density functions (PDF) of coherence samples within the areas of high feature values in the first band (green) and the last band (red) of the time series.

case, we are content with just surmising that the decrease over time of coherence could be caused either by a higher/denser vegetation volume, or to the presence of water caused by flooding.

One more challenging problem would be to find ways of automatizing the feature selection and analysis. In our exercise we were dealing with only seven points in time and data sets of roughly 5 GB. Here, we were just picking prospective interesting areas of change by visual inspection of the features' imagery. Yet, suppose having a longer time series, say one hundred points, and a much wider coverage – a situation that could easily be common with missions such the ESA Sentinels. We could then foresee a typical setting of big data problems, where it would be necessary to ensure timely and consistent extraction of useful information (i.e. important for a specific geophysical process) to develop an expert system, where we could off-load enough knowledge to automatize the chain, from feature extraction to analysis, down to change interpretation. Here we just throw a pebble in the pond, and leave this matter for future endeavors (or perhaps the next book).

REFERENCES

1. R. Touzi, A. Lopez, J. Bruniquel, and P.W. Vachon, "Coherence estimation for SAR imagery", *IEEE Transactions on Geoscience and Remote Sensing*, vol. 37, no. 1, January 1999.
2. S.R. Cloude, "*Polarisation, Applications in Remote Sensing*, Oxford, UK: Oxford University Press, 2010.
3. R. Touzi, and A. Lopes, "Statistics of the Stokes parameters and of the complex coherence parameters in one-look and multilook speckle field", *IEEE Transactions on Geoscience and Remote Sensing*, vol. 34, pp. 519–532, March 1996.
4. J-S. Lee, S.R. Cloude, K.P. Papathanassious, M.R. Grunes, and I.H. Woodhouse, "Speckle filtering and coherence estimation of polarimetric SAR interferometry data for forest applications", *IEEE Transactions on Geoscience and Remote Sensing*, vol. 41, no. 10, 2254–2263, October 2003.
5. J.E.S. Fransson, G. Smith, J. Askne, and H. Olsson (2001), "Stem volume estimation in boreal forests using ERS-1/2 coherence and SPOT XS optical data", *International Journal of Remote Sensing*, vol. 22, mo. 14, 2777–2791.
6. J.I. Askne, P.B.G. Dammert, L.M.H. Ulander, and G. Smith, "C–band repeat-pass interferometric SAR observations of the forest", *IEEE Transactions on Geoscience and Remote Sensing*, vol. 35, no. 1, 25–35, 1997.
7. M. Santoro, J.I. Askne, U. Wegmueller, and C.L. Werner, "Observations, modeling, and applications of ERS-ENVISAT coherence over land surfaces", *IEEE Transactions on Geoscience and Remote Sensing*, vol. 41, no. 8, 2600–2611, August 2007.
8. S. Mallat, *A Wavelet Tour of Signal Processing*, 2nd ed., San Diego, CA: Academic Press, pp. 163–219, 1999.
9. D.L. Donoho, "De-noising by soft-thresholding", *IEEE Transactions on Information Theory*, vol. 41, no. 3, 613–627, May 1995.
10. G.F. De Grandi, M. Leysen, J.S. Lee, and D. Schuler, "*Radar reflectivity estimation using multiple SAR scenes of the same target: techniques and applications*", *Proceedings IEEE IGARSS'97*, August 3–8, 1997, Singapore, pp. 1044
11. J.S. Lee, M.R. Grunes, and S.A. Mango, "Speckle reduction in multipolarization multifrequency SAR imagery", *IEEE Transactions on Geoscience and Remote Sensing*, vol. 29, no. 4, 535–544, July 1991.
12. E.C. De Grandi, E. Mitchard, and D. Hoekman, "Wavelet based analysis of TanDEM-X and LiDAR DEMs across a tropical vegetation heterogeneity gradient driven by fire disturbance in Indonesia", *Remote Sensing*, vol. 8, 641, 2012.
13. Sarmap SA, 2102, available at: http://www.sarmap.ch/pdf/SARscapeTechnical.pdf.
14. S.M. Kay, *Fundamentals of Statistical Signal Processing, Volume II: Detection Theory*, 1 st ed., Englewood Cliffs, NJ: Prentice Hall, 1998, p. 672.

10 Proxies of Forest Volume Loss and Gain by Differencing InSAR DSMs

Fingerprints of Forest Disturbance

In Chapter 6 we saw how the spatial statistic of a single InSAR Digital Surface Model (DSM) reflects the forest structure. This happens because in a vegetation layer the coherence phase center's position with respect to the top of canopy depends on the volume height and on the properties of the scattering elements within the volume, i.e. vertical distribution, number density and moisture content.

So what about taking two DSMs generated by InSAR data obtained on two different dates? Barring configuration changes due to the instrument (e.g. baseline, incidence angle), seasonality and rainfall, the height differences measured by the two DSMs will be related to changes in the vegetation volume. Positive changes may be connected to forest regrowth (volume gain), and negative changes to forest degradation or deforestation (volume loss).

In this chapter we report on a study that, based on this principle, aims at estimating proxies of the forest volume gain and loss in connection with small-scale degradation in a tropical forest, situated around the city of Ouesso, Republic of the Congo. The experiment was carried out within the framework of the DLR Announcement of Opportunity AO TanDEMs-X AO (VEGE6702) [1]. It is based on two DSMs generated by TanDEM-X datasets acquired in a time frame of one year (2012–2013).

10.1 MOTIVATION

Carbon stocks in the Congo Basin are significant and account for about 25% of total tropical carbon stocks [2]. As an important carbon sink and source, the Congo Basin is a priority area of focus in international initiatives aimed at mitigating the effects of climate changes, such as Reducing Emissions from Deforestation and Forest Degradation initiatives (REDD+), set up under the UN Framework Convention on Climate Change (UNFCCC) [3].

Rapid deforestation and forest degradation, driven by agricultural expansion, commercial logging, plantation development, mining, industrialization and urbanization, are widespread across these areas [4]. Recent estimates indicate a tree cover loss equal to 471.38 Kha between 2001 and 2014 (with the highest loss occurring in 2014) and a tree cover gain reaching 559.82 Kha (2001–2012) (considering a 30% canopy density threshold) [5]. Compared to Indonesia and Brazil, the rate of annual forest change for the period 1990–2015 was comparatively low for countries in the Congo Basin (e.g. –0.1% for the Republic of Congo in comparison to –1.1% for Indonesia and –0.4% for Brazil [6]).

Forest degradation is considered an anthropogenically driven process that results in a change of carbon stocks [7] and forest structure [8]. A degraded forest still maintains a canopy cover above a threshold attributable to forest, therefore the process of degradation does not imply a land-use conversion [7]. Forest degradation is more challenging to detect and there are limited approaches to derive it in comparison to deforestation, which makes this process vastly unaccounted for [9]. Recent estimates indicate that between 2005 and 2010, forest degradation made up 25% of the combined emissions from deforestation and forest degradation [10].

In some countries, forest degradation has been identified as a predecessor of deforestation [11] but there has been limited evidence for this in Africa [12], which indicates that degraded forests are not necessarily converted to non-forest. This is widely observed in locations where small-scale disturbance caused by selective logging or shifting cultivation is the prevalent form of forest disturbance (e.g. in the Congo Basin).

Deforestation mapping is accomplished operationally using space-borne optical sensors (e.g. LANDSAT), with the ever-increasing recognition and successful provision of the ability of radar instruments to provide forest area change estimates [13] independently of cloud cover, haze and daylight conditions.

Forest degradation mapping and monitoring are still considered to be comparatively challenging with respect to deforestation mapping, and novel methods are being tested with an increasing interest in the use of SAR remote sensing.

Generally, research has been oriented vastly towards retrieving changes in canopy spectral properties using spectral mixture analysis (SMA) in the Amazon rainforest by INPE (e.g. [14]) with successful results exploiting the widely available and freely accessible LANDSAT mission. More recent approaches have demonstrated the use of SMA techniques using SPOT4 [15]. These approaches are seldom tested in the tropical rainforest of the Congo Basin, which is a critical contributor to forest degradation especially because of industrial selective logging and shifting agriculture, which are, respectively, the first and third largest land uses in the Congo Basin [16].

Changes in forest structure caused by degradation cannot easily be detected by optical sensors or through SAR backscatter; they are essentially measures of a field in two dimensions. In the case of optical instruments, saturation of the optical signal occurs as a result of fast re-growth of vegetation with canopy gaps closure occurring within 50 days after disturbance [15]. In the case of microwave instruments, the changes in backscatter are limited in magnitude and often masked by speckle and effects related to ground moisture (especially at longer wavelengths because of increased penetration into the canopy) [17].

Importantly, SAR backscatter and optical spectral data, even if they do detect degradation, will very rarely have sufficient sensitivity to estimate its magnitude. For example, the size of disturbance patches from small-scale agriculture and selective logging of trees means that some of these changes can be missed by the most commonly available medium-scale forest cover loss products (e.g. Hansen Forest Cover Loss (HFCL) with a 30 m resolution (MMU = 0.09 ha) [5].

Detecting changes in forest structure (canopy height) caused by forest degradation practices over time can best be accomplished using remote sensing observations that provide three-dimensional digital surface models (3D DSMs), such as LiDAR and InSAR.

Demonstration of the potential to detect structural differences resulting from forest degradation using LiDAR is usually carried out for specific sites on demand for commercial purposes, as LiDAR coverage, especially repeated survey, are still relatively uncommon in tropical African countries. Undoubtedly, LiDAR is well suited for fine-scale detection of forest degradation and has been demonstrated in the case of selective logging in the Amazon [18].

InSAR opens new avenues for degradation monitoring. The sensitivity of InSAR to the vertical structure function of the vegetation layer (e.g. the capability of probing the 3D structure of the target) is the key point that enables the detection of forest degradation since this is essentially a process that changes the vegetation's volume height (in both positive and negative directions).

The physics behind the generation of interferometric DSMs for forest monitoring was discussed in Chapter 6. In a nutshell, waves from the interferometer sense a volume within the canopy whose height depends on the medium extinction (penetration depth) [19]. The integral of all the contributions reflected along the scattering density function on the vertical dimension of the volume results in the measured complex coherence. Surface height is computed from the phase center height (PCH) within the scattering volume. The phase center depends on the scattering density function (vertical structure function) and extinction. For zero extinction, the phase center is set at the center of the scattering volume, while in the limit of infinite extinction the center moves to the top of the volume.

For example, in dense tropical forests, at X-band, the high extinction results in the PCH being located closer to the top of the canopy [20]. For forests and woodlands that are less dense (e.g. boreal forests) and in dry conditions, the penetration depth is higher, and the PCH lies lower within the volume [21,22].

In the context of dense tropical forest (assuming low penetration of the signal), since the PCH is located very close to the top of the canopy, the removal of forests through disturbance will lead to the PCH moving lower within the vertical profile, thus revealing a process such as deforestation or forest degradation that is linked to a change in the target structure.

The InSAR sensitivity to the vertical structure of the vegetation layer can be exploited in several ways to arrive at forest deforestation or degradation monitoring. One line of action aims at the estimation of canopy height models (CHM), these providing a cross-section in time of the forest distribution in space. A CHM can be built through the conversion of the phase center that provides a proxy of the top of the canopy, and subtraction with a reference value of the underlying terrain (e.g. through a digital terrain model (DTM) provided by LiDAR). However, DTMs are not as readily available [23], especially for large areas covering tropical forests.

Forest CHMs can also be estimated by InSAR methods that rely on frequency agility (i.e. using dual-wavelength (P-, L- band) airborne InSAR systems that provide DTMs [24,25]), or methods that are based on the estimation of the vertical scattering density function by polarimetry [26].

We move the spotlight to a second line of action: the detection of relative changes between DSMs on two dates from the phase center positions [27]. This approach, being based on relative changes, does not rely on any reference ground elevation, thus avoiding the pitfall of DTM availability. It provides a proxy of the vegetation volume gain and loss at different dates, the true value being dependent on the variables that characterize the specific degradation process (e.g. large-scale forest conversion or small clearings for agricultural practices).

Critical issues of the approach, which call for careful attention, are: (a) error propagation in differencing two variables (DSM height) affected by coherence phase noise; (b) the dependence of the PCH estimates on varying instrument configurations between the two data takes (e.g. incidence angle, baseline); (c) impact of environmental factors (e.g. moisture); and (d) spatial resolution compatible with the detection of small-scale forest disturbance (i.e. ~0.25 ha).

The principle underpinning the proposed method is illustrated in Figure 10.1. The figure shows an example of forest dynamics mapped along a transect from undisturbed lowland forest (1) to forest volume loss (2) followed by forest volume gain (3) and forest volume loss (4). Notice that the fluctuations within each segment are a result of forest structure variability and signal noise, while relative differences between forest volume loss/gain and stable segments are related to the magnitude of the change event. Forest change magnitude in (2) and (4) can be associated with a different degree of forest volume loss.

10.2 STUDY SITE

The study area is located around the town of Ouesso, Republic of the Congo (UL: 15° 56’20.02” E, 1° 44’42.83” N) covering 25 x 40 km^2. The area is dominated by dense, humid, semi-evergreen forest from the Guinean–Congolese region [28] (primary and logged) and extensive swamp forest around the Sangha River, a major tributary of the Congo River. The study site is partitioned according to the World Resources Institute into management units: community development areas, production areas (logging concessions), protection areas, and other unclassified areas (around the city of Ouesso) [29].

Historically, the site has been prone to a high level of forest degradation from commercial selective logging and from extensive shifting cultivation that started to develop during the 1990s and up

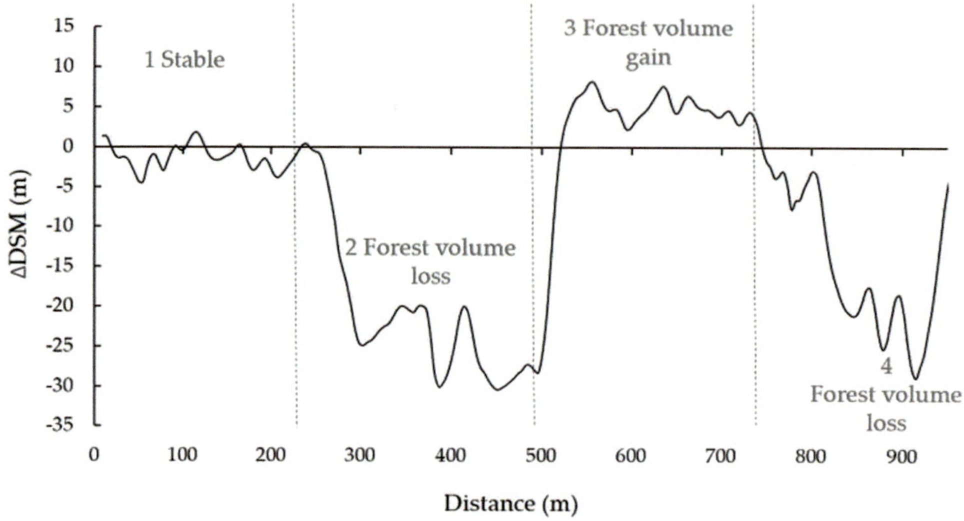

FIGURE 10.1 (a) Three-dimensional TanDEM-X ΔDSM representation showing changes along a 950 m transect across (1) undisturbed lowland forest, (2) forest volume loss, (3) forest volume gain and (4) forest volume loss; (b) the corresponding transect showing the magnitude of changes and (c) ΔDSM image illustrating the transect location.

to 2001 [30]. Forest disturbance is on average no greater than 1 ha and is concentrated in the proximity of Ouesso (department capital and a center for bushmeat trade) and Mboko, along the N2 road and the Sangha River.

The Ngombé and Pokola logging concessions are also situated within the study site, where selective logging of valuable tree species was undertaken on rotation between 1985 and 2008, but is now inactive [30]. Long-term forest disturbance has led to the development of a mosaic landscape (or rural complex) of secondary, degraded forest, re-growth and old clearings spreading outwards from the main road networks. The impact of logging is still visually appreciable by the presence of abandoned and active logging roads, but rapid forest recovery makes it challenging to identify gaps in the canopy and successive re-growth where selective logging occurred in the past.

The climate is classified as Tropical Wet Climate (Am) [31] reaching peak precipitation in November (208 mm), lowest precipitation during July (45.9 mm) [32], and total annual precipitation equal to 1597 mm [32]. The average yearly temperature is 24.8°C and it is relatively constant throughout the year with the highest temperature occurring in March (25.9°C) and lowest temperature in July (23.5°C) [32].

Soil type ranges from Haplic Ferralsols to Umbric Gleysols, which is dominant in swamp forests [33]. Elevation in the study area varies between 320 m and 540 m, with areas of higher elevation located towards the south-west of the study site inside the Ngombé logging concession, while swamp forest is located in areas with the lowest elevation, around the Sangha River.

10.3 TANDEM-X DATA

Two TanDEM-X datasets in bistatic/StripMap mode were acquired on 05/12/2012 and 25/12/2013 at HH polarization in descending mode. Descriptive data on the TanDEM-X acquisitions are reported in Table 10.1. Accumulated precipitation data from the Tropical Rainfall Measuring Mission (TRMM) was used as reference data to ensure no influence by rainfall (Table 10.1).

Data processing was undertaken using ENVI/SARScape software 5.0 interferometric module [34]. The workflow from Co-registered Single Look Slant range Complex (CoSSC) data to a DSM

TABLE 10.1
TanDEM-X StripMap Datasets and Tropical Rainfall Measuring Mission (TRMM) Precipitation Data for a Period of 48 h before the TanDEM-X Acquisition date

Acquisition	Date	$B_\perp$(m)	Inc. Angle (°)	HoA (m)	Range (km)	Rainfall (mm)[1]
t_1	05/12/2012	95.3	48.05	88.6	718.59	0
t_2	25/12/2013	52.0	47.68	164.0	718.51	7.6

[1] Based on TRMM accumulated rainfall for a period of 48 h before the acquisition date. $B_\perp$: Effective baseline, *HoA*: Height of Ambiguity.

included the following steps: (a) multi-looking (2 range and 2 azimuth looks, corresponding to a slant range pixel size of 2.39 x 3.30 m); (b) interferogram generation and flattening; (c) adaptive local frequency filter using the Goldstein filter [35]; (d) interferogram generation; (e) phase unwrapping (the Minimum Cost Flow algorithm); (f) phase to height conversion; and (g) geocoding to Geo-Global Lat/Lon with 3.33 10^{-5} degree pixel size (approximately 4.6 m).

10.4 METHODS

Classical statistical decision theory, when applied to the detection problem, consists basically in deciding whether a signal (deterministic or a random process) is present, or only noise is present. The detector is driven by knowledge of the signal and noise characteristics in terms of their probability density functions (PDFs). In other words, the detector works on random samples and tries to decide to which population (one or many) the sample belongs. For example, in the case of a simple two-hypothesis testing framework for random signal in noise, the likelihood ratio test (Neyman–Pearson theorem) [36] affords a detector where the probability of detection is maximized for a given probability of false alarm [37].

In our case, the signal is composed of objects (spatial random fields), these being defined by the spatial relationship (continuity) of their constituent samples, i.e. by their morphological properties. Objects hold information on the forest height changes Δh. Once an object has been identified and labeled by a clustering algorithm, the detection step (the assignment to signal or noise, in our case change/no-change) is provided by the mean value of the constituent samples (*E*) and the mean value of control plots (*C*) with related standard errors.

Whereas, in dealing with the detection of Δh using a single resolution element, the distribution of the samples would be important, here the distribution of the mean value estimates plays a role. The advantage of dealing with an object-oriented approach is therefore that the variance of the estimates used for detection, because of phase noise and the intrinsic variance of the samples in the object, is much smaller than the one related to the samples in the object. In this way, the criticality at point (a) in the discussion in the previous section is taken care of.

10.4.1 DSM Difference Data Set Generation and Calibration

The object-based change detection method uses two TanDEM-X datasets, acquired at dates 05/12/2012 ($TDXt_1$) and 25/12/2013 ($TDXt_2$). The dates are separated by approximately 1 year, but are in the same season (December), to minimize seasonality and rainfall influences. The difference dataset (ΔDSM) is obtained by differencing between two co-registered DSMs in slant range geometry acquired at times t_1 and t_2.

The ΔDSM dataset presents a trend between far and near range, which was estimated with a tilt angle φ_{tilt}= 5.76 × 10^{-4} degrees, corresponding to a height difference over the swath of 7.1 cm. The tilt is a result of the propagation of small errors in the line of sight baseline determination combined with different effective baselines ($B_{\perp}$) in the two data takes. Theoretically, the φ_{tilt} in a DSM can be estimated from [38]:

$$\varphi_{tilt} = \frac{\Delta B_{los}}{B_{\perp}}$$

For the ΔDSM dataset:

$$\varphi_{tilt} = \Delta B_{los}\left(\frac{1}{B1_{\perp}} - \frac{1}{B2_{\perp}}\right)$$

Where ΔB_{los} is the baseline error in line of sight; $B1_{\perp}$ is the effective baseline at $TDXt_1$and $B2_{\perp}$ is the effective baseline at $TDXt_2$.

Considering the effective baseline values for the TanDEM-X dataset ($B1_{\perp}$= 52 m and $B2_{\perp}$= 95.3 m) and assuming an error of the line of sight baseline of 1 mm (in line with the results in [38] we have: φ_{tilt}= 0.00059°, matching the estimated value from the measurement.

To correct the tilt, the DSM difference as a function of range pixel coordinate j, $\Delta DSM(j)$ is modeled as a zero mean random process $\widehat{\Delta h}$ affected by a systematic error (trend), which is linear in the range coordinate j:

$$\Delta DSM(j) = \widehat{\Delta h}(j) + mj + q$$

The trend is first estimated taking the average of N realizations of the ΔDSM profile in the cross-range direction over a small neighborhood where the statistic is stationary. The average profile is then fitted by a trend line:

$$y(j) = mj + q \tag{10.1}$$

The dataset is then de-trended (brought to zero mean) line by line by subtracting from each pixel the trend line (see Equation (10.2)).

$$p_{detrended}(i,j) = p(i,j) - m\,j - q \tag{10.2}$$

where $p(j)$ is the pixel at position j in line i.

10.4.2 Object-Based Change Detection

The calibrated ΔDSM was used as the starting point for the object-based change detection. Note that ΔDSM does not indicate the "true" vegetation height difference but the difference of the PCH, which is somewhere within the vegetation volume [39] and closer to the true "top of the canopy" at X-band for dense tropical forest.

Object-based change detection was subdivided into two phases: (a) detection of negative changes, these being ascribed to loss of vegetation volume over ground (deforestation/forest degradation); and (b) detection of positive changes, these being ascribed to gain in vegetation volume (re-growth). Accordingly, the calibrated ΔDSM was converted into two images ΔDSM_{neg}, ΔDSM_{pos} holding respectively the absolute values of the negative changes and positive changes.

A Minimum Mapping Unit (MMU) ≅ 0.08 ha (40 TanDEM-X pixels) was chosen to maximize the detection of small-scale changes. The threshold was set, on the one hand, to make use fully of the TanDEM-X DSM resolution and, on the other hand, according to knowledge of the process scale as reported in the literature. The MMU choice was based on the following considerations: (a) clearing sizes, resulting from shifting agriculture in the Congo Basin [40], are usually < 1 ha (lower threshold in the range 0.2–0.3 ha); and (b) size of selective logging gaps, because of the removal of a limited number of trees/hectare, is reported to average 719 m² [41].

An area threshold ($40 \leq T1_{AREA} \leq 10000$ pixels equivalent to 0.08 ha $\leq T1_{AREA} \leq 21.16$ ha) was set to both the ΔDSM_{neg} and ΔDSM_{pos} for filtering the data according to the chosen MMU.

A morphological segmentation algorithm implemented in Wolfram Mathematica 10 [42], was applied to each filtered ΔDSM image. The output is an array where each pixel in the image is labeled by an integer index representing the connected foreground image component in which the pixel lies [42]. Morphological algorithms are well suited for extracting the spatial relationships between groups of pixels rather than the information for a single pixel [43].

10.4.3 Change Objects Refinement

Objects refinement was essential to (a) remove objects within rivers; and (b) remove objects generated by co-registration mismatches.

A river mask was extracted from LANDSAT 8 OLI (01/11/2013) by setting a threshold to the SWIR 1 (Band 6) which enabled a satisfactory delineation of the Sangha River. The river mask was buffered by setting a 20 m outer threshold, and change objects located within the buffered area were removed from the analysis since they did not constitute actual vegetation changes but were linked to problems with phase unwrapping over low coherence (γ) areas at both acquisition dates ($\gamma t_1 = 0.47 \pm 0.17$; $\gamma t_2 = 0.49 \pm 0.19$).

Objects located within a 10 m buffer from the margin of the output ΔDSM image were also removed since they were not considered "true" changes but were related to possible co-registration mismatches between the two TanDEM-X DSMs.

10.4.4 Variance of the Within-Object Mean Height Difference Estimator

Let us first consider one realization of the within-object random field. Sample variation will result from the intrinsic "true" variation of the vegetation height and a noise component given by one realization of the interferometric phase random process.

Height difference(Δh) samples within an object are modeled as independent identically distributed (IID) but correlated random variables. Correlation is a prerequisite for the extraction of objects by the clustering algorithm.

The variance of the mean value estimator $\overline{\Delta h}$ for one realization of the random field is [44]:

$$VAR(\overline{\Delta h}) = VAR\left(\frac{\sum_{i=1}^{N} \Delta h_i}{N}\right) = \frac{\sum_{i=1}^{N} VAR(\Delta h_i)}{N^2} + 2\frac{\sum_{i \neq j}^{N} COV(\Delta h_i, \Delta h_j)}{N^2}$$

where N is the number of samples in the object.

The sample covariance is estimated by an approximation provided by the notion of correlation area, as proposed in [45]. Samples within the correlation area are assumed to be fully correlated, while samples outside are assumed to be uncorrelated. The correlation area is the set where values of the normalized autocorrelation function fall within ½ of the maximum. For N_C samples within the correlation area, there will be $2N \, x \, N_C$ samples where $COV(\Delta h_i, \Delta h_j) = VAR(\Delta h_i)$, therefore:

$$VAR(\overline{\Delta h}) = \frac{(1 + 2N_c) VAR(\Delta h_i)}{N}$$

Considering now M realization of the random field, a component given by the interferometric phase variance within an object is added to the total variance of the mean estimator. The probability density function of the sample phase of the complex coherence resulting from averaging L independent samples (L looks), as derived in [46].

$$f_\varphi(\varphi,\gamma,L) = \frac{\Gamma\left(L+\frac{1}{2}\right)\left(1-\gamma^2\right)^L \beta}{2\sqrt{\pi}\,\Gamma(L)\left(1-\beta^2\right)^{L+1/2}} + \frac{\left(1-\gamma^2\right)^L}{2\pi} F_{21}\left(L,1;\frac{1}{2};\beta^2\right) \quad -\pi < \varphi < \pi$$

With $\beta = \gamma \cos(\varphi - \theta)$ where θ is the mode of the distribution, F_{21} is the hypergeometric function and γ is the sample modulus of the complex coherence, this having the following distribution [47]:

$$f_\gamma(\gamma,\rho,L) = 2(L-1)\left(1-\varrho^2\right)^L \gamma\left(1-\gamma^2\right)^{L-2} F_{21}\left(L,L;1:\gamma^2\rho^2\right) \tag{10.3}$$

Where ρ is the mean coherence modulus within an object.

The variance of the phase, setting $\theta = 0$ is obtained by integrating over the distribution of γ:

$$VAR_\varphi(L,\rho) = \int_0^1\int_{-\pi}^{\pi} \varphi^2 f_\varphi(\varphi,\gamma,L) f_\gamma(\gamma,\rho,L)\,d\varphi\,d\gamma \tag{10.4}$$

The variance of the sample height estimate of the set at time t_1 is:

$$VAR(h1_\varphi) = VAR_\varphi(L,\rho)\frac{\lambda R_1 \sin\vartheta_1}{4\pi B_{1\perp}} \tag{10.5}$$

Where λ is the wavelength, R_1 is the slant range distance, ϑ_1 is the local incidence angle, and $B_{1\perp}$ is the effective baseline for the dataset at time t_1. An analogous expression holds for the dataset at time t_2. The sample variance of the height difference Δh caused by phase fluctuation is:

$$VAR(\Delta h) = VAR(h1_\varphi) + VAR(h2_\varphi)$$

The variance of the mean height difference $\langle\Delta h\rangle$ of an object with N samples is:

$$VAR(\langle\Delta h\rangle) = \frac{VAR(h1_\varphi) + VAR(h2_\varphi)}{N}$$

The total variance of the within-object mean $\langle\overline{\Delta h}\rangle$ estimator is:w

$$VAR\left(\langle\overline{\Delta h}\rangle\right) = \frac{N_c VAR(\Delta h_i)}{N} + \frac{VAR(h1_\varphi) + VAR(h2_\varphi)}{N} \tag{10.6}$$

The total variance of the within-object mean $\langle\Delta h\rangle$ estimator is used to test the statistical hypothesis that the mean Δh of an object associated with a change is shifted significantly with respect to the mean of a control object associated with no change.

10.4.5 Effect Size

To characterize statistically the magnitude of the effect of interest, i.e. the change in time of forest volume with respect to a static situation, we adopt the following measure [48], dubbed the effect size:

$$E_{size} = \frac{\mu_E - \mu_C}{\sigma_{pooled}} \tag{10.7}$$

where μ_E is the Δh mean of the change polygons in the experimental group, μ_C is the Δh mean of the control plots and σ_{pooled} is the pooled standard deviation between the experimental and the control groups [49]:

$$\sigma_{pooled} = \sqrt{\frac{(N_E - 1)SD_E^2 + (N_c - 1)SD_C^2}{N_E + N_C - 2}}$$

where N_E is the number of samples in the experimental group; N_C is the number of samples in the control group; SD_E is the standard deviation in the experimental group; and SD_C is the standard deviation in the control group.

This statistical measure can also be found in the literature, but in a different context [50]. Here, it is defined as the magnitude of the difference between groups and is used as a true measure of the difference between groups.

In a detection and hypothesis testing context, E_{size} is analogous to a measure called "deflection coefficient" in the setting of a mean-shifted Gauss–Gauss problem [37]. Gauss–Gauss statistic is justified because we are dealing with the statistic of the object's mean values. This passage will allow us to interpret E_{size} values in terms of level of significance, i.e. decide $H_1$when H_0 is true (probability of false alarm P_{FA}) and power of the test (probability of detection P_D).

To establish this link, we consider the test statistic $T(x)$:

$$T \rightarrow \begin{cases} \mathbb{N}(\mu_{ct}, \sigma)\, under\, H_0\,(noise) \\ \mathbb{N}(\mu_{obj}, \sigma)\, under\, H_1\,(signal) \end{cases}$$

where $\sigma = \sigma_{pooled}$ and $\mathbb{N}(\mu, \sigma)$ is a normal distribution with mean μ and standard deviation σ.

The problem is to decide between two hypotheses that differ by a shift of the mean of T.

$$P_{FA} = P_r\{T > \gamma'; H_0\} = Q\left(\frac{\gamma' - \mu_{ct}}{\sigma}\right)$$

where $Q(x) = \int_x^{\infty} \mathbb{N}(0,1)\,dx$

$$\gamma' = \mu_{ct} + \sigma Q^{-1}(P_{FA})$$

$$P_D = P_r\{T > \gamma'; H_1\} = Q\left(\frac{\gamma' - \mu_{obj}}{\sigma}\right) = Q\left(Q^{-1}(P_{FA}) - \frac{\mu_{obj} - \mu_{ct}}{\sigma}\right) = Q\left(Q^{-1}(P_{FA}) - E_{size}\right) \tag{10.8}$$

Equation (10.8) gives the dependence of effect size on the detection performance P_D and the levels of significance P_{FA}. Numerical evaluation of $E_{size} = f(P_D)$ parameterized by a set of P_{FA} values is presented in Figure 10.2. For $E_{size} > 5$ the power of test tends asymptotically to probability 1 for all considered values of significance.

10.4.6 Probability of Object Detection by Statistical Decision Theory

An alternative way to characterize the event of the true height change is by means of statistical decision theory. Here the decision is whether the object shows a true change of forest height Δh (the signal), or the object holds a static forest configuration $\Delta h = 0$ (the noise).

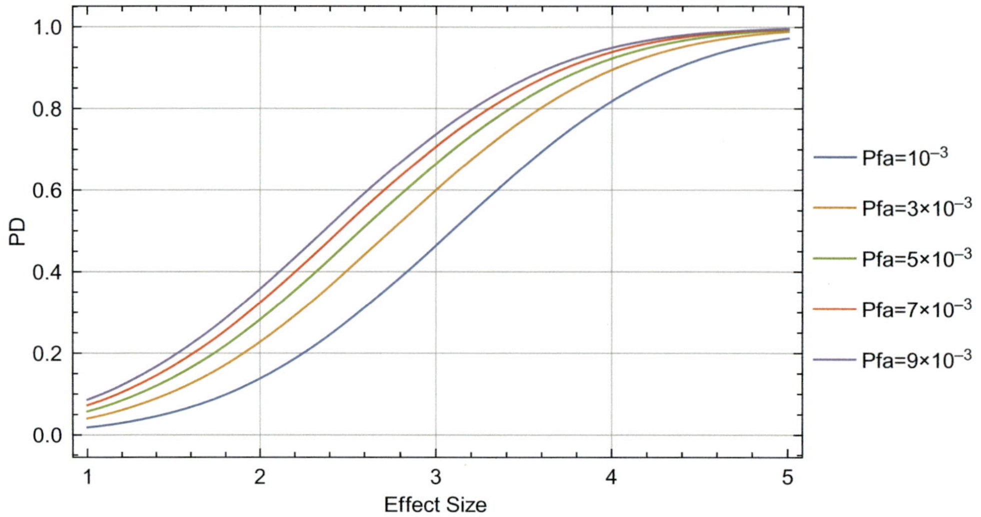

FIGURE 10.2 Dependence of effect size on the detection performance (P_D) and five levels of false alarm P_{FA} (see legend).

In this statistical analysis, each object is considered as the realization of a normal random process, and statistical decision theory is used to assess probabilistically in repeated experiments if the object holds the signal.

We introduce in this context the terms probability of detection and false alarm. However, it must be clear that the purpose of the analysis is not to build a detector that could delineate objects within samples in the dataset. Given the realization of an object and the estimates of its mean Δh and error of the mean Δh, the scope is to estimate the probability in repeated experiments that a new object realization is statistically different from a control object.

10.4.6.1 Neyman–Pearson approach

The test statistic T(x) is:

$$T \to \begin{cases} \mathbb{N}(\mu_{ct}, \sigma_{ct})\, under\ H_0\,(noise) \\ \mathbb{N}(\mu_{obj}, \sigma_{obj})\, under\ H_1\,(signal) \end{cases}$$

where $N(\mu, \sigma)$ is a normal distribution with mean μ and standard deviation σ.

Normal statistic is justified because we are dealing with the statistic of the object's mean values.

To develop the problem, the Neyman–Pearson (NP) theorem is called into play, which states that to maximize the power of the test (probability of detection P_D) given a certain level of significance α (probability of false alarm P_{FA}), H1 is decided when the likelihood ratio:

$$L(x) = \frac{p(x;H_1)}{p(x;H_0)} > \gamma$$

where the threshold γ is derived from the condition:

$$P_{FA} = \int_{L(x)>\gamma} p(x;H_0) = \alpha$$

The likelihood ratio in our setting, after taking logarithms of both sides and assuming $\mu_{ct} = 0$ is:

$$x^2\left(\frac{1}{\sigma_{ct}^2}-\frac{1}{\sigma_{obj}^2}\right)+x\left(\frac{\mu_{obj}}{\sigma_{obj}^2}\right)-\frac{\mu_{obj}^2}{2\sigma_{obj}^2}>\ln(\gamma) \tag{10.9}$$

Since $\frac{1}{\sigma_{ct}^2}\cong\frac{1}{\sigma_{obj}^2}$ at first order

$$L(x)=x\left(\frac{\mu_{obj}}{\sigma_{obj}^2}\right)-\frac{\mu_{obj}^2}{2\sigma_{obj}^2}>\ln(\gamma)=\gamma'$$

The threshold γ is derived by inversion of the condition:

$$P_{FA}=\int_{L(x)>\gamma}p(x;H0)=1-CDF\left(\mathbb{N}(\mu_{ct},\sigma_{ct}),\gamma'\right)=CDF^{-1}\left(\mathbb{N}(\mu_{ct},\sigma_{ct}),1-\gamma'\right)=\alpha \tag{10.10}$$

Where CDF is the cumulative distribution function.

The optimized probability of detection is:

$$P_D=\int_{\gamma'}^{\infty}p(x;H1)=1-CDF\left(\mathbb{N}(\mu_{obj},\sigma_{obj}),\gamma'\right) \tag{10.11}$$

The detection performance ($P_D=f(P_{FA})$) is derived from Equations (10.10) and (10.11):

$$P_D=Q\left(\frac{\sigma_{ct}Q^{-1}(P_{FA})+\mu_{ct}-\mu_{obj}}{\sigma_{obj}}\right)=Q\left(Q^{-1}(P_{FA})+\frac{\mu_{ct}-\mu_{obj}}{\sigma_{obj}}\right)$$

Under the assumption that $\frac{\sigma_{ct}}{\sigma_{obj}}\cong 1$.

Therefore, the deflection performance is monotonic with the term $\frac{\mu_{ct}-\mu_{obj}}{\sigma_{obj}}$, which can be interpreted as signal-to-noise ratio, or as the distance of the object mean from the control mean in standard deviation units.

10.4.6.2 Bayesian Approach

In a detection problem with a Bayesian approach, prior probabilities are assigned to the hypothesis under test. In the Bayesian context, a probability of error P_E is defined as [37]:

$$P_E=P(H_0|H_1)P(H_1)+P(H_1|H_0)P(H_0) \tag{10.12}$$

where $P(H_0|H_1)$ and $P(H_1|H_0)$ are conditional probabilities.

Thus, with this definition P_E is an overall measure of error, with the commission and omission errors being properly weighted.

For a minimum P_E detector, the conditional likelihood ratio is compared to a threshold defined by the prior probabilities. H_1 is decided if:

$$\frac{p(x|H_1)}{p(x|H_0)}>\frac{P(H_0)}{P(H_1)}=\gamma$$

In our case, we assign equal a priori probabilities, $P(H_0)=P(H_1)=0.5$ and the likelihood ratio is:

$$\frac{\mathbb{N}\left(\mu_{obj},\sigma_{obj}\right)}{\mathbb{N}\left(\mu_{ct},\sigma_{ct}\right)} > 1$$

Taking the logarithm of both sides, similar to what is done in Equation (10.9), we have at first order:

$$x\left(\frac{\mu_{obj}}{\sigma_{obj}^{2}}\right) - \frac{\mu_{obj}^{2}}{2\sigma_{obj}^{2}} > 0$$

$$x > \frac{\mu_{obj}}{2} = \gamma \tag{10.13}$$

From Equations (10.12) and (10.13), P_E is computed as:

$$P_E = \frac{1}{2}\left[Q\left(\frac{\gamma - \mu_{obj}}{\sigma_{obj}}\right) + Q^{-1}\left(\frac{\gamma}{\sigma_{ct}}\right)\right]$$

10.4.7 Object Shape

The shape complexity of the objects (patterns of elevation change) identified by the morphological clustering algorithm can make an important connection with the type of process that caused the changes. For example, sharp regular edges and simple geometric objects, such as rectangles or convex polygons, can be ascribed to forest becoming agricultural fields, while more fragmented shapes with irregular borders can be associated with selective logging processes or forest regeneration [51].

To step from a notional concept of shape complexity to a quantitative characterization we adopt a mathematical measure, the fractal exponent D, introduced by [52], by extending the standard dimensional analysis of Euclidean geometry. Here a family of planar shapes is characterized by length–area relationships (e.g. common shapes present the same perimeter-over-area ratio). The perimeter-over-area ratio is extended to become a generalized ratio

$$\frac{Pg^{1/D}}{\sqrt{Ag}} \tag{10.14}$$

where Pg is the linear extent of the shape measured with a yardstick G, and Ag the area measured in units of G^2. It is demonstrated in [52] that this generalized ratio is a constant for a family of shapes whose geometry is fractal or standard.

In our case, we estimate D by solving Equation (10.14) in the logarithmic domain:

$$D = \frac{2\log P}{\log A} \tag{10.15}$$

where P is the polygon perimeter and A is the polygon area measured in units of the resolution element of the differential height dataset.

Objects with very regular boundaries have a low exponent ($D \cong 1$), while objects with highly fragmented boundaries have a high exponent ($D \cong 2$). Examples of objects with different shape complexity and the related shape measures are reported in Figure 10.3.

Multivariate distribution functions of the shape measures and DSM differences are used to infer relationships among these quantities.

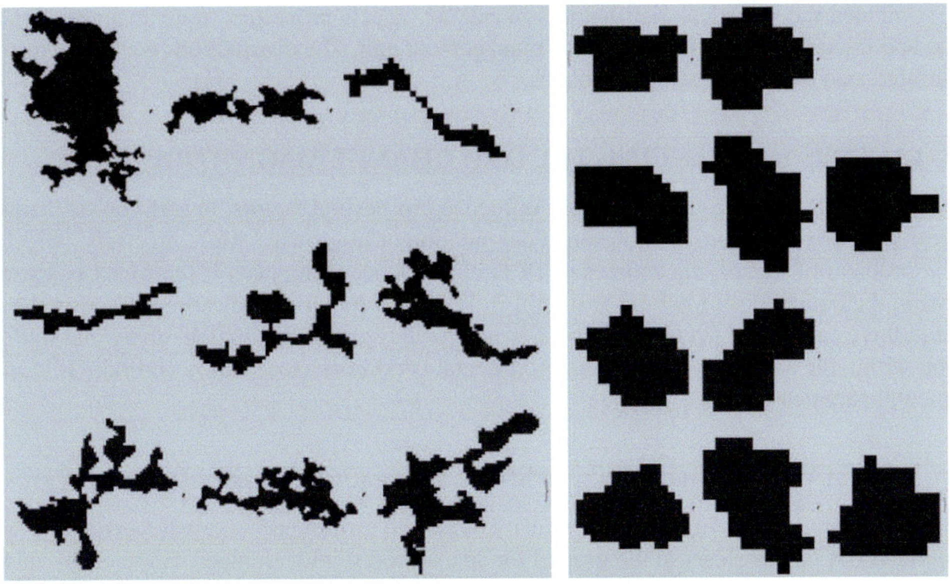

FIGURE 10.3 Examples of high complexity shapes (D = 1.6) (left), and low complexity shapes (D = 1.2) (right).

10.4.8 Characterization of Objects by Contextual Information

10.4.8.1 Distance from Roads

The spatial distribution of objects was characterized by the minimum distance among them and with respect to neighboring roads.

Roads were digitized by the World Resources Institute based on LANDSAT imagery covering a total length equal to 299 km and classified according to type: primary forestry roads (50 km), secondary forestry roads (138 km), operation roads (102 km), and primary public roads (9 km).

The minimum Euclidean distance between $\Delta DSM_{positive}$ and $\Delta DSM_{negative}$ objects was computed between object centroids. The minimum Euclidean distance between ΔDSM and roads was calculated by computing the distance between points distributed every 10 m located along the road networks and the centroid of each ΔDSM object.

10.4.8.2 Attributes by Land Management

To understand forest disturbance dynamics, the location of $\Delta DSM_{negative}$ and $\Delta DSM_{positive}$ objects was labeled by incorporating information from an independent dataset provided by the World Resources Institute (WRI) [29]. The $\Delta DSM_{negative}$ and $\Delta DSM_{positive}$ datasets were partitioned into four land management units: protection (18628.68 ha), production (46178.66 ha), community development (12272.44 ha), and urban (14414.35 ha).

Production areas were allocated to IFO (a company that is part of Danzer GmbH) for forest exploitation (selective logging of < 1 tree/ha) of valuable tree species as part of the Ngombé logging concession. These areas are not allocated for agricultural purposes by the local population [29].

Protection areas are dominated by *Macaranga* forest and pioneer *Macaranga* regeneration based around undisturbed lowland and swamp forest, and are sometimes used by local communities.

Community development areas are reserved for local communities for agriculture and timber harvesting. These are primarily situated around the town of Ngombé and along a primary forestry road connecting Ngombé to the National Road 2 (N2), as well as along secondary logging roads on the east bank of the Sangha River. We assigned the class "urban" to an area around the city of Ouesso in the north of the study site.

We computed the statistical distributions of ΔDSM objects properties (area, magnitude, homogeneity, and fractal dimension) within each management unit. The distributions were then analyzed and characterized by their summary statistics.

10.5 FACTORS INFLUENCING THE DSM CHANGE MAGNITUDE

The estimated DSM changes are inevitably influenced by several factors, in addition to changes in the forest structure (e.g. deforestation/forest degradation or re-growth). More precisely, we address here the problem of identifying sources of PCH changes occurring even if the forest is stationary (i.e. within the boundaries of natural variability). We can summarize the causes affecting the PCH change in three categories: (a) forest vertical structure and spatial distribution, given in a stationary situation within the sampling time; (b) environmental conditions (seasonality and rainfall); and (c) instrumental parameters.

10.5.1 Forest Vertical Structure and Spatial Distribution (Forest Density)

The PCH location depends on the forest density, which in turn impacts on wave extinction within the volume. PCH moves closer to the ground for sparser forest with clumped canopies, because the canopy gaps enable greater signal penetration [53] and increased contribution from surface scattering. In the present work, forest density was considered to be high (highest density and homogeneity is found in swamp forest) but decreasing for degraded forest. Vegetation density and gaps might also have influenced the DSM (and as a consequence ΔDSM) since, in particular in old-growth primary forest, the presence of large emergent trees cast shadows on the surrounding vegetation, thus modulating the local extinction and incidence angle. Shadowing effects were noticed in interferometric coherence derived from airborne sensors over tropical forests [54].

10.5.2 Environmental Conditions (Seasonality and Rainfall)

Seasonality effects can be considered negligible in the tropical evergreen forests of the Congo Basin (as opposed to temperate and boreal forest, where the leaf-off condition leads to higher penetration within the canopy and consequently lowers the PCH [55]).

We also hypothesize that smaller changes (change magnitude between 0 and –7 m and area $\leq$ 100 pixels) could be linked either to disturbance caused by damage from selective logging or to seasonality (e.g. transition from leaf-on/leaf-off conditions) which have been observed in the study area [56]. The study area falls within the so-called Sangha River Interval (located between 14–18° E covering a 400 km wide area) which is reported to host semi-deciduous forest presenting high photosynthetic activity and low endemism ([56,57]). In this area, some trees at least could have more leaves at one date – though both scenes were captured in the same season (dry season, both December).

The effect of rainfall was considered minimal in the study, since both datasets were acquired in December (dry season) ([32,58]). Rainfall 48 h before the data acquisition was low ($t_1 = 0$ mm and $t_2 = 7.6$ mm) [59] (Table 10.1).

10.5.3 Dependence on Instrument Parameters

One more source of uncertainty in the $\langle \Delta h \rangle$ estimator can result from the coherence being measured by different vertical wave numbers at two data takes.

We analyze from the theoretical (modeling) point of view the difference of two estimates of surface height (DSM) of the same natural target, provided by two InSAR observations with slightly different instrumental parameters, these being the baselines, incidence angles and range distances (see Table 10.1).

The problem is split into two scattering scenarios:

1. A dense forest, featuring high extinction at X-band, and therefore amenable to be modeled as a random volume with no return from ground. This situation will correspond to the no-change case in forest change detection, or, in other words, to the primary forest control area. A volume-only model with an exponential structure function is adopted here.
2. A less dense (or sparse) forest, which could represent the case of some forest dynamics, such as regrowth, which can be modeled including a return from the ground. The RVOG model is used in this case.

10.5.3.1 Volume Only

The fractional phase height within the volume is given by:

$$\gamma = f(h,\sigma,z_0,\beta_z) = \frac{p\left(e^{(p+i\beta)h}-1\right)}{(p+i\beta)\left(e^{ph}-1\right)} \quad p = 2\sigma \sec\vartheta \tag{10.16}$$

$$\phi = \arg\left[\gamma\right] \mod 2\pi$$

$$\phi_{center} = \frac{\phi}{h\beta \mod 2\pi} \tag{10.17}$$

where γ is the complex coherence, h is the volume height, σ is the extinction coefficient, ϑ is the incidence angle, $\beta = \dfrac{4\pi bs}{\lambda R \sin\vartheta}$ is the vertical wavenumber, a parameter that summarizes the interferometric sensitivity.

The DSM difference between two observations was computed from Equation (10.17) as a function of extinction, and parametrized by random volume height (Figure 10.4). The maximum difference is of the order of 1 m, it is negative (decrease in phase center height – PCH), and occurs for low extinction and 50 m volume height. Notice that a difference of –1 m corresponds to the mean of the

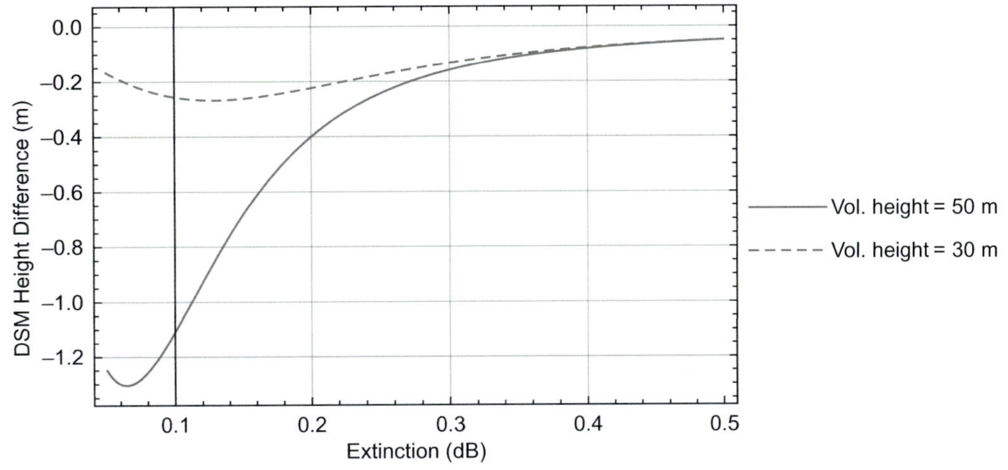

FIGURE 10.4 DSM height difference between two observations as a function of extinction for a volume height of 50 m and 30 m

distribution of samples taken within the primary forest control area. Therefore DSM changes within a stable forest area can be ascribed to differences in instrumental parameters.

For increasing extinction, the PCH moves toward the top of the volume, and the dependence on interferometer sensitivity (β_z) decreases, resulting in lower or negligible DSM difference. The same effect occurs for a lower volume height (30 m), where for all extinction values the range of DSM difference is reduced to only a fraction of a meter.

10.5.3.2 Volume over Ground

In this case, the model for the complex coherence in Equation (10.16) is modified to include the return from the ground in the form of the effective surface-to-volume scattering ratio μ.

$$\gamma = e^{i\beta_z z_0} \frac{\gamma_{vo} + \mu}{1 + \mu}$$

where the volume decorrelation component γ_{vo} is given by Equation (10.16).

The DSM height difference as a function of the surface-to-volume scattering ratio μ is shown in Figure 10.5 for a volume height of 30 m and parametrized by two extinction values: 0.05 dB and 0.3 dB. With lower extinction (0.05 dB), the PCH is set around the middle of volume (15 m), and there is low dependence on β_z for small surface scattering components (small negative DSM difference). However, when the surface component weighs in more, the DSM difference becomes positive and increases. This indicates that for sparser, or lower biomass forest, increase in DSM difference can be caused by the influence of ground return, and is proportional to this return.

For higher extinction, when the PCH is pushed toward the top of the canopy, the DSM difference becomes larger and negative (up to −1 m) when the surface contribution is small. However, it is pushed to positive values when the surface component starts to weigh in, reaching an asymptotic difference of approximately 1.5 m.

All in all, we conclude from this analysis that the range of DSM differences that can be ascribed to instrumental parameters is of the order of ± 1 m, with the positive values attributed to the influence of surface return.

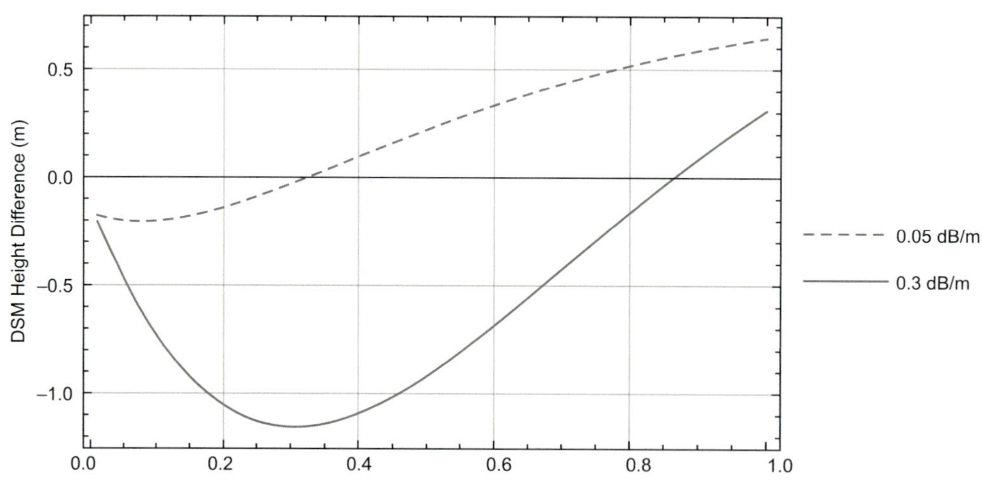

FIGURE 10.5 DSM height difference as a function of the surface-to-volume scattering ratio and parametrized by extinction: 0.05 dB/m (dashed line) and 0.3 dB/m (solid line).

10.6 ANALYSIS

10.6.1 ΔDSM Magnitude and Area Descriptive Statistic

The segmentation algorithm detected 4168 $\Delta DSM_{negative}$ objects and 14534 $\Delta DSM_{positive}$ objects. The frequency distributions (smoothed histograms) of the DSM height difference within each object ($\Delta DSM_{positive}$ solid line, $|\Delta DSM_{negative}|$ dashed line) are shown in Figure 10.6, while the frequency distributions of the objects' area are plotted in Figure 10.7.

All frequency distributions are heavy-tailed (kurtosis $k > 3$) and can be modeled adequately by a log-normal distribution. The statistics include the geometric parameters of the fitted log-normal distribution (mean, standard deviation $\mu^* = e^{\pi}$, $\sigma^* = e^{\sigma}$, and mode $e^{\mu - \sigma^2}$), the kurtosis, the quartiles (Q), and the minimum and maximum values). The log-normal fitted geometric parameters better describe the properties of the distribution with respect to sample mean and standard deviation, these being more affected by the right tail.

Samples in a control population $P_c\Delta h$ (no change area) result in a zero-mean normal distribution with a standard deviation $\sigma_h = 0.19$.

It is also interesting to analyze the occurrence of large values in the tails of the area distributions (outliers). We assume a criterion for the determination of an outlier observation based on the inter-quartile (IQR) range:

$$Q_{out} \rightarrow (\Delta area) > Q_3 + 3(Q_3 - Q_1)$$

Regarding the $\Delta area(-)$ objects, by this criterion, 278 objects with $area > 0.56$ hectares are selected. The median of their distribution is 1 ha. The objects with an area within the interquartile range can be ascribed to two distinct processes: selective logging and shifting cultivation.

Moreover, 24 objects have areas beyond the third quartile ($area > 0.56$ ha). These large objects are likely to correspond to extended deforestation events – for example, clearing for agriculture.

The $\Delta area(+)$ objects present a more even distribution of outliers, with $Q_{out} = 0.23$ ha, and comprised between 0.23 and 1 ha. This relatively more even-sized distribution of outlier objects indicates that these areas may be associated with clusters of regrowth, which are generally large in size.

There is no overall correlation between the area and magnitude of objects. However, there is an interesting connection between outliers of negative changes $\Delta h(-)$ and their corresponding area.

For $\Delta h(-)$, $Q_{out} = 11$ m. Accordingly, there are 20 objects with $\Delta h(-) > 11$ m. For these objects,1 ha $< \Delta area <$ 14 ha. Therefore extreme outliers in negative height change also imply a large change in area.

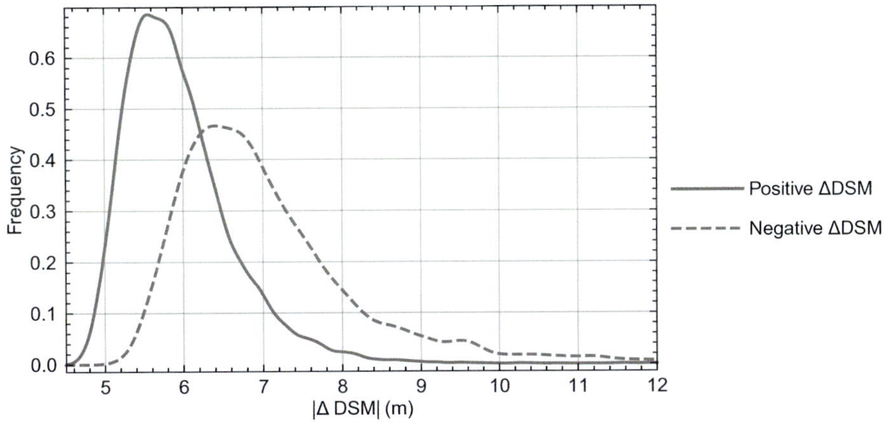

FIGURE 10.6 Frequency distributions of the ΔDSM magnitude. Positive change (solid line), absolute value of negative change (dashed line).

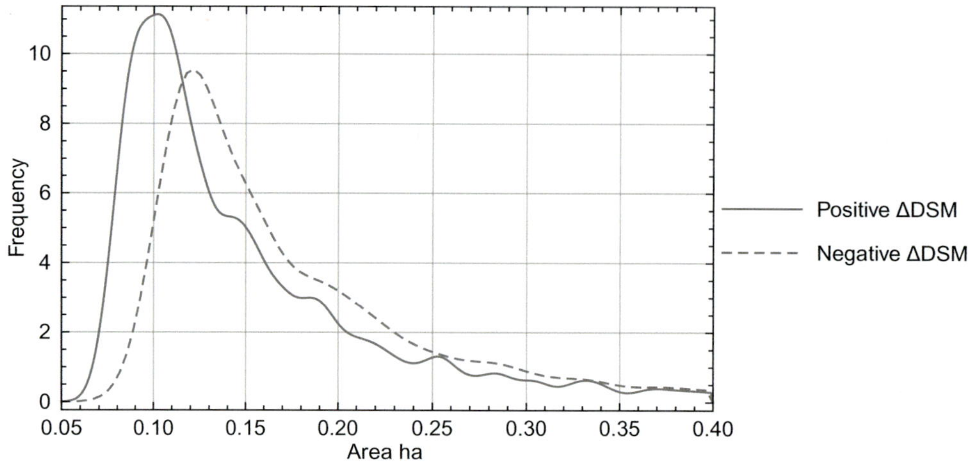

FIGURE 10.7 Frequency distributions of the ΔDSM area. The plot range is up to 0.4 ha to highlight the functional dependence of the distributions. Positive change (solid line), negative change (dashed line).

Finally, notice that the height variation range for $\Delta DSM_{positive}$ objects is narrower with respect to negative variation. This is a consequence of different generating processes. A natural process, i.e. slow vegetation growth within 1 year, in case of positive change, and forest disturbance in case of negative change.

TABLE 10.2
Descriptive Statistics of the Magnitude and Area of the Detected Objects

Object	μ^*	σ^*	$e^{\mu-\sigma^2}$	*Kurtosis*	*Q1*	*Q2*	*Q3*	*min*	*max*
				ΔDSM Magnitude (m)					
$\Delta h(-)$	6.72	1.13	6.62	13.5	6.3	6.8	7.5	5.2	19.7
$\Delta h(+)$	5.77	1.10	5.71	7.8	5.45	5.82	6.27	4.45	13.12
				ΔDSM Area (ha)					
$\Delta area(-)$	0.13	1.41	0.11	275.8	0.12	0.16	0.23	0.1	14.88
$\Delta area(+)$	0.10	1.51	0.08	186.3	0.1	0.13	0.18	0.07	4.78

10.6.2 Standard Error of the Object Mean

We shall follow the path of error propagation from phase noise up to the standard error of the object's mean (SEOM). We start from the mean coherence modulus ρ within the object at acquisition t_1 and t_2, these being the observables that will allow us to estimate the sample phase variance (phase noise), as laid out in Equations (10.4)–(10.6), Section 10.4.4.

The frequency distributions of the mean coherence moduli for the $\Delta DSM_{positive}$ and $\Delta DSM_{negative}$ objects are reported in Figure 10.8. Regarding the $\Delta DSM_{negative}$ objects, there is a notable increase in the distribution mode between t_1 and t_2 (0.84 to 0.89). Moreover, the interquartile deviation decreases from t_1 to t_2 (0.0305 to 0.013). These figures are a tell-tale sign of a physical process consisting of loss of forest volume and of a consequent decrease in volume decorrelation. From data in the literature reporting the dependence of coherence on dense tropical forest biomass [53], we infer that the measured coherence changes in $\Delta DSM_{negative}$ objects may correspond to a loss of forest biomass of the order of 300 Mg/ha.

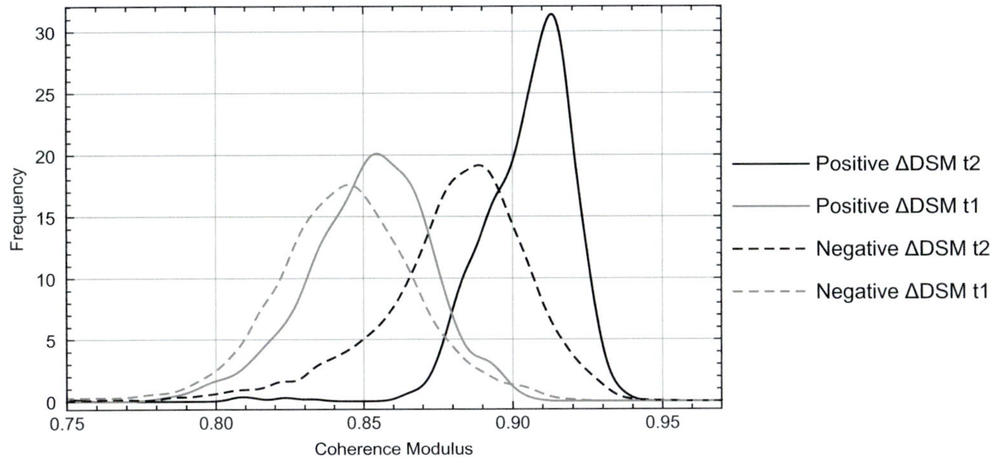

FIGURE 10.8 Frequency distribution of the mean (within object) coherence modulus at acquisition t_1 and t_2 (gray and black lines respectively) for $\Delta DSM_{positive}$ objects and $\Delta DSM_{negative}$ objects (solid and dashed lines respectively).

The reduction in the spread of the per-object coherence indicates that in the final state there is a more homogeneous distribution of scatterers within the area subtended by the object. In other words, because of disturbance, the forest structure's heterogeneity decreased. This point is also confirmed by the within-object coherence variance, this decreasing from t_1 to t_2 (0.03 to 0.02).

Regarding the $\Delta DSM_{positive}$ objects, there is also an increase of the distribution mode (0.87 to 0.91) between the data takes, and a positive shift with respect to the distribution of the negative objects. These figures suggest that, in the case of a shorter vegetation layer (smaller random volume), there is more sensitivity to a change in the vertical wave number, rather than to volume height change. Indeed, the vertical wave number is lower at t_2 (smaller baseline), and therefore the coherence modulus will be higher (see RVOG model in Section 10.5.3).

The phase noise within each object induces the variance of the Δh estimates. The descriptive statistic of the per-object phase noise (standard deviation of height difference $SD(\Delta h)$ caused by phase noise) is depicted by the boxplot in Figure 10.9. The median and the interquartile deviation of the $\Delta DSM_{positive}$ and $\Delta DSM_{negative}$ distributions are very similar ($SD(\Delta h) = 2.44\,\text{m},\ 2.65\,\text{m},\ \Delta Q = 0.16\ m$, $0.17\ m$), albeit the $\Delta DSM_{negative}$ distribution presents outliers over 6 m. These outliers correspond to low correlation values in the acquisition t_2, as indicated by the frequency distribution of the corresponding coherence values (see Figure 10.10). Thus, in the final stage, there is still high volume decorrelation. These objects have a range of Δh between 6 m and 10 m, and a small area (0.1 ha to 0.2 ha).

The frequency distributions of the sample standard error induced by phase noise are shown in Figure 10.11. The median (Q2) for the $\Delta DSM_{positive}$ objects is 95%, and for the $\Delta DSM_{negative}$ objects is 39% (Table 10.3). Even at this stage of error propagation, it becomes clear how pixel-based estimates of the height difference would be problematic.

The object's intrinsic variance statistic is described by the boxplot in Figure 10.12. The $\Delta DSM_{negative}$ objects present a higher variance, with median $SD(\Delta h) = 3.08$ with respect to the

TABLE 10.3
Distribution of the Sample Relative Standard Error Over all Objects

Object	Q1 (%)	Q2 (%)	Q3 (%)	Q3 + 1.5*IQR*
$\Delta DSM_{positive}$	88.7	94.8	100.1	120.6
$\Delta DSM_{negative}$	34.5	38.8	42.7	55.2

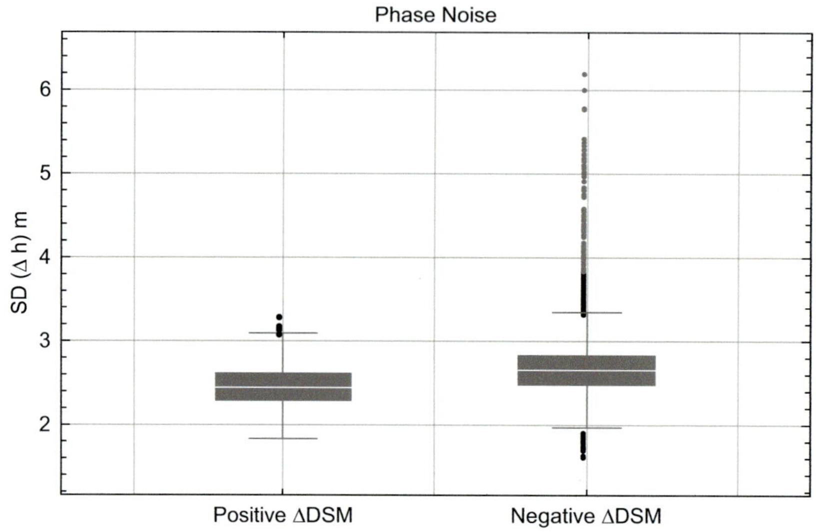

FIGURE 10.9 Boxplot depicting the distribution of $SD(\Delta h)$ induced by phase noise within the $\Delta DSM_{positive}$ and $\Delta DSM_{negative}$ objects.

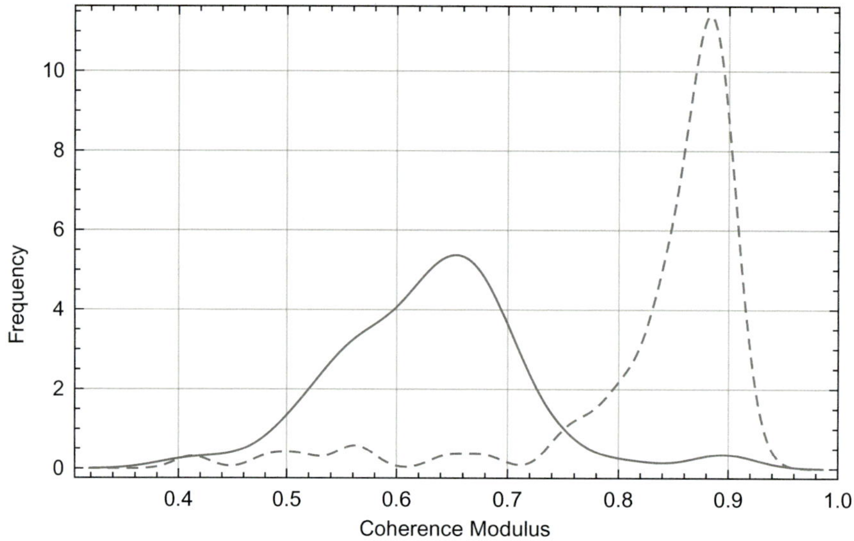

FIGURE 10.10 Frequency distribution of the coherence at t_2 (solid line) and at t_1 (dashed line) corresponding to the outliers of the $\Delta DSM_{negative}$ phase noise.

$\Delta DSM_{positive}$ objects $(\overline{SD(\Delta h)} = 1.44)$. Large variance $\Delta DSM_{negative}$ objects ($2.5 < SD(\Delta h) < 19$) are also those with a higher Δh change, as illustrated in Figure 10.13.

Finally, the within-object sample variance and the phase noise are combined to estimate the SEOM of the object's mean height change $\overline{\Delta h}$, as laid out in Equation (10.7).

The relative SEOM $\left(\frac{SD(\overline{\Delta h})}{\overline{\Delta h}}\right)$ distribution over all objects is illustrated by the boxplots in Figure 10.14 and summarized in Table 10.4, where the quartiles (Q) and the interquartile range (IQR)

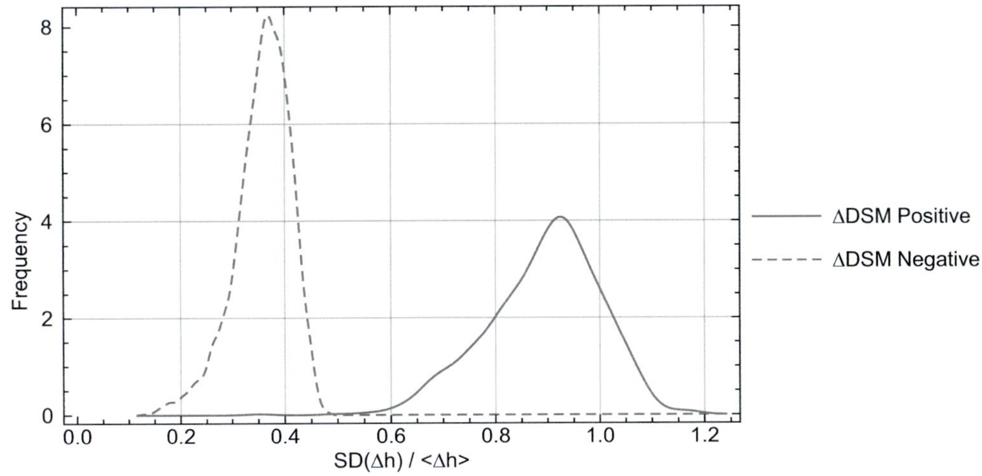

FIGURE 10.11 Frequency distributions of the sample standard error induced by phase noise ($\Delta DSM_{positive}$ is indicated by the solid line and $\Delta DSM_{negative}$ by the dashed line).

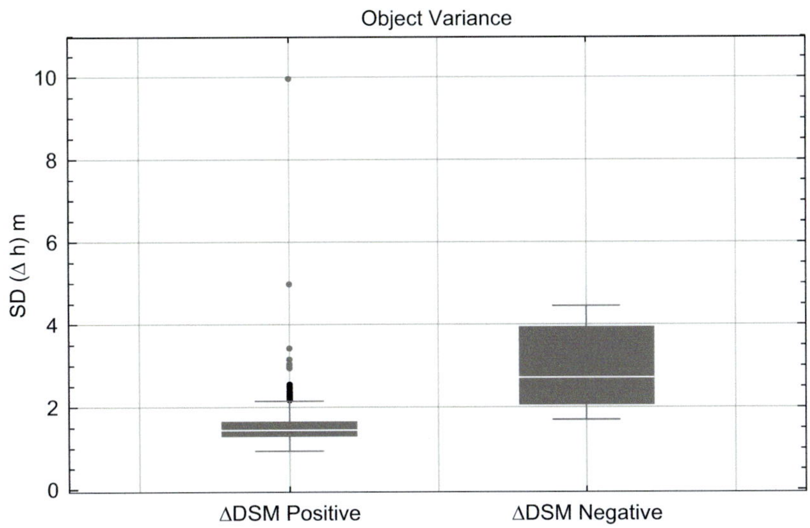

FIGURE 10.12 Boxplots of the object's intrinsic variance.

TABLE 10.4
Statistic of the Standard Error of the Object's mean Height Change (SEOM)

Object	Q1 (%)	Q2 (%)	Q3 (%)	Q3 + 1.5*IQR*
$\Delta DSM_{positive}$	5.2	6.7	7.8	11.8
$\Delta DSM_{negative}$	10.2	12.0	14.3	20.1

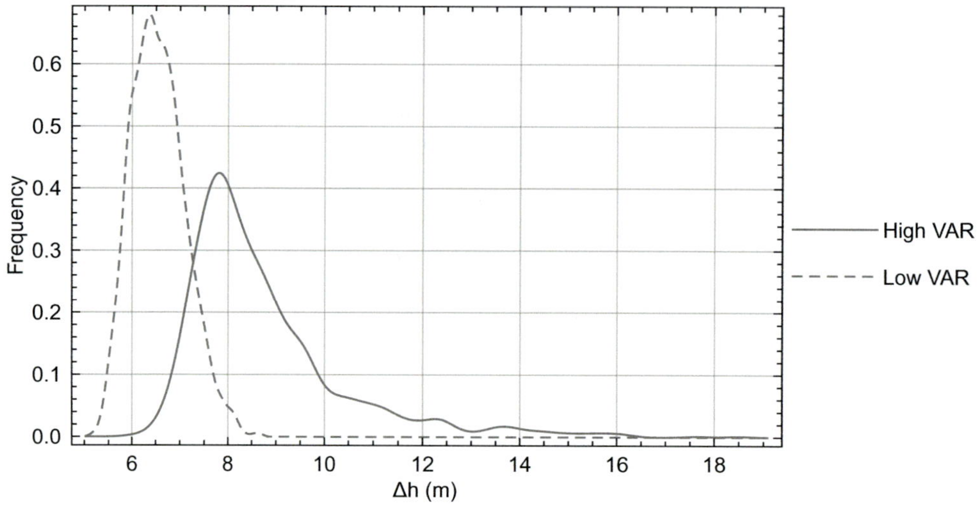

FIGURE 10.13 Height difference Δh (m) frequency distributions corresponding to high variance and low variance $\Delta DSM_{negative}$ objects.

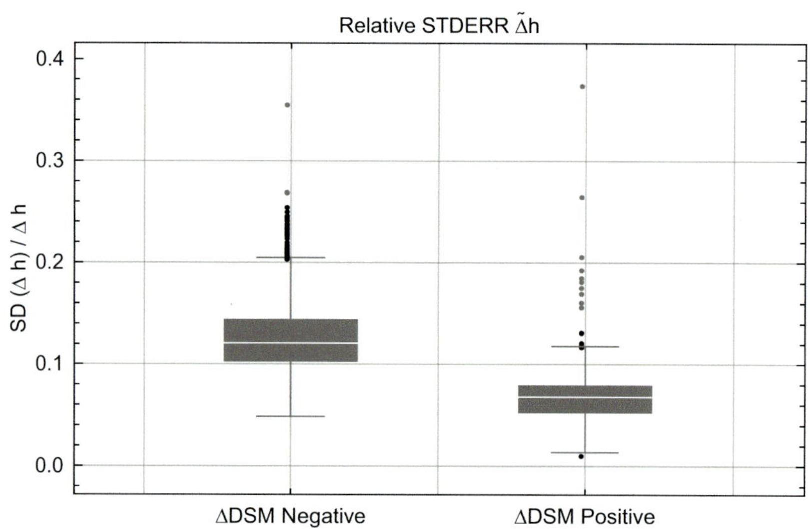

FIGURE 10.14 Boxplot illustrating the distribution of the relative standard error of the object's mean height change over all objects.

are reported. This table should be compared with Table 10.3 to assess the gain in accuracy that is achieved in going from a sample-based to an object-based estimation procedure.

10.6.3 Effect Size

The frequency distribution over all objects of the effect size measure (Equation 10.7) is shown in Figure 10.15. The descriptive statistic is summarized in Table 10.5.

The effect size minimum values, respectively 3.3 and 3.8 for positive and negative objects, when interpreted in a Gauss–Gauss mean-shift problem, give a probability of detection $P_D = 0.83, 0.92$ at probability of false alarm $P_{FA} = 0.01$ (see Figure 10.2). Therefore this test assures that all objects bear a significant height difference with respect to control objects. On the other hand, there are large outlier values beyond the *Q3* quartile, especially belonging to $\Delta DSM_{positive}$ objects.

Further investigation reveals that these outliers correspond to large area objects with a small height change $\Delta h < 4$ m. The large number of samples reduces the variance of the object mean. The median of the variance of the mean of objects with large effect size is 3.5×10^{-3} while the median for all objects is 3×10^{-2}, an order of magnitude larger.

TABLE 10.5
Effect Size Descriptive Statistic

Object	Q1	Q2	Q3	Mean	Std Dev
$\Delta DSM_{positive}$	8.6	10.2	12.3	11.99	2.1
$\Delta DSM_{negative}$	5.6	6.3	7.5	6.88	6.3

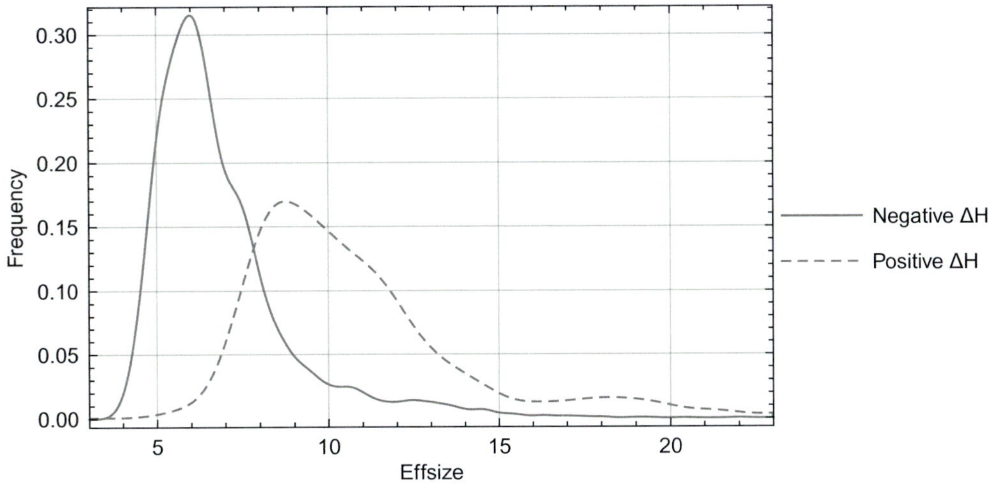

FIGURE 10.15 Effect size frequency distribution for $\Delta DSM_{negative}$ (solid line) and $\Delta DSM_{positive}$ (dashed line) objects. The plot range is set for emphasizing the frequency distribution in the interquartile range.

10.6.4 Object Detection by Statistical Decision Theory

As an alternative path for the characterization of an object's separability from a no-change control object, we consider here a hypothesis test (H_1 is a true change; H_0 is no-change) with a Bayesian approach that would provide the total probability of error, including Type I errors (decide H_1 when H_0 is true), and Type II errors (decide H_0 when H_1 is true).

The distribution of the probability of error P_E over all objects is described by the boxplot in Figure 10.16. The test based on the Bayes criterion illustrates the advantage of the object measures with respect to sample measures. The class labeled "sample" provides the statistic of P_E when the sample variance is taken into account. In this case for ΔDSM_{pos}, median and maximum are: $\bar{P}_E = 34\%$, $max(P_E) = 38\%$. For ΔDSM_{neg}, $\bar{P}_E = 16\%$, $max(P_E) = 24\%$. The class labeled "mean" provides the P_E statistic when the variance of the mean is taken into account. In this case, $\bar{P}_E = 6\%$, $max(P_E) = 10\%$ and $\bar{P}_E = 0.6\%$, $max(P_E) = 7\%$ for the ΔDSM_{pos} and ΔDSM_{neg} objects, respectively.

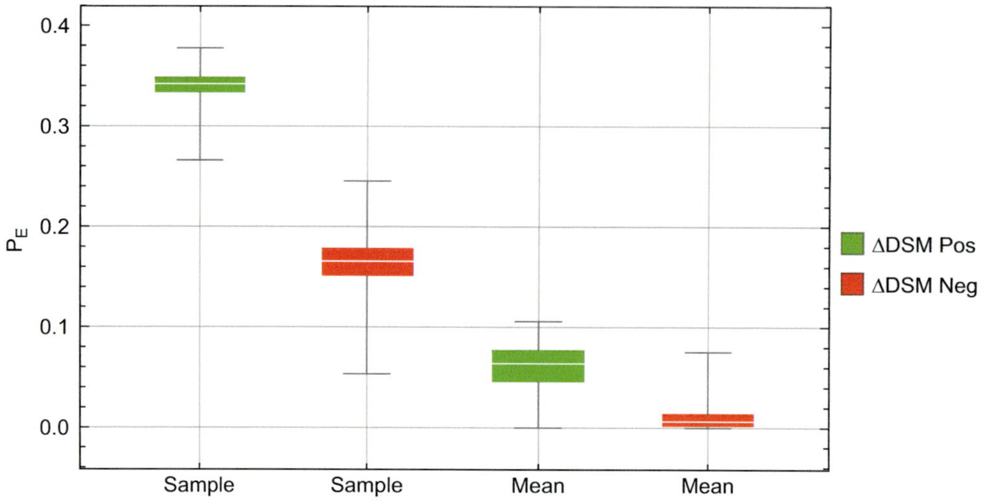

FIGURE 10.16 Boxplot describing the total probability of error P_E in testing the hypothesis that the object holds a true change. The classes labeled "sample" yield P_E when the sample variance is taken into account, while for the classes labeled "mean" the variance of the mean value is considered.

10.6.5 Spatial Location of Objects

The spatial location of $\Delta DSM_{positive}$ objects in relation to $\Delta DSM_{negative}$ objects can give clues about the process driving the change. This can be attributed to: (a) post-disturbance forest re-growth, which is often found in closer proximity to recent deforestation; or (b) isolated $\Delta DSM_{positive}$ events located far from recent deforestation.

$\Delta DSM_{positive}$ objects located far from roads were deemed to be unrelated to re-growth after disturbance. Instead, the proximity of $\Delta DSM_{positive}$ to $\Delta DSM_{negative}$ objects are more likely to indicate re-growth after disturbance caused by smallholders shifting cultivation.

The descriptive statistic of the minimum distance between $\Delta DSM_{positive}$ and $\Delta DSM_{negative}$ objects is reported in Table 10.6. The furthest $\Delta DSM_{positive}$ objects are located 2.93 km away from $\Delta DSM_{negative}$ objects, while the closest objects were a couple of meters away. Distance between $\Delta DSM_{negative}$ and roads ranged between 0.59 m and 6.95 km ($\mu \pm \sigma$: 1.15 ± 1.06 km), while $\Delta DSM_{positive}$ were located between a couple of meters up to 7.25 km from roads ($\mu \pm \sigma$: 1.16 ± 1.19 km) (Table 10.6).

The results are influenced by the road length, with secondary forestry roads extending for 137.5 km, followed by operation roads (101.2 km), primary forestry roads (50 km), and primary public roads covering only a distance of 9 km within the study site.

TABLE 10.6
Minimum Euclidean Distance (km) Between $\Delta DSM_{negative}$ Objects and $\Delta DSM_{positive}$, and Minimum Distance (km) Between Change Objects and Roads (Mean ± Standard Deviation = $\mu \pm \sigma$; median= $\tilde{x}$; Minimum, Maximum; 1st Quartile: Q1, and 3rd Quartile: Q3)

Class	$\mu \pm \sigma$	$\tilde{x}$	min	max	Q1	Q3
$\Delta DSM_{negative} - \Delta DSM_{positive}$	0.73 ± 0.55	0.60	0.03	2.93	0.3	1.03
$\Delta DSM_{negative}$- roads	1.15 ± 1.06	0.84	0.00059	6.95	0.34	1.68
$\Delta DSM_{positive}$- roads	1.16 ± 1.19	0.76	0	7.25	0.32	1.6

10.6.6 Objects' Proximity to Roads

The number of $\Delta DSM_{negative}$ objects is also highest in proximity to secondary forestry roads (38%, n = 1574) followed by operation roads (36%, n = 1515), primary forestry road (21%, n = 862), and 5% in proximity to primary public roads (n = 217). The largest $\Delta DSM_{negative}$ objects were found close to secondary forestry roads (441.96 ha), and the lower area was found in proximity to primary public roads (89.5 ha) (Table 10.7).

Instead, 58% of $\Delta DSM_{positive}$ objects were found closest to secondary forestry road (n = 8371) followed by operation roads (26%, n= 3794), primary forestry road (14%, n= 2003) and primary public road (3%, n= 366). The largest $\Delta DSM_{positive}$ object area was found closest to secondary forestry road (1424.57 ha) whilst the lowest $\Delta DSM_{positive}$ area was closest to primary public roads (52.54 ha).

TABLE 10.7
Road Type, Length (km), Number (*n*) and Area (ha) of $\Delta DSM_{positive}$ and $\Delta DSM_{negative}$ Objects in Proximity to Different Road Types

	Road	$\Delta DSM_{positive}$		$\Delta DSM_{negative}$	
Road type	**Length (km)**	***n* (%)**	**Area (ha) (%)**	***n* (%)**	**Area (ha) (%)**
Operation road	101.9	3794 (26%)	557.28 (23%)	1515 (36%)	312.35 (27%)
Primary forestry road	50	2003 (14%)	356.38 (15%)	862 (21%)	312.8 (27%)
Primary public road	9.1	366 (3%)	52.54 (2%)	217 (5%)	89.5 (8%)
Secondary forestry road	137.5	8371 (58%)	1424.57 (60%)	1574 (38%)	441.96 (38%)
Total	298.6	14534	2390.77	4168	1156.62

10.6.7 Change in Objects by Land Management

Out of the total $\Delta DSM_{negative}$ objects (n = 4168) only a portion (n = 4086; total area = 1135.56 ha) were situated within land management units classified by the WRI. For the $\Delta DSM_{positive}$ objects only 12017 (total area = 1919.71 ha) were considered in the analysis by land management out of the 14534 originally detected. In total, 82 $\Delta DSM_{negative}$ events and 2517 $\Delta DSM_{positive}$ events were not situated inside a classified management area and thus discarded from the land management analysis.

The highest number of $\Delta DSM_{negative}$ objects (58.7%, n = 2448) occurred inside production management followed by 20.9% (n = 873) in community development areas; 10.7% within peri-urban areas (n = 436); and 7.9% (n = 329) inside areas designated for protection. In terms of area, the management unit with the largest $\Delta DSM_{negative}$ area was production (502.05 ha) followed by community (380.44 ha), peri-urban (182.25 ha) and protection (70.82 ha).

Instead, the highest number of $\Delta DSM_{positive}$ objects (48.8%) were located inside production areas (n = 7094); followed by 15.6% inside community development areas (n = 2270); 9.6% within peri-urban areas (n = 1398) and 8.6% in protection areas (n = 1255). The largest $\Delta DSM_{positive}$ area was located in production (1117.75 ha) followed by community (392.68 ha), peri-urban (237.22 ha) and protection (171.93 ha), corresponding to 50%, 16.5%, 9.9% and 7.2%, respectively, of the total area affected by $\Delta DSM_{positive}$ changes (Table 10.8).

The proportion of $\Delta DSM_{negative}$, $\Delta DSM_{positive}$ and no-change areas with respect to the total area of each class is reported in Table 10.9. Note that no-change may indicate stationary forested areas with no estimated height change, rivers, or urban areas. No-change dominates all management units (93.7–98.7%). The no-change category is calculated as the total management area covered by each category minus the sum of the negative and positive changes.

TABLE 10.8

$\Delta DSM_{negative}$ (n = 4086) and $\Delta DSM_{positive}$ (n = 12017) Objects Partitioned Among Land Management Units. (Number of Objects – n, % of Total Objects and Total Area – ha, % of Total Area). The Analysis by Land Management Takes into Consideration a Subset of the Changes Detected within the Study Site

	$\Delta DSM_{negative}$		$\Delta DSM_{positive}$	
Class	***n*** **(%)**	**Area (ha) (%total area)**	***n*** **(%)**	**Area (ha) (% total area)**
Protection	329 (7.9%)	70.82 (6.1%)	1255 (8.6%)	171.93 (7.2%)
Production	2448 (58.7%)	502.05 (43.5%)	7094 (48.8%)	1117.75 (50%)
Community development	873 (20.9%)	380.44 (33%)	2270 (15.6%)	392.68 (16.5%)
Peri-urban	436 (10.7%)	182.25 (1.8%)	1398 (9.6%)	237.22 (9.9%)
Total (WRI)	**4086**	**1135.56**	**12017**	**1919.58**
Unclassified	82 (1.9%)	18.99 (1.6%)	2517 (17.3%)	465.22 (19.5%)
Total (all)	**4168**	**1154.55**	**14534**	**2384.80**

TABLE 10.9

The Proportion of Negative (ΔDSM_{neg}), Positive (ΔDSM_{pos}) and No-change Objects within Management Units

Class	ΔDSM_{neg} (%)	ΔDSM_{pos} (%)	No change (%)
Protection	0.4	0.9	98.7
Production	1.1	2.4	96.5
Community	3.1	3.2	93.7
Peri-urban	1.3	1.6	97.1

Inside all land management units, there is a higher proportion of $\Delta DSM_{positive}$ compared to $\Delta DSM_{negative}$ areas. "Community" is the class where the proportion of $\Delta DSM_{negative}$ and $\Delta DSM_{positive}$ is most balanced, with 3.1% and 3.2%, respectively; it also presents the highest percentage of $\Delta DSM_{negative}$ area among all classes considered.

"Protection" is the class with the highest proportion of no-change (98.7 %). In the "protection" management unit the percentage of $\Delta DSM_{positive}$ (0.9%) is more than double that of $\Delta DSM_{negative}$ (0.4%). This is also true for the "production" class, where $\Delta DSM_{positive}$ accounts for 2.4% and $\Delta DSM_{negative}$ for 1.1%. In the "peri-urban" class the proportion of $\Delta \mathrm{DSM}_{\mathrm{neg}}$ and $\Delta \mathrm{DSM}_{\mathrm{pos}}$ change is similar (1.3% and 1.6%).

10.6.8 Shape Analysis

10.6.8.1 Fractal Exponent

The fractal exponent (D) frequency distribution of all objects is summarized in Figure 10.17. The $\Delta DSM_{positive}$ and $\Delta DSM_{negative}$ objects present significantly different and non-overlapping distributions. The $\Delta DSM_{positive}$ fractal exponent distribution is shifted toward higher values (Q_1, Q_2, Q_3 = 1.53, 1.57, 1.61), with a maximum D = 1.8, thus suggesting more fragmented shapes, which would be typical of natural processes. The negative change fractal exponent distribution presents

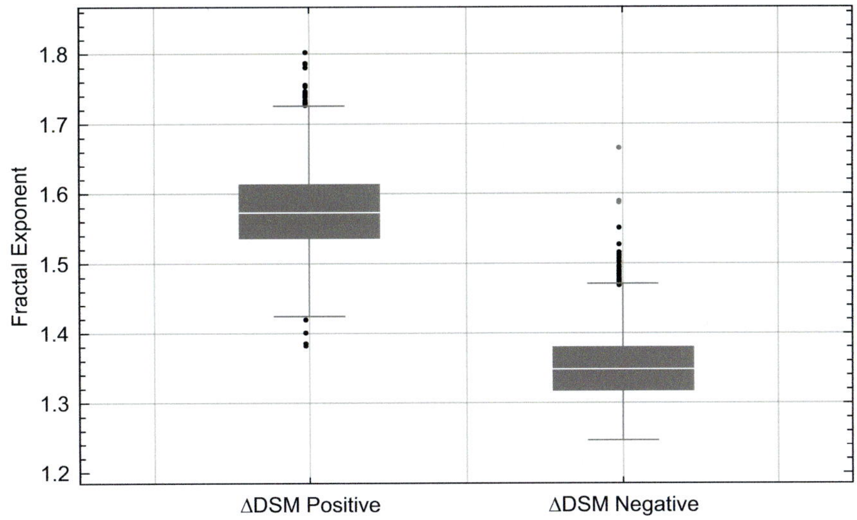

FIGURE 10.17 Fractal exponent box plots for $\Delta DSM_{positive}$ objects and $\Delta DSM_{negative}$ objects.

lower values (Q_1, Q_2, Q_3 = 1.53, 1.57, 1.61), with a minimum of D = 1.2, suggesting more regular shapes, such as those driven by clearing of forest for agriculture.

We investigate the relationship between the regularity of shape and the magnitude of change. The dependence of D on the independent variable (Δh) is shown in Figure 10.18. The $\Delta DSM_{positive}$ cluster is centered around smaller Δh values with respect to the $\Delta DSM_{negative}$ cluster, suggesting that small-change objects are mainly associated with fragmented shapes. Within each cluster, fitting $D = f(\log\Delta h)$ confirms, barring the outliers, the increasing trend between the two variables.

This observation is confirmed by analyzing the dependence on volume change $D = f(\Delta vol)$, as shown in Figure 10.19, and the relative linear fit between the two variables. The Pearson correlation coefficients (R) between the two variables are 0.7 and 0.52 for the negative and positive changes, respectively.

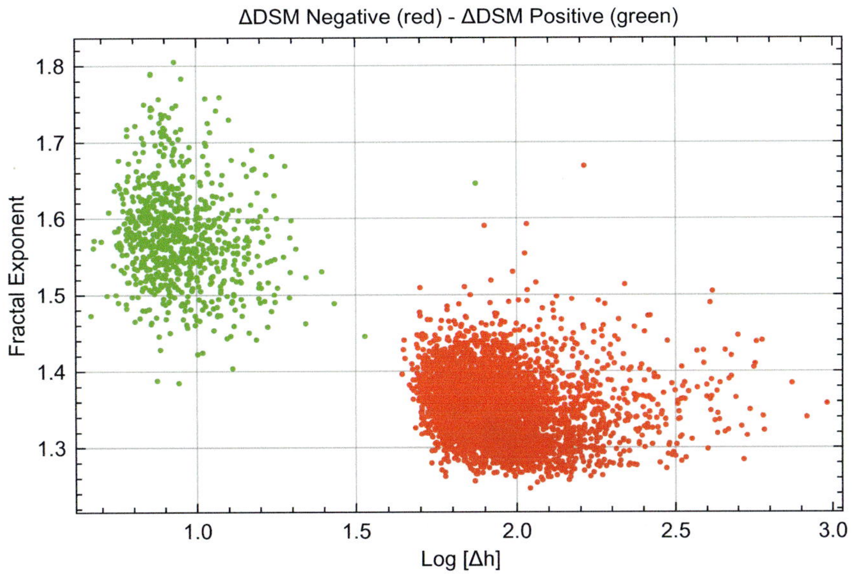

FIGURE 10.18 Relationship between the fractal exponent (D) and the magnitude of change log(Δh).

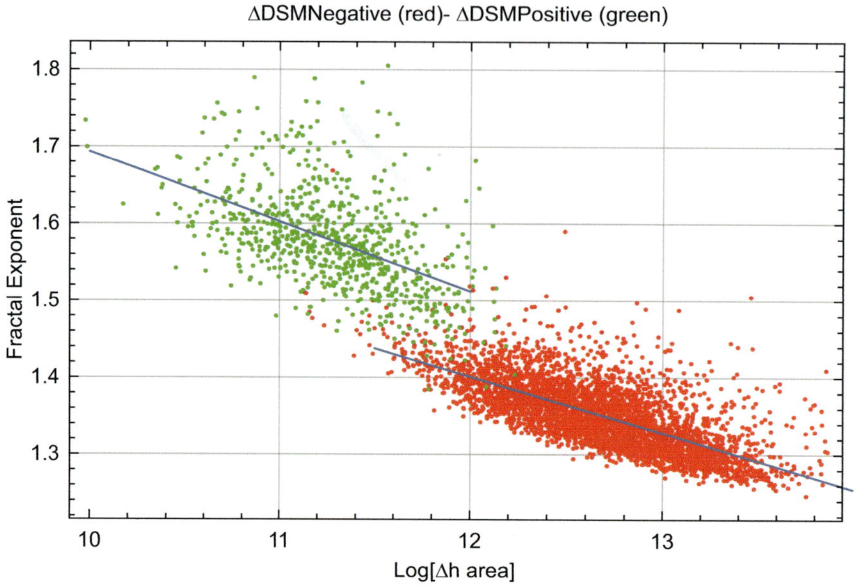

FIGURE 10.19 Dependence of the fractal exponent (D) on volume change log(Δh area).

10.6.8.2 Rectangularity

In this analysis, the object's shape measure is given by the fraction of pixels within the minimal bounding box, this being an indication of the rectangularity of the object. This measure will allow us, in particular, to detect large areas with rectilinear borders, these being a tell-tale sign of deforestation.

A boxplot of the rectangularity $\mathcal{R}$ for all objects is shown in Figure 10.20. In line with the results provided by the fractal exponent, we see here that the $\Delta DSM_{negative}$ objects present more regular and straighter shapes. In particular, the $\Delta DSM_{negative}$ boxplot shows outliers with $\mathcal{R} > 0.8$. These outliers correspond to the shapes illustrated in Figure 10.21. The corresponding areas are 4–6 ha with Δh changes in the order of 7 m. These objects are deemed to reflect clearing for agriculture.

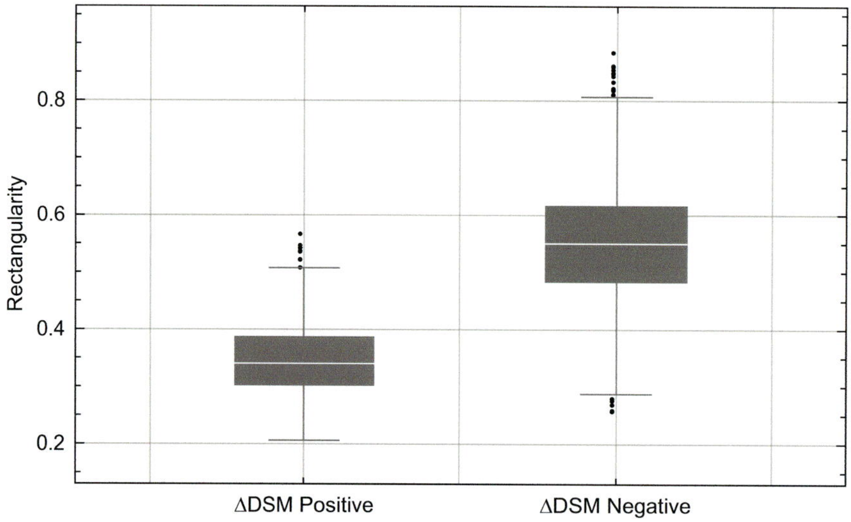

FIGURE 10.20 Boxplot of the rectangularity measures for all objects.

FIGURE 10.21 Shapes corresponding to high rectangularity values ($\mathcal{R} > 0.8$).

Objects with a high rectangularity $\left(0.8 < \mathcal{R} < 0.95\right)$ are located in the community and production land management units. All objects have a large area and Δh changes in the order of 7 m (Table 10.10). This case demonstrates how shape analysis of change objects can supply additional evidence on the nature of specific forest disturbance events.

TABLE 10.10
Statistic of Large Objects with Rectangular Borders in the "Community" and "Production" Land Management Units

		Area (ha)			Δh (m)		
Management Unit	***N***	**Q2**	**Q1**	**Q3**	**Q2**	**Q1**	**Q3**
Production	5	6.4	4.3	6.9	7.3	6.9	7.8
Community	13	4.1	3.1	5.2	8.0	7.1	8.5

10.6.8.3 Regular Boundary Shapes in Land Management Units

The starting point of this analysis is the definition of a right neighborhood of the fractal exponent lower bound $1 < D < \varepsilon$, where all objects' shapes present regular piecewise continuous boundaries and width-to-height ratio of the bounding box greater than 0.5. The neighborhood ε value is found by segmentation of the D space of all negative value objects into four clusters, taking the min–max values of the first cluster, and checking by visual inspection that the corresponding shapes meet the required geometric properties. As in the previous analysis, these objects are meant to represent changes induced by specific forest disturbance processes, such as clearing for agriculture.

Objects with D values within the neighborhood ε, are selected for each land cover class. The related area A and Δh change properties of each set are also selected. The probability density of area conditional to the fractal exponent in the neighborhood ε, $PDF(A \mid D)$ and the joint conditional density of area and height change $PDF(A, \Delta h \mid D)$ are finally computed for each class set. This statistical description attempts to identify area and volumetric measures ($area \times \Delta h$) of forest loss associated with patterns within geometric boundaries.

The conditional probability density $PDF(A \mid D)$ for the classes protection, production, community and peri-urban is shown in Figure 10.22.

The density functions of the four classes overlap for $A < 0.35$ ha, while a significant difference is evident for larger areas. In particular, the community class presents high values of between 0.9 and 1.2 ha.

The joint conditional $PDF(A, \Delta h \mid D)$ for the classes protection, community and peri-urban is represented as a scatter-plot in Figure 10.23. This statistic indicates that clusters of the community and peri-urban classes are present in the large area ($A > 0.8$ ha) region, and also correspond to large Δh in the range of 12 m to 17 m, while probability for protection class clusters around a smaller area (0.45 ha) and Δh (12 m). The interpretation of this shape analysis mirrors the one carried out for the rectangularity measure. The large area objects with regular boundaries in the community management unit can be related to deforestation. While the smaller regular patches with lower change in the protection land management unit cannot be ascribed to a specific disturbance in the absence of additional supporting information, but they could be a result of selective logging or forest die-back.

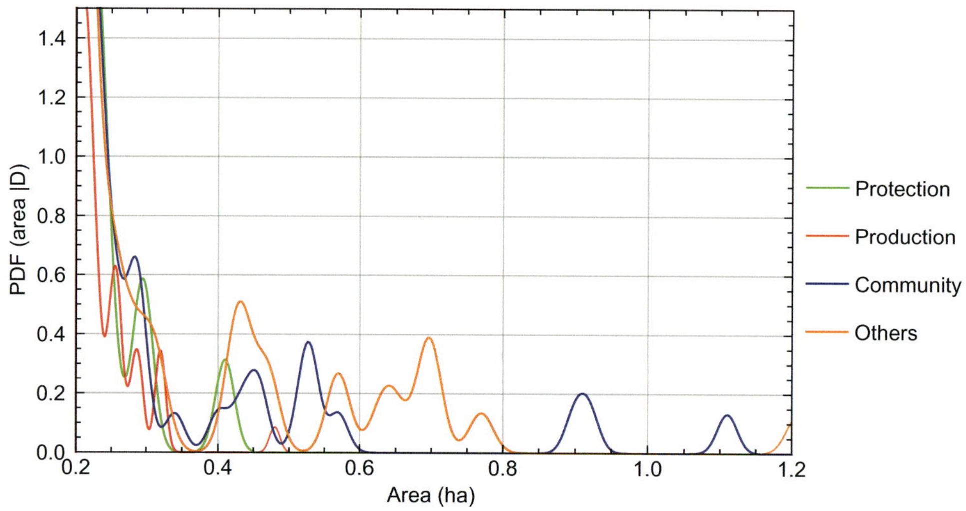

FIGURE 10.22 Conditional probability density function $PDF(A \mid D)$ for classes protection, production, community and others with fractal exponent D in the right neighborhood of 1.

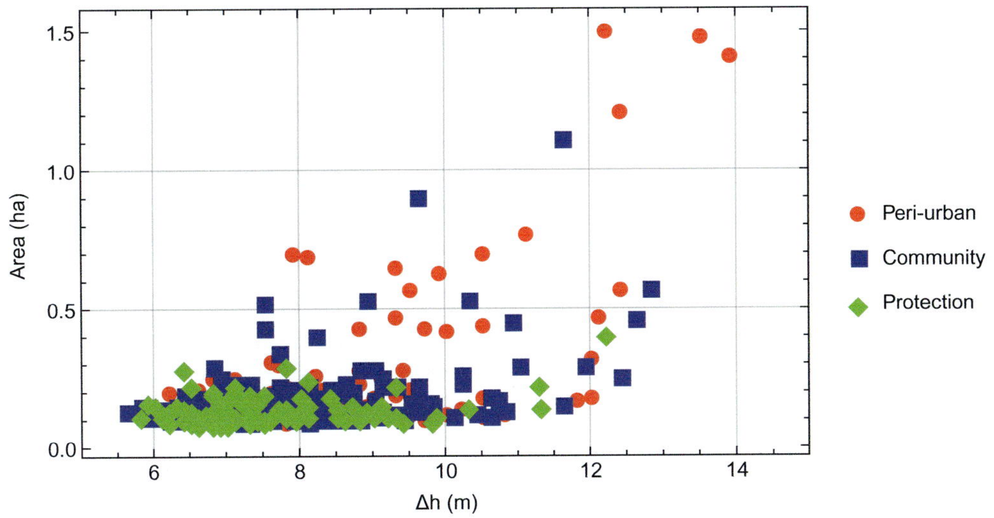

FIGURE 10.23 Joint conditional $PDF(A, \Delta h \mid D)$, represented as a scatter-plot, of protection, community and peri-urban land management units. Community and peri-urban statistics present large area measures (up to 1.5 ha) associated with large height changes (up to 17 m), while the statistics of the protection class clusters around a smaller area (0.45 ha) and height change (12 m).

10.7 COMPARISON BETWEEN OBJECTS DETECTED BY INSAR ΔDSM AND BY OPTICAL IMAGERY

Comparison between observations of the same object by two instruments relying on different physics and measurement techniques supply evidence of truth when the areas overlap and provide clues for justifying the nature of discrepancies when a correspondence does not occur. This comparison procedure should be contrasted with a validation process, where the observation is compared with a piece of information having very low uncertainty, thus being considered "the truth".

The comparison was based on delineation by visual inspection and expert knowledge, of corresponding change objects in the InSAR ΔDSM dataset and very high resolution (VHR) optical imagery. The VHR imagery was acquired by the Pléiades satellite sensor on 22/02/2013 (4 bands pan-sharpened at 0.5 m resolution) [60] and on 17/12/2013 (available from Google Earth at 2 m resolution).

In the object-based comparison, the validity criterion was the hit/miss matching of a spatial geometrical property, this being the intersection of the digitized contour delineating the test object (InSAR ΔDSM) and a reference object (area of change in the VHR imagery). The hit rate is the percentage of objects that are interpreted by the Pléiades scenes as changes, and that geometrically intersect the $\Delta DSM_{negative}$ objects, while the miss rate is the proportion of objects with no intersection between the Pléiades validation dataset and the $\Delta DSM_{negative}$ objects.

Two objects ranging from 0.2 ha to 2.1 ha that were detected by both InSAR and Pléiades imagery are shown in Figure 10.24.

Overall, 77 out of the 85 disturbed areas delineated manually in the Pléiades imagery intersected (some only partially) the InSAR ΔDSM objects (87.6%), while 11 objects (12.3%) did not. This discrepancy can be explained either by an omission error in the $\Delta DSM_{negative}$ detection algorithm, or a commission error during the visual interpretation of the optical data. In some cases, the $\Delta DSM_{negative}$ objects corresponded to either secondary or degraded forest in both Pléiades acquisitions. The reason was ascribed to the acquisition date of the first TanDEM-X scene that occurred about three months before the first Pléiades acquisition.

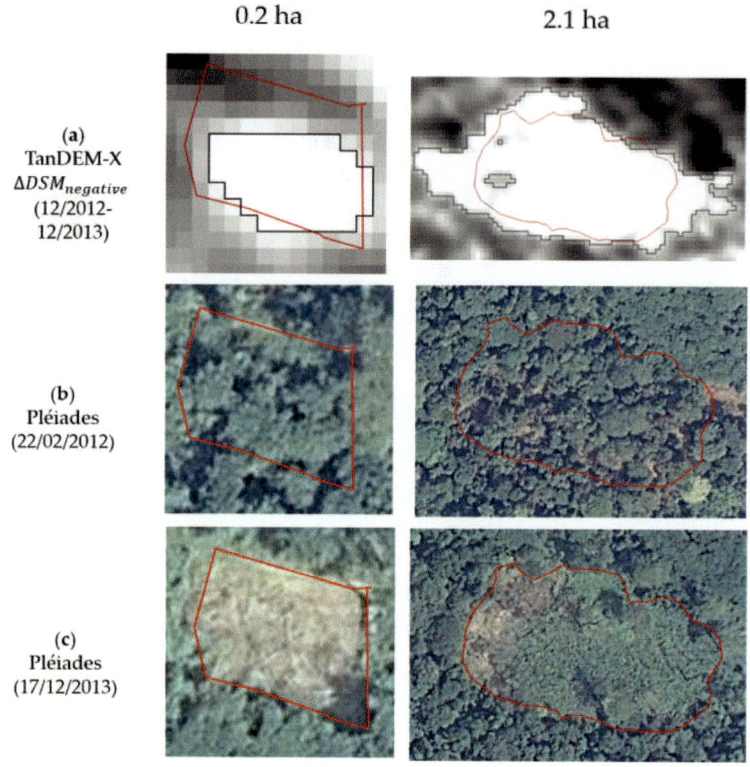

FIGURE 10.24 Sample of manually delineated deforestation/forest degradation objects (red polygon) and $\Delta DSM_{negative}$ objects (black polygon) overlaid on: (a) Digital surface model difference image *(ΔDSM);* (b) Pléiades imagery at t_1 (22/02/2012); and (c) Pléiades imagery at t_2 (17/12/2013).

Change objects detected by TanDEM-X were generally larger in area compared to those observed in the Pléiades images. This can be attributed to two distinct approaches affected by different sets of errors: a computational automatic change detection algorithm for the InSAR DSMs against the user-defined manual delineation of deforestation events using Pléiades.

Pléiades imagery was not suitable for the validation of forest re-growth, since the spectral properties of re-growth at various stages are visually similar after one year. The only way to check these observations would be through height change estimates that are only available from three-dimensional remote sensing datasets (e.g. repeat LiDAR surveys or repeat drone-based digital elevation models) and/or through a time series of ground-based height estimates. These datasets were not available for the study area at the time of our project.

10.8 CONCLUDING REMARKS

In line with the main theme of the book, we have proposed in this chapter a spatial analysis technique that provides objects encapsulating either negative or positive DSM changes occurring between two InSAR acquisitions. Statistical and morphological analysis of the objects' properties (differential surface height, area, and shape) provided links to forest disturbance.

We exploited the potential of using an object-based approach as an alternative to a pixel-based one, which assures, on the one hand, phase noise reduction and good accuracy of the estimates, and on the other, it lays the groundwork for geometrical measures that qualify types of forest disturbance.

The approach provided significant information in terms of (a) forest volume change and (b) area changes linked to deforestation, forest degradation, and to a certain extent post-disturbance re-growth at high resolution (< 5 m) in tropical forest. This is a significant step forward compared to SAR backscatter change detection, which is limited by the inability of scattering physics to sense the forest vertical structure function, and by signal-to-speckle noise ratio precluding the detection of subtle processes (e.g. degradation and regrowth).

The path was traced for the detection of post-disturbance re-growth. We have detected re-growth to a certain extent, despite having only a single year of data, albeit these measures could not be validated because of the lack of an appropriate validation dataset. This ability to map the extent and magnitude of forest re-growth will need longer DSM time series, as the changes in PCH will increase with time and the probability of detection with respect to no-change areas will accordingly increase.

The experiment described in this chapter should be considered as a proof of concept, or a test bed for object-based spatial analysis techniques. The approach requires pairs of InSAR DSMs (or a longer time series) with sufficient vertical and spatial resolution and low phase noise (low temporal decorrelation). These requirements were indeed met by the datasets provided to the project by the TanDEM-X mission.

A space-borne InSAR instrument with a similar configuration and performance as the TanDEM-X, but with greater data availability and more frequent observations, would present the possibility of monitoring tropical forest disturbance and post-disturbance regrowth at an operational level. Moreover, the DSM differencing approach could be extended by incorporating ground-based above-ground biomass estimates, which would allow conversion from volume change to biomass change and thus be used for REDD+ initiatives aimed at estimating above-ground biomass changes.

REFERENCES

1. I. Hajnsek, and T. Busche, "*TanDEM-X Science Activities*", in *10th European Conference on Synthetic Aperture Radar*, Berlin, Germany, 2014.
2. S.S. Saatchi, N.L. Harris, S. Brown, M. Lefsky, E.T.A. Micthard, W. Salas, B.R. Zutta, W. Buermann, S.L. Lewis, S. Hagen, S. Petrova, L. White, M. Silman, and A. Morel, "Benchmark map of forest carbon stocks in tropical regions across three continents", *PNAS*, vol. 108, no. 24, p. 9899–9904, 2011.
3. UNFCCC, "Conference of the Parties (COP): Adoption of the Paris Agreement, Proposal by the President", United Nations Office, Geneva, Switzerland, 2015.
4. G. Duveiller, P. Defourny, B. Desclée, and P. Mayaux, "Deforestation in Central Africa: Estimates at regional, national and landscape levels by advanced processing of systematically-distributed Landsat extracts", *Remote Sensing of Environment*, vol. 112, no. 5, pp. 1969–1981, 2008.
5. M. Hansen, P. Potapov, R. Moore, M. Hancher, S. Turubanova, A. Tyukavina, D. Thau, S. Stehman, S. Goetz, T. Loveland, A. Kommareddy, A. Egorov, L. Chini, C. Justice, and J. Townshend, "High-resolution global maps of 21st century forest cover change", *Science*, vol. 342, no. 6160, pp. 850–853, 2013.
6. FAO, "The Global Forest Resources Assessment (FRA) 2015. Terms and Definitions", Food and Agriculture Organization of the United Nations, Rome, Italy, 2015.
7. M. Herold, M. Román-Cuesta, V. Heymell, Y. Hirata, P. Van Laake, A.G. Souza, C. Avitabile, and V.K. MacDicken, "A review of methods to measure and monitor historical carbon emissions from forest degradation", *Unasylva*, vol. 62, no. 238, pp. 16–24, 2011.
8. A. Grainger, *Controlling Tropical Deforestation*, 1st ed., London.: Earthscan Publications, 1993, p. 310.
9. R.A. Houghton, "How well do we know the flux of CO2 from land-use change?", *Tellus B*, vol. 62, no. 5, p. 337–351, 2010.
10. T.R.H. Pearson, S. Brown, L. Murray, and G. Sidman, "Greenhouse gas emissions from tropical forest degradation: an underestimated source", *Carbon Balance and Management*, vol. 12, no. 3, pp. 1–11, 2017.
11. A. Ahrends, N.D. Burgess, S.A. Milledge, M.T.F.B. Bulling, J.C. Smart, G.P. Clarke, B.E. Mhoro, and S. Lewis, "Predictable waves of sequential forest degradation and biodiversity loss spreading from an African city", *Proceedings of the National Academy of Sciences*, vol. 107, no. 33, pp. 14556–14561, 2010.

12. P. Mayaux, J.-F. Pekel, B. Desclée, F. Donnay, A. Lupi, F. Achard, M. Clerici, C. Bodart, A. Brink, R. Nasi, and A. Belward, "State and evolution of the African rainforests between 1990 and 2010", *Philosophical Transactions of the Royal Society B*, vol. 368, no. 1625, pp. 1–10, 2013.
13. M. Shimada, T. Itoh, T. Motooka, M. Watanabe, T. Shiraishi, R. Thapa, and R. Lucas, "New global forest/non-forest maps from ALOS PALSAR data (2007–2010)", *Remote Sensing of Environment*, vol. 155, pp. 13–31, 2014.
14. E. Matricardi, D. Skole, M. Cochrane, J. Qi, and W. Chomentowski, "Monitoring selective logging in tropical evergreen forests using Landsat: multitemporal regional analyses in Mato Grosso, Brazil", *Earth Interactions*, vol. 9, no. 24, pp. 1–24, 2009.
15. A. Verhegghen, H. Eva, and F. Achard, "*Assessing forest degradation from selective logging using time series of fine spatial resolution imagery in Republic of Congo*", in *IEEE International Geoscience and Remote Sensing Symposium (IGARSS)*, Milan, Italy, 2015.
16. R. Nasi, A. Billand and N. Vanvliet, "Managing for timber and biodiversity in the Congo Basin", *Forest Ecology and Management*, vol. 26, pp. 103–111, 2012.
17. V. Liesenberg, C.R. de Souza Filho, and R. Gloaguen, "Evaluating moisture and geometry effects on L-band SAR classification performance over a tropical rain forest environment", *IEEE Journal of Selected Topics in Applied Earth Observations and Remote Sensing*, vol. 9, no. 12, pp. 5357–5368, 2016.
18. H.-E. Andersen, S. Reutebuch, R. McGaughey, M. d'Oliveira, and M. Keller, "Monitoring selective logging in western Amazonia with repeat lidar flights", *Remote Sensing of Environment*, vol. 11, pp. 157–165, 2014.
19. S. Cloude, *Polarization Applications in Remote Sensing*, Oxford, UK: Oxford University Press, 2010, p. 352.
20. Izzawati, E.D. Wallington and I.H. Woodhouse, "Forest height retrieval from commercial X-band SAR products", *IEEE Transactions on Geoscience and Remote Sensing*, vol. 44, no. 4, pp. 863–870, 2006.
21. F. Garestier, P.C. Dubois-Fernandez and K.P. Papathanassiou, "Pine forest height inversion using single-pass X-band PolInSAR data", *IEEE Transactions on Geoscience and Remote Sensing*, vol. 46, no. 1, pp. 59–68, 2008.
22. I. Hajnsek, F. Kugler, S.-K. Lee, and K. Papathanassiou, "Tropical-forest-parameter estimation by means of Pol-InSAR: The INDREX-II campaign", *IEEE Transactions on Geoscience and Remote Sensing*, vol. 42, no. 7, pp. 481–493, 2009.
23. H. Balzter, C. Rowland, and P. Saich, "Forest canopy and carbon estimation at Monks Wood National Nature Reserve, UK, using dual-wavelength SAR interferometry", *Remote Sensing of Environment*, vol. 108, no. 3, pp. 224–239, 2007.
24. J.C. Mura, L. Sant'Anna Bins, F.F. Gama, C. da Costa Freitas, J.R. dos Santos, and L. Vieira Dutra, "*Identification of the tropical forest in Brazilian Amazon based on the DEM difference from P- and X-band interferometric data*", in *Proceedings of the IEEE Geoscience and Remote Sensing Symposium*, Sydney, Australia, 2001.
25. T. Neef, V. Dutra, Luciano J.R. dos Santos, C. da Costa Freitas, and L. Spinelli Araujo, "Tropical forest measurement by interferometric height modeling and P-band radar backscatter", *Forest Science*, vol. 51, no. 6, pp. 585–594, 2005.
26. K. Papathanassiou, and S. Cloude, "Single-baseline polarimetric SAR interferometry", *IEEE Transactions on Geoscience and Remote Sensing*, vol. 39, no. 11, pp. 2352–2363, 2001.
27. S. Solberg, "Monitoring forest carbon in a Tanzanian woodland using interferometric SAR: a novel methodology for REDD+", *Carbon Balance and Management*, vol. 10, no. 14, pp. 1–14, 2015.
28. F. White, "The vegetation of Africa: a descriptive memoir to accompany the Unesco/AETFAT/UNSO vegetation map of Africa", Unesco, Paris, France, 1983.
29. World Resources Institute, "Congo Basin Forest Atlases", 2013. [Online]. Available: http://www.wri.org/our-work/project/congo-basin-forests/congo#project-tabs. [Accessed January 20 2016].
30. N.T. Laporte, T.S. Lin, J. Lemoigne, D. Devers and M. Honzák, "Towards an operational forest monitoring system for Central Africa", in *Land Change Science: Observing, Monitoring and Understanding Trajectories of Change on the Earth's Surface*, G. Gutman, A.C. Janetos, C.O. Justice, E.F. Moran, J.F. Mustard, R.R. Rindfuss, D. Skole, B.L. Turner, and M. A. Cochrane, editors., The Netherlands: Kluwer Academic Publishers, 2004, pp. 97–110.
31. M.C. Peel, B.L. Finlayson, and T.A. McMahon et al., "Updated world map of the Koppen–Geiger climate classification", *Hydrology and Earth System Sciences Discussions*, vol. 4, no. 2, pp. 439–473, 2007.

32. World Bank, "Climate Change Knowledge Portal", 2016. [Online]. Available: http://sdwebx.worldbank.org/climateportal/countryprofile/home.cfm?page=country_profile&CCode=COG&ThisTab=ClimateBaseline. [Accessed November 15 2016].
33. A. Jones, H. Breuning-Madsen, M. Brossard, A. Dampha, J. Deckers, O. Dewitte, T. Gallali, S. Hallett, R. Jones, M. Kilasara, P. Le Roux, E. Micheli, L. Montanarella, O. Spaargaren, L. Thiombiano, E. Van Ranst, M. Yemefack, and R. Zougmore, *Soil Atlas of Africa*, 1st ed, Luxembourg: European Commission, Publications Office of the European Commission, 2013, p. 176.
34. Sarmap, "SARscape Guide", May 20, 2016. [Online]. Available: https://www.harrisgeospatial.com/docs/pdf/sarscape_help.pdf. [Accessed May 20 2016].
35. R.M. Goldstein, and C.L. Werner, "Radar interferogram filtering for geophysical applications", *Geophysical Research Letters*, vol. 25, no. 21, pp. 4035–4038, 1998.
36. J. Neyman, and E.S. Pearson, "On the problem of the most efficient tests of statistical hypotheses", *Philosophical Transactions of the Royal Society of London. Series A. Containing Papers of a Mathematical or Physical Character*, vol. 231 (694–706), pp. 289–337, 1933.
37. S.M. Kay, *Fundamentals of Statistical Signal Processing, Volume II: Detection Theory*, 1st ed., Englewood Cliffs, NJ, USA: Prentice Hall, 1998, p. 672.
38. M. Wermuth, R. König, Y. Moon, J.M.W. Antony, and O. Montenbruck, "Two years of TanDEM-X baseline determination", *International Journal of Space Science and Engineering*, vol. 2, no. 1, pp. 35–48, 2014.
39. H. Balzter, "Forest mapping and monitoring with interferometric synthetic aperture radar (InSAR)", *Progress in Physical Geography*, vol. 25, no. 2, pp. 159–177, 2001.
40. A. Aweto, *Shifting Cultivation and Secondary Succession in the Tropics*, 1st ed., Ibadan, Nigeria: CABI, 2012, p. 208.
41. T.R.H. Pearson, S. Brown, and F.M. Casarim, "Carbon emissions from tropical forest degradation caused by logging", *Environmental Research Letters*, vol. 9, no. 3, pp. 1–11, 2014.
42. Wolfram, "Morphological Components", 2016. [Online]. Available: https://reference.wolfram.com/language/ref/MorphologicalComponents.html. [Accessed June 18 2016].
43. P. Soille and M. Pesaresi, "Advances in mathematical morphology applied to geoscience and remote sensing", *IEEE Transactions on Geoscience and Remote Sensing*, vol. 40, no. 9, pp. 2042–2055, 2002.
44. B. Picinbono, *Random Signals and Systems*, 1st ed., Englewood Cliffs, NJ, USA: Prentice Hall, 1993.
45. C. Oliver, "The interpretation and simulation of clutter textures in coherent images", *Inverse Problems*, vol. 2, pp. 481–518, 1996.
46. J.-S. Lee, K. Mango, and A. Miller, "Intensity and phase statistics of multilook polarimetric and interferometric SAR imagery", *IEEE Transactions on Geosciences and Remote Sensing*, vol. 32, no. 5, pp. 1017–1028, 1994.
47. R. Touzi, and A. Lopez, "Statistics of the Stokes parameters and of the complex coherence parameters in one-look and multilook speckle fields", *IEEE Transactions on Geoscience and Remote Sensing*, vol. 34, no. 2, pp. 519–531, 1996.
48. J. Cohen, "A power primer", *Psychological Bulletin*, vol. 112, no. 1, pp. 155–159, 1992.
49. S. Nakagawa, and I.C. Cuthill, "Effect size, confidence interval and statistical significance: a practical guide for biologists", *Biological Reviews*, vol. 82, no. 4, pp. 591–605, 2007.
50. G.M. Sullivan, and R. Feinn, "Using effect size – or why the p value is not enough", *Journal of Graduate Medical Education*, vol. 4, no. 3, pp. 279–282, 2012.
51. P. Peralta, and M. Mather, "An analysis of deforestation patterns in the extractive reserves of Acre, Amazonia from satellite imagery. A landscape ecological approach", *International Journal of Remote Sensing*, vol. 21, pp. 13–14, 2000.
52. B. Mandelbrot, *The Factal Geometry of Nature*, New York, USA: W.H. Freeman and Company, 1982.
53. R. Treuhaft, F. Gonçalves, J. R. dos Santos, M. Keller, M. Palace, S.N. Madsen, F. Sullivan, and P.M.L. Graça, "Tropical-forest biomass estimation at X-band from the spaceborne TanDEM-X interferometer", *IEEE Geoscience and Remote Sensing Letters*, vol. 12, no. 2, pp. 239–243, 2015.
54. D.H. Hoekman, and C. Varekamp, "Observation of tropical rain forest trees by airborne high-resolution interferometric radar", *IEEE Transactions on Geoscience and Remote Sensing*, vol. 39, no. 3, pp. 584–594, 2001.
55. J. Praks, O. Antropov, and M. Hallikainen, "LiDAR-aided SAR interferometry studies in boreal forest: scattering phase center and extinction coefficient at X- and L-band", *IEEE Transactions on Geoscience and Remote Sensing*, vol. 50, no. 10, pp. 3831–3843, 2012.

56. V. Gond, A. Fayolle, A. Pennec, G. Comu, P. Mayaux, P. Camberlin, C. Doumenge, N. Fauvet, and S. Gourlet-Fleury, "Vegetation structure and greenness in Central Africa from Modis multi-temporal data", *Philisophical Transactions of the Royal Society B*, vol. 368, no. 20120309, pp. 1–8, 2013.
57. J. Maley, "A catastrophic destruction of African forest about 2,500 years ago still exerts a major influence on present vegetation formations", *IDS Bulletin*, vol. 33, no. 1, pp. 13–30, 2002.
58. N. Philippon, B. de Lapparent, V. Gond, G. Sèze, N. Martiny, P. Camberlin, G. Cornu, B. Morel, V. Moron, S. Bigot, T. Brou, and V. Dubreuil, "Analysis of the diurnal cycles for a better understanding of the mean annual cycle of forests greenness in Central Africa", *Agricultural and Forest Meteorology*, vol. 223, pp. 81–94, 2016.
59. GIOVANNI, "GIOVANNI", 2016. [Online]. Available: http://giovanni.sci.gsfc.nasa.gov/. [Accessed March 2 2016].
60. Astrium, "Pléiades Imagery User Guide", October 2012. [Online]. Available: http://www.cscrs.itu.edu.tr/assets/downloads/PleiadesUserGuide.pdf. [Accessed September 26 2016].
61. S. Solberg, R. Astrup and D.J. Weydahl, "Detection of forest clear-cuts with shuttle radar topography mission (SRTM) and TanDEM-X InSAR data", *Remote Sensing*, vol. 5, no. 11, pp. 5449–5462, 2013.

Appendix
A Wavelet Tour

Wavelets are ubiquitous in the book, and they appear under many guises. In Chapter 2 they appear as orthonormal bases and provide multi-resolution approximations of radar mosaics. Later, in Chapter 6, they are a non-orthogonal basis and work as yardsticks for texture measures. Then in Chapter 7 the trajectory of wavelet maxima in the scale domain are of importance for characterizing discontinuities in radar imagery. In Chapter 8, the possibility of inverting the wavelet representation is exploited together with the maxima trajectories to build smooth, but edge preserving, approximations of the radar image. Finally, in Chapter 9 wavelets appear in a noise-killing role.

Such a dense participation in the drama warrants at least some sketchy presentation of the actors. That's why we propose here to embark together in a quick tour of wavelet signal processing.

The quick tour will try only to explain principles in an intuitive manner, in such a way as to lay the groundwork for understanding the application. The reader will therefore forgive us if the tour looks more like a visit to Disneyland rather than a serious tutorial on the topic. Theory and application of wavelets covers a vast area of knowledge, and a rigorous and exhaustive treatment of the subject is outside the scope of this tour. References are given for the interested reader who would like to explore the subject more deeply.

A.1 SIGNAL REPRESENTATION IN A BASIS

Let us start by considering one fundamental problem that underpins several aspects of theory and practice in signal processing: the representation of a mathematical function in terms of a family of elementary functions, called a basis. A popular toy – the LEGO construction kit – suggests by analogy an intuitive view of the problem (see Figure A.1). A given LEGO object can be constructed – and decomposed – into a series of elementary plastic bricks of different shapes and colors. To do that, we need enough pieces and sufficient elementary shapes and colors to satisfy all the geometric, textural and color rendering requirements of the object we want to build. This is what we call a "complete" representation.

Second, we need to make sure that the laws of statics are satisfied so that our construction does not tumble down. This leads to the second property of our representation: stability. A complete and stable representation would allow us to decompose our object into elementary pieces and reconstruct it starting from the elementary bricks.

Now, if we use all the pieces, we will reassemble an exact copy of the object. We will achieve full reconstruction. On the other hand, if we pick out some pieces, or modify the shape or color of some elementary bricks, we will obtain an approximation of the original object.

Mathematically, this type of representation – analogous to the LEGO toy construction – is studied in the abstract setting of linear vector spaces [1]. Here, our elementary building blocks – the so-called basis – are composed of a set of particular vectors that are: (1) linearly independent and (2) every other vector in the space can be written as a linear combination of the basis vectors (we say that the basis spans the space).

Translating this recipe with our analogy: (1) each brick in our LEGO set has a unique shape and color and cannot be represented in terms of the other ones; and (2) the objects we want to construct (e.g. a house) can be assembled by gathering so many pieces of this brick type, so many pieces of that one, and putting them together.

A particular type of vector space allows us to define a scalar product of two vectors, this being an operation that tells us something about the length of the vectors, but also on the relative orientation,

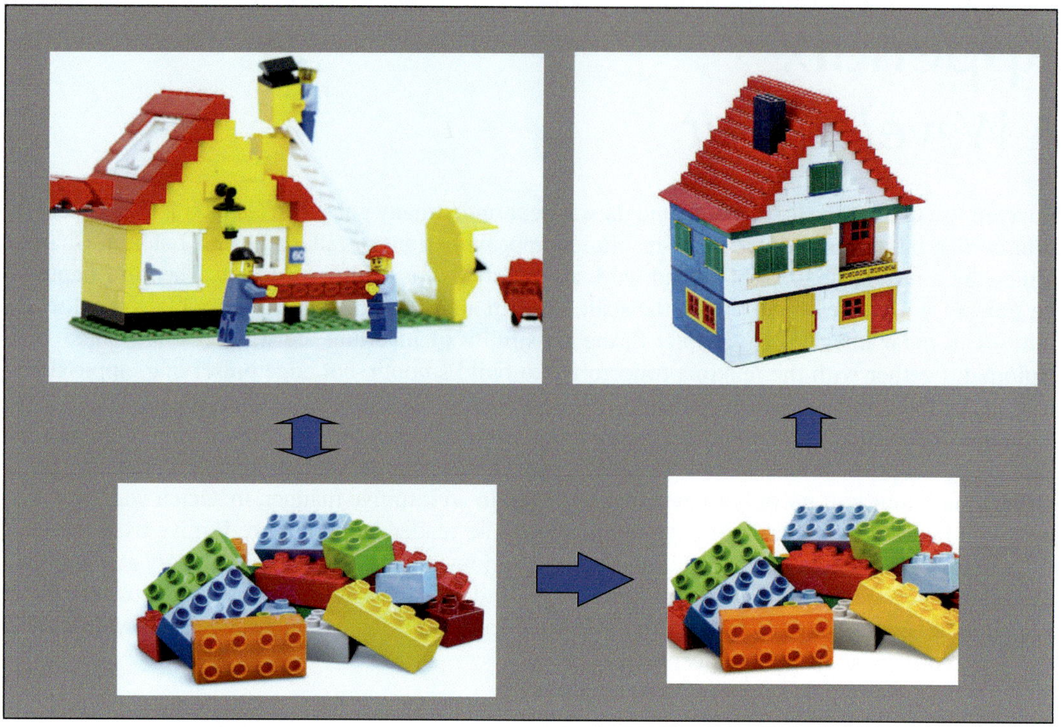

FIGURE A.1 An intuitive view of the signal decomposition into a basis and the signal reconstruction from its projections into the basis.

speaking in terms of the intuitive Euclidean space. Then a basis where all components have unitary length and are mutually perpendicular is called an orthonormal basis.

More precisely, and to be mathematically more serious, we are dealing here with linear spaces, these being vector spaces that obey laws of good conduct. Basically, when you add more elements and multiply by a scalar then you end up with a vector in the same space (see [1], for a rigorous definition).

A linear space becomes an inner product space when there exists a unique scalar (symbolized $<x, y>$) that encapsulates the relationship between two vectors in the space, and that obeys linear laws of summation and scalar multiplication:

$$\langle \alpha x + \beta y, z \rangle = \alpha \langle x, z \rangle + \beta \langle y, z \rangle$$

For example, in the Euclidean space the inner product begets a geometrical meaning:

$$\langle \vec{x}, \vec{y} \rangle = |xy| \cos \vartheta$$

where ϑ is the angle between the two vectors. For $\langle \vec{x}, \vec{y} \rangle = 0$, the vectors are perpendicular. For collinear vectors $\langle \vec{x}, \vec{y} \rangle = |xy|$. $\sqrt{\langle \vec{x}, \vec{x} \rangle} = |x|$ gives the length of the vector $\vec{x}$.

A.2 THE FOURIER KINGDOM

One good and important example of orthonormal basis in a linear inner-product space is afforded by the trigonometric functions [2]. They are a complete orthonormal basis with respect to real valued functions that are piecewise continuous in the interval $0 - 2\pi$. Since it is a complete basis, any function in the space can be expanded in a series of the basis functions. This series – point-wise convergent – is the Fourier series:

$$f(x) = \frac{1}{2}A_0 + \sum_{k=1}^{\infty}\left(A_k \cos kx + B_k \sin kx\right)$$

The coefficients in the expansion are given by:

$$A_k = \frac{1}{\pi}\int_{0}^{2\pi} f(x)\cos kx\, dx$$

$$B_k = \frac{1}{\pi}\int_{0}^{2\pi} f(x)\sin kx\, dx$$

Now the question is, why do we want to decompose our functions? What is the practical use of this effort?

To get an idea, let us have a look at our basis functions in the Fourier series. They look like the one in Figure A.2(a) (Fourier atom). They are like waves that wiggle up and down for ever, from minus infinity to plus infinity.

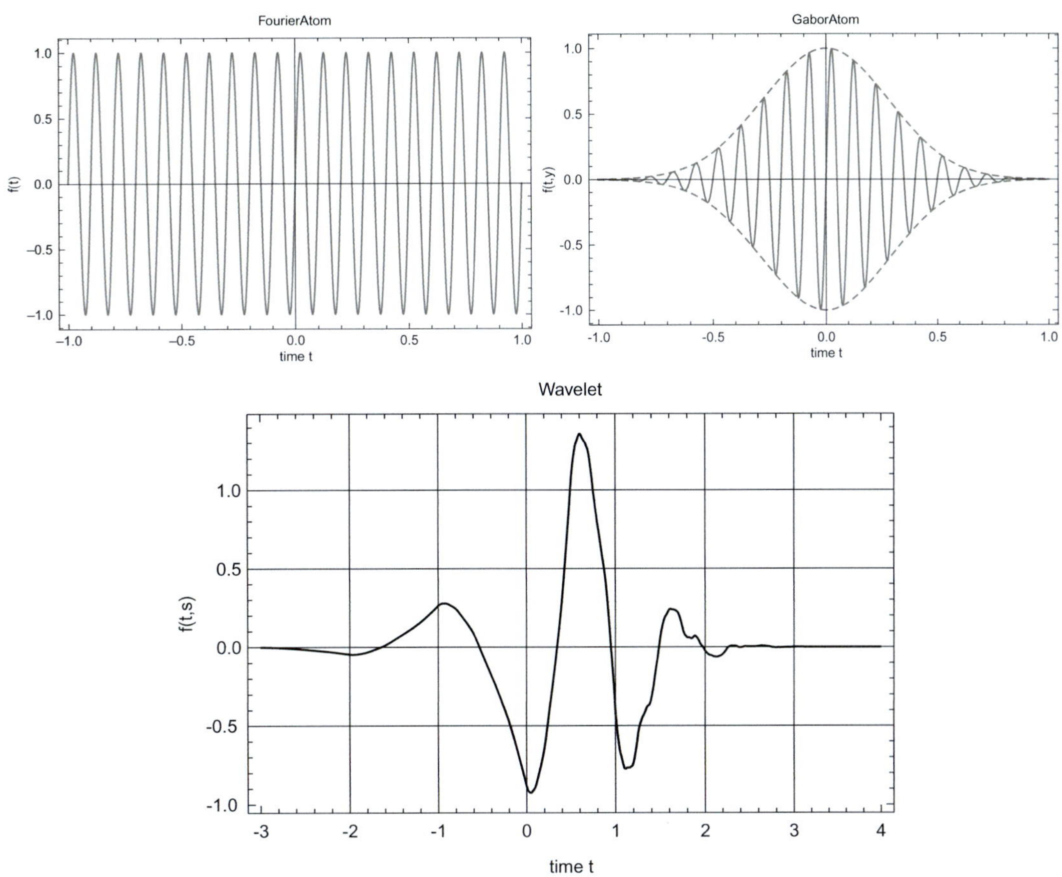

FIGURE A.2 (a) Fourier atom; (b) a Gabor atom; and (c) a wavelet.

Now suppose our function represents some physical quantity, such as an electrical signal. This signal can have quite complicated behavior over time. What we are doing with the Fourier expansion is to break down this complicated signal into elementary components that are far easier to handle: they are periodic functions, or waves, oscillating at a specific frequency. Since we now know how much of each wave is needed to build the original signal – the series coefficients – we have been able to analyze the signal into its frequency components. This is called frequency analysis.

Number two, suppose now to have a linear electrical device with two ports, input and output, a black box whose behavior with respect to some external excitation is not known, except for the fact that the inner components are linear devices. Now if we apply at the input port a series of sinusoidal waves we can record the output, which because of linearity will also be a sinusoidal wave of different amplitude and phase. All these readings will fully characterize the response of our device in the frequency domain. Thanks to the Fourier decomposition, then, we will be able to predict the behavior of our device when any type of complicated periodic excitation is applied to the input port.

The dual problem – synthesis – can also be tackled based on Fourier analysis. Namely, we can construct an arbitrary periodic signal summing up in the proper proportion sinusoidal signals.

This type of analysis–synthesis approach can be applied not only to electrical devices but to any linear physical system, pointing to the great practical importance of the method.

Moving to other vector spaces, the space of integrable function $L^1(\mathfrak{R})$ and the space of square integrable or finite energy functions $L^2(\mathfrak{R})$, we find a relative of the Fourier series, which is the Fourier integral. The analysis part – the Fourier transform – is defined mathematically as:

$$\hat{f}(\omega)=\int_{-\infty}^{+\infty} f(t)e^{i\omega t}dt$$

If $f(\omega)$ is integrable, then the inverse Fourier transform is given by:

$$f(t)=\frac{1}{2\pi}\int_{-\infty}^{+\infty}\hat{f}(\omega)e^{i\omega t}d\omega$$

The inverse transform tells us that the signal is decomposed into sinusoidal waves where the frequency changes continuously.

A.3 EXTENSION TO LINEAR TRANSFORMS WITH MORE INTERESTING ATOMS – WHERE WAVELETS FINALLY APPEAR

As a further extension of this concept, we can define a linear transform where the elementary building blocks – the atoms – are more general functions than the sinusoidal waves of the Fourier kingdom. This transform [3] can be written in the setting of a Hilbert space – a particular type of linear vector space – as a scalar product like this:

$$Tf(\gamma)=\int_{-\infty}^{+\infty} f(t)\varphi_\gamma^*(t)dt=\langle f,\varphi_\gamma\rangle \tag{A1}$$

Now we have the flexibility of choosing the shape of the atoms in such a way that their time–frequency behavior becomes more interesting. This flexibility is summarized in the – possibly – multi-valued index γ.

More interesting in which sense? The Fourier atoms are waves of infinite duration. Therefore they tell us something about the frequency – with infinite resolution – but nothing about the position in time – zero resolution. If there is an event at some time, this will affect a number of sinusoids with different frequencies, but its position in time will be washed out from minus infinity to plus infinity. There is no time localization.

We would like to have a decomposition that establishes a compromise between our ability to see in time and in frequency. This can be accomplished, for example, by windowing the Fourier sinusoids by a function that limits their existence in time. We obtain in this case the Gabor atoms (see Figure A.2b):

Wavelets are a further improvement in this direction [3–7]. As the word says, they are functions that oscillate with zero mean value and fast decay in time (Figure A.2). Mathematically, they are functions belonging to the space of square integrable functions $\psi \in L^2(\Re)$ that satisfy the so-called admissibility condition:

$$C_\psi = 2\pi \int |\omega|^{-1} |\hat{\psi}(\omega)|^2 d\omega < \infty \tag{A2}$$

where $\hat{\psi}(\omega)$ is the Fourier transform of ψ.

If $\boldsymbol{\psi}$ is integrable then this condition can only be satisfied if $\hat{\psi}(0) = 0$ or:

$$\int \psi(t) dt = 0$$

The fact that the Fourier transform is zero at zero frequency, and decays at high frequencies, indicates that the wavelet behaves like a band-pass filter.

To cover the time–frequency plane, the wavelet functions are translated and dilated. We generate the following family of analysis atoms that are dependent on two indexes, translation b and dilation a:

$$\psi^{a,b}(t) = \frac{1}{\sqrt{|a|}} \psi\left(\frac{t-b}{a}\right)$$

Using this atom, our scalar product (A1) becomes:

$$Tf(a,b) = \langle f, \psi^{a,b} \rangle = \int f(t) \frac{1}{\sqrt{|a|}} \overline{\psi\left(\frac{t-b}{a}\right)} dt \tag{A3}$$

Our signal is therefore analyzed by similarities with translated and dilated version of the basic wavelet. This is what is called the continuous wavelet transform (CWT).

It can be proven [5] that a function $f(t)$ can be recovered from its wavelet transform through the so-called resolution of identity, which can be written as:

$$f(t) = \frac{1}{C_\psi} \int_{-\infty}^{+\infty} \int_{-\infty}^{+\infty} Tf(a,b) \psi^{a,b} \frac{da\, db}{a^2} \tag{A4}$$

This is a fundamental fact: the wavelet transform provides a complete and stable representation and our function can be fully reconstructed from its representation in the wavelet domain. Looking at the inversion Equation (A4), we now realize the importance of the admissibility condition (A2). If C_ψ were infinite, no reconstruction would be possible.

Let us now go back to the time–frequency localization problem, and see what we gain from using wavelet atoms in our decomposition. More precisely, what we mean by better time–frequency localization?

An important theorem in the theory of Hilbert spaces states that scalar products and norms are conserved by the Fourier transform up to a factor of 2π. Namely:

$$\int_{-\infty}^{+\infty} f(t)h^*(t)dt = \frac{1}{2\pi}\int_{-\infty}^{+\infty} \hat{f}(\omega)\hat{h}^*(\omega)d\omega$$

$$\int_{-\infty}^{+\infty} |f(t)|^2 dt = \frac{1}{2\pi}\int_{-\infty}^{+\infty} |f(\omega)|^2 d\omega$$

These are called the Parseval and Plancherel relationships. From the physical point of view – if the function is associated, for example, to an electrical signal – the fact that the norm is conserved means that all the energy in the signal goes into the Fourier coefficients, or in other words in the Fourier representation the basis is complete, and nothing gets lost.

Now application of the Parseval relationship to our time–frequency transform proves that:

$$Tf(\gamma) = \int_{-\infty}^{+\infty} f(t)\varphi_\gamma^*(t)dt = \frac{1}{2\pi}\int_{-\infty}^{+\infty} \hat{f}(\omega)\hat{\varphi}_\gamma^*(\omega)d\omega$$

This equation tells us that if our atom is zero outside a certain region around a point t_0 in time (the so called support of the atom) then the scalar product $\langle f, \varphi_\gamma \rangle$ will depend only on values of f within that region around t_0. Symmetrically, if the atom in frequency falls off to zero around a point f_0 then the scalar product will depend only on values of the Fourier transform of f in the neighborhood of frequency f_0. Therefore, indeed, the time–frequency atom is able to give us information about the signal that is localized in bothtime and frequency. In other words, that transform is able to tell us when certain events with specific characteristics in frequency do occur. Typical cases of interest are non-stationary signals, or transients, where a sudden burst of high frequency activity (e.g. a spike) occur at certain points in time. Gabor atoms and wavelets open up an entirely new perspective for this type of analysis.

Back to Equation A1: let us now characterize more precisely and in quantitative terms the time–frequency localization provided by a certain atom. This is done through the concept of Heisenberg boxes. A Heisenberg box delimits a region centered around a point on the time–frequency plane where the information given by the scalar product $\langle f, \varphi_\gamma \rangle$ is available. This measure of the time–frequency localization rests on a fundamental principle in nature, the uncertainty principle. In a nutshell, this principle states that there are pairs of physical variables that we cannot measure with equal infinite accuracy. Zooming in on one variable with more accuracy will make the measurement of the coupled variable sloppier. This principle has important consequences, for example, in quantum mechanics, where the coupled variables are position and momentum. We cannot know exactly where a particle is and its velocity at the same time.

In signal processing, the principle applies to our capability to devise a function that is localized in time and frequency. The Heisenberg theorem in this case states that the temporal variance and the frequency variance satisfy:

$$\sigma_t^2 \sigma_\omega^2 \geq \frac{1}{4}$$

where σ_t^2 is the temporal variance and σ_ω^2 is the frequency variance.

How are the position and the spread of the Heisenberg boxes measured? In our context, $|\varphi_\gamma(t)|^2$can be interpreted as a probability density function since

$$\|\varphi_\gamma\|^2 = \int_{-\infty}^{+\infty} |\varphi_\lambda(t)|^2 dt = 1$$

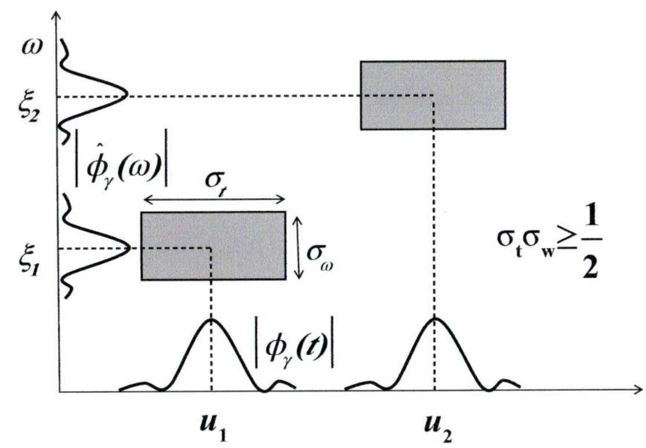

FIGURE A.3 Heisenberg boxes of windowed Fourier atoms.

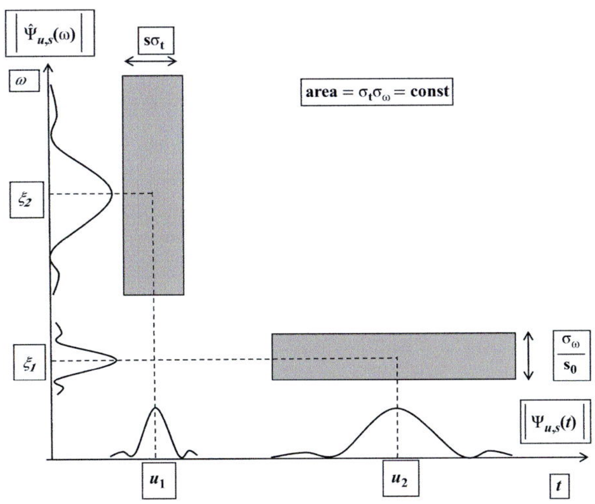

FIGURE A.4 Heisenberg boxes of wavelet atoms.

Then we take the mean value as the position of the box:

$$u_\gamma = \int_{-\infty}^{+\infty} t\left|\varphi_\gamma(t)\right|^2 dt$$

and the variance as a measure of the time–frequency spread (see Figure A.3):

$$\sigma_t^2(\gamma) = \int_{-\infty}^{+\infty} (t-u_\gamma)^2\left|\varphi_\gamma(t)\right|^2 dt$$

Armed with these definitions, let us now see how different atoms behave with respect to time–frequency localization.

For a windowed Fourier atom, we have a situation like the one depicted in Figure A.3. The size of the Heisenberg box is constant and independent of position and frequency. Therefore this transform always has the same resolution in time and frequency across the whole plane.

This is not very desirable, because we would like to have an increased resolution in time to capture high frequency events, and lower resolution in time when dealing with low frequency events. This type of adaptive behavior is afforded by wavelets. Indeed, for a wavelet atom the situation is like the one portrayed in Figure A.4.

Here $\sigma_t \sigma_\omega$ are the standard deviations of the wavelet atom. Now the areas – not the sizes – of the Heisenberg boxes are constant. The resolution in time is directly proportional to the scale s, while the resolution if frequency is inversely proportional to the scale s. Therefore, for low frequency events, the time support gets larger; and for events with a high frequency content the time support becomes smaller. The product of the time and frequency resolution is a constant and this property is called "constant Q" filtering. The window width automatically adapts to the frequency behavior of the portion of the signal that is analyzed.

This characteristic makes wavelets a by far superior tool for time–frequency analysis of a class of signals (e.g. non-stationary) that exhibit a large variation in the frequency spectrum with respect to where in time the analysis is carried out.

A.4 THE WAVELET TRANSFORM IN A DISCRETE TIME SETTING

So far, we have considered functions that are continuous with respect to time. However, we live in a digital world, where sooner or later analogue signals must be sampled and fed into a computer for processing. The passage to the digital world calls for a discretization of the translation and scale parameters in our wavelet.

A natural choice for the discretization of scale is to set $a = a_0^m$ with $m \in Z$ the space of integer numbers. If $a_0 = 2$ we have the so-called dyadic scales. For translation b we can set $b = nb_0$ where with the step b_0 we have to make sure to cover the whole range of values on the time axis.

Now if the wavelet is dilated by a_0^m and becomes $\psi\left(a_0^{-m}t\right)$ then it makes no sense to translate it by fine steps as is done for $\psi(t)$. All points can be covered by the dilated wavelet letting $\psi\left(a_0^{-m}t\right)$ with $a_0 > 1$ and $b_0 > 0$ fixed.

We therefore obtain the following discretization of the wavelet parameters:

$$\psi^{m,n}(t) = a_0^{-m/2}\psi\left(\frac{t - nb_0 a_0^m}{a_0^m}\right)$$

Then our function $f(t)$ would be represented by a set of discrete coefficients given by the inner product:

$$\left\langle f, \psi^{m,n} \right\rangle = \int f(t)\overline{\psi^{m,n}}(t)\,dt \tag{A5}$$

This transform is called in the literature a wavelet series.

The question is: does this discrete setting provide a complete and stable representation of our function, allowing for full reconstruction from the inner product coefficients?

As pointed out in [5], good characterization of the function intuitively implies that if a sequence of coefficients $(\langle f_1, \psi^{m,n} \rangle)$ is "close" to a sequence $(\langle f_2, \psi^{m,n} \rangle)$ then f_1 must also be close to f_2.

If a space is equipped with a norm, the closeness of points is measured by this device. In the case of Hilbert spaces, the norm is induced by the inner product. Now a transform like the one in Equation (A5) maps a space $L^2(\mathfrak{R})$ into a space of discrete square summable sequences $l^2(Z)$. In a similar way, we can measure the distance between two sequences using their norm by:

$$\left\|c^1 - c^2\right\|^2 = \sum_{m,n \in Z} \left|c_1^2 - c_2^2\right|^2$$

But to use this distance we must assume that the sequences $(\langle f, \psi^{m,n}\rangle)$ are themselves in $l^2(Z)$ or, $\sum|\langle f, \psi^{m,n}\rangle|^2 < \infty$.

In reality, it turns out that for any good wavelet with good decay in time and frequency this condition is always satisfied, because:

$$\sum\left|\left\langle f, \psi^{m,n}\right\rangle\right|^2 \leq B\|f\|^2 \tag{A6}$$

What about numerical stability of the reconstruction algorithm? In this case we would require that when $\sum|\langle f, \psi^{m,n}\rangle|^2$is small, $\|f^2\|$ is also small, or seen the other way around, we want to make sure that when there is small energy in the signal, the wavelet coefficients do not go astray, but do also carry small energy. This condition leads to:

$$A\|f\|^2 \leq \sum\left|\left\langle f, \psi^{m,n}\right\rangle\right|^2 \tag{A7}$$

Putting together condition Equation (A5), which imposes on the sequences to be square summable, with condition Equation (A6), which refers to a stability requirement, we have:

$$A\|f\|^2 \leq \sum\left|\left\langle f, \psi^{m,n}\right\rangle\right|^2 \leq B\|f\|^2 \tag{A8}$$

This condition is equivalent in the discrete setting to the resolution of identity in the continuous setting. Now a set of functions $\psi^{m,n}(t)$ satisfying condition Equation (A8) for any f in a Hilbert space constitute a frame. Frames are cousins of bases (and orthonormal bases). But, contrary to a basis that is made up of a set of linearly independent members, the elements of a frame are not independent. There is a certain redundancy. The frame uses more elements to describe a function than is strictly necessary.

However, this waste of resources gives us a gain in other ways. For example, it turns out that frames can be described very sparingly in digital representations. A few bits are often sufficient to store the representation of a signal, a fact that is clearly appealing for data coders. Frames lead to overcomplete decompositions that are translation invariant, a definite advantage in applications such as pattern recognition and image classification.

An example of a frame [5] in a two-dimensional vector space is given in Figure A.5. Define the vectors:

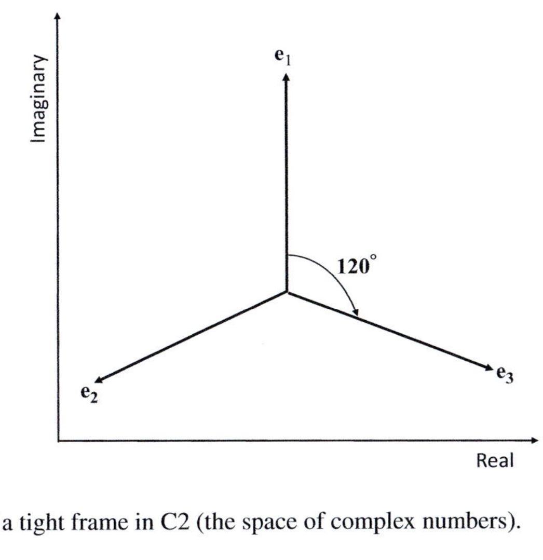

FIGURE A.5 Example of a tight frame in C2 (the space of complex numbers).

$$e_1 = (0,1), \quad e_2 = \left(-\frac{\sqrt{3}}{2}, -\frac{1}{2}\right), \quad e_3 = \left(\frac{\sqrt{3}}{2}, -\frac{1}{2}\right)$$

These vectors are not linearly independent (e.g. $e_1 = -(e_2 + e_3)$). Let us see if they satisfy the frame condition in Equation (A8). For any vector v we have:

$$\sum_{j=1}^{3} \left|\langle v, e_j \rangle\right|^2 = |v_2|^2 + \left|-\frac{\sqrt{3}}{2} v_1 - \frac{1}{2} v_2\right|^2 + \left|\frac{\sqrt{3}}{2} v_1 - \frac{1}{2} v_2\right|^2 = \frac{3}{2}\|v\|^2$$

The condition is therefore satisfied with $A = B = \frac{3}{2}$. In this case the frame is said to be tight and it is demonstrated in [5] that f can be reconstructed from the frame by:

$$f = \frac{1}{A}\sum_j \langle f, \psi_j \rangle \psi_j$$

When the frame is not tight, the situation becomes a bit more complicated. One needs to introduce another family of vectors, the dual frame, given by:

$$\tilde{\psi}_j = \left(F^* F\right)^{-1} \psi_j \tag{A9}$$

where $F = \langle f, \psi_j \rangle$ is the frame operator, F^* is its adjoint.

Then the reconstruction formula becomes:

$$f = \sum_j \langle f, \psi_j \rangle \tilde{\psi}_j \tag{A10}$$

So far we have left the time variable still continuous in our frame representation. Discretizing the integral in (A5) by natural sampling of the wavelet frame $\psi(t) \to \psi[n]$ we obtain:

$$\mathbf{W}^{m,n} = \frac{\mathbf{1}}{\sqrt{\mathbf{a}_0^m}} \sum_k f_k \psi\left(\frac{\mathbf{k} - \mathbf{n}\mathbf{b}_0 \mathbf{a}_0^m}{\mathbf{a}_0^m}\right) \tag{A11}$$

This is called in the literature the discrete wavelet transform.

A.5 COMPUTING THE WAVELET FRAME TRANSFORM: THE "À TROUS" ALGORITHM

When the input signal is sampled, we could then use Equation (A11) to compute the wavelet frame coefficients, and similarly use Equation (A9) and Equation (A10) to reconstruct f from the wavelet frame coefficients. However, this procedure is numerically quite heavy. Indeed, as the scale a^m grows, many more samples of f are required for the computation of the wavelet coefficients.

There is a much more elegant way of computing the wavelet frame representation in the discrete settings. It is an algorithm based on an iterated filter bank, called algorithm "à trous", from the French word that means with holes [8] [6,9].

A beautiful exposition of the algorithm is given in [8], where the author establishes a unified framework that encompasses both wavelet frames and Mallat's multi-resolution analysis (see below).

The basic tenet of the algorithm is to approximate non-integer points in $\psi(t)$ through interpolation by a finite filter h^+. Call g_n the filter obtained by natural sampling of a time-reversed wavelet $\psi(-t)$: $g_n = \psi(-n)$. First, $g_n$is spread apart, leaving space for the points to be interpolated. Next the filter $h^+$is applied to leave the even points fixed and capture the interpolated odd points. Thus an à trous filter satisfies:

$$h_{2k} = \delta(k)/\sqrt{2}$$

where $\sqrt{2}$ is a normalization factor (we assume here we are dealing with a dyadic transform where the scaling factor $a_0 = 2$).

The result of the entire interpolation is that we get an approximation of $2^{-1/2}\psi(n/2)$. Plugging translates of this approximation into Equation (A11) will allow us to compute an approximation of $W(2, 2n)$; that is, the wavelet coefficients at the first level in the decomposition.

Further levels of decomposition require the application the same procedure to an approximation of the function $f$at the next scale. The reader is referred to [8] for the details. Suffice it here to say that, given an à trous filter h and a band-pass filter $g$derived by sampling the wavelet frame, the wavelet frame decomposition can be implemented by the iterated filter bank scheme shown in Figure A.6.

A wavelet frame decomposition implemented by an à trous algorithm is at the core of the texture measures illustrated in Chapter 6.

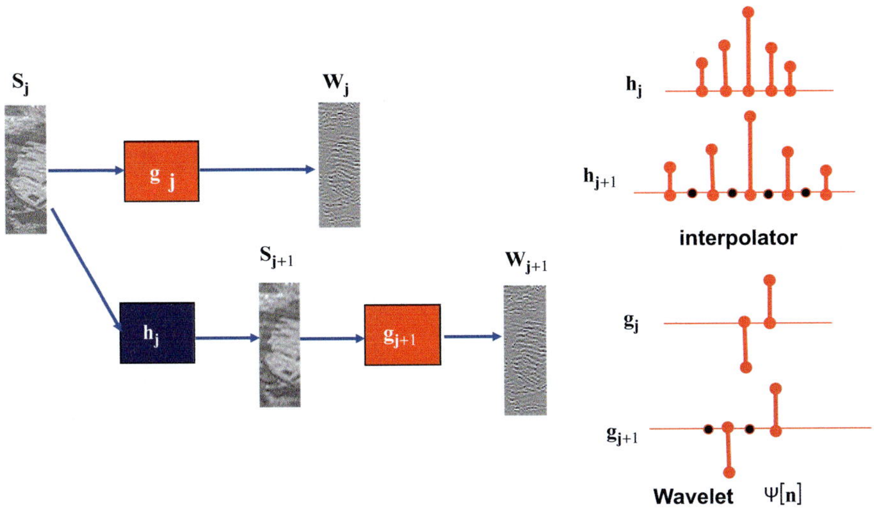

FIGURE A.6 The à trous algorithm implemented as a filter bank.

A.6 THE MULTIRESOLUTION WAVELET REPRESENTATION

Having introduced wavelets as the ultimate step in the evolutionary path of time–frequency analysis, let us now change the standpoint, and look at yet another area where wavelets play a fundamental role: multi-resolution representations [10]. Whereas, in what we said before, the historical roots stemmed from the classical Fourier analysis, multi-resolution analysis has its roots in visual perception studies and in the area of image processing and computer vision.

Multiresolution is the foundation over which our techniques for the radar mosaic product generation and classification are based.

In the vision process (human or machine-based) information that can be extracted from an image depends on the field of view or on the amount of detail that is taken into consideration. In other words, as the local neighborhood that is considered (scale) is enlarged or shrunk, the aggregation of image tones within the neighborhood carries different meaning to the observer.

For example, if we observe an entire continent using a very large field of view and coarse resolution, we can infer from the shape of the coastlines, maybe from the major rivers and mountain ridges, that we are dealing with a specific continent or part of the world. However, precise information on the details – for example, on vegetation patterns, or on town structure and extent – is completely lost. To retrieve information on these features we must restrict our field of view and increase the resolution.

To extract richer information, we therefore need an analysis tool that behaves like a virtual microscope, zooming in and out at several resolutions. Wavelets provide a powerful mathematical setting for such a tool.

Let us begin once again from an intuitive view to gain an understanding of such a tool. Suppose to have an object, for example, an apple, which lives in a certain world, the world of apples. For reference, we call this world the space V_0 (see Figure A.7). Starting from our apple we now create a new object, peeling the apple and leaving only the core of the fruit. The new object, the peeled apple, lives now in a different world that we call space V_1.

In passing from one world to the next, we have lost part of the original object, the skin. We can store this part in a third world, that we call space W_0. The process of taking an object and generate a new representation of the object where we lose some detail is called decomposition. By analogy, in multi-resolution analysis we describe an image or a signal at a certain resolution, then degrade the resolution, and obtain a new description of the image. In this passage we will lose some details of the original image that we save into another description. As will be explained in a moment, wavelets are the mathematical tool that allows us to extract the details that are lost in moving from one resolution level to the next.

Now coming back to our world of apples, suppose we would like to reconstruct the original full apple starting from the peeled one. We could do so by putting together the apple in world V_1 with

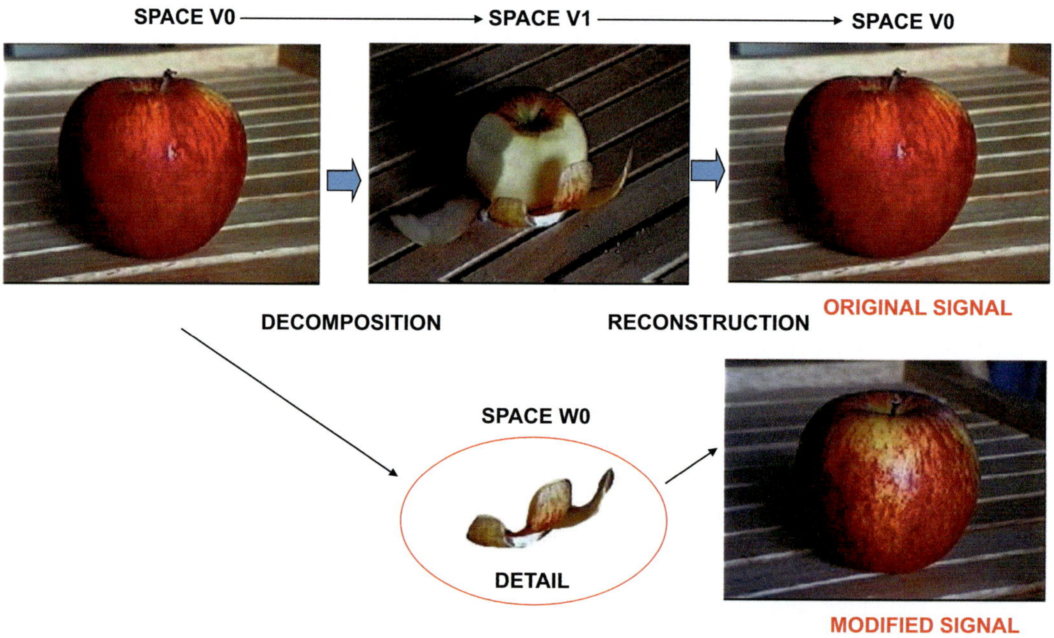

FIGURE A.7 Multi-resolution analysis in the wavelet representation.

the peel that we have saved in world W_0. This process is called signal reconstruction. If we use all the peel that was saved, we will get back our original apple (full reconstruction). If, on the other hand, we manipulate some properties of the peel (for example, the color) we will get back a different apple, or an approximation of the original apple.

This process – the generation of different representations of the original object, extraction of the details lost in going from one representation to the next, and reconstruction of a suitable approximation of the original object – is what wavelet-based multi-resolution analysis is all about. It is also the basic recipe that underpins the signal processing component of the radar mosaic machine – generation of multi-scale imagery, and generation of piecewise constant approximations before the classification step.

Let us now see how these abstract concepts can be put into a mathematical framework. Let us consider the space of square integrable functions $L^2(\Re)$. The first thing to do is to define ways to represent our functions at different resolutions. This step is called multi-resolution representation. For our purpose, we define a nested sequence of vector spaces where the resolution by which we represent our function decreases by a factor of 2 as 2^{-j} (or the sampling interval – the scale – increases as 2^j).

These spaces can therefore be thoughts as grids whose spacing is successively incremented by a factor of 2. If our function corresponds to a physical signal measured by an instrument (e.g. remote sensing image) we can start from the resolution afforded by the instrument, say V_0, and define in succession smaller spaces at lower resolution $V_1 \ldots V_n$. To the limit, we will get an empty space. In the opposite direction, we will have spaces with higher resolution, tending to the ideal space at infinite resolution of mathematical (abstract) functions of a real variable.

The V spaces must obey specific laws if they are to be of any use. These are summarized below.

$$\boldsymbol{V}_{j+1} \subset \boldsymbol{V}_j \tag{A12}$$

$$\boldsymbol{f}(\boldsymbol{t}) \in \boldsymbol{V}_j \Leftrightarrow \boldsymbol{f}\left(\boldsymbol{t} - \boldsymbol{2}^j \boldsymbol{k}\right) \in \boldsymbol{V}_j \tag{A13}$$

$$\boldsymbol{f}(\boldsymbol{t}) \in \boldsymbol{V}_j \Leftrightarrow \boldsymbol{f}(\boldsymbol{t}/\boldsymbol{2}) \in \boldsymbol{V}_{j+1} \tag{A14}$$

$$\lim_{j \to \infty} V_j = {}_{-\infty}^{+\infty}\cap V_j = \{0\} \tag{A15}$$

$$\lim_{j \to -\infty} V_j = closure_{-\infty}^{+\infty} \Delta V_j = L^2(\Re) \tag{A16}$$

Equation (A12) tells us that each space at a lower resolution must be included in the preceding one at a higher resolution. In other words, we do not want to create new things when representing a signal at a lower resolution. All the information necessary to create a space at a lower resolution must be contained in the preceding one.

Equation (A13) tells us that a translated version of the function proportional to the scale must still reside in the same space. Translations do not imply a modification of the shape of the object and therefore must produce something that can be represented in the same space.

Equation (A14) means that dilations by two of a function represented at resolution 2^{-j} will reside in the next space in the sequence at resolution $2^{-(j+1)}$. Indeed, by dilating by 2, the details of the function will be enlarged and the function can be represented on a correspondingly enlarged grid.

Equations (A15) and (A16) describe the two limiting processes for infinite resolution (functions in $L^2(\Re)$ and zero resolution (empty set).

To obtain a representation in an abstract vector space we need a reference system where to project our vectors. In spaces of functions, the reference system is constituted of an orthonormal basis. Let us briefly outline the logical steps that one needs to follow to obtain such a basis for the spaces V_j.

First let us consider the space V_0. This can be thought of as a grid with spacing equal to the scale of our discrete original signal. Recollect that this signal is obtained in practice by sampling an analogue signal (e.g. voltage). Discretization in time is the issue here. A continuous function $f(t)$ can be uniquely decomposed in this discrete setting by a particular set of functions $\{\theta(t-n)\}_{n\in z}$ as:

$$f(t)=\sum_{n=-\infty}^{n=+\infty} a[n]\theta(t-n) \tag{A17}$$

with

$$A\|f\|^2 \le \sum_{n=-\infty}^{n=+\infty} |a[n]|^2 \le B\|f\|^2 \tag{A18}$$

$\{\theta(t-n)\}$ is called a Riesz basis, and it is a generalization of the concept of orthonormal basis, when the condition of orthogonality is relaxed. The $\{\theta\}$ vectors are still linearly independent, but they are not necessarily orthogonal to each other.

Interpreting Equation (A18) in the energy sense, it means that the coefficients in the Riesz basis expansions capture a finite multiple of the total energy of the original signal. This condition assures, therefore, that the basis is stable (it does not diverge in energy). The total energy in the expansion coefficients can be greater than the original energy because the vectors are not mutually orthogonal, and therefore the same amount of energy can be split redundantly among more of them.

Equation (A17) provides a discretization theorem, namely the link between the discrete samples $a[n]$ that are available in our sampled signal, and the underlying continuous function that it is known only through its projection in V_0.

An example of Riesz basis is constituted of unitary pulse functions defined as:

$$p_1(t)=\begin{cases}1\rightarrow|t|<1\\0\rightarrow|t|>1\end{cases}$$

The next step is to orthogonalize the Riesz basis and find an orthonormal basis for all spaces V_j by translation and dilation of a unique function ϕ, which is called the scaling function.

$$\varphi_{j,n}(t)=\frac{1}{2^j}\varphi\left(\frac{t-n}{2^j}\right)$$

It can be proven that the family of functions $\phi_{j,n}$ generates an orthonormal basis for all spaces V_j if its Fourier transform can be expressed as a function of the Riesz basis θ in the following way:

$$\hat{\varphi}(\omega)=\frac{\hat{\theta}(\omega)}{\left(\sum_{k=-\infty}^{\infty}\left|\hat{\theta}(\omega+2k\pi)\right|^2\right)^{1/2}} \tag{A19}$$

The orthogonality condition can be expressed in the Fourier domain as:

$$\sum_{k=-\infty}^{\infty}\left|\hat{\varphi}(\omega+2k\pi)\right|^2=1$$

Exploiting the properties of the nested V spaces, it is finally possibly to associate a discrete filter to the scaling function.

This final step will allow us to work efficiently in a completely discrete setting. It is done as follows.

If $\varphi(t - n)$ is a scaling function in V_0, then by property (A12) the dilated scaling function $\left(1/\sqrt{2}\right)\varphi\left(t/2\right)$ that belongs to space V_1 is also included in V_0. Therefore, it can be decomposed as:

$$\frac{1}{\sqrt{2}}\varphi\left(\frac{t}{2}\right)=\sum_{n=-\infty}^{n=+\infty}h\left[n\right]\varphi\left(t-n\right) \tag{A20}$$

where

$$h\left[n\right]=\left\langle\frac{1}{\sqrt{2}}\varphi\left(\frac{t}{2}\right),\varphi\left(t-n\right)\right\rangle$$

This is the called two-scale equation or refinement equation.

Properties of the filter $h[n]$ can readily be derived in the Fourier domain. They are:

$$\forall\omega\in\mathfrak{R},\left|\hat{\boldsymbol{h}}\left(\omega\right)\right|^2+\left|\hat{\boldsymbol{h}}\left(\omega+\pi\right)\right|^2=\mathbf{2} \tag{A21}$$

$$\hat{h}\left(0\right)=\sqrt{2}$$

A filter whose frequency response obeys Equation (A21) is called a conjugate mirror filter.

Notice that until now wavelets have not entered the picture. Approximations of an image are obtained only through the definition of the scaling function.

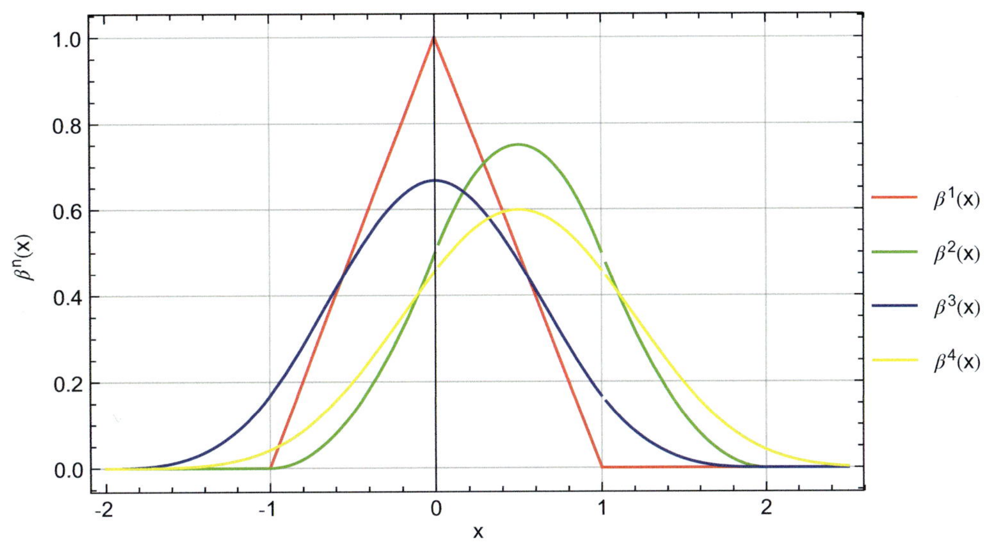

FIGURE A.8 B-splines functions of order $m = 1, 2, 3, 4$. They are obtained by successive m-time convolutions of the unitary pulse function.

A basis can be, for example, constructed out of basic splines functions (B-splines). Let us go through the steps that are needed in this case to define an orthonormal basis for the V_j spaces and to compute the coefficients of the corresponding filter $h[n]$.

B-splines are obtained by successive convolutions of the unitary pulse function (see Figure A.8). More precisely a B-spline of order m β_m is obtained convolving the unitary pulse function $m + 1$ times with itself and centering at 0 or ½ if m is odd or even. Its Fourier transform, since the $m + 1$ – fold convolution in the time domain corresponds to a $m + 1$ product in the frequency domain, is:

$$\beta_m(\omega) = \left(\frac{\sin(\omega/2)}{\omega/2}\right)^{m+1} \exp\left(\frac{-i\varepsilon\omega}{2}\right)$$

With $\varepsilon = 1$ if m is even, $\varepsilon = 0$ if m is odd.

They constitute a Riesz basis in V_0 for $m \geq 0$, as it can be verified using the Fourier equivalent of condition Equation (A18):

$$\frac{1}{B} \leq \sum_{k=-\infty}^{+\infty} \left|\hat{\beta}_m(\omega - 2k\pi)\right|^2 \leq \frac{1}{A}, \forall \omega \in [-\pi, \pi]$$

From Equation (A19) the Fourier transform of the scaling function (orthogonalization of the Riesz basis) is:

$$\hat{\varphi}(\omega) = \frac{\beta_m(\omega)}{\left(\sum_{k=-\infty}^{+\infty} \left|\beta_m(\omega + 2k\pi)\right|^2\right)^{1/2}} \tag{A22}$$

The expression in the denominator can be expanded as follows:

$$\sum_{k=-\infty}^{+\infty} \left|\beta_m(\omega + 2k\pi)\right|^2 = b(\omega) = (2\sin(\omega/2))^{2(m+1)} \sum_k \frac{1}{(\omega + 2k\pi)^{2(m+1)}}$$

$$= (2\sin(\omega/2))^{2(m+1)} S_{2(m+1)}(\omega) \tag{A23}$$

The term $S_{2(m+1)}(\omega)$ can be derived starting from the series:

$$\sum_{k=-\infty}^{+\infty} \frac{1}{(\omega + 2k\pi)^2} = \frac{1}{4\sin^2(\omega/2)} = S_2(\omega)$$

$S_p(\omega)$ is computed by successive derivatives of $S_2(\omega)$. For example:

$$\frac{d^2}{d\omega^2} \sum_{k=-\infty}^{+\infty} \frac{1}{(\omega + 2k\pi)^2} = \sum_{k=-\infty}^{+\infty} \frac{6}{(\omega + 2k\pi)^4} = \frac{d^2}{d\omega^2} \frac{1}{4\sin^2(\omega/2)}$$

$$\sum_{k=-\infty}^{+\infty} \frac{1}{(\omega + 2k\pi)^4} = \frac{2\cos^2(\omega/2) + 1}{48\sin^4(\omega/2)} = \frac{1}{6} \frac{d^2}{d\omega^2} S_2(\omega)$$

Generalizing, we have:

$$S_p(\omega) = \sum_k \frac{1}{(\omega + 2k\pi)^p} = \frac{(-1)^{p-2}}{(p-1)!} \frac{d^{p-2}}{d\omega^{p-2}} S_2(\omega)$$

Using the notation in Equation (A23), Equation (A22) becomes:

$$\widehat{\varphi}(\omega) = \frac{\beta_m(\omega)}{\sqrt{(2\sin(\omega/2))^{2(m+1)} S_{2(m+1)}(\omega)}} \tag{A24}$$

The scaling function $\hat{\Phi}(\omega)$, $\Phi(t)$ for a cubic spline is shown in Figure A.10.
The two scales relation Equation (A18) becomes in the Fourier domain:

$$\widehat{h}(\omega) = \sqrt{2}\,\frac{\varphi(2\omega)}{\varphi(\omega)} \tag{A25}$$

Substituting Equation (A22) into Equation (A23):

$$\hat{h}(\omega) = \sqrt{2}\,\frac{\beta_m(2\omega)}{\beta_m(\omega)} \sqrt{\frac{(2\sin(\omega/2))^{2(m+1)} S_{2(m+1)}(\omega)}{(2\sin(\omega))^{2(m+1)} S_{2(m+1)}(2\omega)}} =$$

$$\sqrt{2}\,\frac{e^{-i\varepsilon\omega}\left(\dfrac{\sin(\omega)}{\omega}\right)^{m+1}}{e^{-i\varepsilon\omega/2}\left(\dfrac{\sin(\omega/2)}{\omega/2}\right)^{m+1}} \sqrt{\frac{S_{2(m+1)}(\omega)}{(2\cos(\omega/2))^{2(m+1)} S_{2(m+1)}(2\omega)}} =$$

$$\sqrt{2}\,\frac{e^{-i\varepsilon\omega/2}(2\cos(\omega/2))^{m+1}}{(1/2)^{m+1}} \sqrt{\frac{S_{2(m+1)}(\omega)}{(2\cos(\omega/2))^{2(m+1)} S_{2(m+1)}(2\omega)}} =$$

$$e^{-i\varepsilon\omega/2} \sqrt{\frac{S_{2(m+1)}(\omega/2)}{S_{2(m+1)}(\omega)}}\, sign(\cos(\omega/2)) \tag{A26}$$

Coefficients of the discrete filter $h[n]$ are the Fourier series coefficients of Equation (A26), namely:

$$h[n] = \frac{1}{2\pi} \int_{-\pi}^{+\pi} \hat{h}(\omega) e^{i\omega n} d\omega$$

Wavelets finally come into play when we consider the details of the function that are lost in passing from the representation at one resolution to the next. To this purpose, we define a complementary abstract space W_j, where the details of the image that were present at scale 2^{j-1} are projected (see Figure A.9).

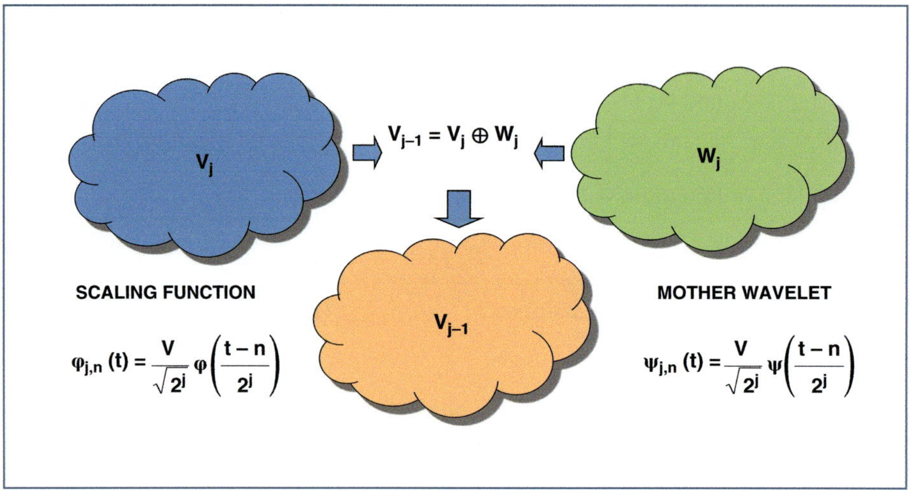

FIGURE A.9 In the multi-resolution representation wavelets provide complementary spaces where the signal details are projected.

$$V_{j-1} = V_j \oplus W_j$$

The mother wavelet $\psi(t)$ plays the same role in these spaces W_j as the scaling function $\varphi(t)$ for the spaces V_j. Therefore, it constitutes an orthonormal basis for all spaces W_j:

$$\psi_{j,n}(t) = \frac{1}{\sqrt{2^j}}\psi\left(\frac{t-n}{2^j}\right)$$

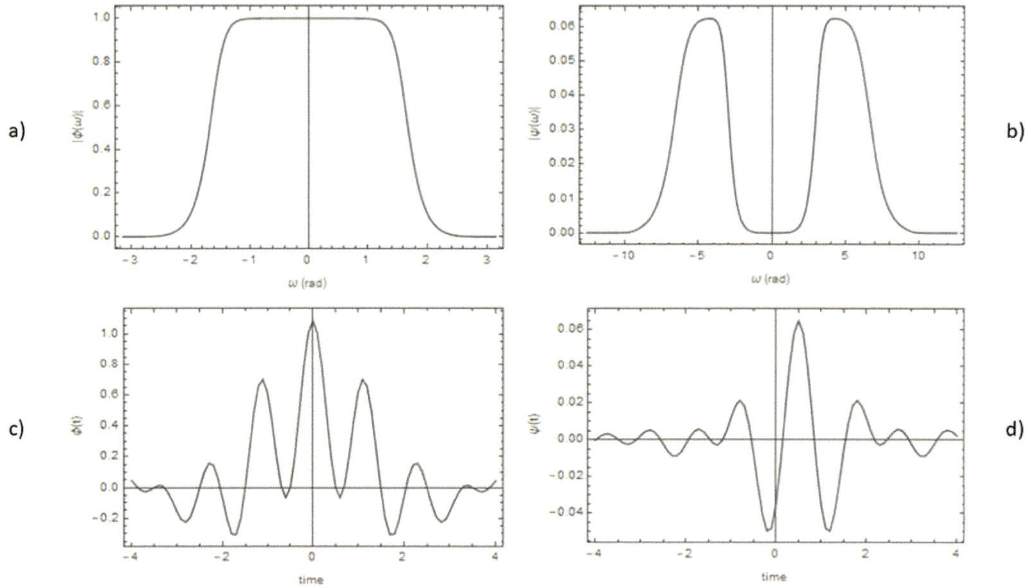

FIGURE A.10 (a) The scaling function $\hat{\Phi}(\omega)$ for a cubic spline in the frequency domain (a low-pass filter) and (c) in the time domain; (b) The mother wavelet in the frequency domain (a band-pass filter); and (d) in the time domain.

The mother wavelet $\Psi(t)$ for a cubic spline is shown in Figure A.10(d).

Again, a discrete filter can be associated with the wavelet ψ. This is done considering an equation similar to the two-scale equation. Indeed $\psi(t/2)$ belongs to W_1, which is comprised in V_0. Therefore:

$$\frac{1}{\sqrt{2}}\psi\left(\frac{t}{2}\right)=\sum_{n=-\infty}^{n=+\infty}g[n]\varphi(t-n)$$

The coefficients $g[n]$ can be computed from the scalar products:

$$g[n]=\left\langle\frac{1}{\sqrt{2}}\psi\left(\frac{t}{2}\right),\varphi(t-n)\right\rangle$$

Also $g[n]$ is a conjugate mirror filter. The family $\{\psi_{j,n}\}$ is an orthonormal basis of W_j if and only if the filter $g[n]$ satisfies the properties:

$$\forall\omega\in\Re,\left|\hat{g}(\omega)\right|^2+\left|\hat{g}(\omega+\pi)\right|^2=2 \tag{A27}$$

$$\hat{g}(\omega)\hat{h}^*(\omega)+\hat{g}(\omega+\pi)\hat{h}^*(\omega+\pi)=0 \tag{A28}$$

Setting:

$$\hat{g}(\omega)=e^{-i\omega}\hat{h}^*(\omega+\pi) \tag{A29}$$

then $\hat{g}(\omega)$ satisfies Equation (A27) and Equation (A28). Taking the inverse Fourier transform of Equation (A29) we have:

$$g[n]=(-1)^{1-n}h[1-n]$$

This relationship links the filter $g[n]$ to the filter $h[n]$.

Finally, we have reached the conclusion that a wavelet-based multi-resolution approximation is characterized by two conjugate mirror filters, one $h[n]$ associated with the scaling function, and the other $g[n]$ associated with the mother wavelet. This paves the way to an efficient implementation of the wavelet multi-resolution approximation using numerical techniques related to filter banks.

For a two-dimensional case (an image) the wavelet decomposition algorithm uses separable filters (rows and columns) and comprises the following steps (see Figure A.11).

The original image S_j is convolved by rows using the mirror filters $H$and G. Each resulting image is then down-sampled by 2.

The image corresponding to the H filter is convolved by columns using the H filter again, and then decimated. This processing chain produces the projection into the V_{j+1} space, or the approximation of the image S_j at lower resolution. S_{j+1} is therefore the approximation image at scale 2^{j+1}.

The image corresponding to the H filter is also convolved by columns using the G filter and decimated. This step produces a projection in the W_{j+1} space corresponding to a spatial orientation along the rows (details occurring along this orientation).

Similarly, the image corresponding to the G filter is convolved with the H filter to generate the detailed image corresponding to spatial orientation along columns.

The image corresponding to the G filter is convolved with the G filter to produce a detailed image corresponding to features oriented at 45° with respect the orthogonal orientation of rows and columns. Note that at each resolution level 2^{-j} the size of the image is reduced by a factor of 2 in rows and columns.

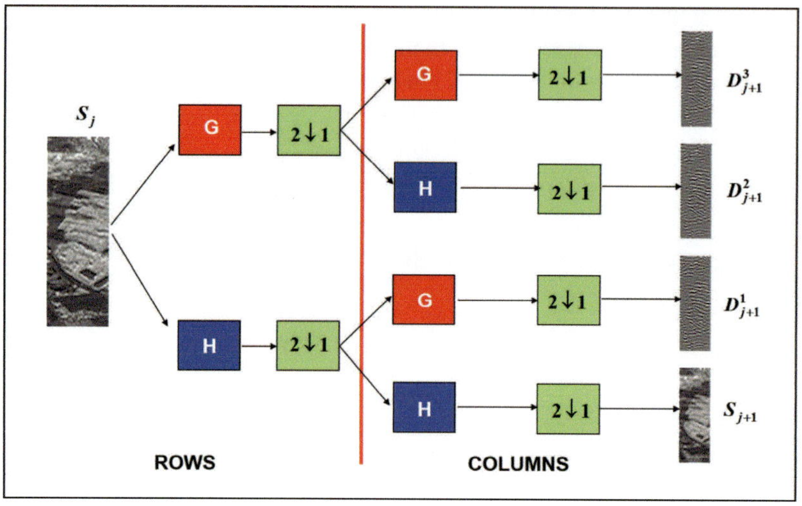

FIGURE A.11 Wavelet multi-resolution representation implemented by a filter bank algorithm.

REFERENCES

1. J.W. Dettman, *Mathematical Methods in Physics and Engineering*, New York: Dover Publications, 1969, ISBN 0-486-65649-7.
2. H.F. Davis, *Fourier Series and Orthogonal Functions*, New York: Dover Publications, 1989, ISBN 0-486-65973-9.
3. S. Mallat, *A Wavelet Tour of Signal Processing*, 2nd ed., San Diego, CA: Academic Press, 1999, ISBN 0-12-466606-X.
4. C.K. Chui, *An Introduction to Wavelets*, San Diego, CA: Academic Press, 1992.
5. I. Daubechies, *Ten Lectures on Wavelets*, Philadelphia, PA: Society for Industrial and Applied Mathematics, 1992, ISBN 0-89871-274-2.
6. M. Vetterli, and J. Kovacevic, *Wavelets and Subband Coding*, Englewood Cliffs, NJ: Prentice Hall PTR, 1995, ISBN 0-13-097080-8.
7. I. Daubechies, "The wavelet transform, time-frequency localization and signal analysis", *IEEE Transactions On Information Theory*, vol. 36, no. 5, pp 961–1005, September 1990.
8. M.J. Shensa, "The discrete wavelet transform: Wedding the à trous and Mallat algorithms", *IEEE Transactions On Signal Processing*, vol. 40, no. 10, pp. 2464–2482, October 1992.
9. O. Rioul, and P. Duhamel, "Fast algorithms for discrete and continuous wavelet transforms", *IEEE Transactions On Information Theory*, vol. 38, no. 2, pp. 569–586, March 1992.
10. S.G. Mallat, "A theory for multi-resolution signal decomposition: The wavelet representation", *IEEE Tranactions on Pattern and Machine Intelligence*, vol. 11, pp. 674, July 1989.

Index

Created with TExtract / www.TExtract.com

A

B

C

D

E

F

M

N

O

P

Q

R

S

For Product Safety Concerns and Information please contact our EU representative GPSR@taylorandfrancis.com Taylor & Francis Verlag GmbH, Kaufingerstraße 24, 80331 München, Germany